Organic Reactions

Organic Reactions

VOLUME 54

ASSOCIATE EDITORS

ROBERT M. MORIARTY
ADRIANA B. PIERINI
OM PRAKASH
ROBERTO A. ROSSI
ANA N. SANTIAGO

JOHN WILEY & SONS, INC.

New York • Chichester • Weinheim • Brisbane • Singapore • Toronto

This text is printed on acid-free paper. ♾

Published by John Wiley & Sons, Inc.

Published simultaneously in Canada.

For ordering and customer service, call 1-800-CALL-WILEY.

Library of Congress Catalog Card Number 42-20265

ISBN 0-471-34888-0

Printed in the United States of America

10 9 8 7 6 5 4 3 2 1

PREFACE TO THE SERIES

In the course of nearly every program of research in organic chemistry the investigator finds it necessary to use several of the better-known synthetic reactions. To discover the optimum conditions for the application of even the most familiar one to a compound not previously subjected to the reaction often requires an extensive search of the literature; even then a series of experiments may be necessary. When the results of the investigation are published, the synthesis, which may have required months of work, is usually described without comment. The background of knowledge and experience gained in the literature search and experimentation is thus lost to those who subsequently have occasion to apply the general method. The student of preparative organic chemistry faces similar difficulties. The textbooks and laboratory manuals furnish numerous examples of the application of various syntheses, but only rarely do they convey an accurate conception of the scope and usefulness of the processes.

For many years American organic chemists have discussed these problems. The plan of compiling critical discussions of the more important reactions thus was evolved. The volumes of *Organic Reactions* are collections of chapters each devoted to a single reaction, or a definite phase of a reaction, of wide applicability. The authors have had experience with the processes surveyed. The subjects are presented from the preparative viewpoint, and particular attention is given to limitations, interfering influences, effects of structure, and the selection of experimental techniques. Each chapter includes several detailed procedures illustrating the significant modifications of the method. Most of these procedures have been found satisfactory by the author or one of the editors, but unlike those in *Organic Syntheses* they have not been subjected to careful testing in two or more laboratories.

Each chapter contains tables that include all the examples of the reaction under consideration that the author has been able to find. It is inevitable, however, that in the search of the literature some examples will be missed, especially when the reaction is used as one step in an extended synthesis. Nevertheless, the investigator will be able to use the tables and their accompanying bibliographies in place of most or all of the literature search so often required.

Because of the systematic arrangement of the material in the chapters and the entries in the tables, users of the books will be able to find information desired by reference to the table of contents of the appropriate chapter. In the interest of economy the entries in the indices have been kept to a minimum, and, in particular, the compounds listed in the tables are not repeated in the indices.

The success of this publication, which will appear periodically, depends upon the cooperation of organic chemists and their willingness to devote time and effort to the preparation of the chapters. They have manifested their interest already by the almost unanimous acceptance of invitations to contribute to the work. The editors will welcome their continued interest and their suggestions for improvements in *Organic Reactions.*

Chemists who are considering the preparation of a manuscript for submission to *Organic Reactions* are urged to write either secretary before they begin work.

CONTENTS

Organic Reactions

CHAPTER 1

AROMATIC SUBSTITUTION BY THE $S_{RN}1$ REACTION

Roberto A. Rossi, Adriana B. Pierini, and Ana N. Santiago

Departamento de Química Orgánica, Facultad de Ciencias Químicas, Universidad Nacional de Córdoba, Córdoba, Argentina

CONTENTS

Organic Reactions, Vol. 54, Edited by Leo A. Paquette et al.
ISBN 0-471-34888-0

ACKNOWLEDGMENTS

Thanks are due to our coworkers at the National University of Córdoba, and to the colleagues who sent us bibliographic material. We are also grateful to CONICET, CONICOR and the Antorchas Foundation for financial support.

INTRODUCTION

The $S_{RN}1$ (Unimolecular Radical Nucleophilic Substitution) is a process through which nucleophilic substitution is achieved on aromatic and aliphatic compounds that do not react or react slowly through polar nucleophilic mechanisms. Eq. 1 shows the generalized reaction with an aryl halide. This transformation takes place through electron transfer steps with radicals and radical anions as intermediates.

$$ArX + Nu^- \longrightarrow ArNu + X^- \qquad \text{(Eq. 1)}$$

The $S_{RN}1$ mechanism was proposed for the first time in 1966 for the substitution of alkyl halides with electron-withdrawing groups[1,2] and in 1970 for the substitution of unactivated aryl halides.[3,4] Since then, the scope of the process has increased considerably, and it constitutes an important synthetic method for achieving the substitution of unactivated aromatic and heteroaromatic substrates, vinyl halides, perfluoroalkyl iodides, and activated and non-activated alkyl compounds. Nucleophiles that can be used include anions derived from carbon or heteroatoms, which react to form a new carbon-carbon or carbon-heteroatom bond.

Several reviews have been published in relation to activated[5,6] and non-activated[5,7] aliphatic $S_{RN}1$ reactions, to aromatic photoinitiated reactions,[8–12] to $S_{RN}1$ reactions at aromatic centers,[5,13–18] to reactions performed with electrochemical techniques,[19–27] and to the synthetic applications of the process.[13,28–32] This review attempts to cover aromatic $S_{RN}1$ reactions including all the relevant literature up to the end of 1996.

MECHANISM

The proposed mechanism of the $S_{RN}1$ reaction is a chain process, whose key steps are presented in Scheme I.

The initiation step (Eq. 2) is electron transfer from an adequate electron source to the substrate (ArX) to form the substrate radical anion, which fragments into an aromatic σ radical Ar· and the anion of the leaving group (Eq. 3). The radical thus formed can react with the nucleophile present in the reaction medium to form the radical anion of the substitution product (Eq. 4). Electron transfer from this radical anion to the substrate reforms the substrate radical anion (Eq. 5), and the propagation cycle continues. Summation of Eqs. 3–5 leads to the net nucleophilic substitution.

Results with radical probes,[33–36] as well as relative reactivities of pairs of nucleophiles toward substrates proposed to give the same radical,[37,38] provide strong

Scheme I

INITIATION	$ArX + \text{Donor} \longrightarrow (ArX)^{-\bullet}$		(Eq. 2)
PROPAGATION	$(ArX)^{-\bullet} \longrightarrow Ar^{\bullet} + X^{-}$		(Eq. 3)
	$Ar^{\bullet} + Nu^{-} \longrightarrow (ArNu)^{-\bullet}$		(Eq. 4)
	$(ArNu)^{-\bullet} + ArX \longrightarrow (ArX)^{-\bullet} + ArNu$		(Eq. 5)
	$ArX + Nu^{-} \longrightarrow ArNu + X^{-}$		
TERMINATION	$(ArNu)^{-\bullet} + Ar^{\bullet} \longrightarrow Ar^{-} + ArNu$		(Eq. 6)
	$Ar^{\bullet} \xrightarrow{SH} ArH + S^{\bullet}$		(Eq. 7)

experimental evidence for the presence of radical intermediates. Other mechanistic evidence includes inhibition by radical traps or radical-anion scavengers.

Because the $S_{RN}1$ reaction is a chain process, its overall rate depends on the efficiency of the initiation, propagation and termination steps. Any of the intermediates (radicals and radical anions) can initiate the chain, but by far the most commonly used is the radical anion of the substrate. Destruction of any of the intermediates can terminate the process. One important practical consequence of the chain mechanism is that $S_{RN}1$ reactions should be carried out under an inert atmosphere to avoid inhibition by oxygen. Although the initiation and termination steps represent only a few reactions compared with the whole, they are decisive in promoting or preventing substitution, depending on the efficiency of the chain propagation cycle. Some terminations depend on the method of initiation, while others depend on the intermediates involved and the solvent used. For instance, electron transfer from any radical anion to an aryl radical (Eq. 6) is one of the proposed termination steps in solvents such as liquid ammonia that are poor hydrogen donors. Hydrogen atom abstraction by the aryl radical (Eq. 7) is a possible termination step in organic solvents. The net effect of these termination reactions is formation of the reduced dehalogenated product. Even though radical dimerization to form biaryls has been observed in some electrochemical systems,[39] this is not generally an important termination step.

The fragmentation reaction of the radical anion of the substitution compound into an unreactive radical is another type of termination. For instance, in the reaction of phenyl halides with nitromethane[40,41] or cyanomethyl anions,[42,43] the main products are derived from benzyl radicals. For example, the radical anion of phenylacetonitrile, formed in the photoinitiated reaction of iodobenzene and cyanomethyl anion, can give the substitution compound by electron transfer to iodobenzene (Eq. 9) or can fragment to afford benzyl radicals (Eq. 10). These radicals are responsible for the formation of toluene and diphenylethane (Eqs. 10, 11).

$$PhI + {}^{-}CH_2CN \xrightarrow[NH_3]{h\nu} (PhI)^{-\bullet} \longrightarrow Ph^{\bullet} + I^{-} \qquad \text{(Eq. 8)}$$

$$Ph^{\bullet} \xrightarrow{{}^{-}CH_2CN} (PhCH_2CN)^{-\bullet} \xrightarrow{PhI} (PhI)^{-\bullet} + PhCH_2CN \ (7\%) \qquad \text{(Eq. 9)}$$

$$(PhCH_2CN)^{-\bullet} \xrightarrow{CN^{-}} PhCH_2^{\bullet} \xrightarrow{NH_3} PhCH_3 \ (6\%) \qquad \text{(Eq. 10)}$$

$$2\ PhCH_2^{\bullet} \longrightarrow PhCH_2CH_2Ph \ (63\%) \qquad \text{(Eq. 11)}$$

However, when the substrates have a low-lying π^* molecular orbital that stabilizes the radical anion of the substitution product,[44] bond fragmentation can be avoided to give α-arylacetonitriles straightforwardly.[45–47] The nitromethane anion gives fragmentation products with all the aromatic substrates studied.[40,41] Fragmentation reactions have also been found with certain alkanethiolate ions.[48,49]

The possibility of an $S_{RN}2$ instead of an $S_{RN}1$ mechanism has been proposed for aromatic substrates.[50,51] However, this proposal has been rejected on the basis of mechanistic considerations,[52] transition state theory[53] and electrochemical evidence.[19,39]

Methods of Initiation

Only a few systems are known to react by the $S_{RN}1$ mechanism in a thermal (or spontaneous) reaction. Most of the systems need to be initiated by different means. Photoinitiation, chemical initiation by alkali metals in liquid ammonia, or electrochemical initiation at a cathode are the most frequently used techniques. However, other reagents to initiate the reactions, such as Fe^{+2}, SmI_2, or Na(Hg), have been reported recently, but their scope and limitations are still unknown.

Thermal (Spontaneous) Reactions. In some systems, the $S_{RN}1$ mechanism occurs without stimulation of any kind other than temperature. In these reactions, spontaneous electron transfer from the nucleophile to the substrate occurs. Spontaneous initiation is possible whenever the substrate is a very good electron acceptor and the nucleophile a very good electron donor. By far the most common substrates for thermal $S_{RN}1$ reactions are aryl iodides. Bromo- or chloro-substituted π-electron-deficient aromatic rings such as pyrimidines,[54] pyrazines,[55] pyridazines,[55] and quinoxalines[56] can also react under these conditions. Arylazo-*tert*-butyl- or phenyl sulfides are also very good electron acceptors that can react spontaneously or under laboratory light with various nucleophiles in organic solvents.[57–59]

When the nucleophiles are carbanions from ketones, in general the higher the pK_a of the ketone the higher the electron donor capability of its conjugate base and so the higher the probability of a spontaneous electron transfer.[60,61] Thermal initiation is more common in dimethyl sulfoxide (at room temperature)[62,63] than in liquid ammonia (at −33°). Even though some reactions in dimethyl sulfoxide occur in the dark, exposure to light is usually recommended to achieve higher yields of substitution in shorter reaction times.

Photoinitiated Reactions. Photoinitiation is one of the most widely used methods, and it has proven extremely suitable for synthetic purposes. One of the mechanisms proposed for initiation is photochemically activated electron transfer within a charge-transfer complex formed between the nucleophile and the substrate.[64–69] Electron transfer from an excited state nucleophile is another possibility since these species are the best candidates for absorption at commonly used wavelengths[67,70] ($\lambda > 350$ nm, see Experimental Considerations).

Even if photoinitation takes place with low quantum yields, it can be successful in achieving substitution whenever it is followed by efficient propagation steps. To improve the yields of these reactions, a 0.02–0.04 M concentration of substrate is recommended. At higher concentrations the reactions are slower, and the product yields decrease. The ratio of nucleophile to substrate concentration also plays an important role in maximizing the yield of substitution in systems involving a chain process of rather low efficiency.

A reaction between a nucleophile and an aryl halide may be slow to initiate, but fast to propagate. The addition of tiny amounts of another substrate (or nucleophile) that is more reactive in the thermal or photoinitiated step increases the level of reactive intermediates and allows the less reactive substrate (or nucleophile) to start its own propagation[71–74] (*entrainment* reactions, see Electrophiles).

Fe^{+2}-Initiated Reactions. Initiation by ferrous ion ($FeSO_4$ in liquid ammonia,[75,76] $FeCl_2$[77] or $FeBr_2$[78] in dimethyl sulfoxide) is growing in importance because of its possible synthetic advantages with respect to the other methods. The reaction, useful mainly with aryl iodides, can be carried out in liquid ammonia or dimethyl sulfoxide without special electrochemical or photochemical devices. Even though the role of ferrous ion in the reaction remains uncertain,[77] this is a promising alternative to traditional methods.

SmI_2-Initiated Reactions. In recent years, SmI_2 has been used as a one-electron reducing agent for many organic halides.[79] However, it has been used to initiate $S_{RN}1$ reactions of haloarenes only once, with the enolate ion of acetophenone in dimethyl sulfoxide.[80] The results obtained indicate that more work in this area is in order.

Electrochemically Initiated Reactions.[23–27,81] Electrochemical initiation is also a widely used approach, though it may have the disadvantage of affording low yields of substitution when the aryl halide radical anion fragments rapidly. In this case, the aryl radical is formed close to the electrode surface, so it may be reduced before it can react with the nucleophile in solution. The situation can be remedied by introduction of a redox mediator that is reduced at a more positive potential than the aryl halide. The radical anion of the mediator lives long enough to form the aryl halide radical anion by an electron transfer reaction away from the electrode. Properties and criteria for selection of adequate redox mediators such as 4,4′-bipyridine, phthalonitrile, 2-phenylpyridine, and benzonitrile have been summarized.[82–85] Even though the yields of substitution are improved, the use of a mediator requires that it be separated from the reaction mixture.

When the aryl halide radical anion fragments slowly, the aryl radical and the radical anion of the substitution compound are formed far from the electrode surface, and this radical anion can transfer its odd electron to an aryl halide in the solution to maintain the propagation cycle. In this way, the complete conversion of aryl halide into the substitution product with a catalytic amount of electrons can be achieved.

When the product radical anion is stable relative to the aryl halide radical anion, the electron transfer step in Eq. 5 can become rate limiting and the $S_{RN}1$ cycle can be sluggish. Such is the case in the reaction of 4-bromobenzophenone with cyanide ions, which under electrochemical initiation requires one electron per molecule to react (noncatalytic system).[86–89]

Electrochemical initiation is quite useful in preparative reactions. The main advantage is the possibility to achieve substitution with nucleophiles that fail to react, or react very poorly, under other conditions. It also provides quantitative information about the mechanism, such as the fragmentation rate constants of radical anions[86,90–93] and the absolute rate constants for the coupling reaction of a wide number of nucleophiles with radicals.[23,86,91,92]

Alkali Metals in Liquid Ammonia. Although a popular method, the use of alkali metals in liquid ammonia is often not the best choice. The main disadvantages of this type of initiation are reductive dehalogenation of the substrate and reduction of the substitution product. For example, in the reaction of the enolate ion of acetone with a phenyl halide under sodium or potassium metal initiation, not only phenylacetone but also 1-phenyl-2-propanol and benzene are formed.[94–97] In addition to reductive dehalogenation and substitution, reductive hydrogenation of the aromatic ring can take place as in the reaction of 1-iodo or 1-chloronaphthalene with the enolate ion of acetone initiated by potassium metal.[45]

Despite the limitations, this method of initiation can be valuable in those cases in which amide products of a benzyne mechanism are to be avoided, as in the reaction of *o*-bromoanisole with amide ion,[4] or the reaction of *p*-halobenzonitriles with the anion of *N,N*-dimethylacetamide.[98]

Sodium Amalgam [Na(Hg)] in Liquid Ammonia. Sodium amalgam selectively dehalogenates haloarenes in liquid ammonia,[99] and it can be used to induce $S_{RN}1$ reactions of substrates with redox potentials close to or more positive than its redox potential.[100] With substrates whose radical anions fragment rapidly and consequently close to the Na(Hg) surface, substitution is achieved only when the reactions are performed in the presence of a redox mediator.[101,102]

SCOPE AND LIMITATIONS

Electrophiles

A great variety of aromatic compounds with appropriate leaving groups have been shown to react as electrophiles in $S_{RN}1$ reactions. Common partners include phenyl and substituted phenyl derivatives, and aromatic polycyclic and heterocyclic compounds. Even though halogens are by far the most commonly used leaving groups, other leaving groups have been used including $(EtO)_2P(O)O$, RS (R = Ph, alkyl), PhSO, $PhSO_2$, PhSe, Ph_2S^+, RSN_2 (R = *t*-Bu, Ph), N_2BF_4, and Me_3N^+.

For the $S_{RN}1$ reaction to work efficiently, good electron acceptors are preferred as substrates in order to favor the initiation step (Eq. 2) as well as to facilitate electron transfer from the radical anion of the substitution product to the aryl halide (Eq. 5).[87,103,104] Besides being good electron acceptors, the substrates must form radical anions that cleave into aromatic radicals and the leaving group at an appropriate rate. There is a rough correlation between the rate constant for the fragmentation reaction of the radical anion and the aryl halide/aryl halide radical anion standard potential.[90–93] Thus, for the same aromatic group, the average rate constants of fragmentation follow the order I > Br > Cl. The order of reduction potential in liquid ammonia, PhI > PhBr > $PhNMe_3I$ > PhSPh > PhCl > PhF > PhOPh,[86] coincides with the reactivity order determined under photoinitiation conditions.

With some highly reactive heteroaromatic and polycyclic aromatic substrates, chlorides are just as good as bromides or iodides in photoinitiated reactions, and even better under electrode initiation. Under the latter conditions, chlorides offer the possibility of avoiding the use of a redox mediator. Heteroaromatic halides are good electron acceptors; however, because they are π-electron deficient, competition with an S_NAr process can occur in some cases mainly when more than one heteroatom is present.

Even though unsubstituted aryl halides react with most nucleophiles, there are examples of such reactions that fail in unsubstituted cases but succeed when the ring has particular substituents. For example, *o*-, *m*- and *p*-cyano, *o*-carbonyl, *o*-methoxy, or *o*-amino functions have an important activating effect on the reaction of aryl halides with nucleophiles such as phenoxides or the enolate ions of aldehydes and β-dicarbonyl compounds.[105–107] Electron-withdrawing groups usually do not interfere with substitution as long as they do not react with nucleophiles by other mechanisms; neither do electron-donating groups. The reaction is quite insensitive to steric hindrance.[30,108] A wide range of *o*-substituents and a large number of substituents in other positions of the aromatic halide are generally compatible. Groups that are tolerated include OR, OAr, SAr, CF_3, CO_2R, NH_2, NHCOR, NHBoc, NR_2, SO_2R, CN, COAr, COMe, $CONH_2$, and F. Even though the reaction is not inhibited by the presence of some negatively charged substituents such as carboxylate ions, other charged groups such as oxyanions hinder the process.

Nitro-substituted phenyl halides produce radical anions that fragment at a rather low rate ($\cong 10^{-2} - 10^{2}\ s^{-1}$).[109–112] For this reason, the nitro group is not a suitable substituent for most $S_{RN}1$ reactions. However, exceptions are found with *o*-iodonitrobenzene[113] and nitroaryldiazo phenyl or *tert*-butyl sulfides,[59,114,115] whose radical anions have a relatively high rate of fragmentation.

When the substitution is performed with a carbanion and an aromatic compound that bears an appropriate substituent ortho to the leaving group, cyclization to give five-, six-, and seven-membered ring products can take place. Another interesting approach to cyclization is the intramolecular $S_{RN}1$ reaction of substrates that have both a nucleophilic center and a nucleofugal group in the molecule. This has been used in the synthesis of different natural products. Here an example of entrainment can be presented: the anion of *o*-iodothioacetanilide is unreactive in dimethyl sulfoxide under photoinitiation. However, the photoinitiated intramolecular substitution of iodine occurs in the presence of an excess of acetonate enolate ion, a good electron donor under $S_{RN}1$ conditions.[7]

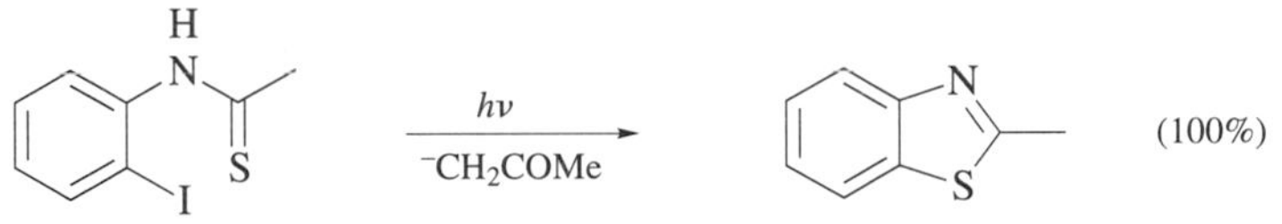
H
N
S
I
hν
⁻CH₂COMe
N
S
(100%)

(Eq. 12)

Compounds with two leaving groups also react to form products of monosubstitution and/or disubstitution, depending on the leaving groups, their relative positions in the molecule, and on the nucleophile. In the case of dihalobenzenes, for example, the occurrence of mono- and/or disubstitution is interpreted based on the following mechanistic proposal (Eqs. 13, 14). The radical anion **1**⁻˙ of dihalobenzene **1** fragments at the more labile carbon-halogen bond to give a haloaryl radical **2**, which in turn forms the radical anion of the monosubstituted

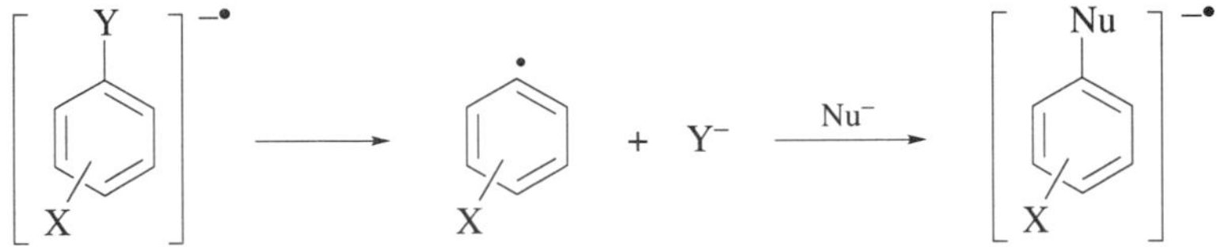
Y
X
X
+ Y⁻
Nu⁻
Nu
X

1⁻˙ **2** **3**

(Eq. 13)

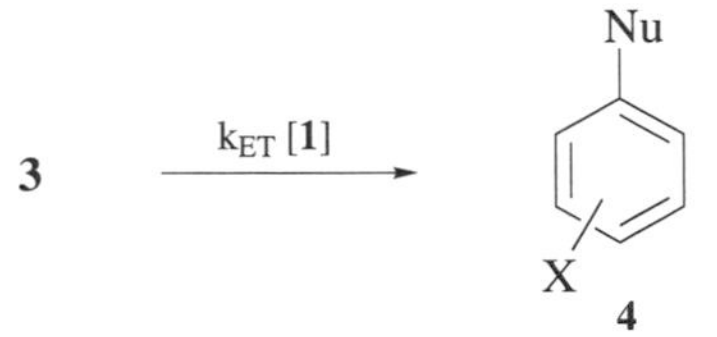
3
kET [1]
Nu
X

4

(Eq. 14)

compound **3** by reaction with the nucleophile. Radical anion **3** can transfer its extra electron to the substrate and in this case the monosubstitution product **4** is formed with retention of halogen.

Alternatively, intramolecular electron transfer to the second carbon-halogen bond results in fragmentation to form radical **5**, which by coupling with a second molecule of nucleophile, affords the disubstitution compound **6** (Eq. 15). In this case, the monosubstitution product **4** is not an intermediate in the formation of **6**.[116]

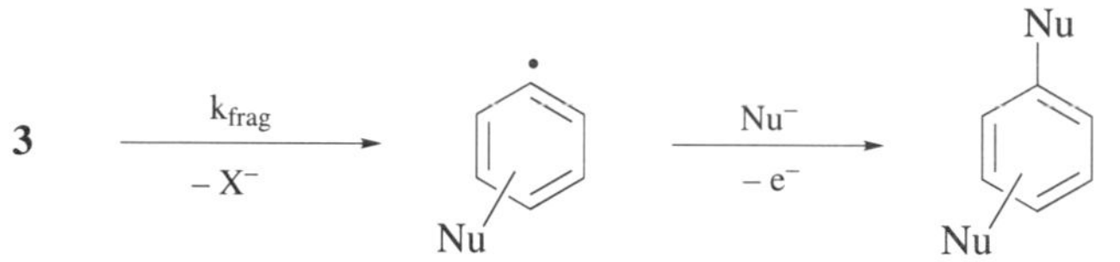

5 **6**

(Eq. 15)

Product **4** is formed in a bimolecular reaction that depends on the concentration of **1** while radical **5** (the precursor of **6**) is formed in a unimolecular reaction. However, competition between the two processes is almost independent of the concentration of **1**. The ratio between **4** and **6** depends on the halogen, its electron affinity in relation to the substrate that acts as another acceptor, and its position on the aromatic ring, as well as on the nature of the nucleophile.

Nucleophiles

Carbanions from hydrocarbons, nitriles, ketones, esters, *N,N*-dimethylacetamides, and mono and dianions from β-dicarbonyl compounds are among the most commonly used nucleophiles for carbon-carbon bond formation. The only oxygen nucleophiles known to react are phenoxides. In these reactions, carbon-carbon instead of carbon-oxygen bond formation occurs. Similar carbon-carbon bond formation is obtained with nitrogen nucleophiles from aromatic amines, pyrrole, indole, and diazoles.

Carboxylation of aromatic halides to form the corresponding acid and/or ester is possible through the use of cobalt carbonyl species. Anions from tin, phosphorus, arsenic, antimony, sulfur, selenium, and tellurium react at the heteroatom to form new carbon-heteroatom bonds in good yields.

A special behavior is shown for diphenylarsenide, diphenylstibide, benzeneselenenate, and benzenetellurate anions. With these nucleophiles, scrambling of the aromatic ring has been reported.[117–120] The reaction of *p*-iodoanisole with benzeneselenenate ions is illustrated (Eq. 16).[117]

$$p\text{-}IC_6H_4OMe + PhSe^- \xrightarrow[NH_3]{h\nu} \underset{(20\%)}{Ph_2Se} + \underset{(25\%)}{PhSeC_6H_4OMe\text{-}p} + \underset{(19\%)}{Se(C_6H_4OMe\text{-}p)_2}$$

(Eq. 16)

According to the proposed mechanism (Eq. 17), the first radical anion **7** formed in the coupling of *p*-anisyl radicals and benzeneselenenate ions can undergo three competitive reactions: reversion to starting materials, electron transfer to the substrate to give the expected substitution product, and fragmentation at the Ph-Se bond to form phenyl radicals and the new nucleophile *p*-anisoleselenenate anion.

$$p\text{-MeOC}_6\text{H}_4^{\bullet} + \text{PhSe}^- \rightleftharpoons \underset{\mathbf{7}}{(p\text{-MeOC}_6\text{H}_4\text{SePh})^{-\bullet}} \rightleftharpoons p\text{-MeOC}_6\text{H}_4\text{Se}^- + \text{Ph}^{\bullet}$$

$$(p\text{-MeOC}_6\text{H}_4\text{SePh})^{-\bullet} \xrightarrow{k_{ET}\,[p\text{-IC}_6\text{H}_4\text{OMe}]} p\text{-MeOC}_6\text{H}_4\text{SePh}$$

(Eq. 17)

The *p*-anisoleselenenate anion can couple with *p*-anisyl radicals to give the symmetrical *p*-dianisyl selenide, while phenyl radicals can react with benzeneselenenate ions to form diphenyl selenide. The fragmentation reaction of radical anion **7** indicates that the C-Se σ^* molecular orbital is of similar energy to the π^* molecular orbital of the system. However, when the π^* molecular orbital is lower in energy than the C-Se σ^* molecular orbital (for example with 1-naphthyl, 2-quinolyl, 4-biphenylyl, and 9-phenanthryl), the straightforward substitution product aryl phenyl selenide is formed in good yields (52–98 %).[16,121,122] This fragmentation process explains the formation of scrambled products of diphenylarsenide ions with some aromatic halides and the straightforward substitution products with others.[118,119] On the other hand, only scrambling of aryl rings is realized with all the substrates studied and diphenylstibide anion.[119]

The absolute rate constants for the reaction of nucleophiles with aromatic radicals (Eq. 4) are close to the diffusion limit.[23] One important factor that favors this step is the thermodynamic stability of the radical anion formed, as this has been proposed to explain the relative reactivity of carbanions toward phenyl radicals under photoinitiation.[41,123] This factor can also explain the regiochemistry of the reaction with conjugated carbanions[44,124] and with anions from aromatic amines or aromatic alkoxides.[125] Relative reactivities of pairs of nucleophiles toward the same radical have been determined from the ratio of the two substitution products in competition reactions.[37,38,117,126]

Under photostimulation, certain nucleophiles such as nitrite and cyanide ions fail to achieve substitution with aryl halides, owing to the formation of a stable radical anion of the substitution product. The situation can be remedied, as in the photoinitiated reaction of cyanide ion, by the use of aryldiazo sulfides as substrates.[115,127]

Carbanions From Hydrocarbons

A relatively limited set of carbanions derived from hydrocarbons has been studied.[70,124,128–133] One of the limitations is that the parent hydrocarbon must be acidic enough to be ionized in liquid ammonia ($pK_a = 33$). In these reactions, diarylation and in some cases triarylation operates in addition to arylation.[70,124,128] For instance, both 9-phenylfluorene and 9,9-diphenylfluorene are formed in the reaction of bromobenzene with the anion from fluorene (Eq. 18).[124]

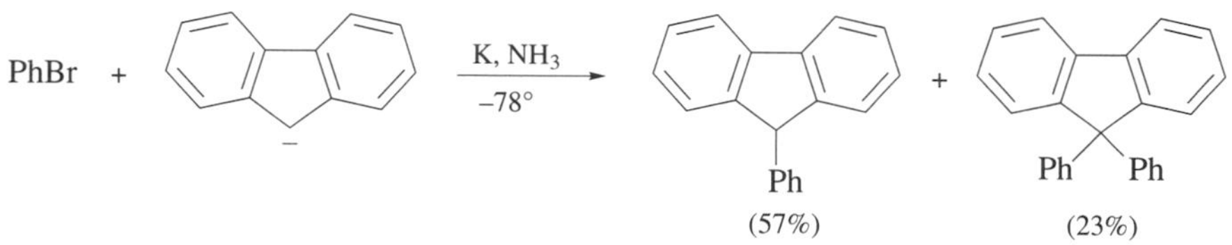

(Eq. 18)

Reactions with anions that have more than one possible coupling position are not regiospecific. For example, the carbanion from 1-(*p*-anisyl)propene gives about three times as much 3-phenyl- (36%) as 1-phenyl-1-(*p*-anisyl)propane (13%) after hydrogenation.[124] Products of further phenylation are also formed in this reaction.[124] Other carbanions such as the anions from acetylene and phenylacetylene fail to give substitution under $S_{RN}1$ conditions.[134,135]

Enolate Ions from Ketones

The enolate ions of ketones are probably the most widely studied nucleophiles in $S_{RN}1$ reactions. Anions of aliphatic, cyclic, and aromatic ketones generally give substitution products in good yield.

Reactions with Substrates with One Leaving Group. The enolate ions of acyclic aliphatic ketones such as acetone,[45,108,136,137] 3,3-dimethyl-2-butanone (pinacolone),[137–141] 2-butanone,[124] 3-methyl-2-butanone (isopropyl methyl ketone),[124,142] 3-pentanone,[142,143] 4-heptanone,[143] 2,4-dimethyl-3-pentanone (diisopropyl ketone),[55,137,142,143] and also the anion of the diacetal from pyruvic aldehyde ($^-CH_2COCH(OMe)_2$)[106] can react through the $S_{RN}1$ mechanism with aryl and heteroaryl halides to afford α-arylation products under photoinitiation. A representative example is shown in Eq. 19.

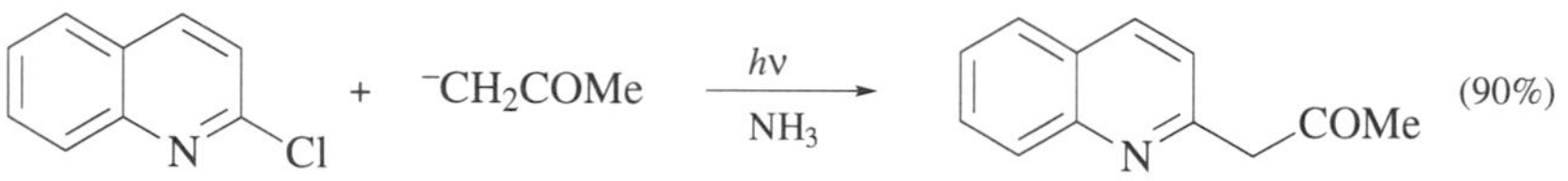

(Eq. 19)

In some reactions, α,α-diarylation can also occur but generally in low yields (<15%). α,α,α-Triarylation has never been reported in these systems. α,α-Diarylation is caused by ionization in the basic medium of the monoarylated product. In order to minimize diarylation, a nucleophile/substrate ratio of 3–5 is recommended. Use of excess *t*-BuOK is also a convenient tactic to avoid the possibility of aldol condensation of the ketone.

The yield of substitution is highly dependent on the solvent. For instance, in the reaction of the enolate ion of acetone with 2-chloroquinoline (Eq. 19) for the same elapsed times, lower yields are obtained by changing the solvent from liquid ammonia (90%) to THF (82%), DMF (74%), dimethoxyethane (28%), diethyl ether (9%), or benzene (4%).[144] For this reason, the reactions are usually performed in liquid ammonia or dimethyl sulfoxide under irradiation. In dimethyl sulfoxide, some reactions can take place by thermal initiation.[62,63,73] The reactions in liquid ammonia can also be initiated electrochemically, by dissolution of alkali metals[45,94,96,97] or by Na(Hg).[101] Iron(II) initiation is also possible with $FeSO_4$ in liquid ammonia,[75] or with $FeCl_2$ in dimethyl sulfoxide.[77]

Isomeric enolate ions can be formed from unsymmetric dialkyl ketones, and it appears that the distribution of the two possible arylation products of this mixture of ions is determined principally by the equilibrium concentration of the various possible enolate ions. However, the selectivity also depends on the structure of the attacking radical. In reactions with the enolate ions from 2-butanone, arylation occurs preferentially at the more substituted α carbon giving about twice as much 3-phenyl-2-butanone as 1-phenyl-2-butanone,[124,145] but in the reaction with the anion from isopropyl methyl ketone, the 1-phenyl derivative is formed predominantly.[124] However, when this anion reacts with 2-chloroquinoline, the ratio of tertiary to primary substitution is 4.77.[142] When there is a substituent *ortho* to the leaving group, the attack at the primary α carbon is enhanced.[145–149]

β-Hydrogen abstractions from the enolate of diisopropyl ketone diminish the substitution reaction with halobenzenes,[143] but do not prevent the high-yield reactions of this anion with heteroaryl halides, such as 2-bromopyridine,[137] 2-chloroquinoline,[142] and 2-chloro- and 4-chloropyrimidines[55,140] under irradiation. The heteroarylation of ketones with 5-bromo-4-*tert*-butylpyrimidines can be performed in the dark, under irradiation or potassium metal initiation (Eq. 20).[54] However, when the 4-*tert*-butyl group is replaced by a 4-phenyl group, irradiation or potassium metal initiation is recommended in order to avoid the dark addition of pinacolone anion to the N_1-C_6 azomethine bond.[54]

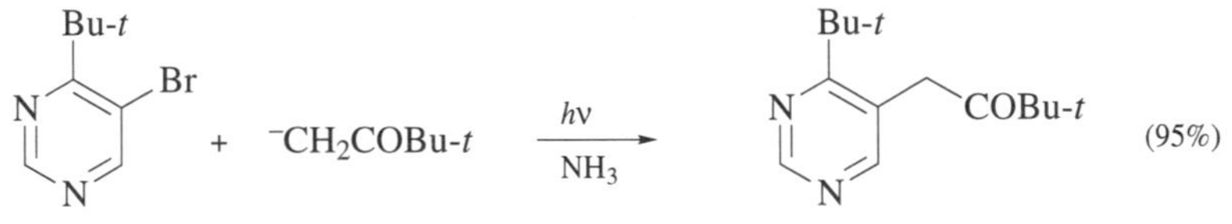
Bu-t
Br
N
N
+ ⁻CH2COBu-t
hν
NH3
Bu-t
COBu-t
N
N
(95%)

(Eq. 20)

2-Chloropyrazine,[55] chloropyridazines,[55] and 2-chloroquinoxaline[56] are other substrates that can react in the dark. In the thermal reaction of the last compound with the anion of pinacolone or diisopropyl ketone, products ascribed to a competing ionic cyclization process can also be formed along with the expected substitution compound (Eq. 21).[56]

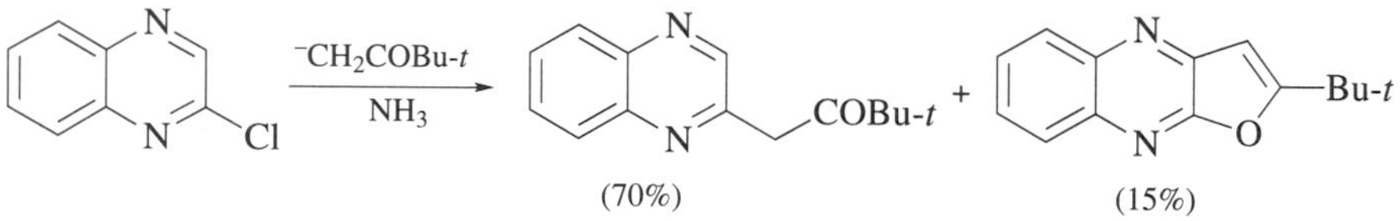

N
N
Cl
⁻CH2COBu-t
NH3
N
N
COBu-t
+
N
N
O
Bu-t
(70%)
(15%)

(Eq. 21)

Poorly reactive compounds include 3-halo (X = Cl, Br, I)-2-aminobenzo[*b*]thiophene (the amino group protected as an amide or as a carbamate),[150] 3-iodobenzo[*b*]thiophene,[151] and 2-chloro- and 2-bromothiophene.[152] However, 3-bromothiophene reacts with the enolate ion of acetone to give arylation (51%) and diarylation (25%) products.[152] Substitution succeeds under irradiation with halo-substituted five-membered heterocycles containing two heteroatoms such as 2-bromo- and 2-chlorothiazoles.[141] The $S_{RN}1$ reaction is also a useful method of C-C bond formation in 6-iodo-9-ethylpurines, as indicated in Eq. 22.[153–154]

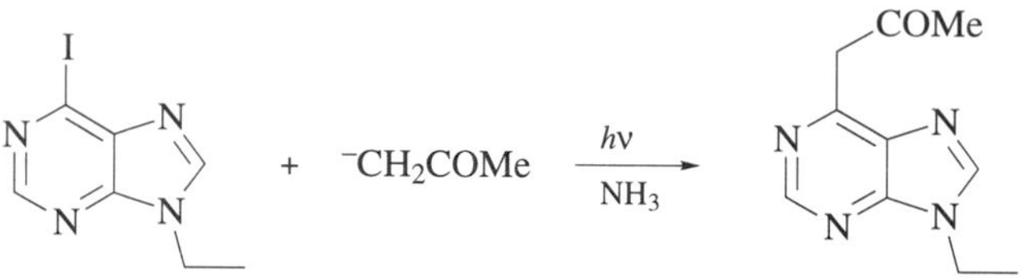

I
N
N
N
N
+
⁻CH2COMe
hν
NH3
COMe
N
N
N
N
(70%)

(Eq. 22)

The enolate ions of cyclic ketones **8** (from cyclobutanone to cyclooctanone) give α-phenylation products **9** (Eq. 23).[143] The enolate ions of cyclohexanone,[137,153] 2-methylcyclohexanone[153] (mainly arylated at the more substituted carbon enolate), cyclopentanone,[142,153] and tetralone[153,155] can also be heteroarylated. No substitution product is formed with the enolate ion of cyclohex-2-en-1-one.[143]

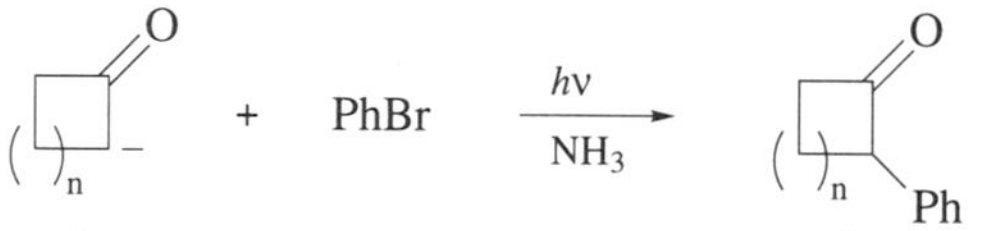

O
n
+
PhBr
hν
NH3
O
n
Ph

(Eq. 23)

8 **9**

An interesting reaction is the arylation of the enolate ion of (+)-camphor.[156] The almost exclusive *endo*-arylation at C_3 in excellent yield opens a new stereospecific C_3-arylation route of (+)-camphor by this mechanism (Eq. 24).[156]

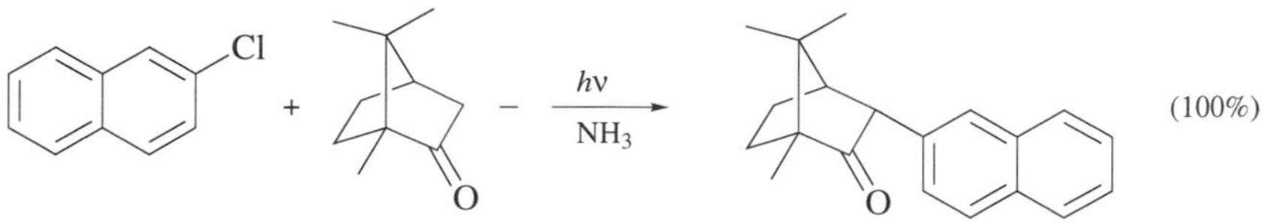

(Eq. 24)

A special behavior is shown by the enolate ions of aromatic ketones. For example, the enolate anion of acetophenone fails to react with halobenzenes or halonaphthalenes in liquid ammonia under photoinitiation.[45,143] Heteroarylation[54,153,157] is possible in liquid ammonia under photoinitiation, and even in the dark.[55] On the other hand, arylation and heteroarylation can be achieved in this solvent with solvated electrons[45,54] or Na(Hg)[101] initiation. Phenylation, unsuccessful under photoinitiation in liquid ammonia, and heteroarylation are possible under photoinitiation in dimethyl sulfoxide.[41,155] The anion of acetophenone also reacts with different aryl and heteroaryl halides in dimethyl sulfoxide under SmI_2 initiation.[80] Other anions of this family are the enolate ions of 2-naphthyl methyl ketone,[158] 2-furanyl[153,155,159] and 2-thienyl methyl ketone.[159]

Phenylation of ketones is also possible with the compounds Ph_2S, $Ph_3S^+Cl^-$, Ph_2Se, $PhNMe_3^+I^-$, Ph_2O, $PhOPO(OEt)_2$,[67,94,136] and aryldiazo *tert*-butyl sulfides.[57,58] Diaryl sulfides and the related sulfoxides and sulfones react with pinacolone enolate ion under irradiation with fragmentation of one carbon-sulfur bond to form arylmethyl *tert*-butyl ketones.[67] Unsymmetrical diaryl sulfides can fragment at both carbon-sulfur bonds, affording a mixture of aryl ketones.[160]

Although good yields of arylation are obtained by reaction of (*Z*)-arylazo *tert*-butyl sulfides with enolates of ketones in dimethyl sulfoxide under laboratory light (Eq. 25),[57,58] competition with an ionic mechanism is observed when the substrates have benzylic hydrogens in the 2- or 4-position with respect to the azothio group.[57,58,161–163]

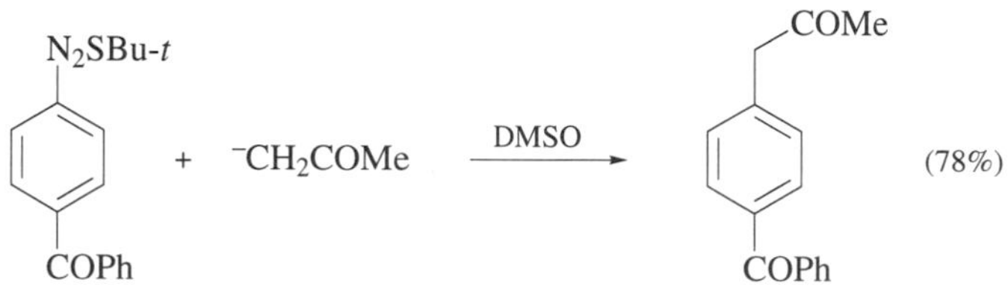

(Eq. 25)

Substrates with Two Leaving Groups. Dihalobenzenes react with enolate ions of ketones to afford monosubstitution or disubstitution products. *m*-Fluoroiodobenzene[108] and 2-fluoro-3-iodopyridine[139] give monosubstitution compounds with retention of fluorine under photoinitiation. Since fluoride ion is not a good leaving group in $S_{RN}1$ reactions, the synthesis of the antiinflammatory drug fluorobiprofen can be achieved by reaction of 4-bromo-2-fluorobiphenyl with the enolate ion of acetone followed by methylation and oxidative demethylation (Eq. 26).[164]

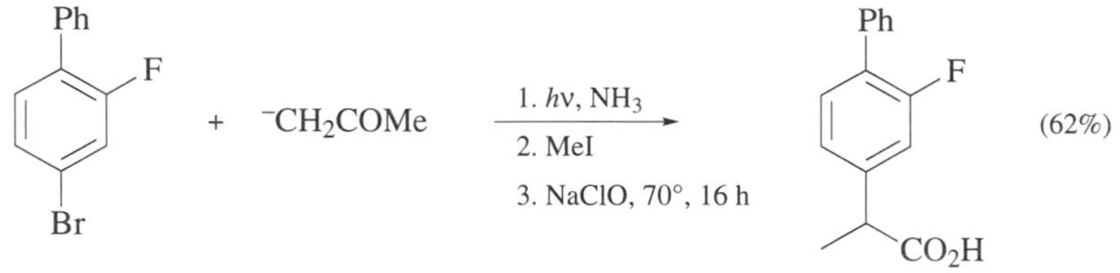
Ph
F
Br
+
$^-CH_2COMe$
1. hν, NH_3
2. MeI
3. NaClO, 70°, 16 h
Ph
F
CO_2H
(62%)

(Eq. 26)

The products of the reaction of *o*-dibromobenzene depend on the nucleophile used. Reaction with the anion of pinacolone leads to the disubstitution product in 62% yield with no evidence of monosubstitution,[165] but with acetone enolate ion, cyclization compounds are formed from aldol condensation of the disubstitution product.[165] Monosubstitution of *p*-dichlorobenzene can be achieved electrochemically in the presence of a redox mediator.[85] On the other hand, disubstitution of this compound and of *p*-dibromobenzene is possible under photoinitiation (Eq. 27).[166]

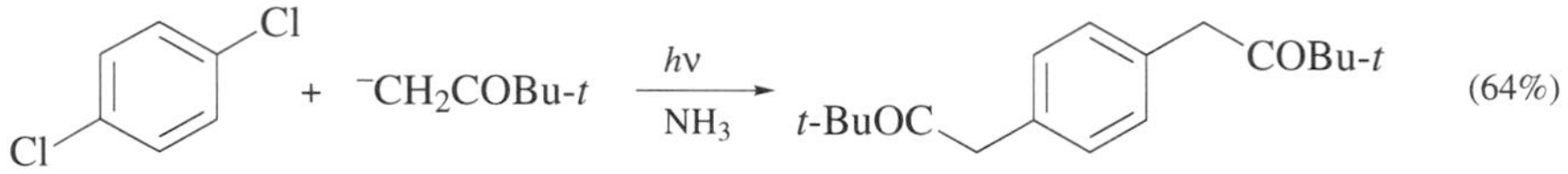
Cl
Cl
+
$^-CH_2COBu$-*t*
hν
NH_3
t-BuOC
COBu-*t*
(64%)

(Eq. 27)

Disubstitution also occurs in the reaction of 2,6- and 2,5-dibromopyridines, and 2,3-, 3,5-, and 2,6-dichloropyridines with the enolate ion of pinacolone.[137,167] A monosubstitution product (70%) resulting from the selective displacement of chlorine from C_4 is obtained in the reaction of the same anion with 4,7-dichloroquinoline.[167] These reactions are not recommended for the synthesis of diketones from dichloropyrazines[55] and dichloropyridazines,[167] which afford mixtures of products.

Reactions of *o*-Functionalized Substrates Followed by Ring Closure. Benzo[*b*]furan,[145] furo[3,2-*h*]quinolines,[155] and furo[3,2-*b*]pyridines[155] can be synthesized through the reaction of enolate ions of ketones or aldehydes with aryl halides substituted with an ortho alkoxide group. The *o*-hydroxyaryl ketones, formed after deblocking the alkoxide function of the $S_{RN}1$ product, undergo spontaneous cyclodehydration to give the furan derivatives quantitatively (Eq. 28).[155]

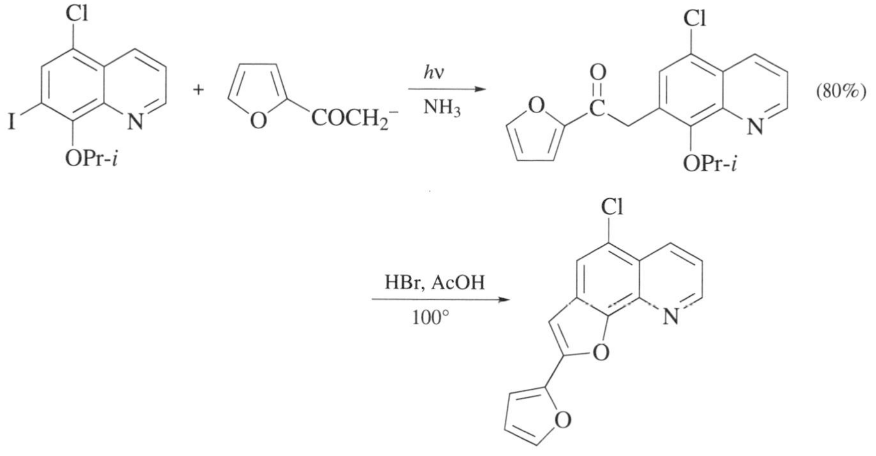
Cl
I
N
OPr-i
+
O
$COCH_2^-$
hν
NH_3
O
O
C
Cl
N
OPr-i
(80%)
HBr, AcOH
100°
Cl
N
O
O

(Eq. 28)

Indoles can be synthesized by the reaction of *o*-amino haloarenes with enolate ions. The uncyclized $S_{RN}1$ substitution product is not isolated (see Ring Closure Reactions). However, when the amino group is protected as the pivaloyl derivative, aminoiodopyridines react with the enolate ions of acetone or pinacolone to give substitution products in yields ranging from 90 to 98% (Eq. 29).[139] Hydrolysis of the pivaloyl moiety, cyclization, and dehydration are simultaneously achieved under acidic conditions to afford 7-, 6- and 5-azaindoles in almost quantitative yields.[139]

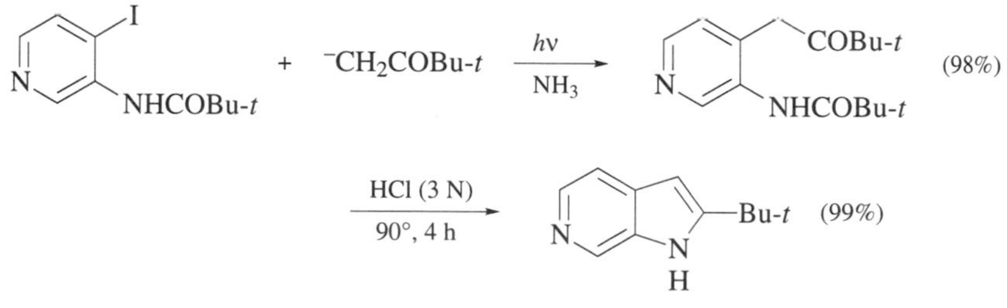
I
N
NHCOBu-t
+
$^-CH_2COBu$-t
hν
NH_3
COBu-t
N
NHCOBu-t
(98%)
HCl (3 N)
90°, 4 h
N
N
H
Bu-t
(99%)

(Eq. 29)

Isocoumarins and benzazepines are some of the interesting compounds synthesized by the $S_{RN}1$ reaction. For example, the reaction of *o*-bromobenzoic acid and its derivatives **10** with the anion of acetone can be used as a synthetic route to esters of type **11** and isocoumarin **12** by treatment of the substitution product with diazomethane or acids, respectively (Eq. 30).[168]

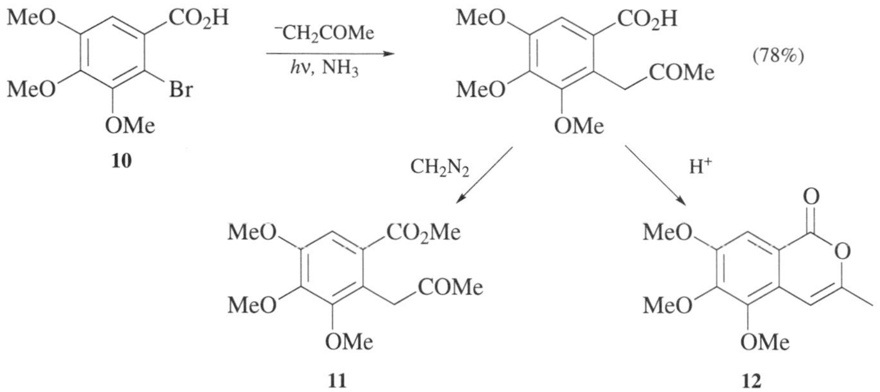
MeO
CO2H
MeO
Br
OMe
10
⁻CH2COMe
hν, NH3
MeO
CO2H
MeO
COMe
OMe
(78%)
CH2N2
H⁺
MeO
CO2Me
MeO
COMe
OMe
11
MeO
O
O
MeO
OMe
12

(Eq. 30)

Benzazepines can be obtained through the photoinitiated reaction of (*o*-iodophenyl)acetic acid derivatives with different enolate ions in 50–60% yields (Eq. 31).[149]

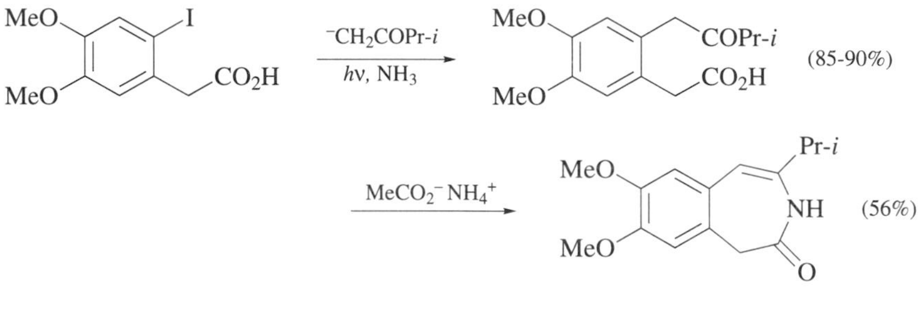
MeO
I
MeO
CO2H
⁻CH2COPr-i
hν, NH3
MeO
COPr-i
MeO
CO2H
(85-90%)
MeCO2⁻ NH4⁺
Pr-i
MeO
NH
MeO
O
(56%)

(Eq. 31)

Carbanions from Esters and Carboxylate Ions

α-Arylacetic esters can be obtained through the reaction of aryl and heteroaryl halides with *tert*-butyl acetate ions. The reaction can be initiated by light[135,169,170] or Fe^{+2} ions.[76] Arylation and diarylation products are formed in yields that depend on the carbanion:substrate ratio. Arylation is favored under Fe^{+2} initiation with a six-fold excess of nucleophile.[76] In most cases, *tert*-butyl acetoacetate is also formed (10–15%). In these reactions, the acids are generally obtained by hydrolysis with aqueous *p*-toluenesulfonic acid (Eq. 32).[76]

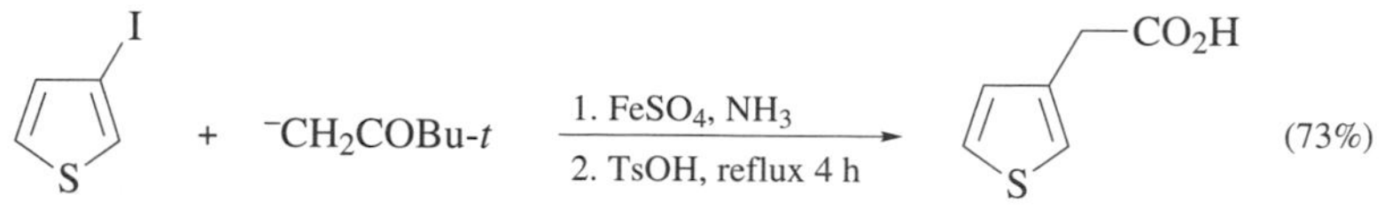

(Eq. 32)

Bromobenzene and electron-rich substrates react sluggishly, and use of a full equivalent of $FeSO_4$ is necessary to achieve complete reaction in a reasonable time.[76] Other nucleophiles reported to react under photoinitiation are the anions of ethyl phenylacetate[167,169] and methyl diphenylacetate.[167] With these anions, substitution of halopyridines[169] and disubstitution of dihalopyridines[167] can be achieved (Eq. 33).

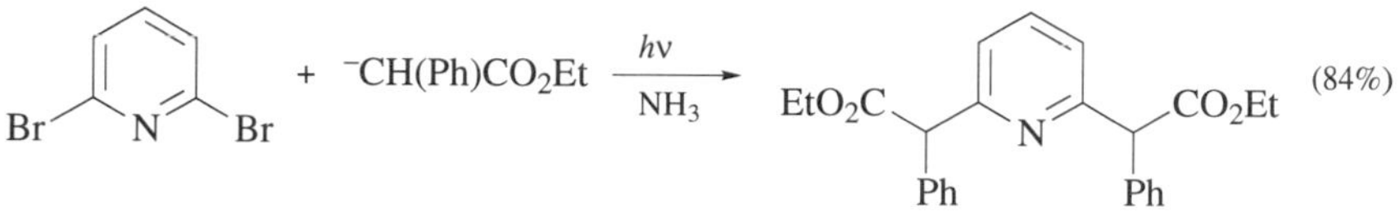

(Eq. 33)

Low yields of substitution products are obtained with anions of tertiary esters that bear β hydrogens. In these reactions, reductive dehalogenation of the substrates competes with the substitution reaction.[135,170]

The only example known of carbanions from carboxylate ions is the reaction of dianion **13**, formed when phenylacetic acid is treated with amide ions in liquid ammonia. This dianion, when irradiated in the presence of haloarenes, can afford products of *p*-arylation and α-arylation depending on the counterion used. For example, with potassium as counterion, only *p*-arylation occurs (Eq. 34), but with lithium only α-arylation is observed, and with sodium both products are formed in comparable yields.[140]

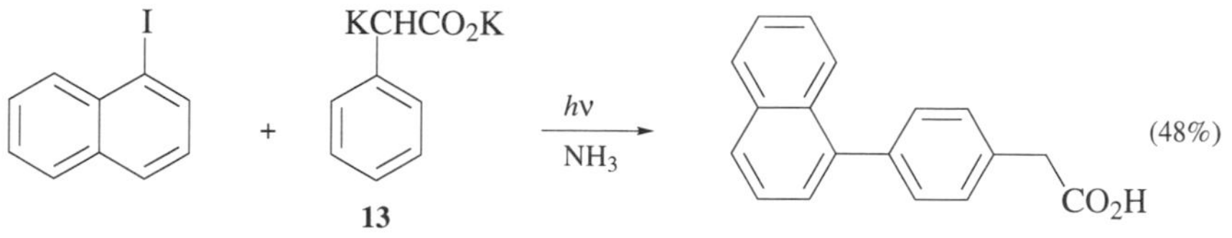

(Eq. 34)

Carbanions from *N,N*-Disubstituted Amides and Thioamides

The anions from acetamide and *N*-methylacetamide are unreactive toward aryl halides under $S_{RN}1$ conditions.[171] However, the carbanions of *N,N*-disubstituted amides react with certain aromatic halides in liquid ammonia under light[98,164,171–173] or Fe^{+2} initiation[76] to form the expected α-arylated compounds in good yield. The insolubility of the potassium salt of *N*-acetylpiperidine in liquid ammonia prevents its satisfactory reaction.[171] A series of *N,N*-dimethyl α-aryl and α,α-diarylacetamides, some of them herbicides, can be prepared by this procedure. In these reactions, the monoarylated compound is favored when the nucleophile:substrate ratio is 10–15.[98,164] Under these conditions, 9-bromoanthracene reacts with the anion of *N,N*-dimethylacetamide to give the arylated compound in 70% yield (Eq. 35).[164] The arylated and diarylated compounds (66% and 12% yields respectively) are formed with a nucleophile:substrate ratio of 5.[98] When the ratio is 2, diarylation becomes more important.[98]

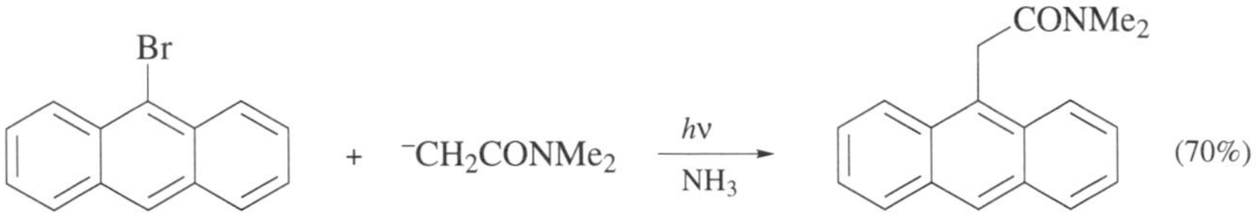
Br
+ ⁻CH2CONMe2
hν
NH3
CONMe2
(70%)

(Eq. 35)

The synthesis of α-arylpropanamides is possible when the product anion is quenched by methyl iodide after irradiation.[164,76] Unsymmetrical α,α-diaryl amides can be formed by reaction of aryl halides with the anion of the α-aryl-*N,N*-dimethylacetamides, as shown in Eq. 36.[98]

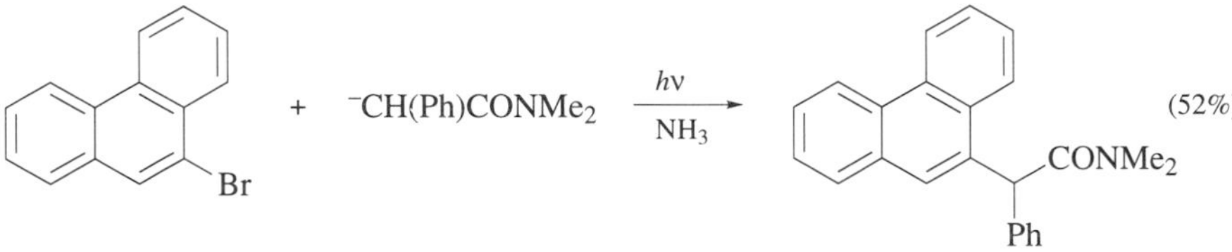
Br
+ ⁻CH(Ph)CONMe2
hν
NH3
CONMe2
Ph
(52%)

(Eq. 36)

When an electron-withdrawing group is attached to the aryl halide, most reactions take place by a benzyne mechanism to yield substituted anilines. However, the reaction of *p*-halobenzonitriles (X = Cl, Br, I) with *N,N*-dimethylacetamide ion with an excess of potassium metal provides high yields (68–98%) of the substitution compound uncontaminated by benzyne products.[98]

Disubstitution is possible by reaction of the anion of *N*-acetylmorpholine with *p*-diiodo- and *p*-chloroiodobenzene in the presence of $FeSO_4$, and fluoroiodobenzenes react as usual to give monosubstitution with retention of fluorine.[76]

Lactam nucleophiles, such as the anion of 1-methyl-2-pyrrolidinone, react with aryl halides under irradiation.[172] When the reaction with 1-iodonaphthalene is quenched with methyl iodide, the 3-(1-naphthyl)-3-methyl product is obtained (Eq. 37).[172]

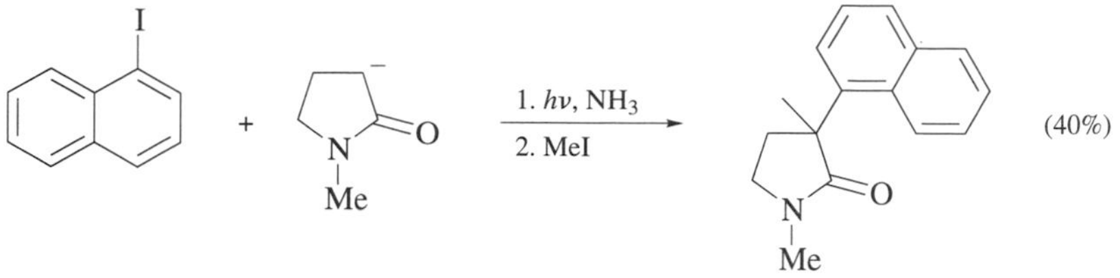
I
+
N
Me
O
1. hν, NH3
2. MeI
N
Me
O
(40%)

(Eq. 37)

An interesting example of stereoselective coupling of an aromatic radical with a nucleophile is found in the reaction of 1-iodonaphthalene with an imide ion containing a chiral auxiliary. In this reaction, the diastereomeric isomers of the substitution compound are formed.[173] The selectivity in this reaction is highly dependent on the metal counterion used. All the ions studied [Li, Na, K, Cs, Ti(IV)] are selective, but the highest stereoselection is reached with Li at low temperature (−78°) and with Ti(IV) (ca. 99%).[173]

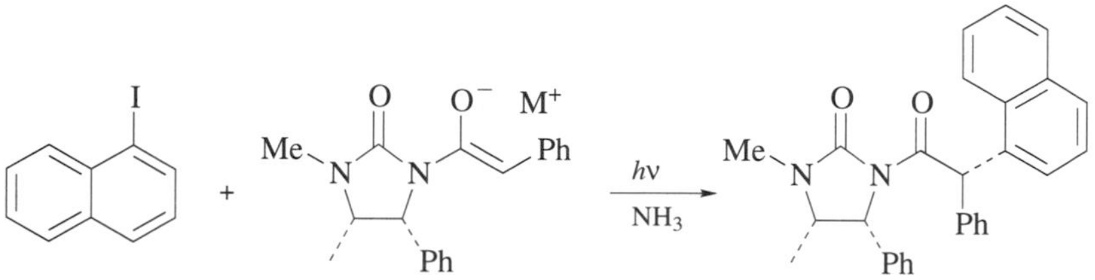
I
+
O
O⁻ M⁺
Me
N
N
Ph
Ph
hν
NH3
O
O
Me
N
N
Ph
Ph

(Eq. 38)

Recently, it has been reported that the anion from *N*-acetylthiomorpholine reacts with aryl iodides in dimethyl sulfoxide under irradiation or Fe^{2+} initiation to give good substitution yields (60–87%).[78]

Carbanions from Nitriles

The main feature of nitrile nucleophiles is that, depending on the aromatic substrate involved, two different outcomes of the substitution reaction are possible: formation of the substitution compound or formation of products from elimination of the cyano group as is the case with phenyl halides[42,43] and halothiophenes.[174] The α-arylated nitrile is formed exclusively by reaction of the anion with compounds that have a π^* molecular orbital of low energy to stabilize the radical anion intermediate of the substitution product as in the reaction of halogen derivatives of naphthalene, benzophenone, quinoline, pyridine, phenanthrene, and biphenyl with acetonitrile,[45–47,134] and of naphthalene and biphenyl with propionitrile anions.[66,175,176] The products with the latter nucleophile give α-arylpropionic acids in excellent yield on hydrolysis. 2-Phenylbutyronitrile[167] and phenylacetonitrile[134,177] also react by the $S_{RN}1$ mechanism. The nitriles formed in the reaction of the latter nucleophile with halopyridines and haloquinolines can undergo oxidative decyanation under phase transfer catalytic conditions to afford phenyl heteroaryl ketones in excellent yield (Eq. 39).[177]

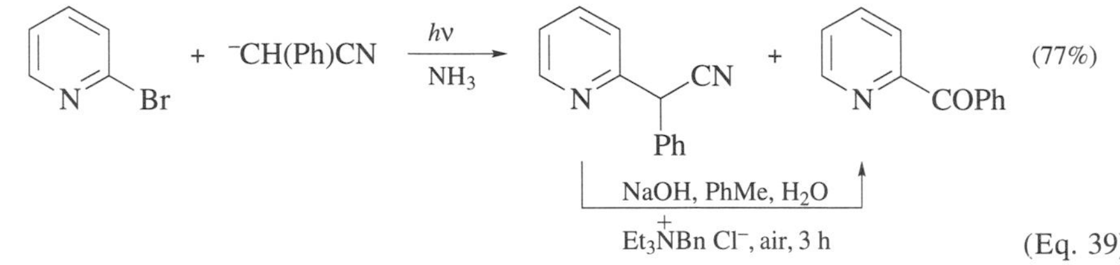
+ ⁻CH(Ph)CN
hν
NH3
Br
N
CN
Ph
+
COPh
(77%)
NaOH, PhMe, H2O
Et3N⁺Bn Cl⁻, air, 3 h

(Eq. 39)

β-Dicarbonyl and Related Carbanions

1,3-Dianions from β-dicarbonyl compounds are suitable nucleophiles and react quite well at the terminal carbon under photoinitiation. For example, arylation at C_1 of the dianion of 2,4-pentanedione with 2-bromomesitylene affords 1-mesityl-2,4-pentanedione in 82% yield.[143] Dialkali salts of benzoylacetone react with 2-chloroquinoline; the best yield of C_1 arylation (71%) is obtained with Li as counterion.[178]

On the other hand, efforts to phenylate a number of monoanions of β-dicarbonyl compounds have been unsuccessful.[41,135,143] Such monoanions do not react with 2-bromopyridine,[137] 2-chloroquinoline[142] or 2-bromobenzamide, but they do react with more electrophilic substrates such as halobenzonitriles,[107,179,180] *p*-bromobenzophenone,[180] bromocyanopyridines,[107] 2-chloro-(trifluoromethyl)-pyridines,[181] and iodoquinolines.[182] In these reactions, excellent yields of arylation products are obtained (Eq. 40).[107] Depending on the nucleophile used, the reaction can be further modified, for example by basic workup, to give the product resulting from a retro-Claisen reaction (Eq. 41).[107]

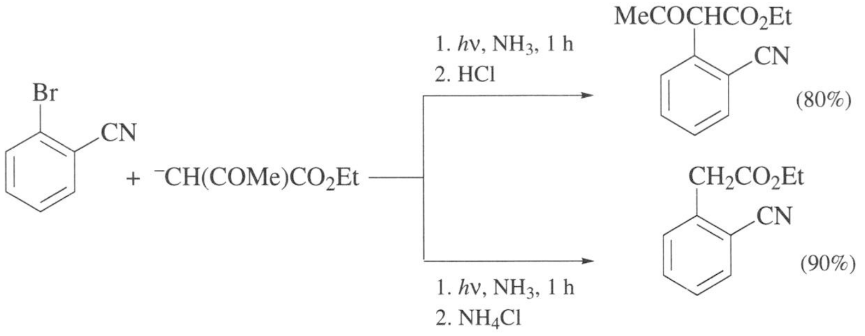
Br
CN
+ ⁻CH(COMe)CO2Et
1. hν, NH3, 1 h
2. HCl
MeCOCHCO2Et
CN
(80%)
CH2CO2Et
CN
(90%)
1. hν, NH3, 1 h
2. NH4Cl

(Eq. 40)

(Eq. 41)

(*E*)-Arylazo phenyl sulfides and (*Z*)-arylazo *tert*-butyl sulfides are also good substrates for these nucleophiles in dimethyl sulfoxide.[59] In the reactions with the potassium monoanion of 2,4-pentanedione, good yields of 3-aryl-2,4-pentanediones are obtained when electron-withdrawing groups are present in the aryl moiety (Eq. 42).[59,183]

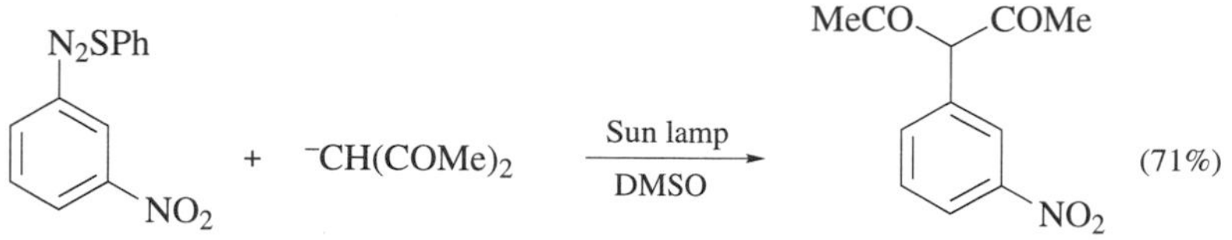
N2SPh
NO2
+ −CH(COMe)2
Sun lamp
DMSO
MeCO
COMe
NO2
(71%)

(Eq. 42)

Carboxylation Reactions

Cobalt carbonyl species allow the catalytic carboxylation of aryl halides through a proposed $S_{RN}1$ process.[184,185] For example, in the reaction of substituted phenyl bromides and iodobenzene with the species CoCRACO (NaH/*tert*-Amyl-ONa/Co(OAc)$_2$ /CO) in the presence of carbon monoxide at atmospheric pressure, the corresponding benzoic acid and ester derivatives are generally formed.[186,187] When amines are present in the reaction media, they compete efficiently with the alkoxide ion to yield benzamides.[186] Among the solvents used (THF, DME, DMF, anisole, benzene), THF gives the best results.

Cobalt carbonyl-catalyzed carboxylation is also possible with $Co_2(CO)_8$ under phase-transfer catalysis conditions provided that the reaction mixture is irradiated (Eq. 43).[188,189] Similar yields of carboxylation products are obtained with

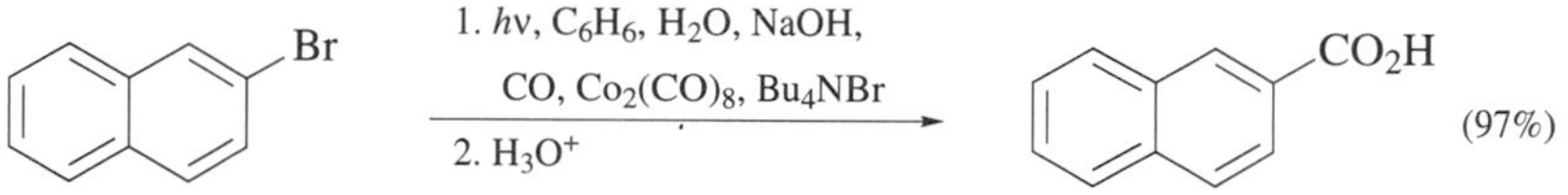
Br
1. hν, C6H6, H2O, NaOH,
CO, Co2(CO)8, Bu4NBr
2. H3O+
CO2H
(97%)

(Eq. 43)

350-nm lamps as with a sunlamp.[189] Tetrabutylammonium bromide is used as the phase transfer catalyst since benzyltriethylammonium chloride is carbonylated under the reaction conditions.[188,190] Carboxylation does not occur with chlorobenzene, a fact that is used to perform the selective carboxylation of *p*-chlorobromobenzene to *p*-chlorobenzoic acid.[188,189] However, a good yield of carboxylation both at the benzylic and the aromatic sites of compound **14** is reported to afford product **15** (Eq. 44).[190]

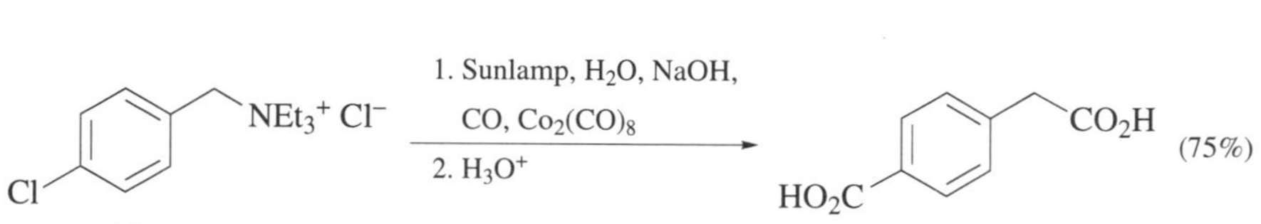
NEt3+ Cl−
Cl
1. Sunlamp, H2O, NaOH,
CO, Co2(CO)8
2. H3O+
CO2H
HO2C
(75%)

14 **15** (Eq. 44)

Cobalt carbonyl-catalyzed polycarboxylation of the less reactive polychlorobenzenes can be achieved under photoinitiation in aqueous NaOH under pressurized carbon monoxide (2 atm) in the absence of an organic solvent and a phase-transfer catalyst.[191] Polycarboxylations of aryl chlorides situated meta or para to another halogen atom (X = Cl, Br) or carboxy group are successful under these conditions (Eq. 45).[191] The reaction with *o*-dichloro compounds can be easily achieved in NaOMe/MeOH (CO, 2 atm) but probably through a different mechanism.[192,193]

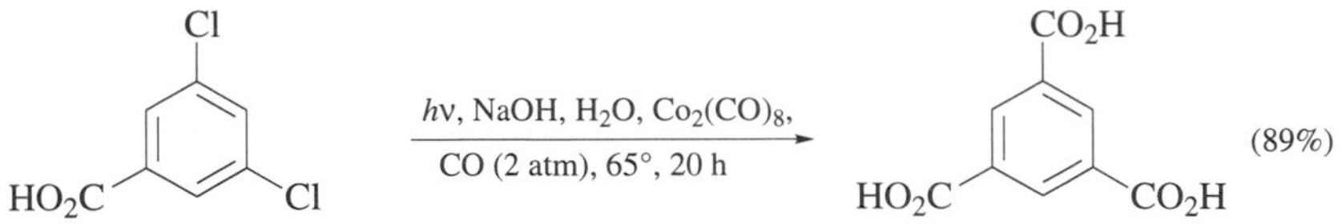

(Eq. 45)

When the reactions are performed with $Co(OAc)_2$, not only halides bearing an electron-withdrawing group, but also simple chlorides are carboxylated.[194] When the aryl halide bears an electron-releasing group, the reaction is less effective.[194]

Carboxylation of aryl halides can also be achieved in the presence of a mixture of $Fe(CO)_5$ and $Co_2(CO)_8$.[195,196] When the reaction is performed in the absence of $Co_2(CO)_8$, benzophenone is the major product, given adequate H_2O/benzene and iodobenzene/$Fe(CO)_5$ ratios as well as NaOH concentration.[197–200]

Phenoxide and Related Ions

Phenoxide and substituted phenoxide ions, mainly di-*tert*-butyl-substituted phenoxides and 1- and 2-naphthoxide ions, react with aromatic substrates under $S_{RN}1$ conditions. These anions couple with radicals at carbon. This reaction is an important route to biaryls unsymmetrically substituted by electron-withdrawing groups and electron-donor groups, as well as to cyclic compounds. The best yields of arylation and heteroarylation are obtained with aryl halides that have electron-withdrawing groups ortho or para to the leaving group and with phenoxide anions substituted by electron-donating groups.

The 2,6- and 2,4-di-*tert*-butyl phenoxide ions have the advantage that the *tert*-butyl groups block two of the three possible coupling positions, thus leading to the selective synthesis of either the ortho or the para isomer. The *tert*-butyl substituents can be easily removed later by transalkylation.[201,202] Phenoxide anions are known to react under electrochemical conditions,[84,85,202–208] usually in the presence of a redox mediator (Eq. 46), or under irradiation.[105,181,209] Monosubstitution of *p*-dichlorobenzene[85,210] as well as the 2,5- and 3,5-dichloropyridines[85] with 2,6-di-*tert*-butyl phenoxide can be achieved electrochemically.

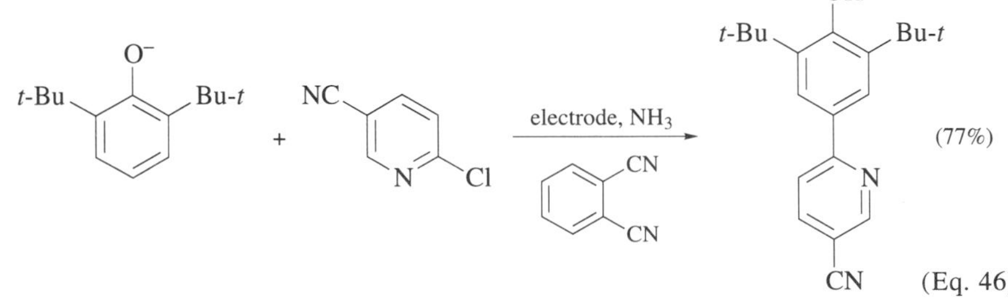

(Eq. 46)

The 1- and 2-naphthoxide ions react with aryl halides under light[105,211,212] or electrochemical initiation.[213] In the reaction of 1-naphthoxide ions, a mixture of 2- and 4-monosubstituted products together with 2,4-disubstituted compounds are formed under irradiation in liquid ammonia.[105,211] Only 4-substitution is obtained (50–85%) with the 2-methyl or 2-*tert*-butyl derivatives of the anion,[211,213] whereas with 4-methoxy-1-naphthoxide anion, the 2-substitution and 4-addition compounds can be formed depending on the solvent used.[211] 2-Naphthoxide ions react to give substitution at C_1 (Eq. 47).[181,209,212,214] The reactivity of the naphthoxide system allows the synthesis of hydroxy derivatives of naphthylquinolines and naphthylisoquinolines via a coupling reaction with haloquinolines (Eq. 48).[215] Another option is the reaction of iodonaphthalenes with the anion of hydroxyquinolines, but this occurs in lower yield (15–51%).[215]

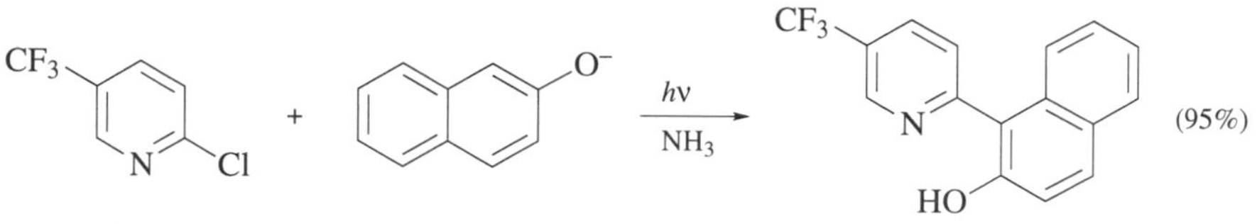
CF3
N
Cl
+
O⁻
hν
NH3
CF3
N
HO
(95%)

(Eq. 47)

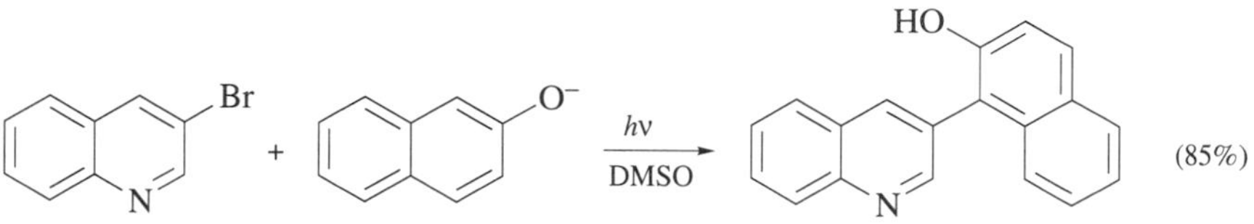
Br
N
+
O⁻
hν
DMSO
HO
N
(85%)

(Eq. 48)

The reaction of aryldiazo phenyl sulfides and aryldiazo *tert*-butyl sulfides with aromatic alkoxides in dimethyl sulfoxide gives moderate to good yields of substitution products.[216,217] The absence of electron-withdrawing groups on the aromatic ring of the azo compounds hampers the arylation process, although yields increase again when the base strength of the nucleophile is increased.[217]

Cyanide and Nitronate Ions

The cyanide anion is one of the less reactive nucleophiles toward aromatic radicals and thus shows low reactivity under $S_{RN}1$ conditions.[86,87,218] Even so, it reacts under controlled potential scale electrolysis with 4-bromobenzophenone (95% substitution), but it fails to react under similar conditions with 1-bromonaphthalene.[88] The recommended substrates to be employed with cyanide ion are diazosulfide derivatives, which by reaction with excess tetrabutylammonium cyanide in dimethyl sulfoxide under irradiation or electrode initiation give nitriles and phenyl aryl sulfides (Eq. 49).[114,115,219]

The yields of nitriles can be improved with higher cyanide/diazosulfide ratios.[114] When the aryldiazo substrate has bromine or chlorine as a substituent, the introduction of two cyano groups is achieved with satisfactory results.[115,127]

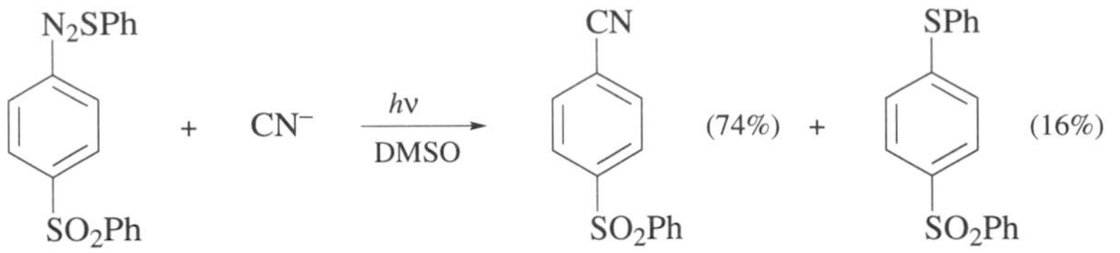
(Eq. 49)

Nitronate anions are poor reaction partners because of fragmentation reactions of the radical anion intermediate of the substitution product.[40,41] For example, whereas 4-isopropylbenzophenone (50%) and benzophenone (26%) are formed in the reaction of 2-nitropropane anion with 4-bromobenzophenone under electrochemical initiation, benzene (71–96%) is the main product in the reaction with iodobenzene.[40]

Other Carbanions

Picolyl anions can be used as nucleophiles of the $S_{RN}1$ reaction (Eq. 50).[134,220] In their reactions with bromobenzene and iodobenzene, benzyne and $S_{RN}1$ processes are formed competitively.[220]

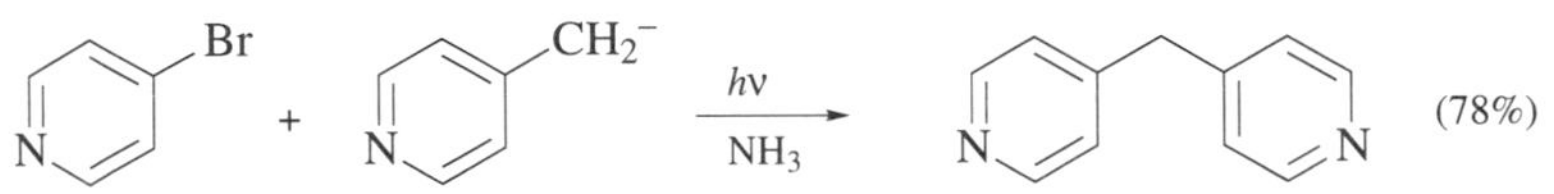
(Eq. 50)

Other carbanions from alkyl-substituted oxazolines and thiazoles are also known to react. Thus, the carbanion from 2,4,4-trimethyl-2-oxazoline is phenylated (56%) and diphenylated (28%) by iodobenzene under light initiation.[169] With bromomesitylene, monoarylation prevails (Eq. 51).[169]

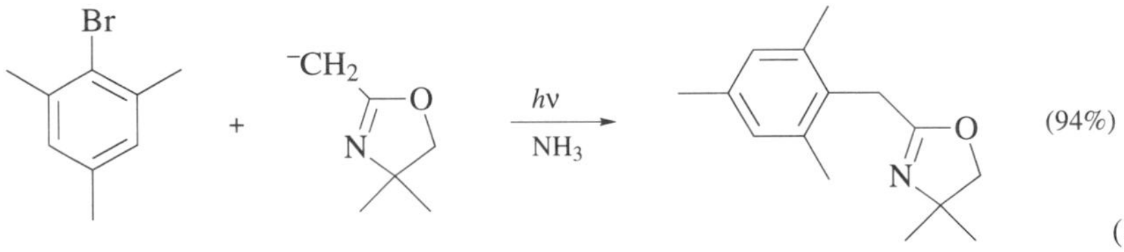
(Eq. 51)

Nucleophiles from Tin

Although reactions of triorganostannyl ions with haloarenes are known, there are few reports concerning their mechanism.[221,222] Recently it has been proposed that aryl chlorides react with trimethylstannyl anions through the $S_{RN}1$ mechanism to give the substitution compounds in high yield (88–100%).[223] Disubstitution is achieved with *p*-dichlorobenzene (Eq. 52).[223] On the other hand, bromo and iodo arenes are not recommended because they react with trimethylstannyl ions by a halogen metal exchange reaction to give dehalogenated products.[223]

Different behavior is observed with the triphenylstannyl anion. The substitution products are formed (62–80% yield) in the photoinitiated reaction of this

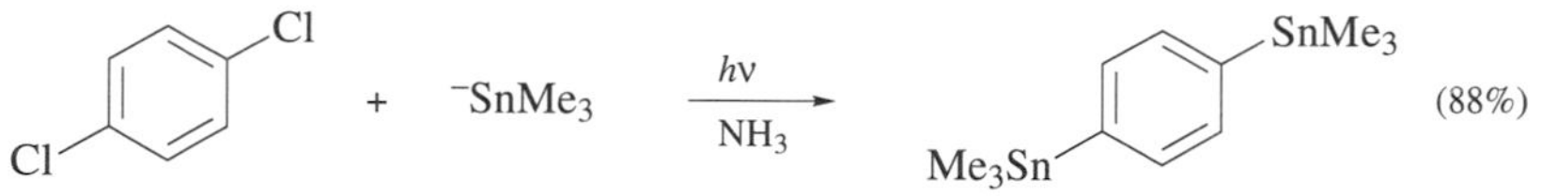

(Eq. 52)

anion with *p*-chloro- and *p*-bromotoluenes, 1-chloro- and 1-bromonaphthalenes, and 2-chloroquinoline in liquid ammonia.[223] On the other hand, with aryl iodides reduction to arenes by a halogen metal exchange reaction predominates.[223]

Other nucleophiles derived from tin can be prepared by Sn-alkyl bond fragmentation from a tetralkyltin compound. Thus, *p*-anisyltrimethyltin (**16**) reacts selectively with sodium metal in liquid ammonia to afford nucleophile **17**, which reacts with *p*-chlorotoluene upon irradiation to give unsymmetrical dimethyldiaryltin **18** in high yield.[224]

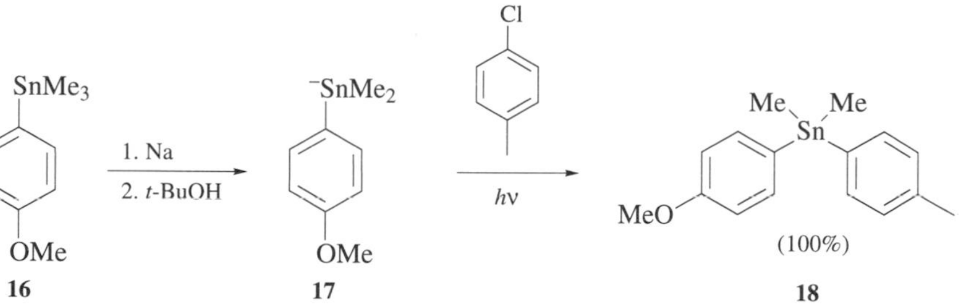

(Eq. 53)

Arylated product **18** can also be obtained by a one-pot reaction starting from trimethylstannyl ions and *p*-chloroanisole, which gives intermediate **16** upon irradiation. This compound is treated without isolation with sodium metal to form nucleophile **17**, which by photoinitiated reaction with *p*-chlorotoluene affords **18** (89%) in a one-pot reaction.[224]

Nucleophiles from Nitrogen

Amide ions are the simplest nitrogen nucleophiles that can react through the $S_{RN}1$ process, and the possibility of a benzyne mechanism can be avoided by reaction in the presence of excess solvated electrons. For example, the reaction of amide ions with *o*-haloanisoles (X = I, Br) in the presence of excess potassium metal provides only *o*-anisidine uncontaminated by the meta isomer.[4] Aniline (73%) can be obtained by the potassium metal-initiated reaction of potassium amide with phenyl diethyl phosphate in liquid ammonia.[225] The photoinitiated reaction of amide anion with 2-bromomesitylene, in which the benzyne mechanism is precluded, affords 2-aminomesitylene and mesitylene in 70% and 6% yields, respectively.[126] 2-Bromopyridine fails to react under irradiation.[47]

In the reaction of aryl halides with the anions of aromatic amines, coupling at carbon instead of at nitrogen usually takes place. For example, 2-naphthylamide ions react under irradiation with aryl iodides to render mainly the 1-aryl-2-naphthylamines (45–63%) (Eq. 54).[226]

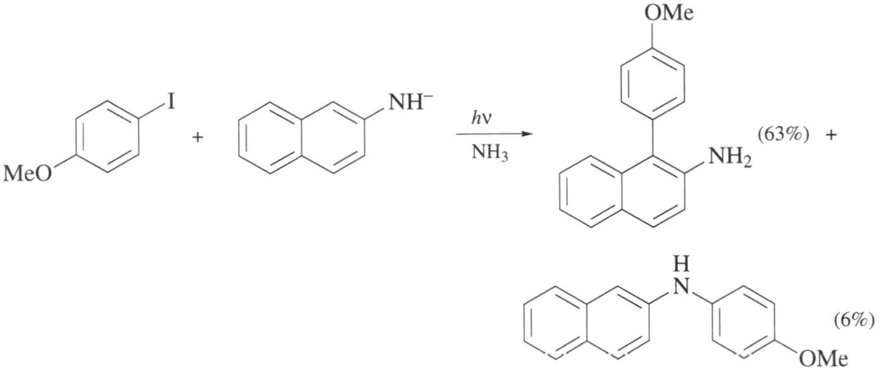

(Eq. 54)

Arylpyrroles, arylindoles, and arylimidazoles can be synthesized electrochemically under $S_{RN}1$ conditions by reaction of pyrrolyl, indolyl, or imidazolyl anions in liquid ammonia.[227–229] For instance, the electrochemically initiated reaction of pyrrolyl anion with 4-chloropyridine in the presence of a redox mediator provides the 2-substituted product together with small yields of the 3-substituted isomer (Eq. 55).[227]

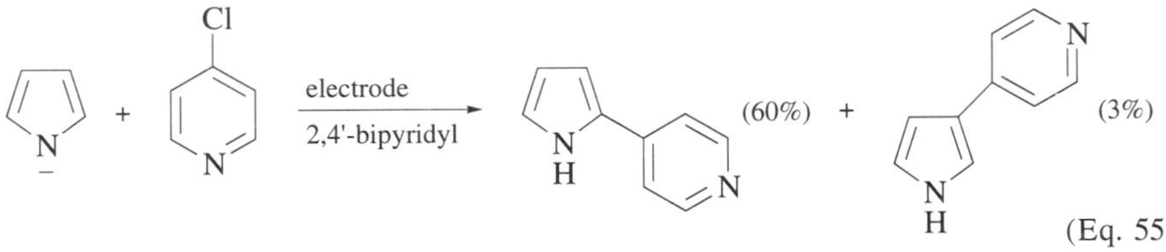

(Eq. 55)

With the anion from 2,5-dimethylpyrrole, the product corresponding to coupling at C-3 is obtained in moderate yield (40%),[228] and addition products at position 2 are also formed.[229] Substitution of the 3-position is the main path in the reaction of indolyl anion with 4-chloropyridine and *p*-chlorobenzonitrile.[227,229]

Direct electrochemical reduction of 4-chlorobenzonitrile in the presence of uracil anion in dimethyl sulfoxide gives nucleoside **19** (Eq. 56).[230] Nucleosides substituted at the 5 position represent an important class of biologically active compounds.

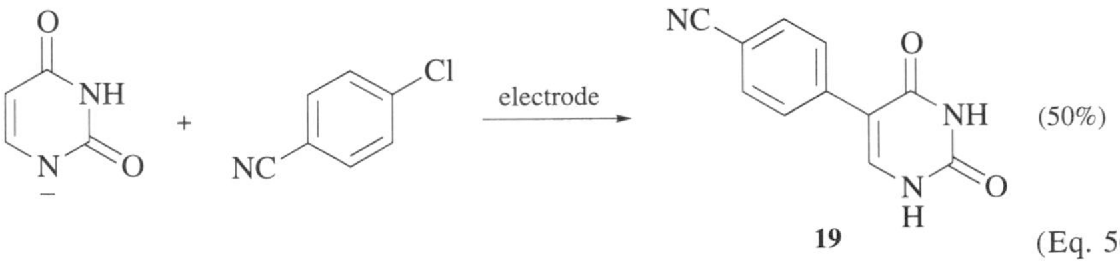

19

(Eq. 56)

Nucleophiles from Phosphorus

Aryl iodides react with diethyl phosphite anion [$(EtO)_2PO^-$] in liquid ammonia,[231–234] and acetonitrile/THF[235] under photoinitiation to form diethyl aryl phosphonates in 70–98% yield. Aryl bromides react sluggishly,[231,236] but the addition of sodium iodide accelerates the photoinitiated reaction in acetonitrile/THF[65] and in liquid ammonia.[64] Consequently, these compounds become suitable for preparative use. The anion can also react by Fe^{+2} catalysis in liquid ammonia[75] but fails to react under these conditions with iodobenzene in dimethyl sulfoxide.[77]

Substitutions of bromo- and iodoanilines,[233,234] bromoaminonaphthalenes,[234,237] and bromoaminopyridines[234] offer interesting synthetic possibilities (Eq. 57).[234] Product **20**, after iodination via the diazonium salt, gives the substrate of a second $S_{RN}1$ reaction, which affords disubstituted compound **21** (Eq. 58).[234] Disubstitution by diethyl phosphite anion can be achieved upon reaction with appropriate haloiodobenzenes.[231,238,239]

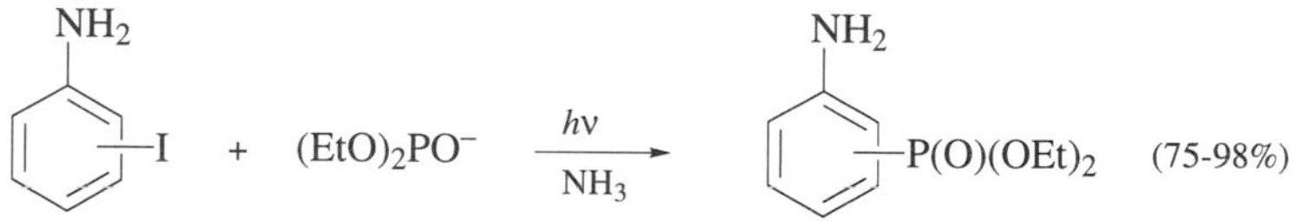

NH2
I
+
(EtO)2PO⁻
hν
NH3
NH2
P(O)(OEt)2
(75-98%)

20

(Eq. 57)

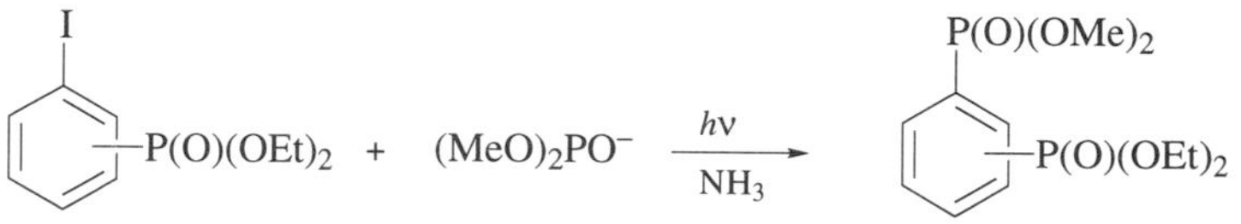

I
P(O)(OEt)2
+
(MeO)2PO⁻
hν
NH3
P(O)(OMe)2
P(O)(OEt)2

21

o- (60%), *m*- (70%), *p*- (99%)

(Eq. 58)

The anions butyl phenylphosphinate [$PhP(OBu)O^-$],[72] *O,O*-diethyl thiophosphite [$(EtO)_2PS^-$],[72] *N,N,N',N'*-tetramethylphosphonamide [$(Me_2N)_2PO^-$],[72] and dibenzylphosphinite [$(PhCH_2)_2PO^-$][240] can also be used as nucleophiles. Reactions of the first two anions with iodobenzene occur in nearly quantitative yield.[72] Another nucleophile that reacts with different aryl halides is diphenylphosphide ion, which gives aryldiphenylphosphines, generally isolable as the oxides.[71,240] This nucleophile reacts in the dark with *p*-iodotoluene in liquid ammonia and in dimethyl sulfoxide (89% and 78%, respectively).[71] Reaction of the anion with 1-chloronaphthalene and 2-chloroquinoline affords substitution products (83% and 96%, respectively) when initiated by Na(Hg).[102] Initiation by sonication under pressurized liquid ammonia is also possible.[241]

Another route to symmetrical and unsymmetrical triarylphosphine oxides is the reaction of elemental phosphorus with sodium metal in liquid ammonia which affords a "P^{3-}" species that reacts with phenyl halides under irradiation to form triphenylphosphine oxide in fair to good yield after oxidation.[242] When the triphenylphosphine thus formed is further treated with sodium metal, it gives a diphenylphosphide ion, which reacts with *p*-bromoanisole to form the unsymmetrical phosphine oxide (55%) in a one-pot reaction (Eq. 59).[242]

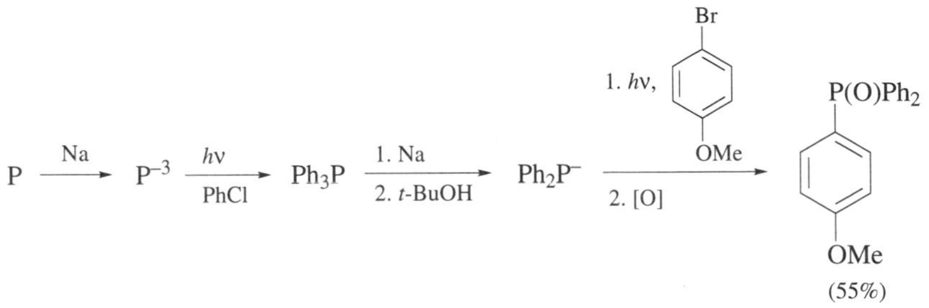
Br
1. hν,
OMe
2. [O]
P(O)Ph2
P
Na
P-3
hν
PhCl
Ph3P
1. Na
2. t-BuOH
Ph2P−
OMe
(55%)

(Eq. 59)

Nucleophiles from Arsenic and Antimony

As in the case of phosphorus, the reaction of elemental arsenic (or antimony) with sodium metal in liquid ammonia forms an "As^{3-}" (or "Sb^{3-}") species that reacts with chloro- or bromobenzene under irradiation to form triphenylarsine (or triphenylstibine) in fair to good yield.[242] Diphenylarsenide ions react in liquid ammonia under photoinitiation with *p*-halotoluenes to afford four products: triphenylarsine, *p*-tolyldiphenylarsine, di-*p*-tolylphenylarsine, and tri-*p*-tolylarsine.[118] Similar product distributions are obtained in reactions with *p*-haloanisoles,[118] 1-bromonaphthalene,[119] and 9-bromophenanthrene.[119] This product distribution is explained on the basis of the fragmentation reaction of the radical anion intermediates.[118,119] Unsymmetrical aryldiphenylarsines can be prepared by reaction of diphenylarsenide ions with 2-chloroquinoline[119] and 4-chlorobenzophenone[118] without scrambling of the aryl rings (Eq. 60).

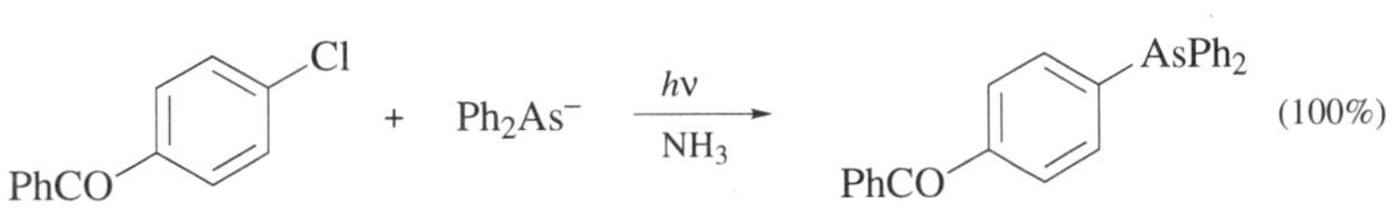
Cl
PhCO
+
Ph2As−
hν
NH3
AsPh2
PhCO
(100%)

(Eq. 60)

The diphenylstibine ion reacts under photoinitiation to afford scrambled products even with 4-chlorobenzophenone.[119]

Nucleophiles from Sulfur

Alkanethiolate Ions. The photoinitiated reaction of iodobenzene with ethanethiolate ion in liquid ammonia affords not only ethyl phenyl sulfide but also benzenethiolate ion, trapped as the benzyl derivative, and diphenyl sulfide (Eq. 61).[160] This product distribution is attributed to fragmentation of the radical anion intermediate at the S-C alkyl bond. This reaction competes with the electron transfer step that leads to the substitution product.

$$\mathrm{PhI} + \mathrm{EtS}^- \xrightarrow[\text{2. BnCl}]{\text{1. } h\nu,\ \mathrm{NH_3}} \mathrm{PhSBn}\ (44\%) + \mathrm{PhSEt}\ (30\%) + \mathrm{Ph_2S}\ (3\%) \qquad \text{(Eq. 61)}$$

However, straightforward substitution takes place in the reaction of the ethanethiolate anion with bromobenzenes substituted by electron-withdrawing groups (70–95% yield),[49] halopyridines, even in the presence of electron-releasing groups (35–87%),[49] and 2-chloroquinoline (85%).[49]

A variable pattern of substitution relative to fragmentation is also observed with the anions from methanethiol,[48,88,243] *n*-[45,48] and *tert*-butylthiol,[48,88,89] benzyl mercaptan,[48,49,244] ethyl 3-mercaptopropionate [$EtO_2C(CH_2)_2SH$],[49,244] and ethyl mercaptoacetate (EtO_2CCH_2SH),[49] depending on the substrate involved.

Straightforward substitution is achieved by reaction of the anion of mercaptoethanol [$HO(CH_2)_2S^-$] with different aryl halides.[45,49,139,245] Thiazine and benzo[*e*]- and benzo[*g*]thiazines can be synthesized from the substitution product obtained with *o*-aminobromo compounds, after replacement of the hydroxy group by chloride (Eq. 62).[233,234,237]

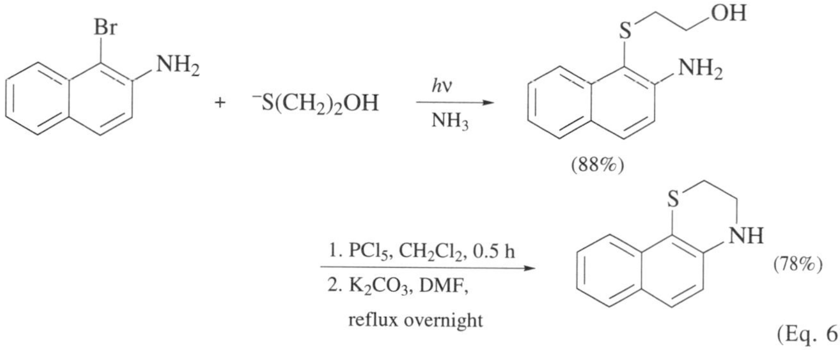
Br
NH2
+ ⁻S(CH2)2OH
hν
NH3
OH
S
NH2
(88%)
1. PCl5, CH2Cl2, 0.5 h
2. K2CO3, DMF,
reflux overnight
S
NH
(78%)

(Eq. 62)

Polymethylenedisulfides can be obtained by reaction of 1- or 2-halonaphthalenes (X = Br, I) with the dianions $^-S(CH_2)_nS^-$ (n = 2–4) (Eq. 63).[246]

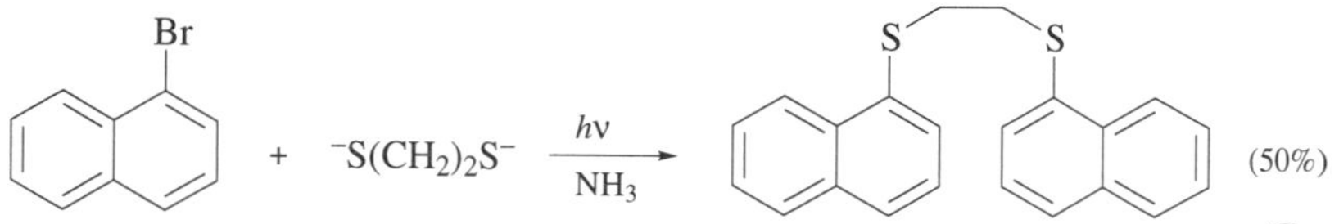
Br
+ ⁻S(CH2)2S⁻
hν
NH3
S
S
(50%)

(Eq. 63)

Arene- and Heteroarenethiolate Ions. Aryl iodides react with benzenethiolate ions in liquid ammonia under irradiation to afford aryl phenyl sulfides in good yields.[160,247] The reaction with different bromo-, chloro- and iodoaryl derivatives can be achieved under photo- or thermal initiation (usually in dimethyl sulfoxide, DMF, or acetonitrile)[248–253] or under electrochemical initiation,[40,88,254–260] as well as by reaction of arenediazonium tetrafluoroborates in dimethyl sulfoxide at room temperature.[127,261,262]

2-Chloropyridines bearing a CF_3 substituent exhibit good reactivity toward a number of heteroarenethiolate ions in liquid ammonia under irradiation.[245] The synthesis of unsymmetrical disubstituted phenyl, naphthyl, and pyridyl derivatives is possible in this solvent through the reaction of haloarylamino compounds with heteroarenethiolate ions. This affords the substituted arylamine **22** (Eq. 64).[234] Compound **22**, after iodination via the diazonium salt, gives product **23** which is the substrate of a second $S_{RN}1$ process that ultimately affords the unsymmetrical disubstitution product **24** (Eq. 65).[234]

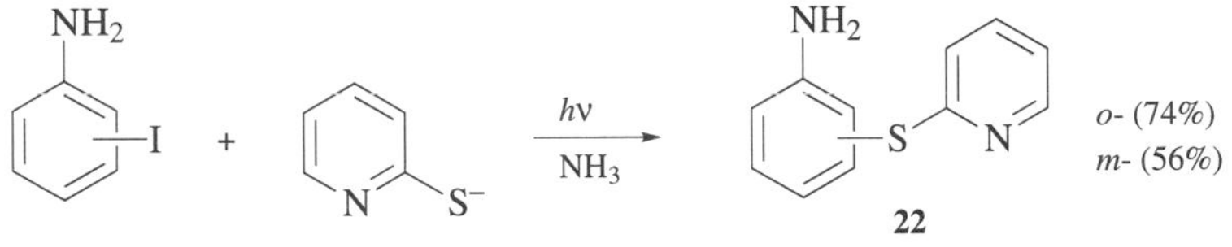
NH2
I
S⁻
N
hν
NH3
NH2
S
N
22
o- (74%)
m- (56%)

(Eq. 64)

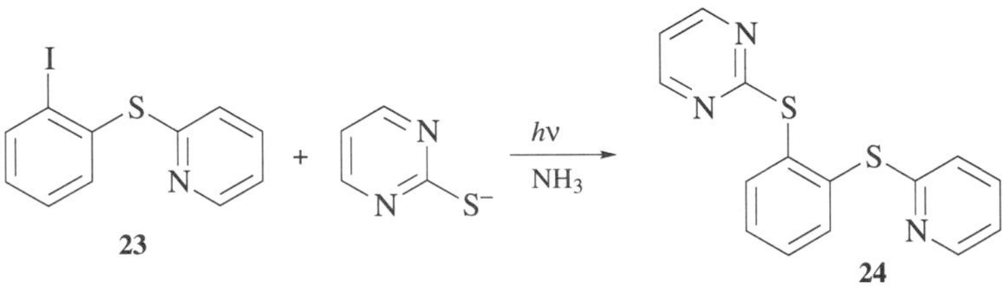
I
S
N
23
N
N
S⁻
hν
NH3
N
N
S
S
N
24
(90%)

(Eq. 65)

Disubstitution by the same sulfur nucleophile can be achieved under light or electrode stimulation with appropriate dihaloarenes[85,263,264] or with phenyldiazonium salts bearing a halogen in the benzene ring.[127,262]

Trisubstituted phenyl, naphthyl, and pyridyl rings can be obtained by $S_{RN}1$ substitution of a haloarylamino- or haloheteroarylamino compound. Halogenation of the product affords the substrate of a second $S_{RN}1$ process which is followed by replacement of the amino group by halogen in order to obtain the substrate of a third $S_{RN}1$ substitution. Compounds of type **25**, **26**, and **27** are obtained by this procedure.[234]

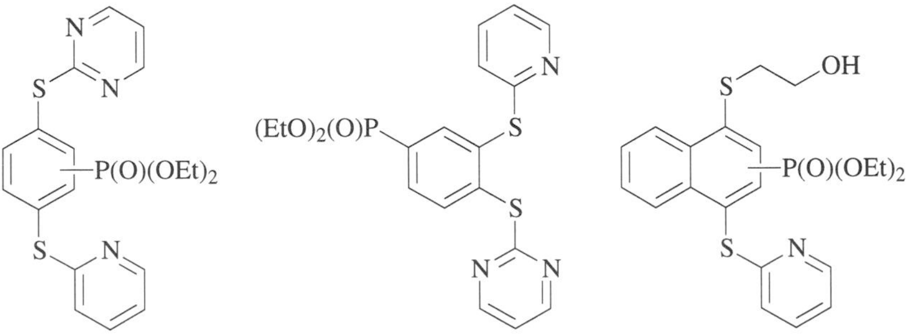
N
N
S
P(O)(OEt)2
S
N
(EtO)2(O)P
N
S
S
N
N
S
OH
P(O)(OEt)2
S
N

25 **26** **27**

The protected *p*-mercaptophenylalanine methyl ester ion can be coupled with a variety of iodinated aryl amino acid derivatives.[265] For example, diiodotyrosine methyl ether gives disubstitution products in excellent yields (Eq. 66).[265] The

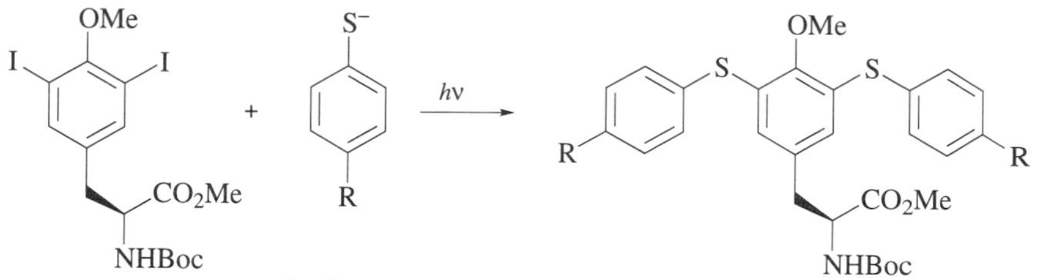
OMe
I
I
+
S⁻
R
hν
CO2Me
NHBoc
S
S
R
R

R = H (>90%)

R = $CH_2CH(NHBoc)CO_2Me$ (>95%) (Eq. 66)

phenylalanine derivatives do not racemize under these conditions. On the other hand, the substrate and product racemize rapidly in the reaction of *N*-Boc protected (3,5-diiodo-4-methoxyphenyl) glycine methyl ester with benzenethiolate anion in liquid ammonia at −33°. Racemization can be avoided by use of the unprotected amino acid.[265]

Thiocarboxylate Ions. The reaction of arenediazonium tetrafluoroborates with potassium thioacetate[34] or sodium thiobenzoate ions[266] in dimethyl sulfoxide leads to aryl thioesters, which can either be isolated (Eq. 67) or react further, providing a convenient one-pot access to other aromatic sulfur derivatives.[267]

N2+BF4−
+ PhCOS−
DMSO
SCOPh
(70%)

(Eq. 67)

Nucleophiles from Selenium and Tellurium

Selenide and Telluride Ions. Na_2Se, formed by reaction of selenium metal with sodium metal in liquid ammonia, reacts under irradiation with iodobenzene to give diphenyl selenide (12%) and diphenyl diselenide (78%), after oxidation of the benzeneselenenate ions formed as products.[268] However, if sodium metal is added after irradiation, benzeneselenenate ion is the only product formed. This is isolated by oxidation to diphenyl diselenide (Eq. 68), or trapped with methyl iodide to give methyl phenyl selenide (Eq. 69).[268] This is a convenient route either

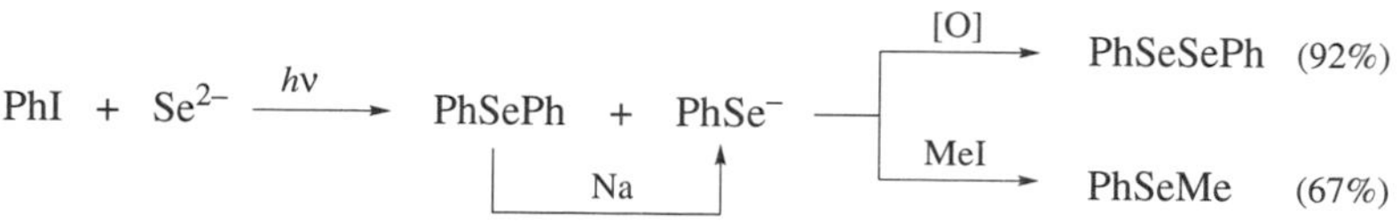
PhI + Se2−
hν
PhSePh + PhSe−
Na
[O]
PhSeSePh (92%)
MeI
PhSeMe (67%)

(Eq. 68)

(Eq. 69)

to the areneselenenate nucleophiles used to synthesize diaryl selenides by $S_{RN}1$ reaction with aryl halides, or to diaryl diselenides used as precursors of the areneselenenate nucleophiles.

Alkali metals react directly with elemental selenium or tellurium in polar aprotic solvents to give, depending on their relative concentration, M_2Z_2 or M_2Z (Z = Se, Te) thus avoiding the use of liquid ammonia.[269–272] The Na_2Te can also be prepared from tellurium and Rongalite ($HOCH_2SO_2Na$) in dilute aqueous alkali and from tellurium and sodium hydride in dry DMF.[273–275] The nucleophiles formed by any of these procedures undergo a direct thermal substitution reaction with aryl halides. There is not enough mechanistic evidence to confirm the existence of an $S_{RN}1$ process in these reactions.

Another approach to diaryl diselenides or ditellurides is the reaction of an aryl halide with the nucleophiles Z_2^{2-} (Z = Se,Te) electrogenerated from the elements by using sacrificial Z electrodes in acetonitrile (Eq. 70).[276,277]

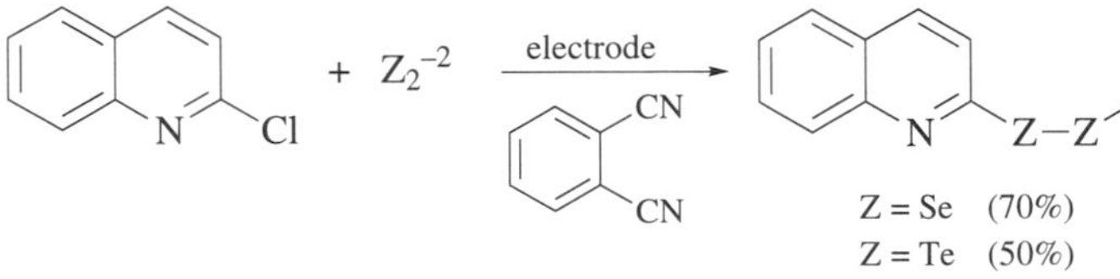

Cl
N
+ Z2-2
electrode
CN
CN
N
Z–Z
N
Z = Se (70%)
Z = Te (50%)

(Eq. 70)

An improvement in this technique is the use of an undivided cell equipped with a Mg sacrificial anode, with addition of fluoride ions to avoid precipitation of the Z_2Mg.[257,278,279] Under these conditions, 2-chloroquinoline reacts with Se_2^{2-} ions to give substitution product in 79% yield. Other diselenides can also be prepared by this methodology (Eq. 71).[278] The yields with Te_2^{2-} are lower.

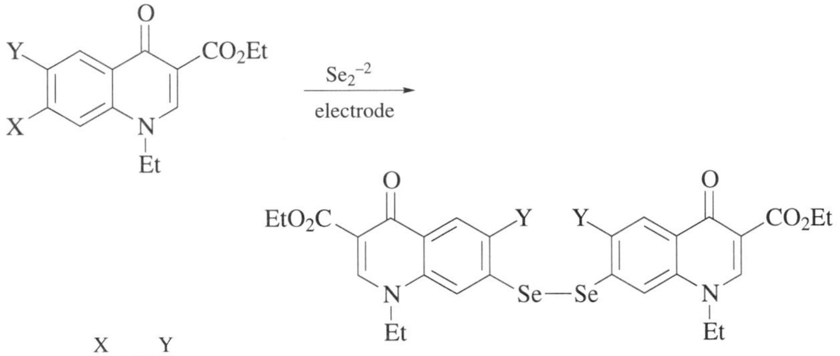

O
Y
CO2Et
X
N
Et
Se2-2
electrode
O
O
EtO2C
Y
Y
CO2Et
N
Se—Se
N
Et
Et

X	Y	
Br	H	(70%)
Cl	F	(82%)

(Eq. 71)

Benzeneselenenate and Benzenetellurate Ions. These anions, formed by deprotonation of their conjugated acids or by any of the previously mentioned methods, react with aryl halides under photoinitiation in liquid ammonia. Straightforward and scrambled products are obtained with both nucleophiles depending on the substrate employed.[120,122] The degree of fragmentation is lower with benzeneselenenate ions. Disubstitution is obtained in the photoinitiated reaction of *p*-bromoiodobenzene with benzeneselenenate ions (Eq. 72),[122] whereas lower yields (40%) are obtained with benzenetellurate ions.[122]

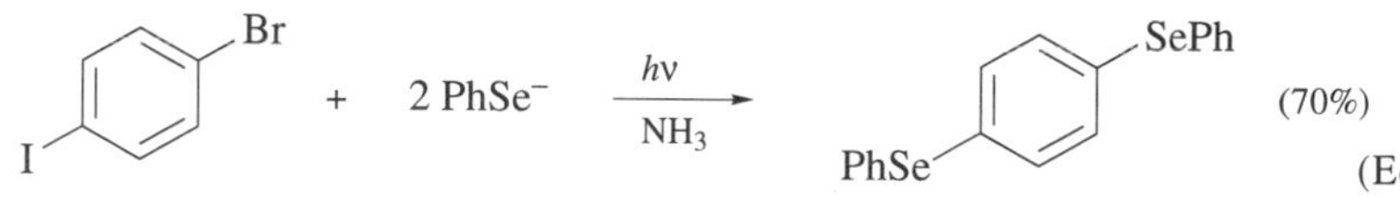

(Eq. 72)

Bromobenzophenones,[280,281] bromo-[282,283] or chlorobenzonitriles,[257,281,282,284] and other aryl bromides and dibromides[285] react with benzenetellurate or benzeneselenenate ions prepared by electrochemical reduction of diphenyl diselenide or ditelluride in acetonitrile by direct or mediated cathodic reduction.

Ring Closure Reactions

$S_{RN}1$ Substitution Followed by a Ring Closure Reaction Involving an *ortho* Substituent. In this field, an important example of substitution followed by spontaneous ring closure is the synthesis of substituted indoles by the photoinitiated reaction of *o*-iodo- and *o*-bromoanilines with carbanions from ketones in liquid ammonia (Eq. 73).[106,286] This reaction can also be initiated electrochemically[287] or by Fe^{2+} ions.[75]

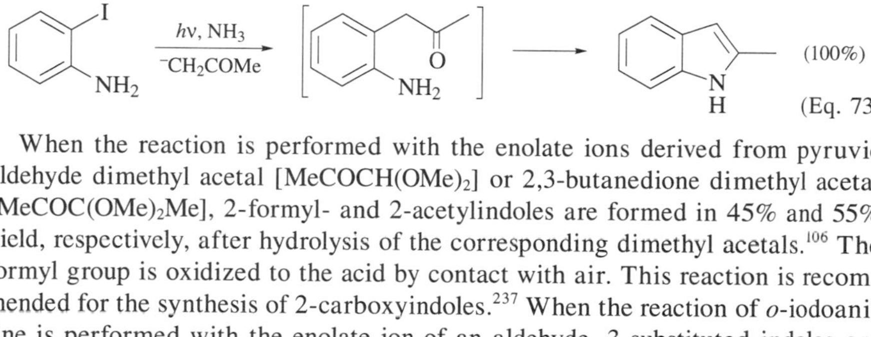

(Eq. 73)

When the reaction is performed with the enolate ions derived from pyruvic aldehyde dimethyl acetal [$MeCOCH(OMe)_2$] or 2,3-butanedione dimethyl acetal [$MeCOC(OMe)_2Me$], 2-formyl- and 2-acetylindoles are formed in 45% and 55% yield, respectively, after hydrolysis of the corresponding dimethyl acetals.[106] The formyl group is oxidized to the acid by contact with air. This reaction is recommended for the synthesis of 2-carboxyindoles.[237] When the reaction of *o*-iodoaniline is performed with the enolate ion of an aldehyde, 3-substituted indoles are obtained.[106] The yields (26–75%) are lower than with enolate ions from aliphatic ketones. Benzo[*e*]- and benzo[*g*]indoles can be synthesized through the reaction of 1-bromo-2-amino- or 1-amino-2-bromonaphthalene with the anions of pinacolone (67% and 82% yields of cyclization, respectively).[237] *o*-Haloaminopyridines react with carbanions to afford azaindoles.[139,146,148,288]

The $S_{RN}1$ mechanism can be a route to thienopyridines by reaction of the anion of ethyl mercaptoacetate (EtO_2CCH_2SH) with 2-bromo-3-cyanopyridine (Eq. 74).[49]

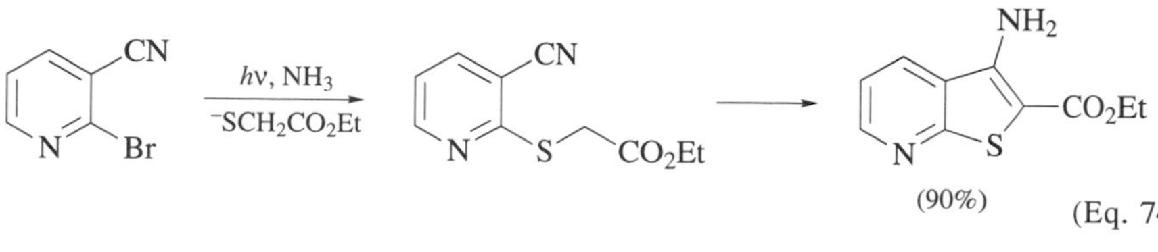

(Eq. 74)

Isocarbostyrils **28** are accessible by the reaction of *o*-bromo- or *o*-iodobenzamides with carbanions (Eq. 75).[147]

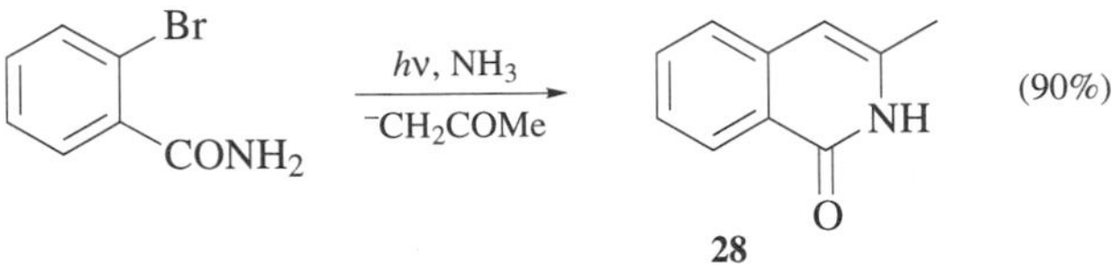

28

(Eq. 75)

The yields in the cyclization are better when the benzamides are not *N*-methylated. The *N*-alkylated isocarbostyrils can be obtained by phase-transfer alkylation of product **28**.[168] The reaction of substituted *o*-iodobenzamides and the enolate ion from 2-acetylhomoveratric acid leads to key tricyclic compounds that can be easily converted into either the berberine or benzo[*c*]phenanthridine ring systems (Eq. 76).[289]

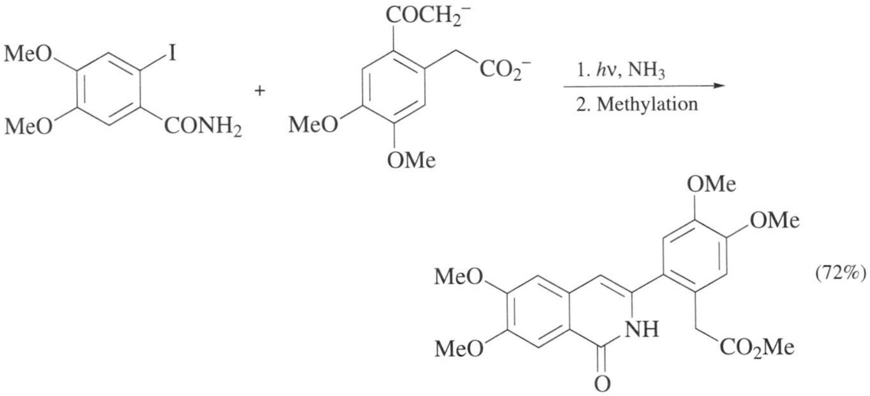

(Eq. 76)

Compounds of the isoquinoline family can also be obtained through the $S_{RN}1$ reaction. The substitution product formed in the irradiated reaction of *o*-iodobenzylamine with ketone enolate ions cyclizes spontaneously to give an intermediate product that affords isoquinolines by Pd/C dehydrogenation or tetrahydroisoquinolines by reduction with $NaBH_4$ (Eq. 77).[290,291] 11,12-Dihydrobenzo[*c*]-

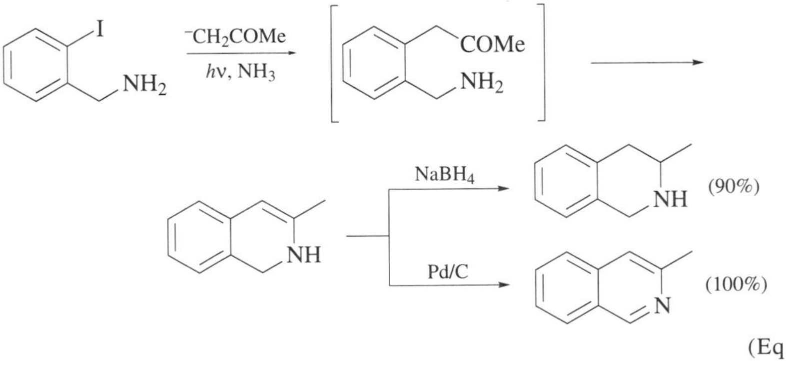

(Eq. 77)

phenanthridines and phenanthridones are obtained by reaction of tetralone enolate ion with *o*-iodobenzylamines and *o*-iodobenzoic acids, respectively.[292]

2,2′-Binaphthyls can be obtained directly through the reaction of the enolate ion of 2-naphthyl methyl ketone with *o*-bromoacetophenone under irradiation in liquid ammonia or dimethyl sulfoxide (Eq. 78).[158,293] Unsymmetrically substituted binaphthyls can also be synthesized by this procedure.[158] Naphthylquinolines and naphthylisoquinolines can be obtained by reaction of the anions of 2-naphthyl and 3-naphthyl methyl ketones with appropriate *ortho*-substituted aryl halides.[215]

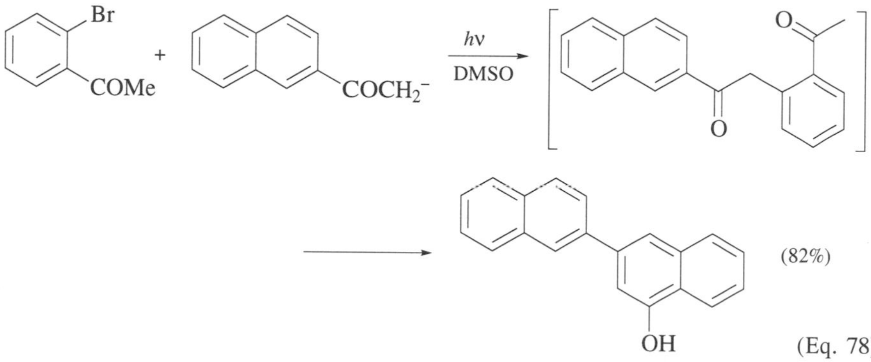
Br
COMe
+
COCH2−
hν
DMSO
O
O
OH

(82%)

(Eq. 78)

The reaction of phenoxides with *o*-substituted aryldiazo sulfides is an important route to the benzopyranone family. The dibenzo[*b*,*d*]pyran-6-one skeleton of benzocoumarins and related compounds can be obtained via a rather straightforward two-step route by 2-cyanoarylation of a phenol in dimethyl sulfoxide, followed by lactonization under very mild conditions. There is no need to isolate the intermediate hydroxy biaryl (Eq. 79).[294]

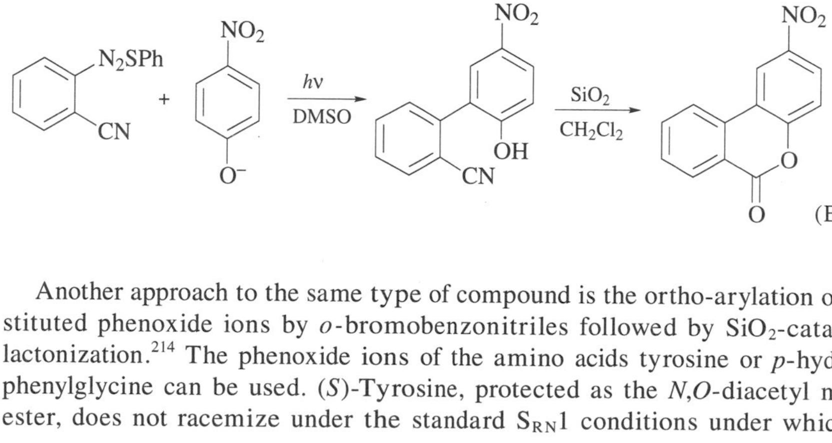
N2SPh
CN
+
NO2
O−
hν
DMSO
NO2
OH
CN
SiO2
CH2Cl2
NO2
O
O

(59%)

(Eq. 79)

Another approach to the same type of compound is the ortho-arylation of substituted phenoxide ions by *o*-bromobenzonitriles followed by SiO_2-catalyzed lactonization.[214] The phenoxide ions of the amino acids tyrosine or *p*-hydroxyphenylglycine can be used. (*S*)-Tyrosine, protected as the *N*,*O*-diacetyl methyl ester, does not racemize under the standard $S_{RN}1$ conditions under which the

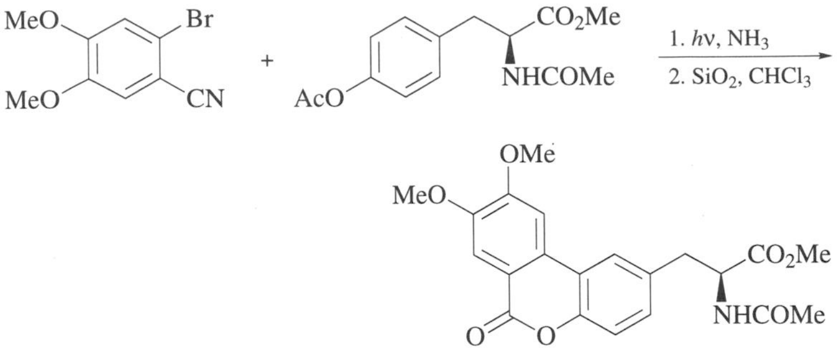

(52%)

(Eq. 80)

O-acetyl group is readily removed to afford the phenoxide nucleophile (Eq. 80). Racemic dibenzopyranones are obtained by reaction with the anion from the *N*-acetyl methyl ester of (*R*)-hydroxyphenylglycine.[214] The oxazolidine derivative of (*R*)-*p*-hydroxyphenylglycine does not racemize in the reaction media.[214] The reaction of 2-naphthoxide ion with *o*-bromobenzonitrile is a route to benzonaphthopyranone.[214]

Another route to five- and six-membered ring benzolactams and benzolactones is the carboxylation under phase-transfer catalysis conditions of aryl halides bearing amino or hydroxy groups on a side chain α to the halogen.[189]

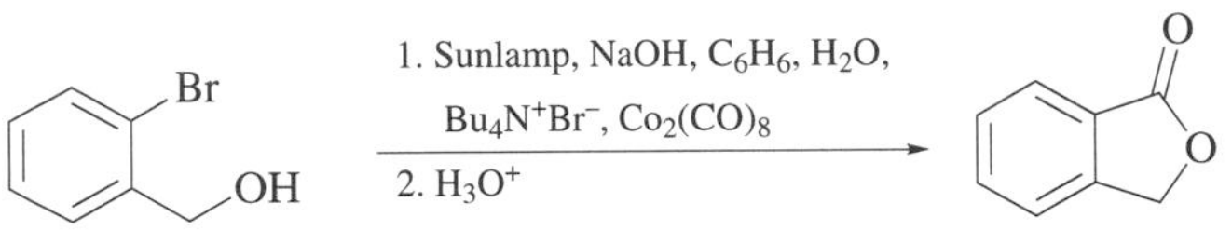

(95%)

(Eq. 81)

Intramolecular $S_{RN}1$ Reactions. One of the first reports of an intramolecular $S_{RN}1$ reaction is the synthesis of cephalotaxinone (**30**). The iodoketone **29** cyclizes in liquid ammonia with excess potassium amide under sodium or potassium metal initiation to afford product **30** in 45% yield. The yield increases to 94% when the reaction is initiated by light (Eq. 82).[295–297] The studies of this

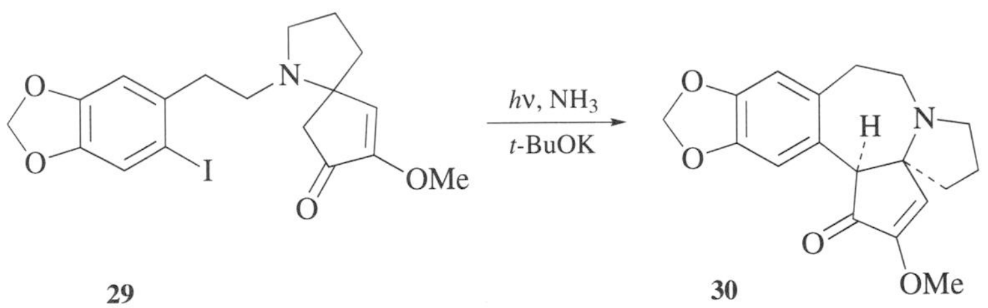

(94%)

29 **30** (Eq. 82)

system were extended to other analogs in order to determine the ring size preferences in the cyclizations and the regioselectivity with ketones that can give two enolate ions. For systems in which only the terminal enolate is possible, the degree of cyclization is high, but decreases as the number of methylene groups α to the ketone functionality increases.[135,170]

Cyclization of *N*-acyl and *N*-alkyl-*N*-acyl-*o*-haloanilines to afford oxindoles can be achieved in the presence of lithium diisopropylamide in THF or potassium amide in liquid ammonia (Eq. 83).[298,299]

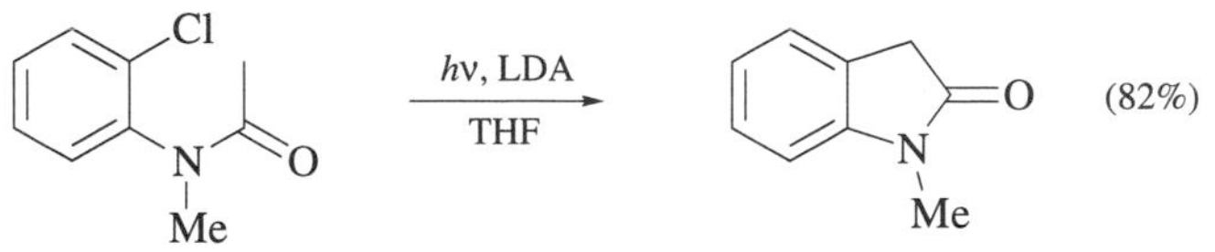
Cl
N
Me
O
hν, LDA
THF
O
N
Me
(82%)

(Eq. 83)

N-Methyl α,β-unsaturated anilides undergo intramolecular arylation exclusively at the α position to afford 3-alkylideneoxindoles.[299] The best results are obtained with potassium amide in liquid ammonia under photoinitiation (Eq. 84).[299]

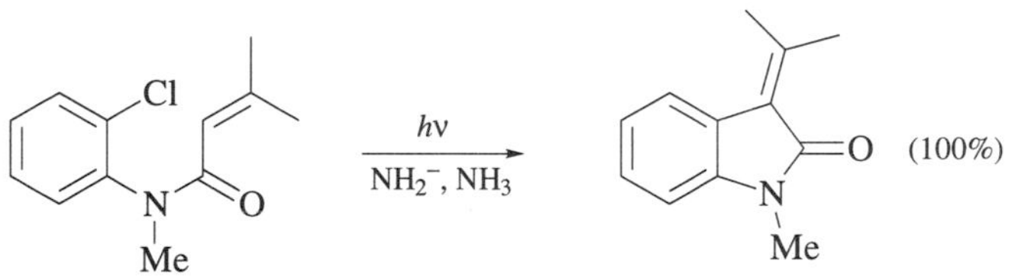
Cl
N
Me
O
hν
NH2−, NH3
O
N
Me
(100%)

(Eq. 84)

Photocyclization of the carbanions from *N*-acyl-*N*-methyl-*o*-chlorobenzylamines formed by reaction with potassium amide in liquid ammonia gives 1,4-dihydro-3-(2*H*)-isoquinolinones in fair to good yield (Eq. 85).[299] The corresponding *N*-trimethylsilylamide serves as precursor to *N*-unsubstituted isoquinolines, which are not accessible by direct photocyclization.[299]

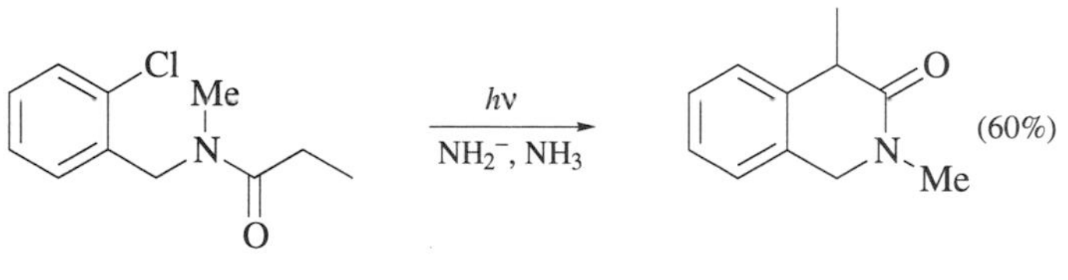
Cl
Me
N
O
hν
NH2−, NH3
O
N
Me
(60%)

(Eq. 85)

An intramolecular $S_{RN}1$ reaction is the key step in the synthesis of several precursors of alkaloids[300–304] such as the azaphenanthrene alkaloid eupoulauramine shown in Eq. 86.[300]

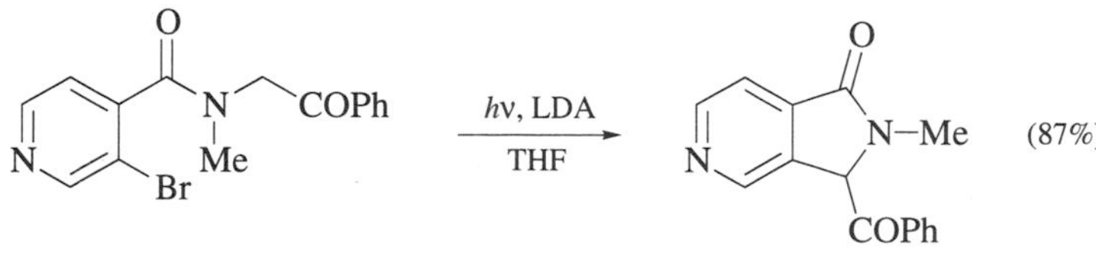
O
N
COPh
N
Me
Br
hν, LDA
THF
O
N−Me
N
COPh
(87%)

(Eq. 86)

Miscellaneous Ring Closure Reactions. Another approach to ring-closed products is the reaction of a substrate bearing two leaving groups with a dinucleophile having the required geometry. For instance, *o*-diiodobenzene reacts under irradiation with 3,4-toluenedithiolate ion in liquid ammonia to afford 2-methylthianthrene in good yields (Eq. 87).[305]

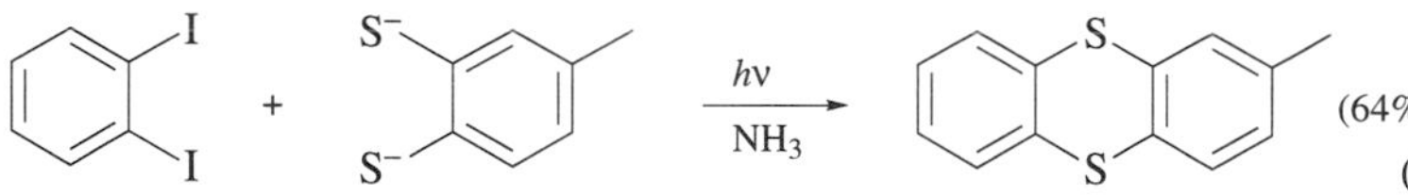
I
I
S⁻
S⁻
hν
NH₃
S
S
(64%)

(Eq. 87)

The synthesis of *meta*- and *para*-cyclophanes is achieved through the reaction of *m*- or *p*-dibromobenzene with dianion **31** under irradiation in liquid ammonia, although in low yield (Eq. 88).[306,307]

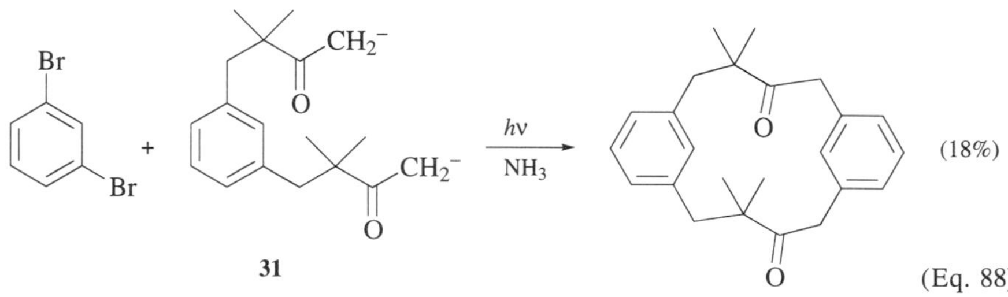
Br
Br
CH₂⁻
O
CH₂⁻
O
hν
NH₃
O
O
(18%)

31

(Eq. 88)

In the reaction of *o*-dihalobenzenes with 2-naphthoxide ion under irradiation in liquid ammonia, mixtures of monosubstitution and cyclization products are formed. The best yield of the latter is obtained with *o*-diiodobenzene and a threefold excess of the nucleophile under dilute conditions (Eq. 89).[308] This is an example in which a phenoxide reacts at oxygen rather than carbon because of geometric constraints. In the reaction of 2-naphthalenethiolate ion with *o*-diiodobenzene, the only compound formed is the cyclized one (Eq. 90).[308]

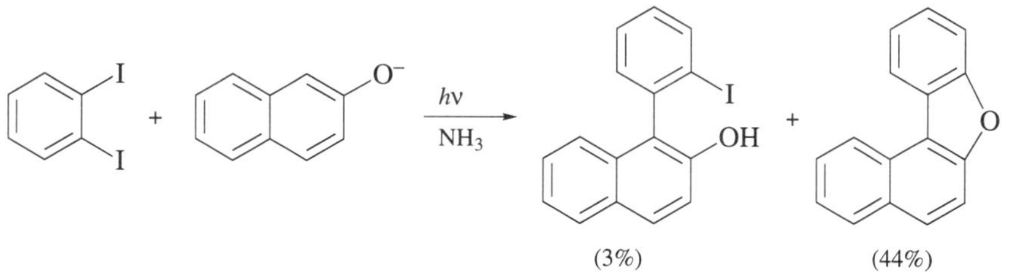
I
I
O⁻
hν
NH₃
I
OH
O
(3%)
(44%)

(Eq. 89)

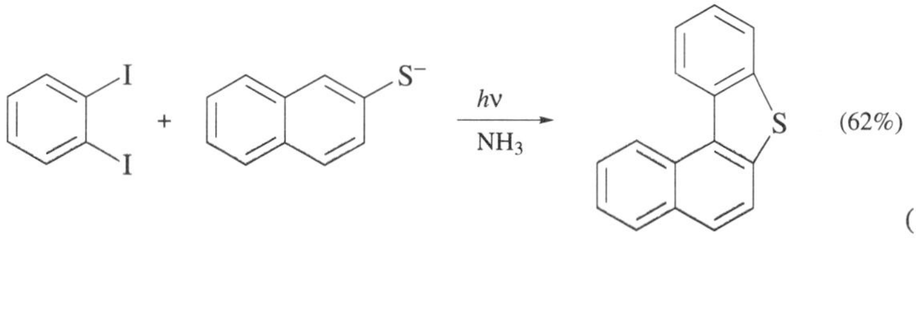
I
I
S⁻
hν
NH₃
S
(62%)

(Eq. 90)

The radical probe *o*-(3-butenyloxy)iodobenzene reacts with different nucleophiles such as benzenethiolate ions in liquid ammonia to give both the cyclized and the straightforward substitution products (Eq. 91).[33]

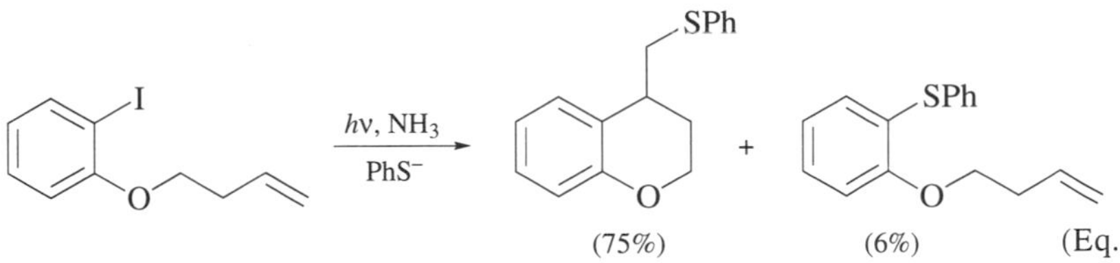
I
hν, NH3
PhS⁻
O
SPh
O
+
SPh
O
(75%)
(6%)

(Eq. 91)

COMPARISON WITH OTHER METHODS

It is difficult to make a detailed comparison between the $S_{RN}1$ reaction and other methods that can be used to obtain the same products. However, an attempt is made to present some of the most relevant unique features of the $S_{RN}1$ in comparison with other methods mainly in relation to the formation of new carbon-carbon and carbon-heteroatom bonds and ring closure reactions.

In general, the synthetic utility of aryl halides is rather limited because of the inertness of the carbon-halogen bond. For this reason, the attention of many research groups has been focused on activation, cleavage, and functionalization of the carbon-halogen bond of these compounds. In addition to nucleophilic substitution reactions, which are limited to aryl halides with strong electron-withdrawing substituents, the employment of transition metal complex catalysis can be an important and efficient approach.[309–312] Other possibilities are the complexation between aryl halides and some transition metals (e.g., the chromium(0) tricarbonyl complexes[313]), the use of stoichiometric amounts of transition metals,[314–317] and the substitution of diaryl halonium salts.[318,319]

Formation of C-C bonds. The $S_{RN}1$ is a powerful method for arylating carbanions. The reaction is regiospecific, is not highly sensitive to steric effects, and a number of different substituents on the substrate are tolerated. These advantages can be compared with the limitations of the substitution of nonactivated aryl halides through the benzyne mechanism, which requires strong bases, in some cases drastic conditions, and often leads to mixtures of isomers.

On the other hand, the S_NAr process is possible mainly with aromatic substrates activated by strong electron-withdrawing groups, especially in positions ortho and para to the leaving group, and hindered by electron-donating groups.

The $S_{RN}1$ arylation of β-dicarbonyl compounds is possible whenever the substrate is activated by electron-withdrawing groups such as NC-, PhCO-, and CF_3-, especially with substituted arylazo phenyl or *tert*-butyl sulfides to give, in most cases, high yields of arylation. In this field, the Hurtley reaction performed with copper halide catalysts is mostly limited to *o*-bromobenzoic acids and to other compounds in which the carboxy and the leaving groups are close to each other.[320] Even though attempts to utilize other *o*-haloaryl compounds have been unsuccessful, the coupling of unactivated aryl bromides and aryl iodides with diethyl malonate or cyanoacetate and copper iodide has been reported.[321] The reaction can also be performed with a palladium complex catalyst mainly with aryl iodides.[322]

Tetraphenylbismuth esters, other bismuth reagents, and aryllead tricarboxylates[323] permit arylation of β-diketones starting from an arene by replacement of hydrogen. In these reactions, the arylation agents are usually phenyl rings without the activating groups required in the $S_{RN}1$ reaction. The arylation of enolate ions of ketones and nitronate ions can also be performed with these reagents.[323] Aryl halides can arylate enolate ions of ketones under Ni[295] catalysis and can cross-couple with α-acetonitrile, acetonyl-, and cyano-tin compounds under palladium catalysis.[310] The metal catalyzed reactions take place mainly with aryl iodides and aryl bromides, and few heteroaryl halides are reported to participate in these reactions.

Aryldiazo phenyl or *tert*-butyl sulfides are the best substrates for arylation of cyanides. However, the reaction of aryl halides with these anions can be performed in the presence of stoichiometric amounts of Cu(I) compounds,[324] Ni-,[325,326] Pd-,[327] or Co-[328] complex catalysis. The $S_{RN}1$ reaction fails to arylate alkyne anions, a reaction that occurs with the copper salt of the anion[324] or can be catalyzed by Pd with CuI as co-catalyst.[311,329]

The $S_{RN}1$ reaction offers the possibility of obtaining not only products of α-arylation but also of α,α-diarylation. Another interesting advantage is the possibility of achieving the disubstitution of aromatic rings, which occurs mainly through the reaction of a wide family of carbon nucleophiles with substrates bearing two leaving groups.

The transition metal-catalyzed carboxylation of aryl halides is an efficient and simple method for the synthesis of various aromatic carbonyl compounds. Fe, Co, Ni and Pd complexes are commonly used as catalysts for the carboxylation of aryl iodides and aryl bromides.[312] The $S_{RN}1$ Co-carbonyl-carboxylation of aryl halides is an important advance in the carboxylation of the less reactive mono- and polychloroarenes. With appropriate reagents, the reaction can be performed in the presence of substituents that need to be protected in reactions that proceed by other mechanisms. Another possibility of obtaining carboxylic acids from aryl chlorides is the one-pot electro-assisted reaction; however, polyacids are not obtained.[330]

Aromatic nitrogen and oxygen nucleophiles react with aryl halides by $S_{RN}1$ pathways to afford hydroxy- and aminobiaryls. The reaction with phenoxide ion is an interesting route to biaryls substituted by electron-donating and electron-withdrawing groups. Different methods such as Ni, Pd, and other metal catalysts have been used to obtain good yields in cross-coupling of aryl halides to unsymmetrical biaryls.[309–311] Other possibilities are the phenylation of phenols and different stabilized indoles with tetraphenylbismuth esters and bismuth reagents that are among the best and most selective reagents for ortho phenylation of phenols under basic reaction conditions.[323] Yet there are few reports of the applications of either of the previous methods to the direct synthesis of unsymmetrical biphenyls bearing electron-withdrawing groups in one ring and electron donors in the other.

Formation of Carbon-Heteroatom Bonds. A number of different carbon-heteroatom bonds can be formed by the $S_{RN}1$ reaction. The generation of alkyl phenyl sulfides may have limitations because of the formation of fragmentation products. The $S_{RN}1$ substitution by benzenethiolate ions can be applied mainly to aryl iodides in liquid ammonia under irradiation, and aryl iodides, aryl bromides and even aryl chlorides in acetonitrile/THF, and dimethyl sulfoxide under irradiation or electrochemically.

Similar types of substitutions can be achieved by the S_NAr reaction between aryl halides with arenethiolate ions if polar aprotic solvents and elevated temperatures are used.[331] Good results are also obtained with activated aryl halides, but in this case side reactions can occasionally be important. Other approaches from aryl halides are the use of the copper salt of the arylthiolate anion as the nucleophile,[315] the alkali metal salt of the anion in the presence of CuI,[315] or catalytic amounts of $(Ph_3P)_4Pd$.[332,333] Conversion of Grignard reagents is another possibility.[334] However, the recent application of the $S_{RN}1$ reaction to substitution by heteroarenethiolate anions and the possibility of obtaining mono-, di-, and even tri-substitution with these anions has to be particularly mentioned. The reaction is also an important alternative to the Ziegler method where an aryldiazonium tetrafluoroborate is used as substrate.[261,262,266]

In the organoselenium family, the $S_{RN}1$ reaction has the advantage of affording areneselenenate ions by reaction of aryl halides with selenium ions formed by treatment of selenium metal with an alkali metal in liquid ammonia. The areneselenenate anion reacts in the presence of an aryl halide to give the unsymmetrical or symmetrical selenenide in a one-pot reaction. Another possibility for the areneselenenate anion is its oxidation to diaryl diselenide compounds used to generate the nucleophile in photoinitiated or electrochemical reactions. This sequence avoids the Grignard reaction[335] or the use of alkyllithium reagents.[336]

A wide range of dialkyl arylphosphonates can be obtained by reaction of unactivated aryl iodides in the presence of copper derivatives or nickel catalysts.[337–339] Aryl bromides and aryl iodides are readily arylated by phosphine oxides of the type Ph(alkylO)P(O)H and $(alkylO)_2P(O)H$ in the presence of $Pd(PPh_3)_4$. However, comparative studies showed that $(EtO)_2P(O)H/\ Pd(PPh_3)_4$ affords good results with some substrates, but fails mainly with aryl iodides, with which good yields of substitution are obtained under $S_{RN}1$ conditions.[235]

Ring Closure Reactions. Different ring sizes and different ring types such as indoles, azaindoles, benzofurans, isocarbostyrils, isocumarones, and isoquinolines can be obtained with the $S_{RN}1$ reaction, making difficult its comparison with other methods. In this field, other possibilities are offered by Cu- and Pd-mediated arylations. However, some features of the $S_{RN}1$ reactions can be mentioned such as the easily available starting materials, the usually fast and clean reactions, the regiospecific site of cyclization, and the possibility of achieving cyclization when other processes, such as the benzyne mechanism, are precluded.[299] Semmelhack and coworkers studied various cyclization methods to obtain cephalotaxinone.[295–297] The benzyne approach leads to 12–15% cyclization; the ring closure via σ-aryl nickel intermediates leads to 30–35% product, and the reaction activated by copper enolates fails to effect the desired ring closure, whereas the photoinitiated $S_{RN}1$ reaction proceeds in 94% yield.[296] The last has been successfully applied as the key cyclization step in synthesis of a number of alkaloids such as eupolauramine,[300] tortuosamine,[301] and rugulovasines[302] where the benzyne reaction failed. The regioselectivity of the intramolecular aryl-aryl coupling reaction of 2′-bromoreticulines depends on the method of cyclization. The Pd(0)-catalyzed reaction leads preferentially to the salutaridine derivative with formation of a quaternary sp^3 carbon center para to the phenolic group, whereas the $S_{RN}1$ reaction leads to the aporphine skeleton through cyclization ortho to the phenolic group.[303]

EXPERIMENTAL CONDITIONS

Recommended Solvents

The most widely used solvent has traditionally been liquid ammonia, but dimethyl sulfoxide can also be a solvent of choice. The photoinitiated reactions with certain carbanions, which do not take place in liquid ammonia or do so only under drastic conditions, can occur more easily in dimethyl sulfoxide at room temperature.[41] The main requirements for a solvent for $S_{RN}1$ reactions are that it: 1) dissolve both the organic substrate and the ionic alkali metal salt of the nucleophile, 2) does not have hydrogen atoms that can be readily abstracted by aryl radicals, 3) does not have protons that can be ionized by the bases or the basic nucleophiles and radical anions involved in the reaction, and 4) does not undergo electron transfer reactions with the various intermediates in the reaction. In addition to these characteristics, the solvent should not absorb significantly in the wavelength range normally used in photoinitiated reactions (300–400 nm), should not react with solvated electrons and/or alkali metals in reactions initiated by these species, and should not undergo reduction at the potentials employed in electrochemically promoted reactions, but should be sufficiently polar to facilitate electron transfer processes.

The effect of different solvents has been studied for the photoinitiated reaction of diethylphosphite ion with PhI[340] and 2-chloroquinoline with acetone enolate ion.[144] However, the marked dependence of solvent effect on the nature of the aromatic substrate, the nucleophile, its counterion, and the temperature at which the reaction is carried out often makes comparisons difficult. For example, in the photoinitiated reaction of 2-chloroquinoline with acetone enolate ion,[144] THF gives almost the same results as liquid ammonia. The reactions occur in low yields in 1,2-dimethoxyethane, and even lower yields in diethyl ether or benzene.[144] THF has proved to be an efficient solvent in intramolecular ring closure reactions.[298,299] In the reaction of pinacolone enolate ion with bromobenzene under irradiation, DMF leads to formation of benzene (28%) in a fast reaction while ammonia gives no benzene. Reaction in dimethyl sulfoxide produces little benzene, but the monophenylation/diphenylation ratio (6:1) is less favorable than in ammonia (20:1). On the other hand, THF almost completely inhibits this photoinitiated $S_{RN}1$ reaction, leading slowly to benzene as the major product. Coupling of the aryl radical with the solvent can also be observed in THF.[135]

In the electrochemically induced $S_{RN}1$ reactions between aryl halide and PhZ^- (Z = S, Se, Te) ions, it was concluded that even though acetonitrile is a better hydrogen donor than dimethyl sulfoxide, the aryl radicals couple faster with the nucleophiles, giving better yields of the substitution product in acetonitrile.[258] For preparative electroinitiated reactions, liquid ammonia is more useful than DMF, while at the laboratory scale DMF is an alternative for $S_{RN}1$ reactions.[341]

Reactions in Liquid Ammonia

Anhydrous ammonia is supplied in cylinders from which it can be removed in liquid or gaseous form. All the operations with this solvent must be conducted in an efficient fume hood because of the toxicity and pungent odor of this gas. Owing to ammonia's low boiling point (−33.4°), an efficient condenser is required to perform these reactions. A Dewar condenser containing a slurry of liquid nitrogen in EtOH or CO_2/EtOH will normally suffice. In order to eliminate iron impurities and water, the liquid ammonia has to be distilled from a preliminary vessel in which it was dried with sodium metal until a blue color persists (about 15 minutes).

Apparatus

Reactions are ordinarily conducted in a three-necked round-bottomed flask with a gas inlet and magnetic stirring bar. A positive pressure of an inert gas such as nitrogen or argon is also required. When metal solutions are used, glass or polyethylene stir blades or stir bars are preferred since Teflon darkens with repeated use in metal/ammonia solutions. However, the use of blackened stirring bars does not seem to interfere with the reactions. They can be treated with hot concentrated nitric acid to recover the white color. A detailed description of the experimental equipment has been published.[15,342]

Photochemical reactors are commercially available or can be custom-built. The commercial Rayonet RPR 204 apparatus equipped with four RUL 3000 tubes is available from the S.O New England Ultraviolet Co. Another type of reactor consists of an oval mirror-type wall of ca. 30-cm maximum radius equipped with two Hanovia 450-W or Philips HPT 400-W high-pressure mercury lamps inserted into a water-refrigerated Pyrex flask placed ca. 20 cm from the reaction vessel, or equipped with a 300-W Osram sunlamp placed ca. 14 cm from the reaction vessel (Pyrex flask). In the latter case, an appropriately positioned fan serves to maintain the reaction temperature around 25°.

EXPERIMENTAL PROCEDURES

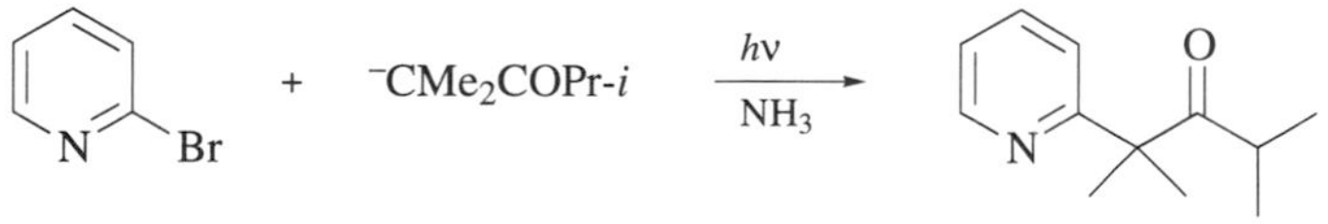
N
Br
+
⁻CMe₂COPr-i
hν
NH₃
N
O

2,4-Dimethyl-2-(2-pyridyl)-3-pentanone (Photoarylation of a Ketone).[137] 2-Bromopyridine (3.16 g, 20 mmol) was added to an enolate solution prepared from 2,4-dimethyl-3-pentanone (8.56 g, 75 mmol) and potassium amide (4.12 g,

75 mmol) in liquid ammonia (300 mL). After the mixture had been irradiated for 1 hour, ether was added, the ethereal suspension remaining after evaporation of the ammonia was decanted through a filter, and the residual salts were washed with ether (4 × 75 mL). This reaction afforded 5% of 2,4,4,5,5,7-hexamethyloctane-3,6-dione and 97% (3.70 g) of 2,4-dimethyl-2-(2-pyridyl)-3-pentanone. The latter was isolated as a colorless oil by preparative GC. ^{1}H NMR (CCl_4) δ 0.85 (d, J = 6.6 Hz, 6 H, isopropyl methyl), 1.45 (s, 6 H, CMe_2), 2.61 (septet, J = 6.6 Hz, 1 H, CH) 6.97–7.22 (m, 2 H, PyH-3,5), 7.44–7.64 (m, 1 H, PyH-4) and 8.38–8.48 (m, 1 H, PyH-6), Anal. Calcd. for $C_{12}H_{17}NO$: C, 75.35; H, 8.96; N, 7.32. Found: C, 75.47; H, 9.08; N, 7.39.

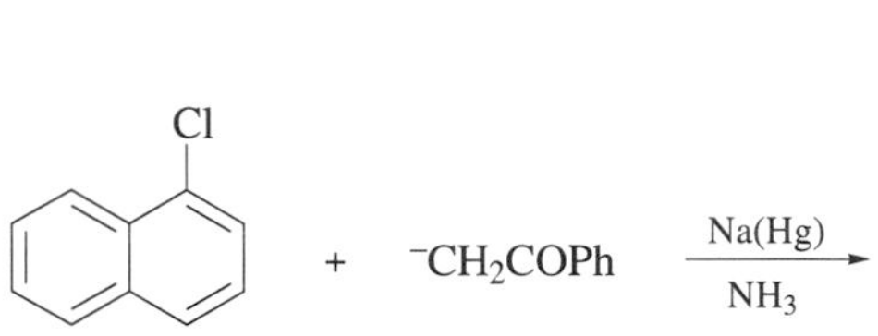
Cl
+
⁻CH2COPh
Na(Hg)
NH3
O
Ph

***α*-(1-Naphthyl)acetophenone [Na(Hg)-Initiated $S_{RN}1$ Reaction].**[101] To 200 mL of distilled liquid ammonia (previously dried under nitrogen with sodium metal) and under nitrogen were added *t*-BuOK (2.24 g, 20 mmol), acetophenone (1.44 g, 12.0 mmol), and 1-chloronaphthalene (0.65 g, 4.0 mmol). The reaction vessel was wrapped with aluminum foil to exclude light, and Na(Hg) (9.2 g, 12 mmol) (3% w/w) was added under magnetic stirring. After 2 hours, the reaction mixture was quenched by the addition of NH_4NO_3 (1.76 g, 22 mmol) in excess, and the ammonia was allowed to evaporate. The residue was dissolved in water and the solution was extracted with CH_2Cl_2. The CH_2Cl_2 solution was dried, and the solvent was removed under reduced pressure. Purification of the residue by radial chromatography on silica gel (elution with petroleum ether:diethyl ether 7:3) gave α-(1-naphthyl)acetophenone (0.96 g, 98%), mp 102–103°; MS (70 eV) m/z (relative intensity) 246 (9, M^+), 215 (7), 202 (15), 105 (30). ^{1}H NMR (CCl_4) δ 4.73 (s, 2 H), 7.3–8.1 (m, 12 H).

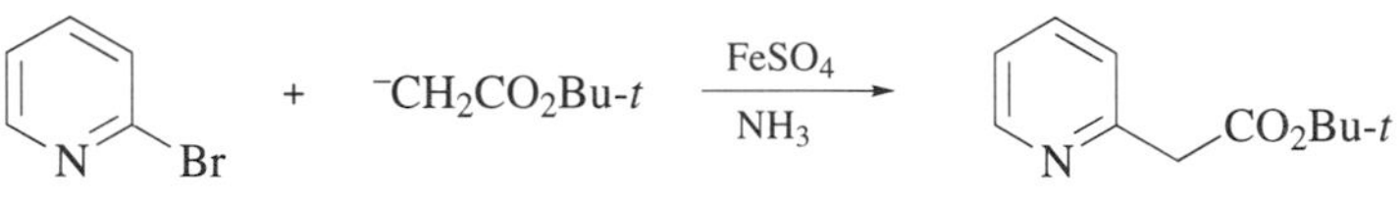
N
Br
+
⁻CH2CO2Bu-t
FeSO4
NH3
N
CO2Bu-t

***tert*-Butyl 2-Pyridylacetate (Fe^{2+}-Initiated $S_{RN}1$ Reaction).**[76] A 250-mL three-necked flask was equipped with a nitrogen inlet, a solid CO_2 condenser and a magnetic stirrer, and was flame-dried under a stream of nitrogen. The coolant well was filled with a dry ice-acetone mixture and the flask was immersed in a dry ice-acetone bath. Ammonia (150 mL) was distilled into the flask and $NaNH_2$ (4.8 g, 0.12 mol) was added. *tert*-Butyl acetate (13.4 g, 0.12 mol) was introduced slowly by syringe to the stirred suspension, and the resulting mixture was stirred at −70° for 0.5 hour. The cooling bath was removed and $FeSO_4$ and 2-bromopyridine (3.16 g, 0.02 mol) were added in quick succession. The reaction mixture was stirred rapidly for 0.75 hour and quenched by the gradual addition of small portions of NH_4Cl (20 g, 0.37 mol). The ammonia was allowed to evaporate and the residue was extracted with diethyl ether (3 × 75 mL). The combined extracts were washed with water (50 mL), 5% hydrochloric acid (3 × 50 mL) and brine (50 mL), and then dried ($MgSO_4$), filtered, and evaporated under reduced pressure to give the crude product mixture. *tert*-Butyl 2-pyridylacetate (3.44 g, 89%) was isolated as a colorless oil by standard column or flash chromatography using graded mixtures of light petroleum ether (bp 40–60°), methylene chloride, and ethyl acetate. IR (film) ν_{max} 1735 cm^{-1}; ^1H NMR δ 1.43 (s, 9 H, CMe_3), 3.76 (s, 2 H, $PyrCH_2CO$), 7.00–7.81 (m, 3 H, 3,4,5-Pyr-H) and 8.45–8.64 (m, 1 H, 6-Pyr-H); MS *m/z* 193 (M^+), 138 (M^+-55), 120 ($PyrCH_2CO^+$) and 92 (Pyr-CH_2^+). Anal. Calcd. for $C_{11}H_{15}NO_2$: C, 68.37; H, 7.82; N, 7.25. Found: C, 68.3; H, 7.8; N, 7.1.

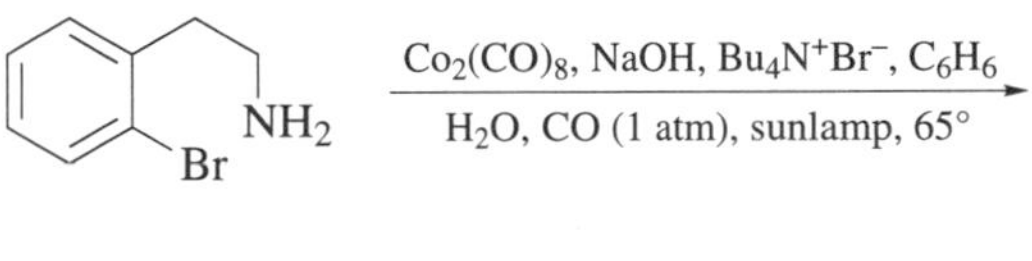
NH2
Br
Co2(CO)8, NaOH, Bu4N+Br−, C6H6
H2O, CO (1 atm), sunlamp, 65°
NH
O

1-Oxo-1,2,3,4-tetrahydroisoquinoline (Carboxylation of *o*-Bromo(2-aminoethyl)benzene, Synthesis of a Benzolactam).[189] To a 250-mL Pyrex flask were added benzene (25 mL), $Co_2(CO)_8$ (0.34 g, 1 mmol), aqueous 5 N NaOH (50 mL), $Bu_4N^+Br^-$ (0.64 g, 2 mmol), and *o*-bromo(2-aminoethyl)-benzene (40 g, 20 mmol). The reaction mixture was heated to 65° under a slow stream of carbon monoxide. Irradiation over 5.5 hours was achieved with a sunlamp placed about 30 cm from the reaction flask. The organic phase was washed several times with water, dried over $MgSO_4$, and purified by column chromatography on a short silica gel column or by recrystallization. The 1-oxo-1,2,3,4-tetrahydroisoquinoline was isolated in 72% yield (2.12 g). IR(film) ν_{max} 3250, 1665 (s), 1605, 1575 cm^{-1}; ^1H NMR ($CDCl_3$) δ 2.90 (t, 2 H), 3.3–3.7 (m, 2 H), 7.0–7.6 (m, 3 H, aromatic), 7.7–8.4 (m, 2 H, aromatic and NH).

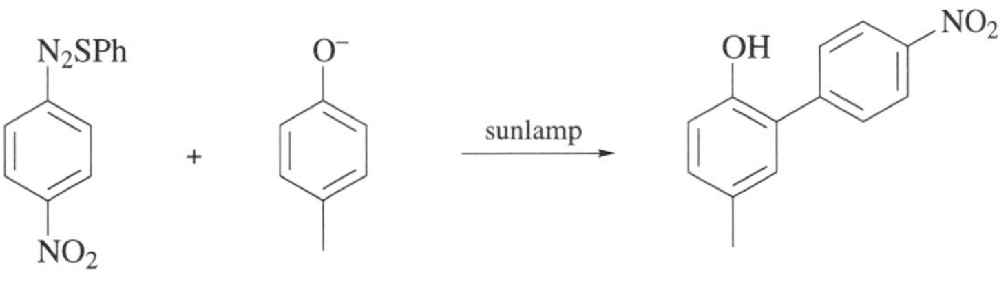
N2SPh
NO2
O−
+
sunlamp
OH
NO2

4-Methyl-2-(4-nitrophenyl)phenol (Photoarylation of a Phenol by an Aryldiazo Phenyl Sulfide).[217] The experiments were carried out under argon, the apparatus being deaerated. Reactions were started by dropping a dimethyl sulfoxide (9 mL) solution of (4-nitrophenyl)azo phenyl sulfide (0.5 g, 1.93 mmol)[343] into a magnetically stirred solution (18 mL) of the nucleophile (19.3 mmol prepared in situ from 4-methylphenol (2.1 g, 19.3 mmol) and *t*-BuOK (2.16 g, 19.3 mmol)). The initial substrate concentration was 0.07 M. Irradiation was performed with a sunlamp during 0.5 hour until gas evolution ceased. Dilution with 3–5% HCl (150 mL) and 4-fold extraction with diethyl ether (50 mL) was followed by washing of the combined extracts with brine. The organic layer was dried (Na_2SO_4) and the solvent removed under reduced pressure at room temperature. Column chromatography on silica gel (hexane-CH_2Cl_2, followed by hexane-ethyl acetate mixtures as eluants) yielded pure products. 4-Methyl-2-(4-nitrophenyl)phenol ($C_{13}H_{11}NO_3$): (70%, 0.31 g) mp 117.5–119.0°, ^{1}H NMR ($CDCl_3$) δ 2.33 (3 H, s), 4.96 (1 H, br, s), 6.9–7.1 (3 H, m), 7.70 and 8.29 (2 H each, AA′BB′, J = 8.9 Hz); IR: 3494 cm^{-1} (OH). 4-Methyl-4-(4-nitrophenyl)-2,5-cyclohexadienone ($C_{13}H_{11}NO_3$) (10%, 0.04 g): mp 110.3–111.6°; ^{1}H NMR ($CDCl_3$) δ 1.75 (3 H, s), 6.34 and 6.89 (2 H each, AA′BB′, J = 10.2 Hz), 7.49 and 8.21 (2 H each, AA′BB′, J = 8.9 Hz); IR: 1664 cm^{-1} (CO).

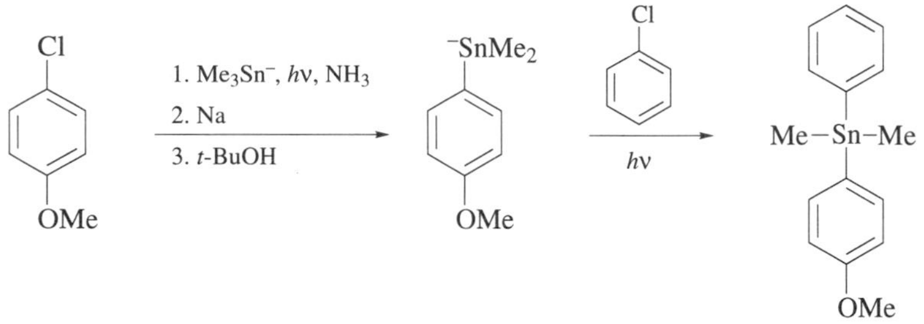
Cl
OMe
1. Me3Sn−, hν, NH3
2. Na
3. t-BuOH
−SnMe2
OMe
Cl
hν
Me−Sn−Me
OMe

***p*-Anisyldimethylphenyltin (One-pot Preparation of a Diaryldimethyltin).**[224] To distilled dry liquid ammonia (500 mL) were added trimethyltin chloride (3.98 g, 0.020 mol) and sodium metal (1.06 g, 0.046 mol) to form trimethylstannylsodium in about 1.5 hours. To this solution was added *p*-chloroanisole (0.020 mol), and the reaction mixture was irradiated for 1 hour. With the lamp off, sodium metal (1.06 g, 0.046 mol) and then *t*-BuOH (1.80 mL, 0.02 mol) were added to the solution to form the second nucleophile. Chlorobenzene (2.5 g, 0.022 mol) was added with a syringe (without solvent) and the reaction mixture was irradiated for 1 hour and quenched with NH_4NO_3 (3.5 g, 0.044 mol); diethyl ether (50 mL) was added and the ammonia was allowed to evaporate. Water was added, the phases were separated, and the aqueous phase was extracted twice with diethyl ether. The dried organic extract was distilled under vacuum using a Kugelrohr apparatus. Trimethylphenyltin was obtained as a byproduct (less than 5% yield), and it was distilled before the *p*-anisyldimethylphenyltin at 77° (0.4 mm Hg). *p*-Anisyldimethylphenyltin: colorless liquid, purified by vacuum distillation at 160–170° (0.5 mm Hg) using a Kugelrohr apparatus; 2.2 g (66% yield) of pure product (97%, GLC) was obtained. ^{1}H NMR

(CD_3COCD_3, 200 MHz), δ 0.479 [s, 6 H, $^2J(^{119}SN, ^1H)$ = 56.2 Hz], 3.78 (s, 3 H), 6.95 [*p*-anisyl BB′ system, 2 H, ^3J(H, H) = 8.4 Hz], 7.65 − 7.20 (m, 7 H). ^{13}C NMR (CD_3COCD_3), δ [$^nJ(^{119}Sn, ^{13}C)$ in Hz] − 10.32 (CH_3, ^1J $SnCH_3$ not detected), 55.22 (OCH_3), 115.01 [^3J(Sn, C) = 53.0 Hz, C-3 An], 129.02 [^3J(Sn, C) = 48.2 Hz, C-3 Ph], 129.24 [^4J(Sn, C) = 10.9 Hz, C-4 Ph], 131.24, 136.91 [^2J(Sn, C) = 37.4 Hz, C-2 Ph], 138.04 [^2J(Sn, C) = 42.8 Hz, C-2 An], 141.26, 161.26. Anal. Calcd. C, 54.10; H, 5.45. Found: C, 54.47; H, 5.68.

Column chromatography seems to be a purification method unsuitable for these anisyltin derivatives, since they decompose on long contact with silica gel.

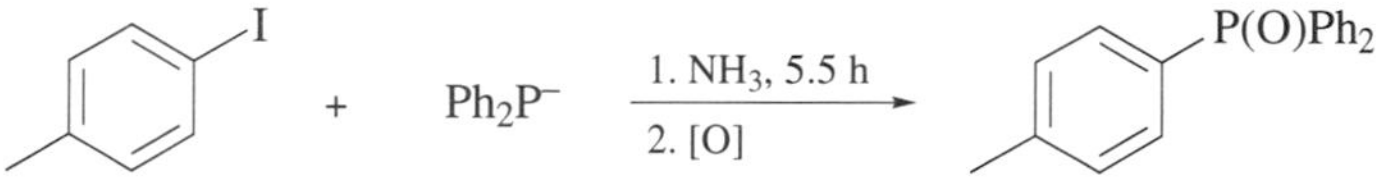
I
+ Ph_2P^-
1. NH_3, 5.5 h
2. [O]
P(O)Ph_2

Diphenyl-*p*-tolylphosphine Oxide (Preparation of an Aryldiphenylphosphine Oxide).[71] To ammonia (250 mL) previously distilled under nitrogen was added *t*-BuOK (1.18 g, 0.0105 mol) and diphenylphosphine (1.86 g, 0.01 mol). *p*-Iodotoluene (2.18 g, 0.01 mol) was added, and the mixture was stirred for 5.5 hours. Solid NH_4NO_3 was added until a colorless solution was obtained, and then diethyl ether (100 mL) was added and the ammonia was allowed to evaporate. Water (100 mL) was added, and the layers were separated. The water layer was extracted with diethyl ether (3 × 75 mL), and the combined ether extracts were dried with $MgSO_4$. The ether was evaporated from the main fraction of the reaction mixture, the residue was dissolved in CH_2Cl_2 (50 mL), and the solution was washed with 5% H_2O_2 (2 × 50 mL), 10% NaOH (2 × 50 mL), and water (50 mL), dried ($MgSO_4$) and evaporated to give 2.59 g (89%) of diphenyl-*p*-tolylphosphine oxide, mp 121–124°. Recrystallization from diethyl ether and petroleum ether yielded 1.33 g, mp 129.5–130°; ^1H NMR ($CDCl_3$) δ 8.07–7.23 (m, 14 H), 2.40 (s, 3 H); IR (KBr) cm^{-1} 3050, 2990 (C-H), 1180 (P=O), 818 (*p*-disubstituted benzene), 750, 698 (monosubstituted benzene); MS *m/z* 292, 291, 215, 213, 199, 183, 165, 152.

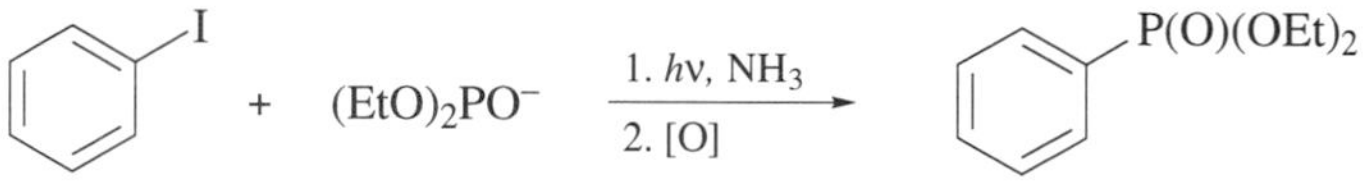
I
+ $(EtO)_2PO^-$
1. $h\nu$, NH_3
2. [O]
P(O)$(OEt)_2$

Diethyl Phenyl Phosphonate (Preparation of an Aryl Diethyl Phosphonate).[344] A 2-L three-necked flask was equipped with a nitrogen inlet and magnetic stirrer. Sodium metal (11.8 g, 0.51 mol) was added to freshly distilled liquid ammonia, and the mixture turned blue. Diethyl phosphonate (70.4 g, 0.51 mol) was cautiously added dropwise until the color changed from blue to colorless. Some white foam formed as the diethyl phosphonate was added. Because water in the ammonia consumes some of the sodium, not quite all the diethyl phosphonate was required to reach the endpoint. Iodobenzene (52.4 g, 0.26 mol) was added slowly, and the solution took on a slight yellowish tint. The dropping funnel was replaced by a stopper and the whole system was mounted in a photochemical reactor and irradiated for 1 hour. After irradiation, solid NH_4NO_3 (about 50 g, 0.61 mol) was added with stirring to acidify the mixture, followed by ether (200 mL). The nitrogen flow was stopped, the condenser was removed and the ammonia was allowed to evaporate. Water (300 mL) and diethyl ether (300 mL) were then added. The ether layer was separated, the water layer was extracted twice with diethyl ether, and the combined ether extracts were dried (Na_2SO_4). After evaporation of the ether, the residue was distilled under reduced pressure through a short Vigreux column. After a small forerun, diethyl phenyl phosphonate was collected at 73–74° (0.02 mm). The yield was 50.3–56.1 g (90.4–92.5%). IR (neat) cm^{-1}: 1440 (P-C aryl), 1250 (P=O), 1020 (POEt), and 3060 (H-C aryl); ^1H NMR ($CDCl_3$) δ 1.3 (t, Me, J = 7 Hz), 4.13 (q, CH_2, J = 7 Hz), 7.33–8.06 (m, 5 H, phenyl *H*).

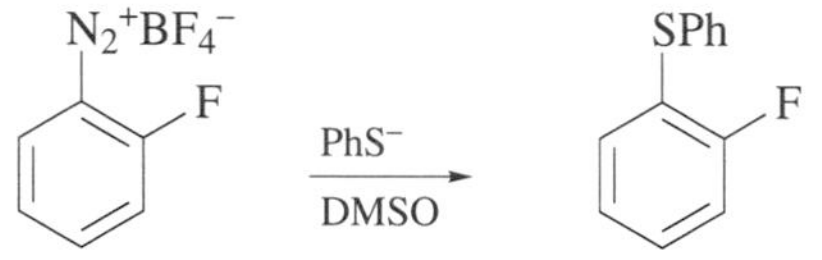

$N_2^+BF_4^-$
F
PhS⁻
DMSO
SPh
F

2-Fluorophenyl Phenyl Sulfide (Preparation of an Unsymmetrical Aryl Phenyl Sulfide from a Diazonium Salt).[262] The experiment was carried out under argon and the equipment was deaerated and left under positive pressure (*ca.* 30 mm Hg regulated with a mercury bubbler). The reaction was started by dropping the dimethyl sulfoxide solution of 2-fluoro-1-phenyldiazonium tetrafluoroborate (1 g in 10 mL, 4.77 mmol) into a magnetically stirred solution (20.0 mL) of sodium benzenethiolate (1.88 g, 14.3 mmol) in a flask that was kept in a water bath at 25° and wrapped in aluminum foil. Gas evolution and darkening of the solution were immediately observed. The progress of reaction was followed by TLC (aliquots being diluted with brine and extracted with diethyl ether) and completion was judged by cessation of nitrogen evolution and by disappearance of the diazosulfide by TLC. The mixture was worked up by dilution with brine (150 mL) and 4-fold extraction with diethyl ether (50 mL), followed by washing of the combined extracts with 10% NaOH and brine. The organic layer was dried (Na_2SO_4) and the solvent was removed under reduced pressure at room temperature. Column chromatography on silica gel (hexane-CH_2Cl_2 as eluant) yielded pure product (80%, 0.78 g). The sulfide was an oil and was identified by spectroscopic analysis of the sulfone. 2-Fluoro-1-(phenylsulfonyl)benzene: mp 105–106° (petroleum ether, bp 80–100°). Anal. Calcd. for $C_{12}H_9FO_2S$ C, 61.0; H, 3.8. Found: C, 61.1; H, 4.0. ^1H NMR: δ 8.22–7.95 (3 H, m) and 7.66–6.97 (6 H, two partly overlapping multiplets).

MeO
$CONH_2$
MeO
Br
OMe
$h\nu$, NH_3
$^-CH_2COMe$
O
MeO
NH
MeO
OMe

5,6,7-Trimethoxy-3-methyl-1(2*H*)-isoquinolone (Substitution of an *o*-Halobenzamide Followed by Cyclization).[168] To liquid ammonia (25 mL), prepared by condensing ammonia (gas) under argon in a three-necked flask fitted with a dry-ice condenser, was added acetone (0.23 g, 4 mmol) and freshly sublimed *t*-BuOK (0.45 g, 4 mmol). To the solution of the acetone enolate thus formed was added 2-bromo-3,4,5-trimethoxybenzamide (0.29 g, 1 mmol), and the reaction mixture was irradiated in a photochemical reactor. The reaction was monitored by the removal of aliquots, which were analyzed by TLC. Quenching was accomplished by addition of NH_4Cl (0.5 g) when all the substrate had reacted. The ammonia was evaporated and slightly acidified water (50 mL containing 1 mL of 2 M HCl) was added. Extraction with CH_2Cl_2 (4 × 30 mL) followed by purification yielded 5,6,7-trimethoxy-3-methyl-1(2*H*)-isoquinolone (80%, 0.2 g), mp 198–200°, MS *m/z* 249 (M^+); ^1H NMR ($CDCl_3$) δ 2.40 (3 H, s), 4.0 (9 H, s), 6.55 (1 H, s) and 7.60 (1 H, s).

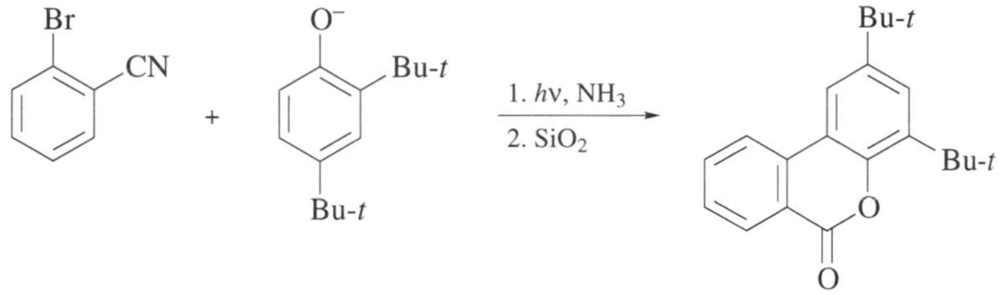
Br
CN
O⁻
Bu-t
+
Bu-t
1. hν, NH3
2. SiO2
Bu-t
Bu-t
O
O

2,4-Di-*tert*-butyldibenzo(*b,d*)pyran-6-one (Substitution of an *o*-Halobenzonitrile Followed by Cyclization).[214] Ammonia (50 mL) was condensed into a 100-mL two-necked Pyrex flask containing freshly sublimed *t*-BuOK (0.34 g, 3 mmol) through a dry ice condenser cooled at −78°. Under an argon atmosphere, 2,4-di-*tert*-butylphenol (0.6 g, 3 mmol) and 2-bromobenzonitrile (0.18 g, 1 mmol) were successively introduced. The reaction mixture was irradiated in a photochemical reactor for 1 hour (−33°) and the solvent was evaporated. Acidic water (100 mL) was added and the aqueous phase was extracted with ethyl acetate (2 × 50 mL). After evaporation, the residue was dissolved in CH_2Cl_2, SiO_2 (1 g) was added, and stirring was continued overnight at room temperature. The silica gel was removed by filtration and washed with acetone. The product 2,4-di-*tert*-butyldibenzo(*b,d*)pyran-6-one was isolated in 78% yield (0.24 g) by TLC, mp 163°, MS *m/z* 308 (M^+), 293; IR 1730 cm^{-1} (C=O); ^1H NMR δ 1.40 (s, 9 H, 3 Me), 1.50 (s, 9 H, 3 Me), 7.43 (m, 2 H), 7.67 (t, 1 H), 7.87 (s, 1 H), 8.03 (d, J = 8 Hz, 1 H), 8.40 (d, J = 8 Hz, 1 H). Anal. Calcd. for $C_{21}H_{24}O_2$; C, 81.82; H, 7.79; O, 10.38; Found: C, 81.84; H, 7.83; O, 10.27.

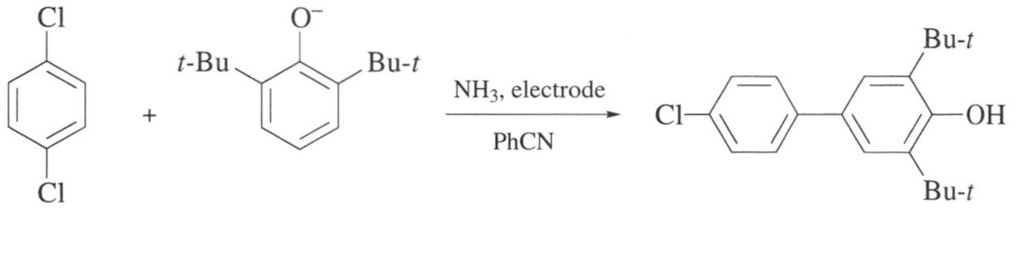
Cl
Cl
+
t-Bu
O⁻
Bu-t
NH3, electrode
PhCN
Cl
Bu-t
OH
Bu-t

3′,5′-Di-*tert*-butyl-4-chloro-4′-hydroxy-1,1′-biphenyl (Electrochemically Initiated $S_{RN}1$ Reaction in the Presence of a Redox Mediator).[210] 2,6-Di-*tert*-butylphenol (10.63 g, 51.5 mmol) and KOBu-*t* (8.93 g, 79.5 mmol) were successively introduced into a single-compartment electrochemical cell containing liquid NH_3 (200 mL) and KBr (6.2 g, 52 mmol) as the supporting electrolyte. The temperature of the solution was maintained at −40° by a cryocooler. The excess base (28 mmol) was neutralized by H_2O (700 mg, 39 mmol). To the reaction suspension, 1,4-dichlorobenzene (26 mmol, 5.3 g) dissolved in benzonitrile (5.02 g, 36.5 mmol) was added slowly. This gave a finely divided suspension; 1,4-dichlorobenzene, which is nearly insoluble in liquid ammonia, could then pass quickly into solution as the electrolysis proceeded. The electrolysis was performed under galvanostatic conditions (i = 100 mA), using a stainless steel grid as the cathode (60 cm^2, 364 mesh cm^{-2}) and a sacrificial Mg rod as the anode. It was stopped when 1614 Coulombs (0.23 Faraday per mole of 1,4-dichlorobenzene) had passed through the circuit. The solution was then neutralized by NH_4Br (10 g, 102 mmol). After NH_3 evaporation, the organic products were extracted by CH_2Cl_2 (2 × 250 mL). The solvent was evaporated. Benzonitrile and the unreacted di-*tert*-butylphenol were distilled at 120° under partial vacuum (5–10 Torr). After cooling, the remaining solid was dissolved in a minimum volume of CH_2Cl_2. 3′,5′-Di-*tert*-butyl-4-chloro-4′-hydroxy-1,1′-biphenyl was precipitated upon the addition of pentane. From the amounts of unreacted 1,4-dichlorobenzene (6 mmol, determined by HPLC) and 3′,5′-di-*tert*-butyl-4-chloro-4′-hydroxy-1,1′-biphenyl (6.36 g, purity higher than 98%), the reaction yield relative to the reacted starting aromatic dihalide could be estimated to be about 67%. The product was recrystallized from hexane; mp 141°. ^{1}H NMR ($CDCl_3$) δ = 1.49 (s, 18 H), 5.28 (s, 1 phenolic H), 7.36 (s, 2 H), 7.40 and 7.46 (AA′BB′, J_{app} = 8 Hz, 4 H). MS (CI, NH_3): *m/z* = 316 (M), 301, 300, 284, 283, 282, 267.

TABULAR SURVEY

The computer search of *Chemical Abstracts* covered the literature to the end of 1995, although most of the papers that appeared during 1996 as well as some papers in 1997 are also included.

The $S_{RN}1$ reactions of aromatic and heteroaromatic compounds with different nucleophiles to form C-C and C-heteroatom bonds are presented, as well as those reactions that lead to ring closure products. Most of the systems presented are explained on the basis of firmly established concepts and experimental demonstration of the $S_{RN}1$ mechanism, but in same cases, in which not enough experimental evidence is available, the interpretation is speculative.

The Tables are presented in the order of the discussion in the Scope and Limitations Section. Within each table the aromatic substrates are listed according to the following ring system ordering: carbon six-membered rings, polynuclear carbon aromatic compounds ordered by increasing carbon number; heteroaromatic compounds of six-membered rings bearing one, two or more heteroatoms;

polynuclear heteroaromatic compounds and finally five-membered heteroaromatic rings.

Within each ring system the following ordering has been followed: compounds bearing one leaving group; one leaving group and one substituent in *ortho-*, *meta-*, or *para-* position with respect to the leaving group. Compounds with higher degrees of substitution follow, and immediately after, compounds with two leaving groups.

The leaving groups are ordered following the periodic table, i.e. from right to left and increasing atomic number within a row.

The nucleophiles are ordered by complexity. First alicyclic aliphatic nucleophiles, then cyclic aliphatic and aromatic nucleophiles of the same family.

Table IX, in which ring closure reactions are presented, is ordered by the size and type of the ring formed. For instance, carbon five-membered rings are followed by increasing numbers of heteroatoms (with one nitrogen, two nitrogens, etc.), and the same for increasing ring size.

The following abbreviations are used in the text and in the Tables.

)))	ultrasonic irradiation
Ac	acetyl
Bn	benzyl
Boc	*tert*-butoxycarbonyl
DME	1,2-dimethoxyethane
DMF	*N,N*-dimethylformamide
DMSO	dimethyl sulfoxide
e^-	electrode
HMPA	hexamethylphosphoramide
LDA	lithium diisopropylamide
rt	room temperature
TBDMS	*tert*-butyldimethylsilyl
THF	tetrahydrofuran
TMS	trimethylsilyl
Ts	*p*-toluenesulfonyl (tosyl)

TABLE I. CARBANIONS DERIVED FROM HYDROCARBONS

Substrate	Nucleophile	Conditions	Product(s) and Yield(s) (%)	Refs.
PhBr	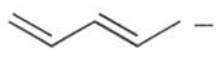	1. K/NH_3, –78° 2. H_2 (cat.)	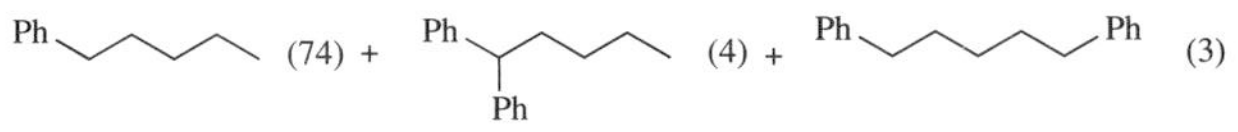 (74) + (4) + (3)	124
	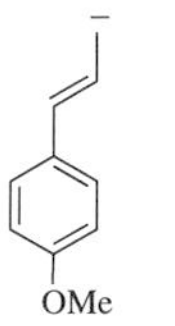	1. K/NH_3 2. H_2 (cat.)	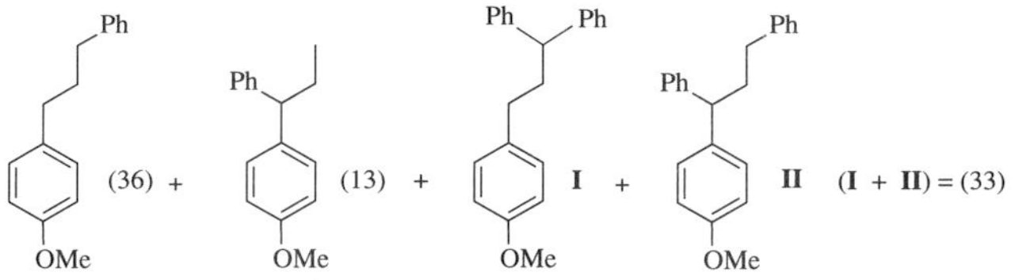 (36) + (13) + **I** + **II** (**I** + **II**) = (33)	124
	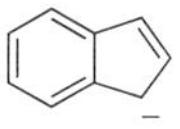	K/NH_3, -78°	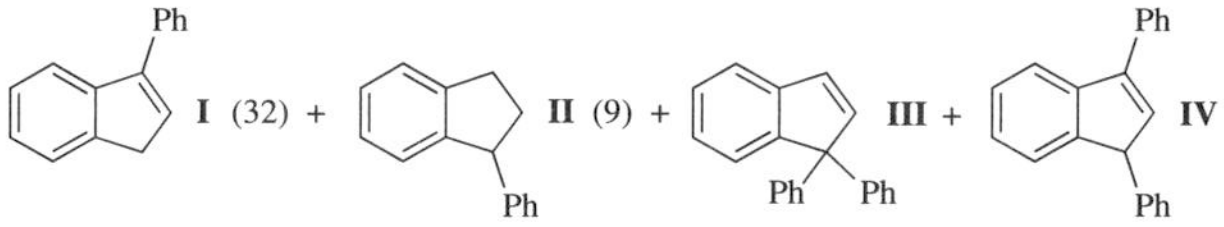 **I** (32) + **II** (9) + **III** + **IV** (**III** + **IV**) = (12)	124
		$h\nu$, DMSO, 0.5 h	**I** (25) + **III** (8) + **IV** (2)	70
	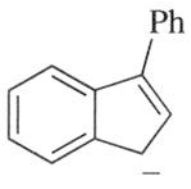	$h\nu$, DMSO, 0.5 h	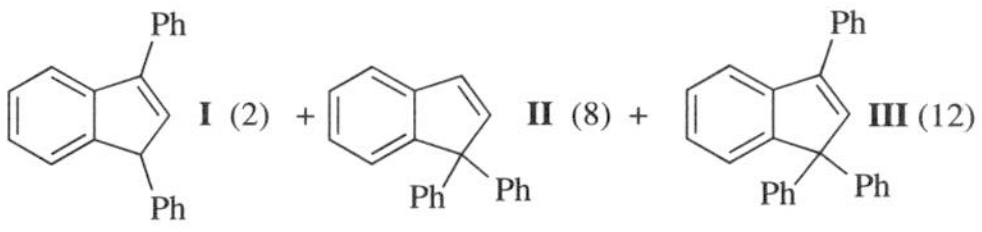 **I** (2) + **II** (8) + **III** (12)	70
		$h\nu$, NH_3, 0.5 h	**I** (20) + **II** (1) + **III** (9)	70
	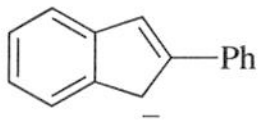	$h\nu$, DMSO, 1.3 h	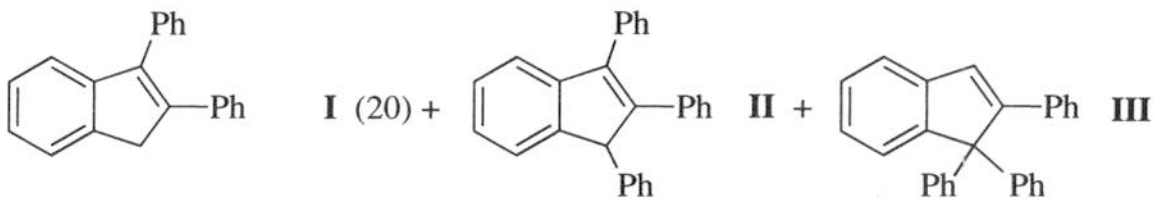 **I** (20) + **II** + **III** (**II** + **III**) = (10)	70
		$h\nu$, NH_3, 1 h	**I** (38), (**II** + **III**) (16)	70

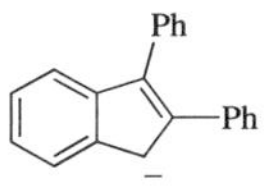Ph Ph	$h\nu$, DMSO, 4 h	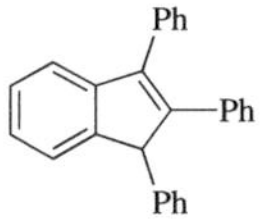Ph Ph Ph **I** (20) + 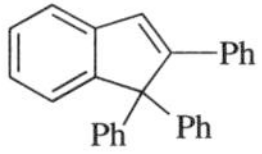Ph Ph Ph **II** (7)	70
	$h\nu$, NH_3, 0.5 h	**I** (30) + **II** (10)	70
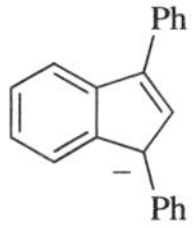Ph Ph	$h\nu$, DMSO, 10 h	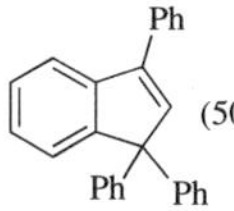Ph Ph Ph (50) + 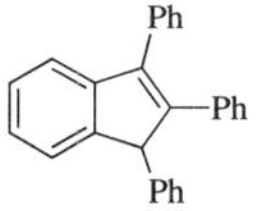Ph Ph Ph (9)	70, 128
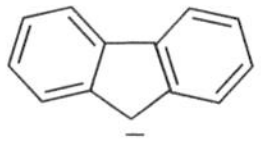	K/NH_3, –78°	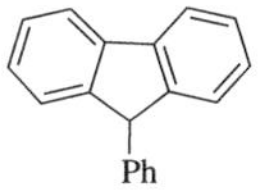Ph (57) + 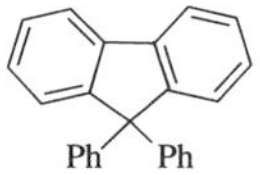Ph Ph (23)	124

TABLE II. CARBANIONS DERIVED FROM KETONES

A. Enolate from Acetone

Substrate	Nucleophile	Conditions	Product(s) and Yield(s) (%)	Refs.
PhF	$^{-}CH_2COMe$	K, NH_3, –78°	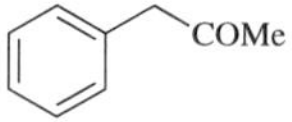 **I** (3) + 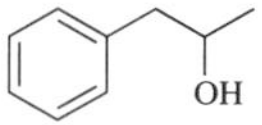 **II** (46)	94
		*h*ν, NH_3, 3.3 h	**I** (60) + C_6H_6 **III** (31)	136
PhCl		1. K, NH_3, –78° 2. [O]	**I** (68)	94
		*h*ν, NH_3, 3 h	**I** (61) + **III** (31) + 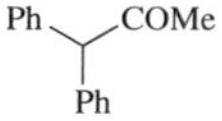 **IV** (5)	136
PhBr		K, NH_3, –78°	**I** (67) + **II** (10) + **IV** (14)	94
		*h*ν, NH_3, 11 min	**I** (85) + **IV** (14)	136
		*h*ν, NH_3:Et_2O 1:1, 0.5 h	**I** (41) + **III** (13)	108
		*h*ν, DMSO, 1 h	**I** (61) + **IV** (6)	108
PhI		1. K, NH_3, –78° 2. [O]	**I** (71)	94
		*h*ν, NH_3, 5 min	**I** (67) + **III** (20) + **IV** (10)	136
		*h*ν, DMSO, 1 h	**I** (88) + **IV** (4)	138, 340, 41
		DMSO, 1 h	**I** (50)	62
		$FeCl_2$, DMSO, 10 min	**I** (60)	75
PhOPh		K, NH_3, –78°	**II** (5)	94

	$h\nu$, NH_3, 4.2 h	**I** (14) + PhOH (20)	136
$PhOP(O)(OEt)_2$	K, NH_3	**I** (5) + **II** (56) + **III** (27)	108, 94
	$h\nu$, NH_3, 4.2 h	**I** (13) + **III** (11) + PhOH (71)	136
PhSPh	K, NH_3, –78°	**I** (18) + **II** (71) + PhSH (84)	94
	$h\nu$, NH_3, 0.5 h	**I** (66) + **III** (26) + **IV** (5) + PhSH (97)	136
$Ph_3S^+Cl^-$	$h\nu$, NH_3, 1.2 h	**I** (75) + PhSH (52) + Ph_2S (22)	136
PhSePh	$h\nu$, NH_3, 3.3 h	**I** (95) + PhSeH (83)	136
$Ph\overset{+}{N}Me_3I^-$	K, NH_3	**I** (46) + **II** (18) + **IV** (7)	94
	$h\nu$, NH_3,1 h	**I** (57) + **III** (37)	136
PhN_2SBu-*t*	Lab. light, DMSO, 1.7 h	**I** (75)	58
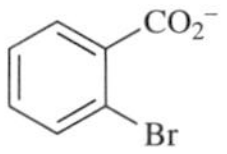	1. $h\nu$, NH_3, 1.5 h 2. CH_2N_2	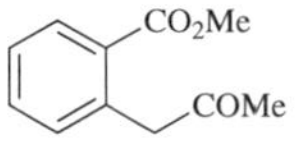 (85) + 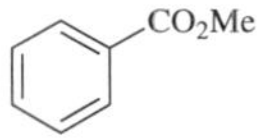 (10)	168, 108
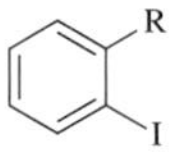	$h\nu$, NH_3	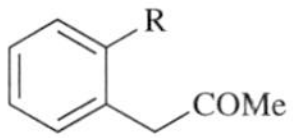 **I** + PhR **II**	
R		**I** / **II**	
OMe	5-15 min	(67) / (—)	291
CH_2NHCO_2Et	0.8 h	(58) / (11)	291, 290
$CH_2NHCOMe$	0.7 h	(30) / (28)	291
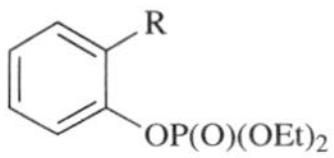	K, NH_3	**I** + 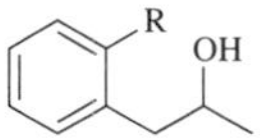 **II** + PhR **III**	108
R		**I** / **II** / **III**	
Pr-*i*		(6) / (24) / (54)	
Bu-*t*		(1) / (4) / (73)	

TABLE II. CARBANIONS DERIVED FROM KETONES (*Continued*)

A. Enolate from Acetone (*Continued*)

Substrate	Nucleophile	Conditions	Product(s) and Yield(s) (%)	Refs.
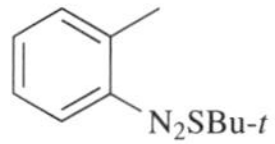	$^{-}CH_2COMe$	Lab. light, DMSO, 1.5 h	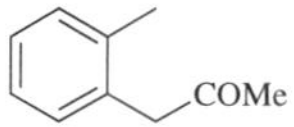 (10) + 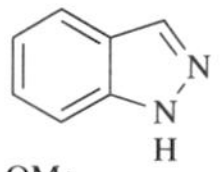 (70)	58
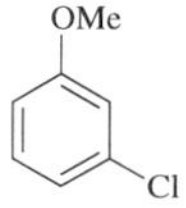		K, NH_3	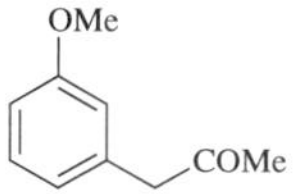 (6) + 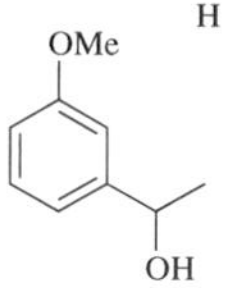 (12) + 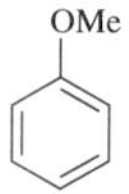 (9)	96
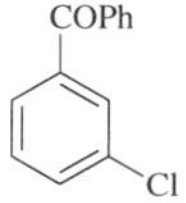		1. *h*ν, NH_3, 2 h 2. MeI	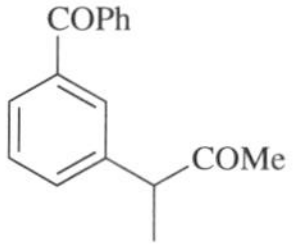 (12) + 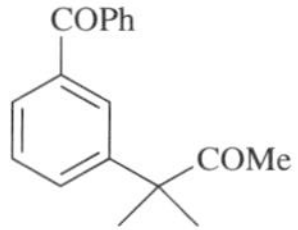 (32)	164
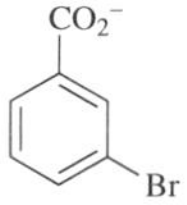		1. *h*ν, NH_3, 2.3 h 2. CH_2N_2	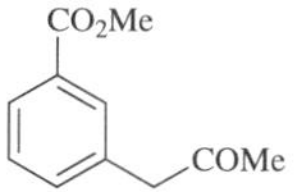 (80) + 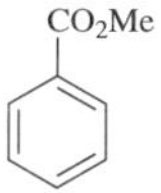 (8)	108
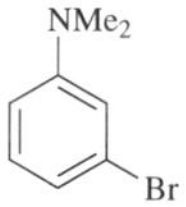		*h*ν, NH_3	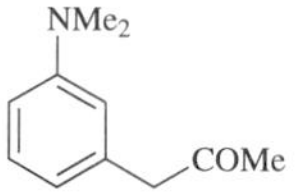 (82)	146
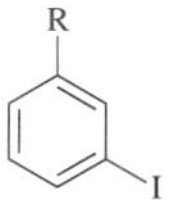		*h*ν, NH_3	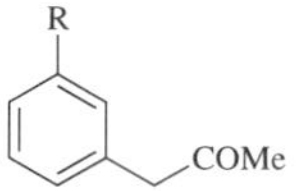	
R				
F		1.2 h	(56)	108
CF_3		1.5 h	(35) + $PhCF_3$ (14)	108
NH_2		4 min	(66)	106

Reactant	Conditions	Product(s) and Yield(s) (%)	Refs.
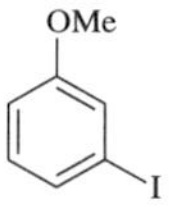	K, NH_3	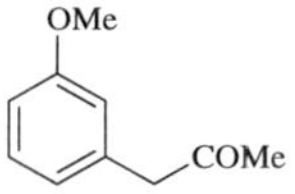 (54) + 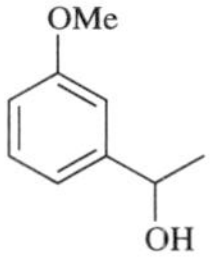 (8)	96
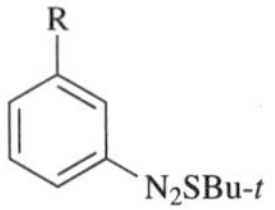	Lab. light, DMSO,	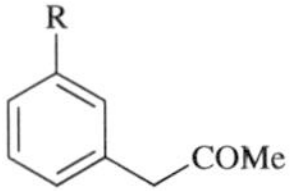	58
R			
Me	1.5 h	(81)	
OMe	1.5 h	(79)	
COPh	0.7 h	(44) + PhCOPh (36)	
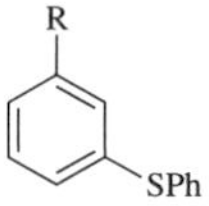	$h\nu$, NH_3	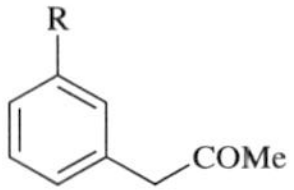 **I** + 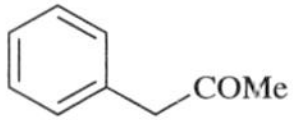 **II**	160
R		**I** / **II**	
Me	1.75 h	(48) / (39)	
OMe	1.25 h	(49) / (28)	
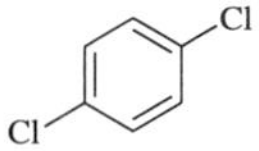	e^-, NH_3, –78°, mediator	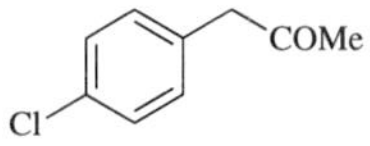 (48)	85
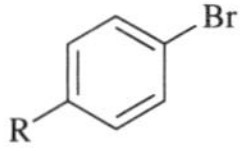		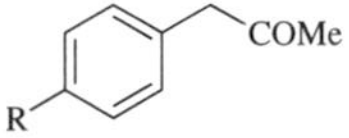 + 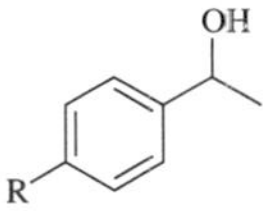 + 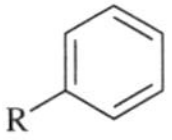	
R = Me	K, NH_3, –78°	(57) + (17) + (26)	94
R = OMe	K, NH_3, –78°	(36) + (10) + (44)	94
R = OMe	Na(Hg), NH_3, mediator	(57) + (—) + (21)	101
R = COPh	Na(Hg), NH_3	(78) + (—) + (—)	101
R = Ph	$h\nu$, NH_3, 2.5 h	(69) + (—) + (—)	108
R = NEt_2	$h\nu$, NH_3, 1.5 h	(—) + (—) + (8)	108

TABLE II. CARBANIONS DERIVED FROM KETONES (*Continued*)

A. Enolate from Acetone (*Continued*)

Substrate	Nucleophile	Conditions	Product(s) and Yield(s) (%)			Refs.
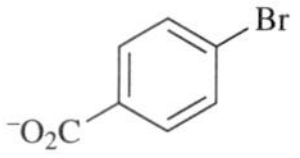	$^{-}CH_2COMe$	1. *h*ν, NH_3, 1.5 h 2. H_3O^+	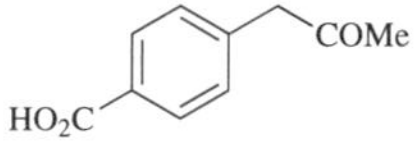 (70)			108
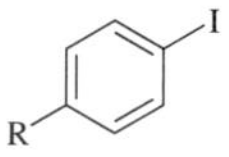		*h*ν, NH_3	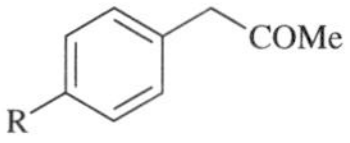 R: NH_2 (33)[a]; NMe_2 (90)			106 146
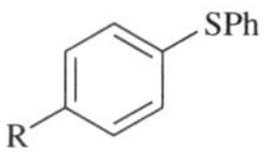		*h*ν, NH_3	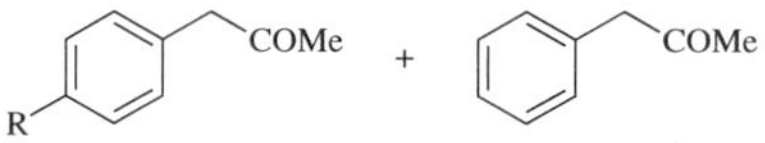			160
R						
Me		1.75 h	(38)	(28)		
OMe		1.5 h	(50)	(16)		
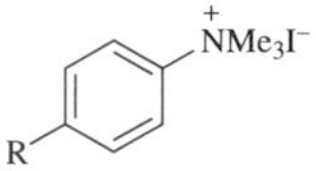		K, NH_3, –78°	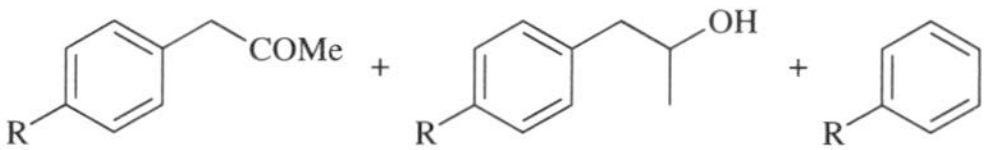			94
R						
Me			(30)	(42)	(6)	
OMe			(20)	(39)	(40)	
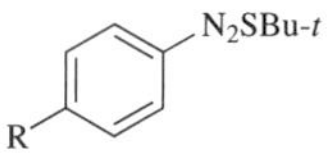		Lab. light, DMSO,	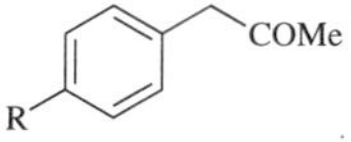			58
R						
Bu-*i*		0.5 h	(42)			
Bu-*t*		1.5 h	(86)			
OMe		1.5 h	(69)			
COPh		0.75 h	(78)			

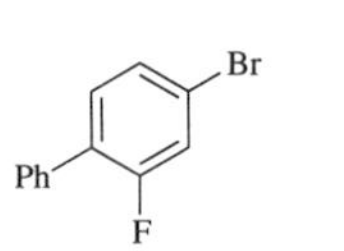	1. $h\nu$, NH_3, 2 h 2. MeI	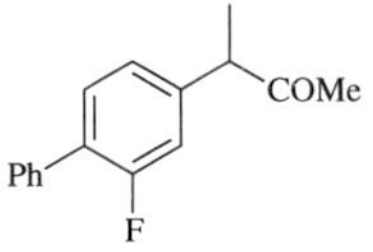 (62)	164
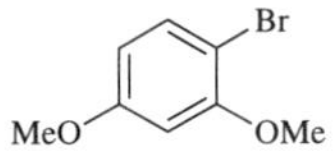	$h\nu$, NH_3, 2 h	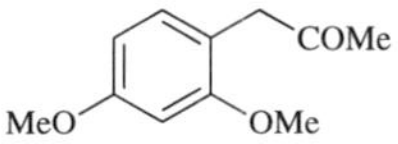 (76)	108
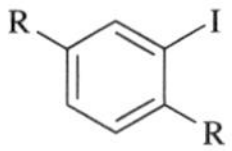	$h\nu$, NH_3, 2.2 h	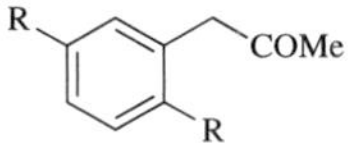 R: Pr-*i*- (78); Bu-*t* (26)	108
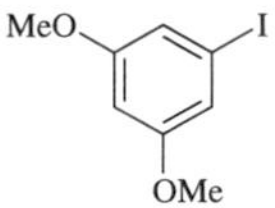	$h\nu$, NH_3, 1 h	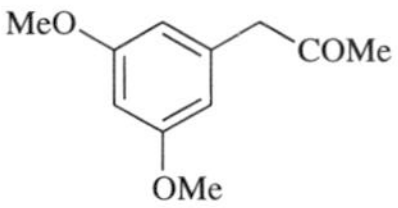 (68)	108
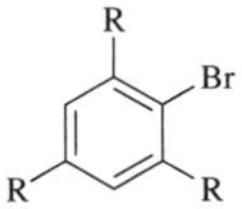	$h\nu$, NH_3	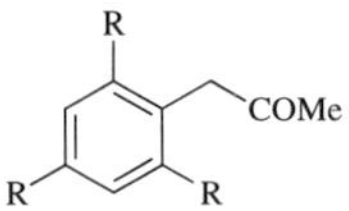 + 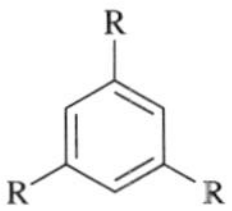	108
R: Et	2 h	(70) + (22)	
R: Pr-*i*	2.5 h	(2) + (—)	
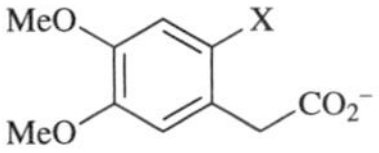	1. $h\nu$, NH_3 2. H_3O^+	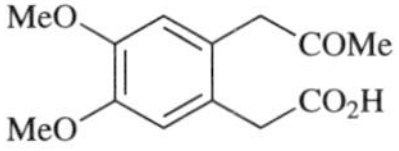 + 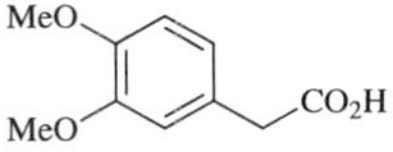	149
X: Br	3 h	(60) + (40)	
X: I	25 min	(75-80) + (20-25)	

TABLE II. CARBANIONS DERIVED FROM KETONES (*Continued*)

A. Enolate from Acetone (*Continued*)

Substrate	Nucleophile	Conditions	Product(s) and Yield(s) (%)	Refs.
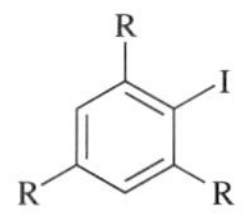	$^{-}CH_2COMe$	$h\nu$, NH_3	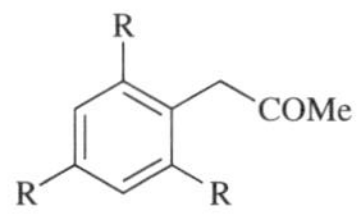 + 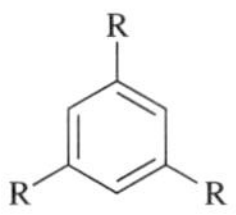	108
R: Me		0.5 h	(82) (10)	
R: OMe		1.5 h	(92) (—)	
R: Pr-*i*		3 h	(16) (37)	
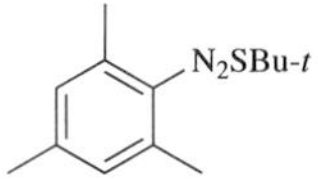		Lab light, DMSO, 2 h	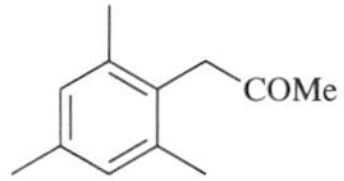 (30) + 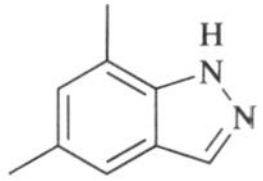 (50)	58
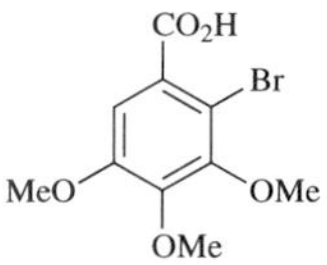		1. $h\nu$, NH_3 2. H_3O^+	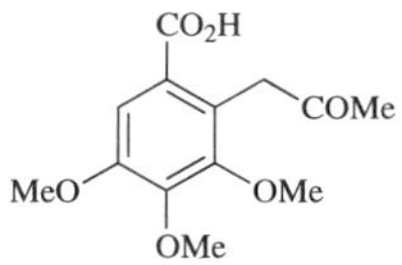 (78) + 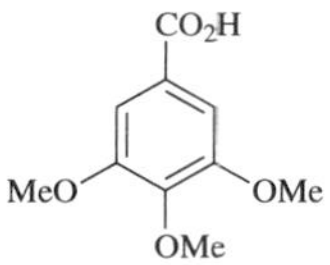 (11)	168
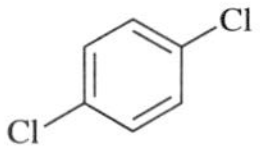		$h\nu$, NH_3, 3 h	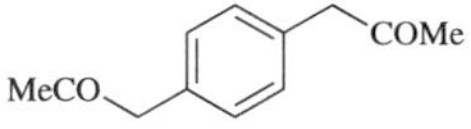 (42)	166
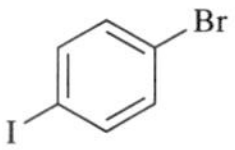		$h\nu$, NH_3, 4 h	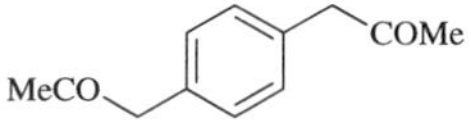 (32)	166
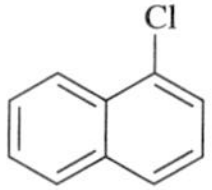		K, NH_3	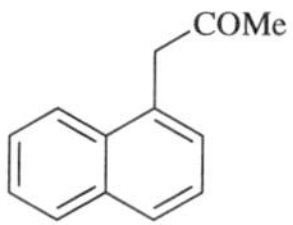 **I** (23) + 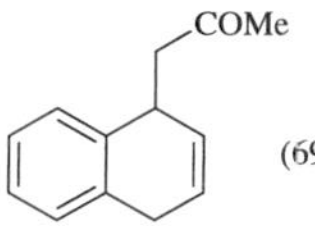 (69)[b]	45

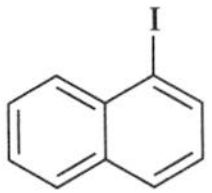	$h\nu$, NH_3, 1.5 h	**I** (88)	45
	Na(Hg), NH_3	**I** (98)	101
	K, NH_3	**I** (6) + **II** (84)[b]	45
	$h\nu$, NH_3, 1 h	**I** (76)	108
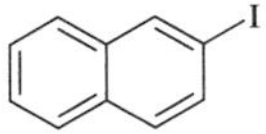	$h\nu$, NH_3, 1.5 h	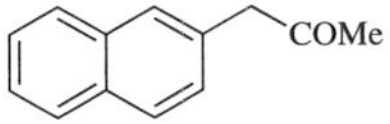 (75)	108
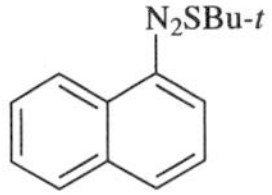	Lab light, DMSO, 1 h	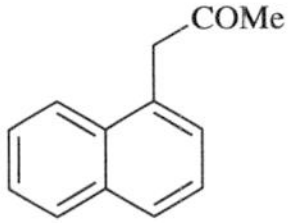 (75)	58
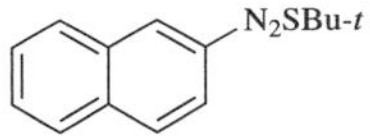	Lab light, DMSO, 0.5 h	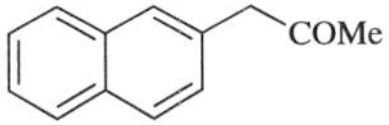 (76)	58
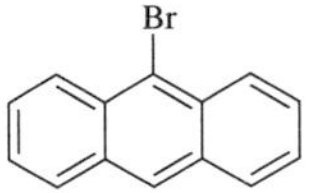	$h\nu$, NH_3, 1 h	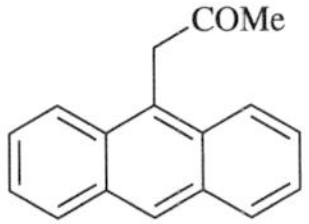 (98)	108
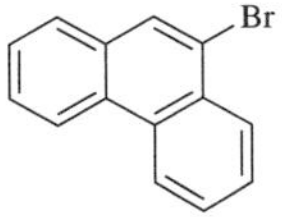	$h\nu$, NH_3, 1.3 h	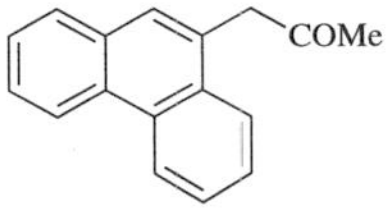 (62)	108
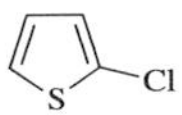	Na, NH_3, –78°	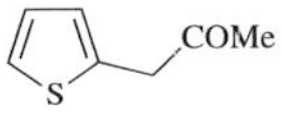 **I** (2) + 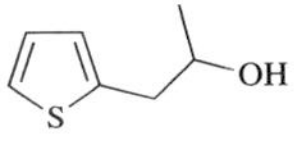 (22) +  (46)	152
	$h\nu$, NH_3, 1 h	**I** (17)	152

TABLE II. CARBANIONS DERIVED FROM KETONES (*Continued*)

A. Enolate from Acetone (*Continued*)

Substrate	Nucleophile	Conditions	Product(s) and Yield(s) (%)	Refs.
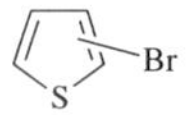	$^{-}CH_2COMe$		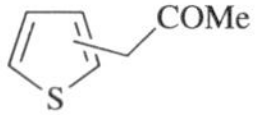 **I** + 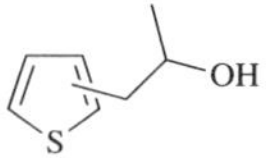 **II** +  **III**	
		Na, NH_3, –78°	2-Br, **I** (4) + **II** + (3) + **III** (70) + 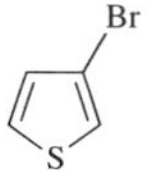 (10)	152
		*h*ν, NH_3, 1 h	2-Br, **I** (31)	152
		K, NH_3, –78°	3-Br, **I** (29) + **II** (23) + **III** (29) + 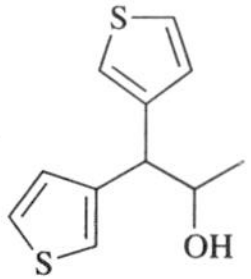 **IV** (10)	152
		*h*ν, NH_3, 1 h	3-Br, **I** (51) + **IV** (25)	152
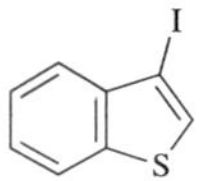		DMSO, 1 h	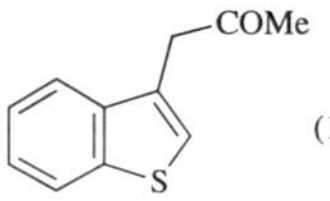 (12) + 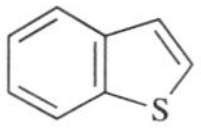 (23)	151
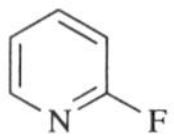		*h*ν, NH_3, 2 h	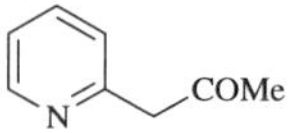 (40)	137
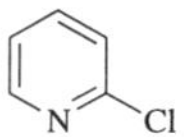		*h*ν, NH_3, 1 h	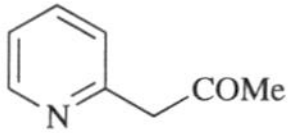 **I** (85)	137
		Na(Hg), NH_3	**I** (15) + 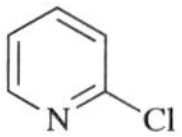 (10) + 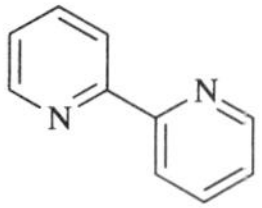 (55)	101

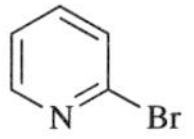	K, NH_3	**I** (4)	137
	$h\nu$, NH_3, 15 min	**I** (100)	137
	$h\nu$, THF, 15 min	**I** (64)	144
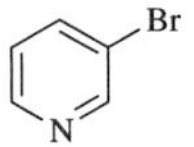	$h\nu$, NH_3, 15 min	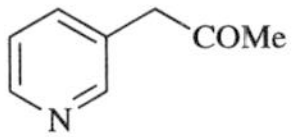 (65)	137
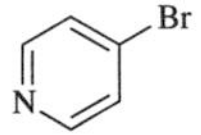	$h\nu$, NH_3, 15 min	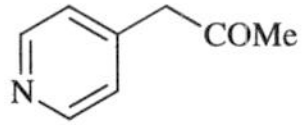 (28)	137
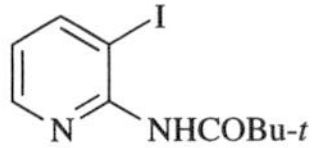	$h\nu$, NH_3	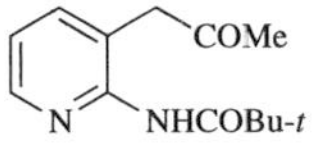 (90)	139
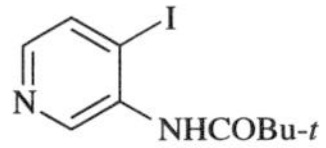	$h\nu$, NH_3	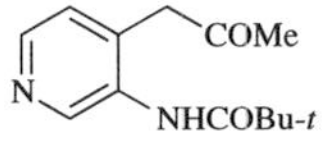 (95)	139
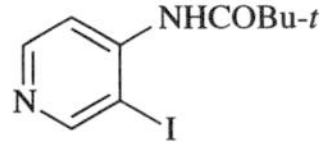	$h\nu$, NH_3	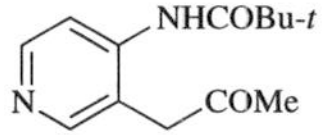 (90)	139
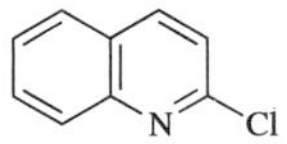	K, NH_3	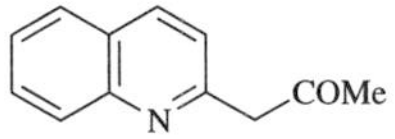 **I** (43) + 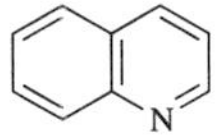 (29)	142
	$h\nu$, NH_3,[c] 1 h	**I** (90)	142
	$h\nu$, NH_3,[d] 1 h	**I** (62)	157
	Na(Hg), NH_3	**I** (49) + 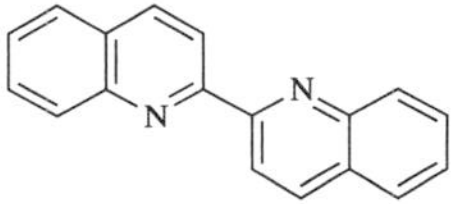 (50)	101

TABLE II. CARBANIONS DERIVED FROM KETONES (*Continued*)

A. Enolate from Acetone (*Continued*)

Substrate	Nucleophile	Conditions	Product(s) and Yield(s) (%)	Refs.
	$^{-}CH_2COMe$	*h*ν, 1 h		144
		THF	**I** (82)	
		DMF	**I** (74)	
		DME	**I** (28)	
		Et_2O	**I** (9)	
		C_6H_6	**I** (4)	
		DMSO, 5 min	**I** (37)	
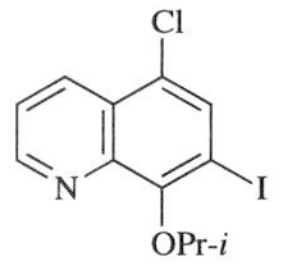		*h*ν, NH_3, 1 h	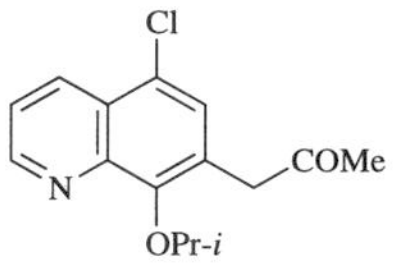 (73)	155
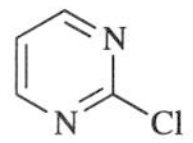		*h*ν, NH_3, 15 min	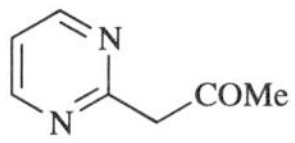 **I** (15) + 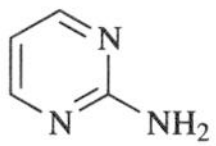 (4)	55
		*h*ν, THF, 15 min, 0°	**I** (61)	55
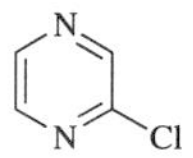		NH_3, 15 min	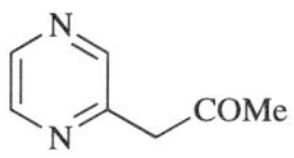 (98)	55
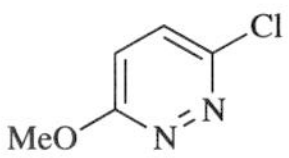		NH_3, 15 min	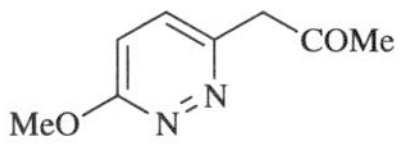 (60)	55
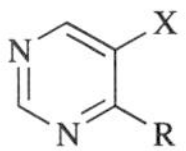			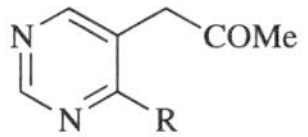	54

Substrate		Conditions	Product and Yield (%)		Refs.
X	R				
Cl	Bu-*t*	K, NH_3	(42)		
Cl	Bu-*t*	*h*ν, NH_3, 1.25 h	(60-65)		
Cl	Ph	NH_3, 16 h	(39)		
Cl	Ph	K, NH_3	(47)		
Cl	Ph	*h*ν, NH_3, 1.25 h	(20-25)		
Br	Bu-*t*	NH_3, 16 h	(20-25)		
Br	Bu-*t*	K, NH_3	(30)		
Br	Bu-*t*	*h*ν, NH_3, 1.25 h	(70-75)		
Br	Ph	NH_3, 16 h	(30)		
Br	Ph	K, NH_3	(42)		
Br	Ph	*h*ν, NH_3, 1.25 h	(25-30)		
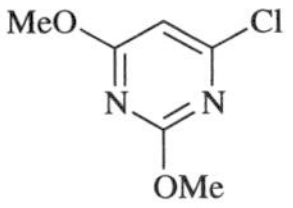		*h*ν, NH_3, 15 min	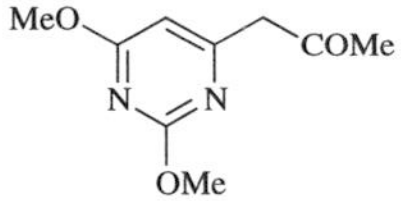	(75)	140
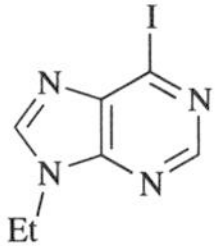		*h*ν, NH_3, 0.5 h	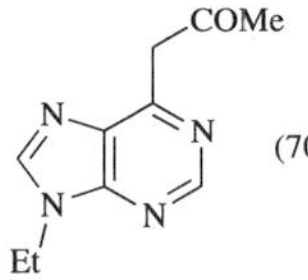	(70)	153, 154

[a] This compound decomposes in contact with air.

[b] This product is a mixture of dihydro and tetrahydro derivatives.

[c] The potassium salt of the nucleophile was used.

[d] A $LiNH_2$:nucleophile ratio of 3 was used.

TABLE II. CARBANIONS DERIVED FROM KETONES (*Continued*)

B. Enolate from Pinacolone

Substrate	Nucleophile	Conditions	Product(s) and Yield(s) (%)	Refs.
PhBr	$^{-}CH_2COBu\text{-}t$	$h\nu$, DMSO, 1.5 h	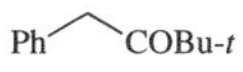 **I** (87) + 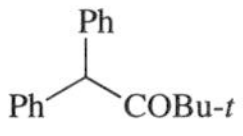 **II** (9)	138
		$h\nu$, NH_3, 1.5 h	**I** (90) + **II** (10)	143
		$FeCl_2$, DMSO, 20 min	**I** (99)	77
		$FeSO_4$, NH_3, 1.25 h	**I** (58)	75
PhI		$h\nu$, DMSO, 1 h	**I** (99)	138
		$h\nu$, NH_3, 15 s	**I** (83)	345
		DMSO, 1 h	**I** (60)	62
		$FeCl_2$, DMSO, 20 min	**I** (74)	77
		$FeSO_4$, DMSO, 20 min	**I** (34)	77
		$FeSO_4$, NH_3, 20 min	**I** (87)	75
$Ph_2I^+Br^-$		$FeCl_2$, DMSO, 20 min	**I** (97)	77
$PhN_2SBu\text{-}t$		Lab. light, DMSO, 2 h	**I** (80)	58
Ph_2S		$h\nu$, DMSO, 3 h	**I** (67)	67
		$h\nu$, NH_3, –76°, 2.5 h	**I** (98)	67
$PhSO_2Ph$		$h\nu$, DMF, 14 h	**I** (70) + C_6H_6 (11)	67
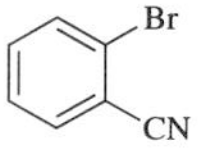		$FeCl_2$, DMSO, 20 min	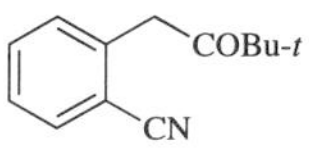 (46) + 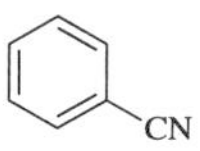 (6)	77

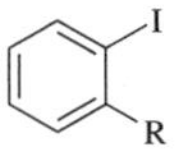	$h\nu$, NH_3	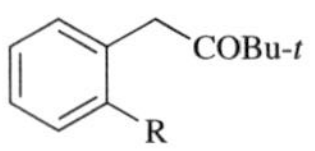	
R			
OMe	5-15 min	(100)	145
NO_2	8 min	(66) + $PhNO_2$ (20)	113
SO_2NHBn	0.5 h	(85)	140
Ph	0.5 h	(83)	30
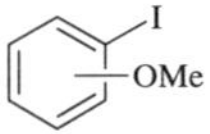	DMSO, 1 h	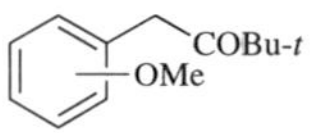 **I** + 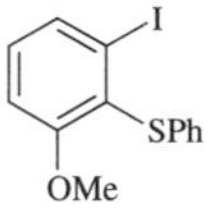 **II**	
		m-OMe, **I** (58) + **II** (11)	63
		p-OMe, **I** (81)	73
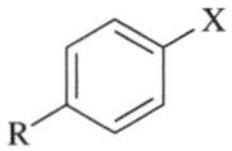		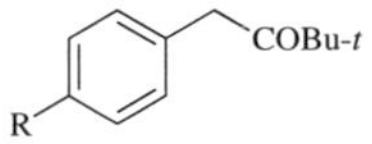	
R X			
Cl Br	$FeCl_2$, DMSO, 20 min	(20)	77
OMe Br	$FeCl_2$, DMSO, 20 min	(13)	77
F I	DMSO	(71)	73
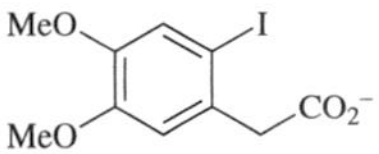	1. $h\nu$, NH_3, 15 min 2. H_3O^+	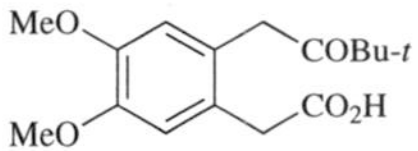 (85-90) + 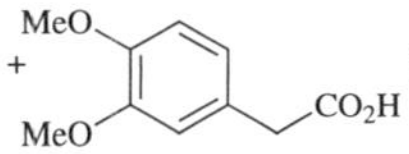 (15-10)	149
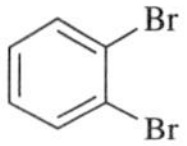	$h\nu$, NH_3, 2 h	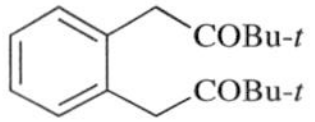 (62)	165

TABLE II. CARBANIONS DERIVED FROM KETONES (*Continued*)

B. Enolate from Pinacolone (*Continued*)

Substrate	Nucleophile	Conditions	Product(s) and Yield(s) (%)	Refs.
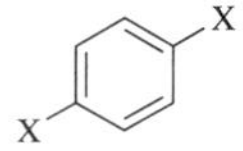	$^{-}CH_2COBu$-*t*	*h*ν, NH_3	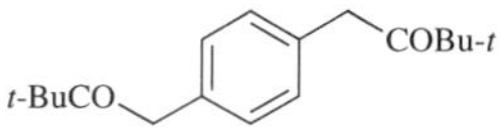 I	166
X				
Cl		4 h	(64)	
Br		5 h	(38)	
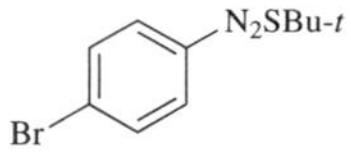		Lab light, DMSO, 1.5 h	**I** (49) + 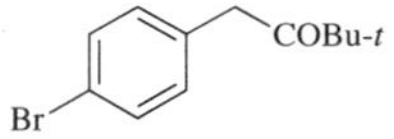 (10)	58
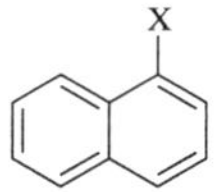		$FeCl_2$, DMSO, 20 min	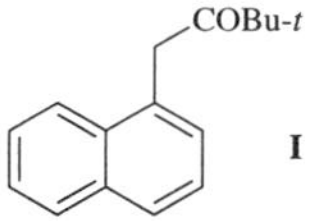 **I** + 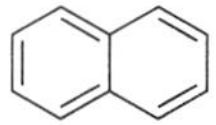 **II**	77
X				
Br			(41) (25)	
I			(60) (26)	
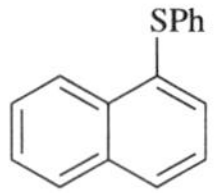		*h*ν, DMSO, 5 h	**I** (18) + 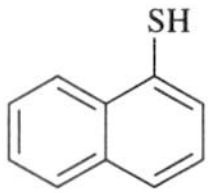 **II** (64) + 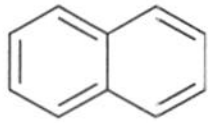 **III** (9) + 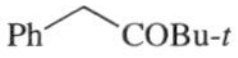 (64) + PhSh (9)	67
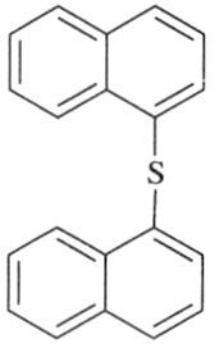		*h*ν, DMF, 19 h	**I** (28) + **II** (67) + **III** (39) + 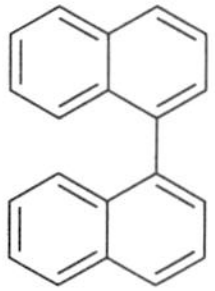 (11)	67

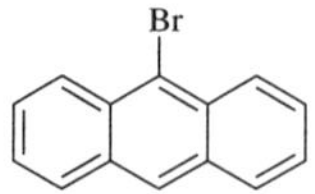Br	$FeCl_2$, DMSO, 20 min	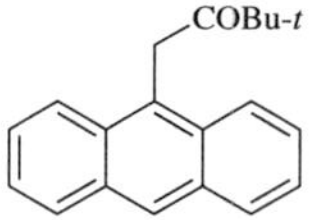COBu-*t* (46) + 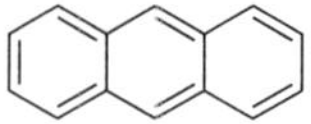 (30)	77
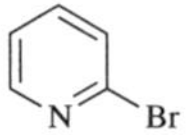N Br	$FeCl_2$, DMSO, 20 min	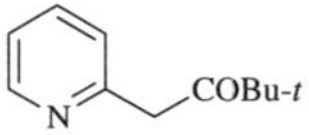N COBu-*t* **I** (80)	77
	*h*ν, NH_3, 1.5 h	**I** (94)	137
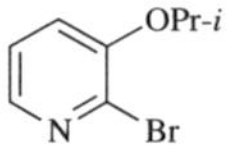OPr-*i* N Br	*h*ν, NH_3, 15 min	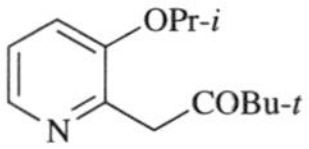OPr-*i* N COBu-*t* (70)	155
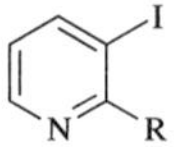I N R	*h*ν, NH_3	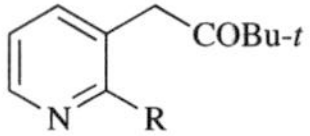COBu-*t* N R	139
R			
F	2 h	(92)	
OMe	2 h	(84)	
NHMe	2 h	(50)	
NH_2	2.5 h	(87)	
NHCOBu-*t*	2 h	(99)	
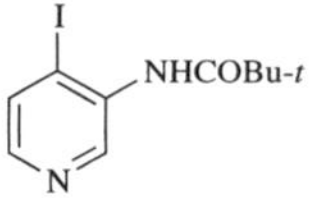I NHCOBu-*t* N	*h*ν, NH_3	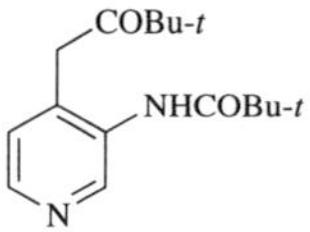COBu-*t* NHCOBu-*t* N (98)	139
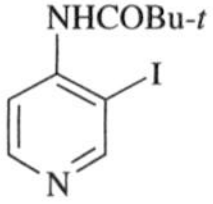NHCOBu-*t* I N	*h*ν, NH_3	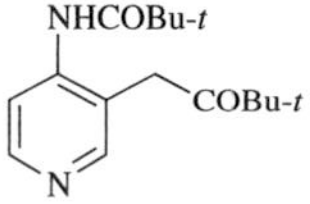NHCOBu-*t* COBu-*t* N (99)	139

TABLE II. CARBANIONS DERIVED FROM KETONES (*Continued*)

B. Enolate from Pinacolone (*Continued*)

Substrate	Nucleophile	Conditions	Product(s) and Yield(s) (%)	Refs.
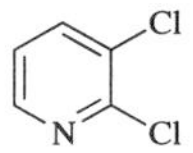	$^{-}CH_2COBu$-*t*	*h*ν, NH_3, 15 min	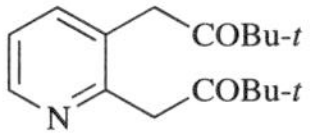 (63)	167
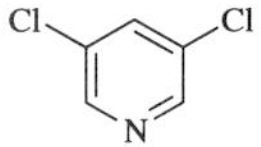		*h*ν, NH_3, 15 min	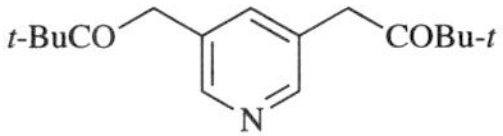 (43)	167
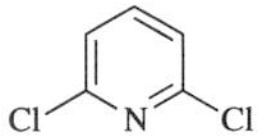		*h*ν, NH_3, 1 h	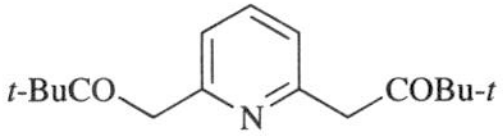 (86)	137
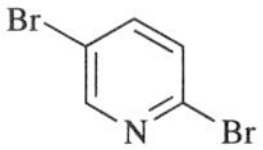		*h*ν, NH_3, 15 min	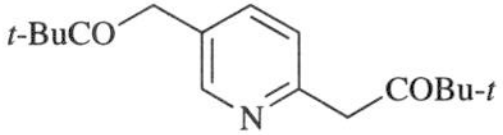 (85)	167
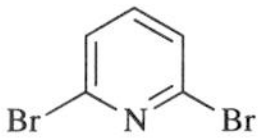		*h*ν, NH_3, 1 h	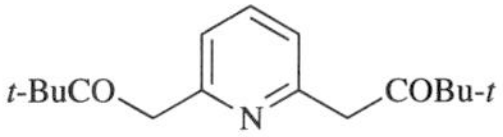 (89)	137
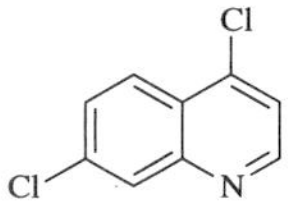		*h*ν, NH_3, 0.5 h	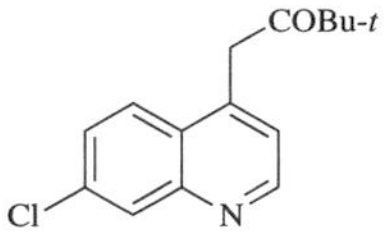 (70)	167
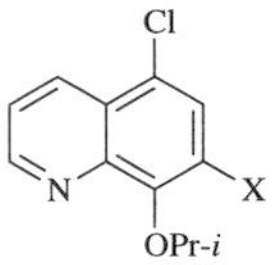		*h*ν, NH_3, 1 h	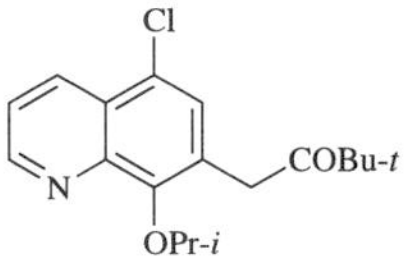 X: Cl (70); I (70)	155

Substrate	Conditions	Products (yield %)	Refs.
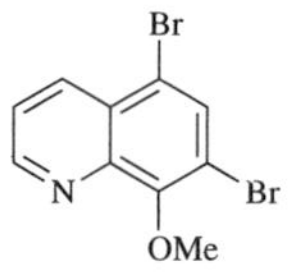	$h\nu$, NH_3	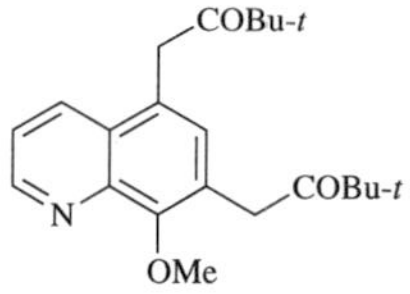 (60) + 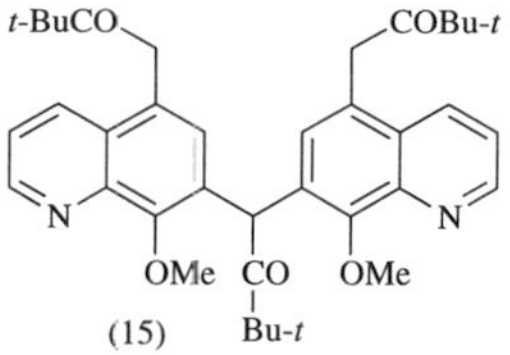 (15)	155
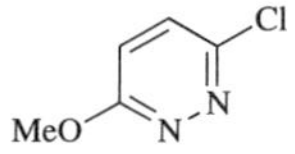	$h\nu$, NH_3, 15 min	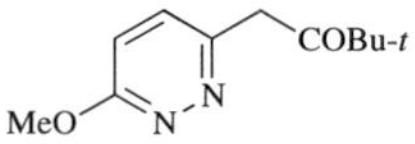 (72)	55
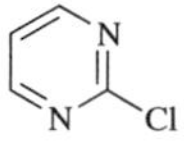	$h\nu$, NH_3, 15 min	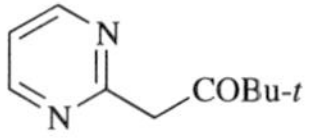 (32) + 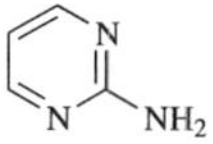 (4)	55
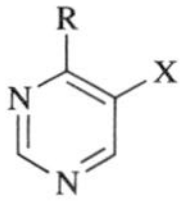		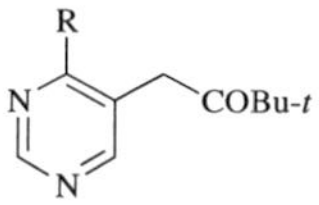 **I**	54
R X			
t-Bu Cl	K, NH_3	**I** (65)	
t-Bu Cl	$h\nu$, NH_3, 15 min	**I** (90-95)	
Ph Cl	$h\nu$, NH_3, 1.25 h	**I** (45)	
Ph Cl	K, NH_3	**I** (50) + 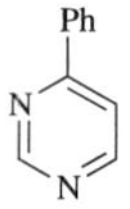 **II** (16)	
t-Bu Br	K, NH_3	**I** (65)	
t-Bu Br	$h\nu$, NH_3, 1.25 h	**I** (95)	
Ph Br	$h\nu$, NH_3, 1.25 h	**I** (60-65)	
Ph Br	K, NH_3	**I** (45-50) + **II** (22)	

TABLE II. CARBANIONS DERIVED FROM KETONES (*Continued*)

B. Enolate from Pinacolone (*Continued*)

Substrate	Nucleophile	Conditions	Product(s) and Yield(s) (%)	Refs.
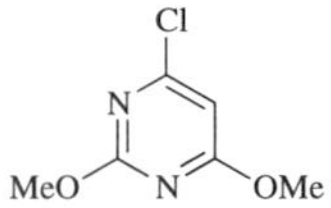	$^{-}CH_2COBu$-*t*	*hν*, NH_3, 15 min	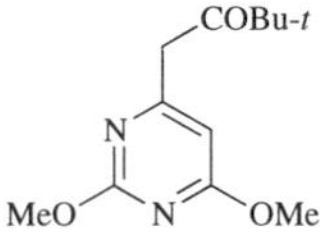 (98)	55
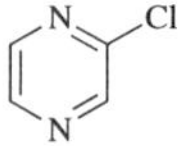		NH_3, 15 min	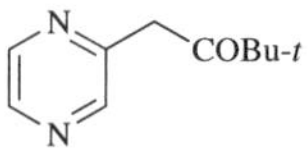 (95)	55
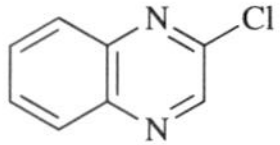		NH_3, 15 min	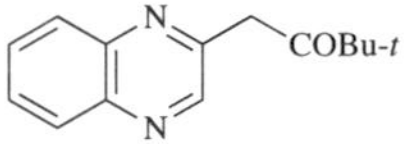 (70) + 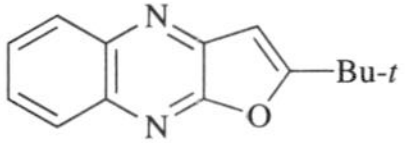 (15)	56
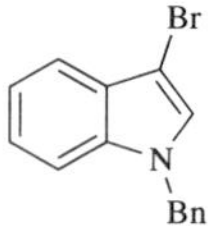		*hν*, NH_3, 2 h	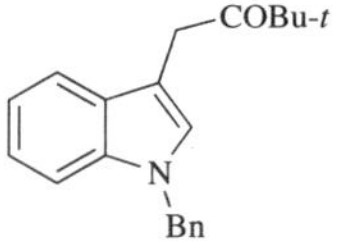 (56) + 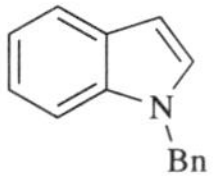 (15)	140
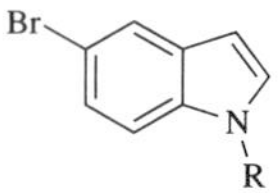		*hν*, NH_3, 15 min	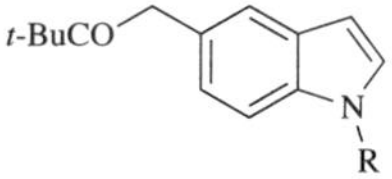 R: Me (62); Bn (72)	140
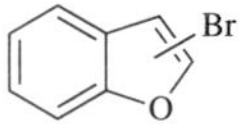		*hν*, NH_3	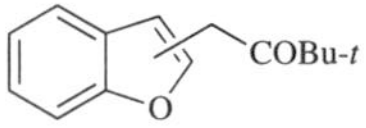 2-Br, 1.5 h (42); 3-Br, 3 h (42)	140
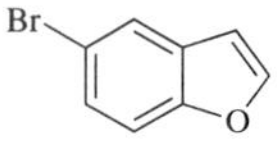		*hν*, NH_3, 15 min	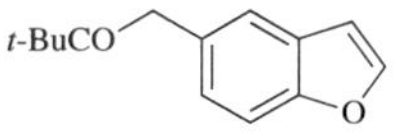 (78) + 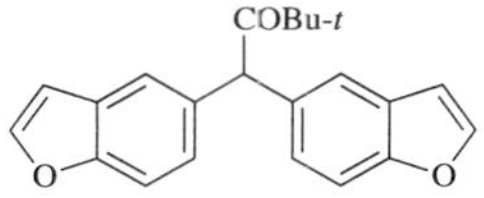 (9)	140

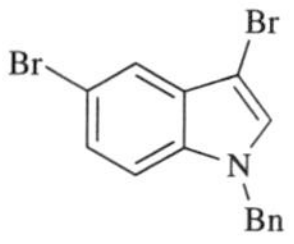	*hν*, NH_3, 2 h	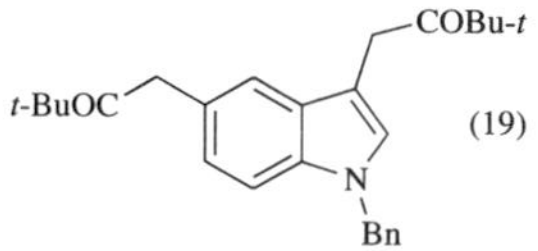 (19) + 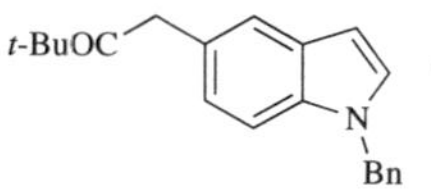 (30)	140
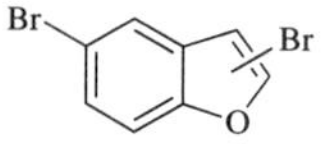	*hν*, NH_3	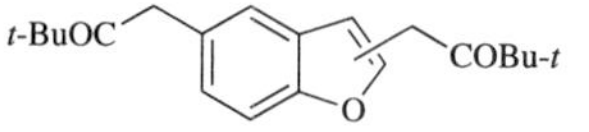 + 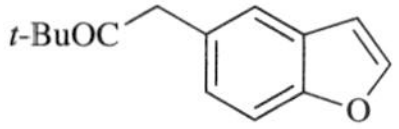	140
2-Br	15 min	(8)	(11)	
3-Br	1 h	(37)	(27)	
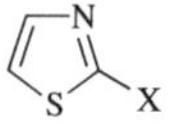	*hν*, NH_3	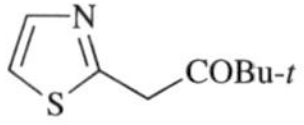 + 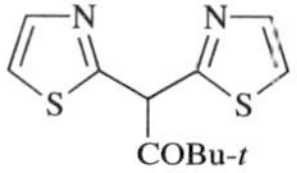	141
X			
Cl	1 h	(53) (25)	
Br	15 min	(44) (—)	
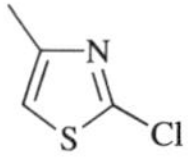	*hν*, NH_3, 1 h	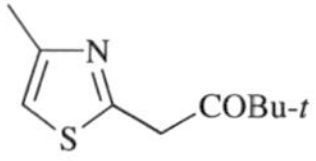 (64) + 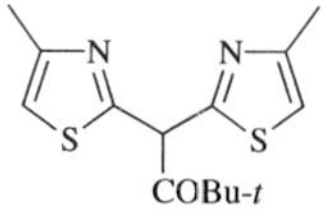 (10)	141
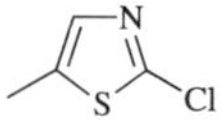	*hν*, NH_3, 1 h	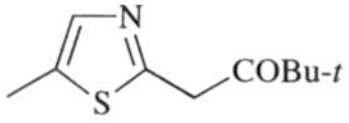 (67) + 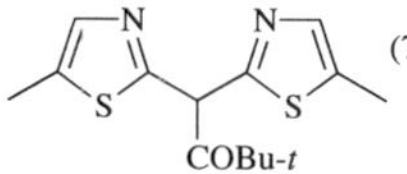 (7)	141
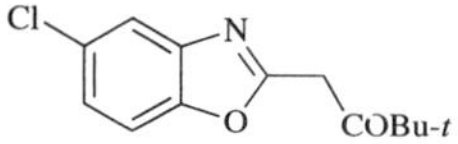	*hν*, NH_3, 1 h	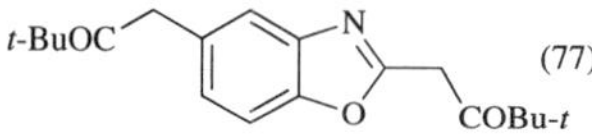 (77) + 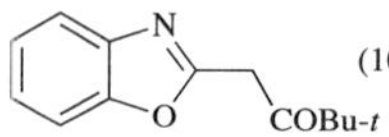 (10)	141

TABLE II. CARBANIONS DERIVED FROM KETONES (*Continued*)

C. Enolate Ions from Other Alicyclic Ketones

Substrate	Nucleophile	Conditions	Product(s) and Yield(s) (%)	Refs.
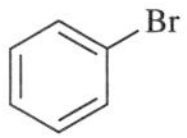Br	$^{-}CH_2COEt$ ⇌ $^{-}CHMeCOMe$	DMSO, 1 h	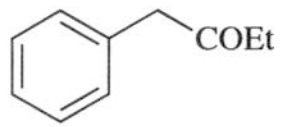COEt **I** (18) + 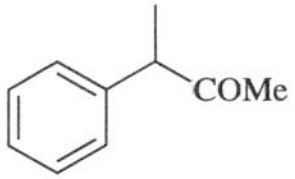COMe **II** (52)	62
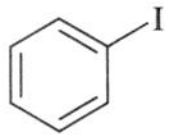I		$h\nu$, NH_3, 12 min	**I** (23) + **II** (41)	124
		$h\nu$, NH_3, 0.8 h	**I** (19) + **II** (61)	124
		1. K, NH_3 2. $K_2Cr_2O_7$, H_2SO_4	**I** (18) + **II** (52)	124
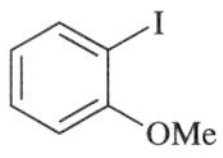I OMe		$h\nu$, NH_3, 5-15 min	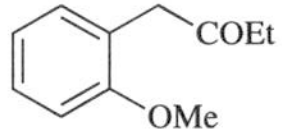COEt OMe (23) + 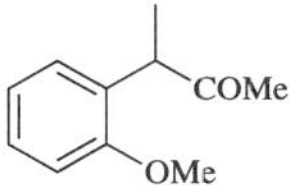COMe OMe (34)	145
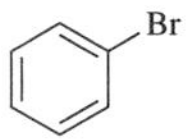Br	$^{-}CH_2COPr$-*i* ⇌ $^{-}CMe_2COMe$	$h\nu$, NH_3, 2 h	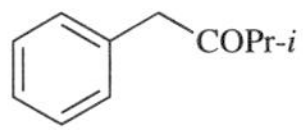COPr-i **I** (81) + 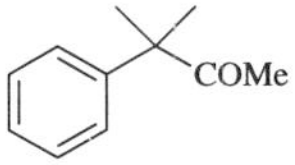COMe **II** (9)	124, 135
		1. K, NH_3 2. $K_2Cr_2O_7$, H_2SO_4	**I** (56) + **II** (7)	124
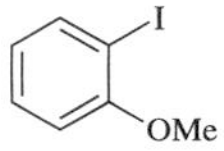I OMe		$h\nu$, NH_3, 5-15 min	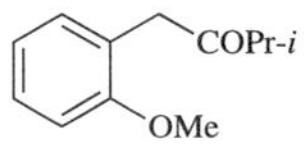COPr-i OMe (66)	145
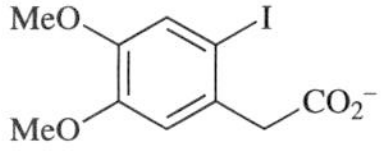MeO I MeO CO2−		1. $h\nu$, NH_3 2. H_3O^+	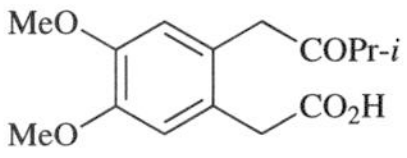MeO COPr-i MeO CO2H (85-90) + 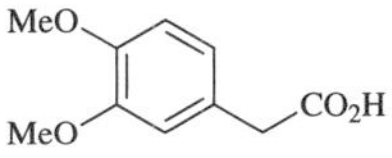MeO MeO CO2H (15-10)	149

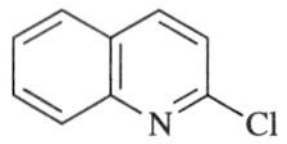		$h\nu$, NH_3, 1 h	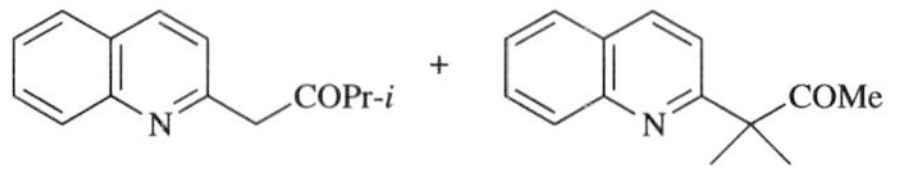 (13) + (62)	142
	$^{-}CH_2COCH(OMe)_2$	$h\nu$, NH_3, 1 h	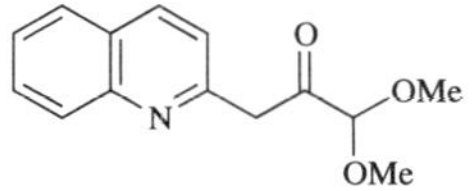 (88)	106
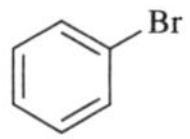	$^{-}$CHMeCOEt	$h\nu$, NH_3, 1.2 h	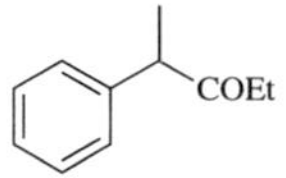 **I** (80) + 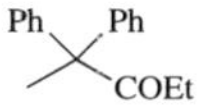 (19)[a]	143
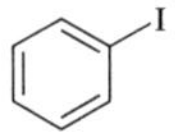		DMSO, 1 h	**I** (70)	62
		$FeCl_2$, DMSO, 20 min	**I** (55)	77
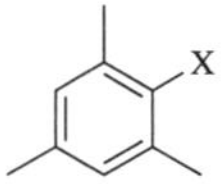		$h\nu$, NH_3, 2.2 h	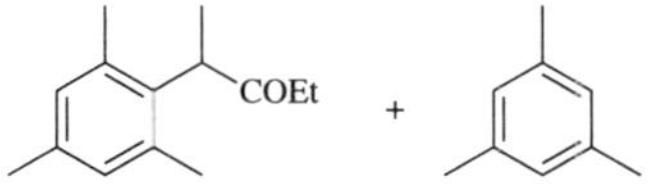	143
X = Br			(14) + (20)	
X = I			(24) + (25)	
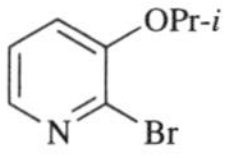		$h\nu$, NH_3, 1 h	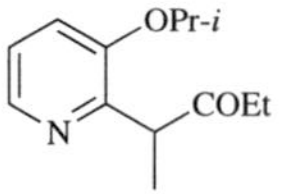 (86)	155
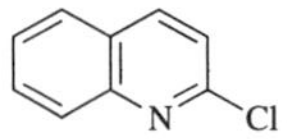		$h\nu$, NH_3, 1 h	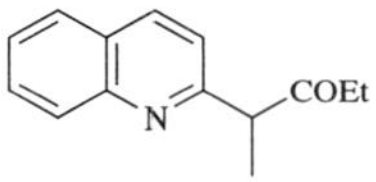 (68)	142

TABLE II. CARBANIONS DERIVED FROM KETONES (*Continued*)

C. Enolate Ions from Other Alicyclic Ketones (*Continued*)

Substrate	Nucleophile	Conditions	Product(s) and Yield(s) (%)	Refs.
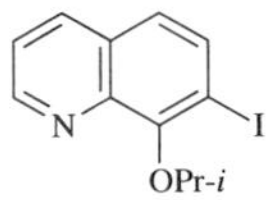	$^{-}$CHMeCOEt	$h\nu$, NH_3, 1 h	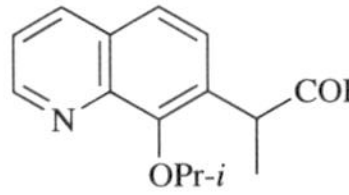 (70)	155
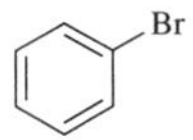	$^{-}$CHEtCOPr-*n*	$h\nu$, NH_3, 2 h	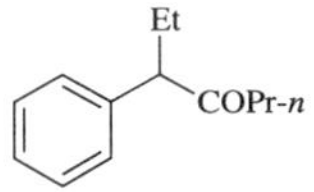 (80) +  (11)	143[b]
	$^{-}$CH(OMe)COMe	$h\nu$, NH_3, 2.2 h	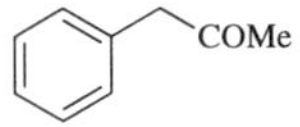 (17)	143
	$^{-}CMe_2$COPr-*i*	$h\nu$, NH_3, 1.5 h	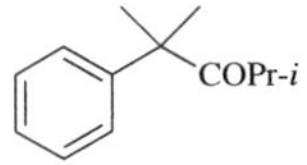 **I** (6) +  **II** (8)	143[c]
		$h\nu$, NH_3, 2 h	**I** (15)	135
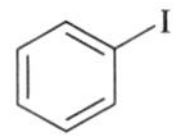		$h\nu$, NH_3, 3 h	**I** (32) + **II** (19)	143[c]
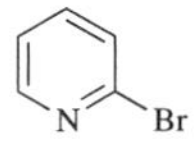		$h\nu$, NH_3, 1 h	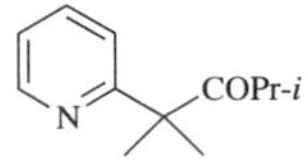 (97)	137
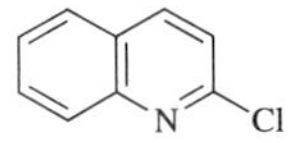		$h\nu$, NH_3, 15 min	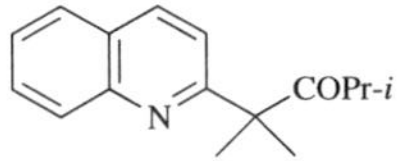 **I** (98)	142
		$h\nu$, THF, 1 h	**I** (65) + 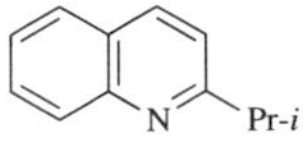 (25)	144

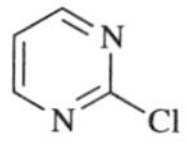	$h\nu$, NH_3, 15 min	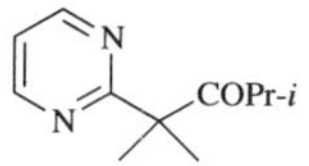 (88)	55
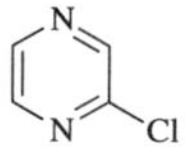	NH_3, 15 min	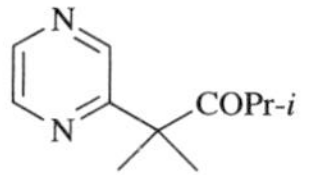 (85)	55
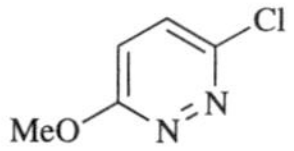	$h\nu$, NH_3, 1 h	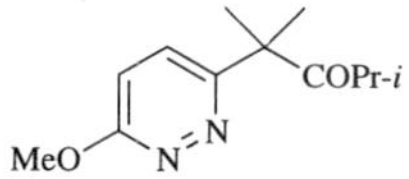 (20)	55
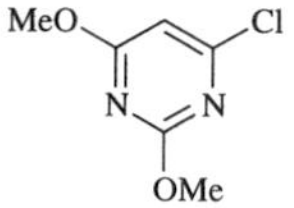	$h\nu$, NH_3, 15 min	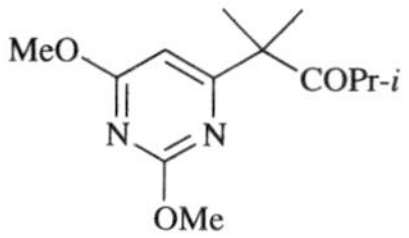 (52)	140
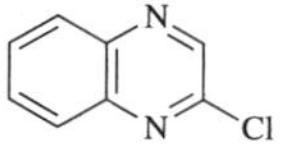	NH_3, 15 min	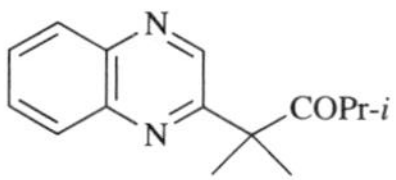 **I** + 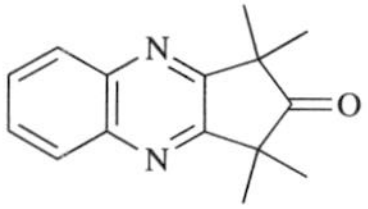 **II** Enolate ion:Substrate = 3.75 **I** (31) + **II** (28) Enolate ion:Substrate = 1.00 **I** (43)	56

[a] The product was a mixture of isomers.

[b] Diphenylheptan-4-one isomers were also formed in 10% yield.

[c]  was also formed.

TABLE II. CARBANIONS DERIVED FROM KETONES (*Continued*)

D. Enolate Ions from Cyclic Ketones

Substrate	Nucleophile	Conditions	Product(s) and Yield(s) (%)	Refs.
PhBr	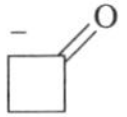	$h\nu$, NH_3, 2.5 h	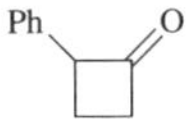 (90)	143
	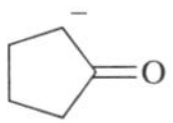	$h\nu$, NH_3, 2.5 h	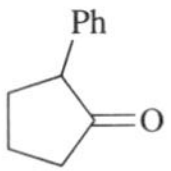 (64) +  (28)	143
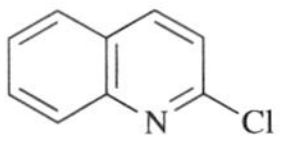		$h\nu$, NH_3, 1 h	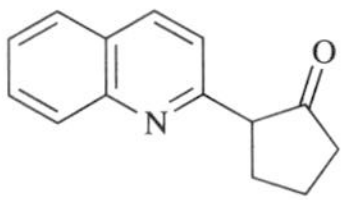 (63)	142
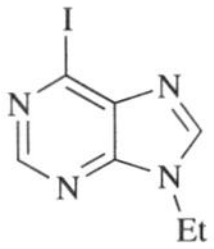		$h\nu$, NH_3, 0.5 h	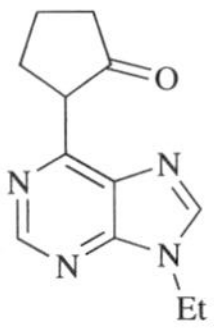 (65)	153, 154
PhBr	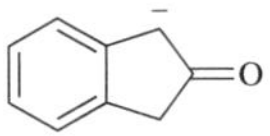	$h\nu$, NH_3, 2.5 h	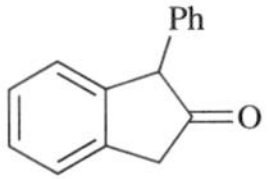 (90)	143
	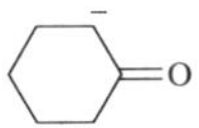	DMSO, 1 h	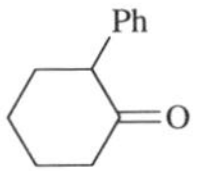 **I** (75) +  **II** (25)	62
PhI		$h\nu$, NH_3, 1 h	**I** (72) + **II** (6)	143
		$FeCl_2$, DMSO, 20 min	**I** (39)	77

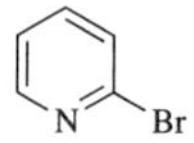		$h\nu$, NH_3, 1 h	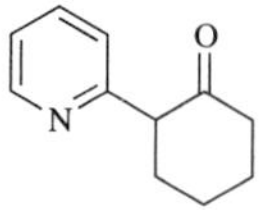 (47)	137
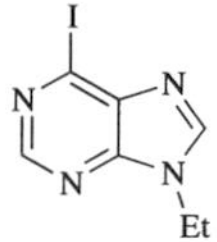		$h\nu$, NH_3, 0.5 h	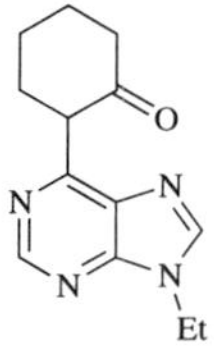 (50)	153, 154
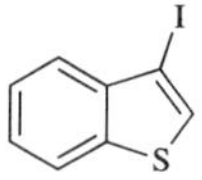		DMSO, 1 h	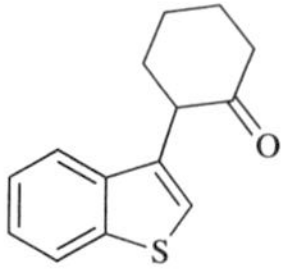 (6) + 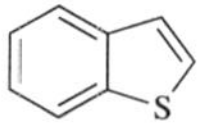 (22)	151
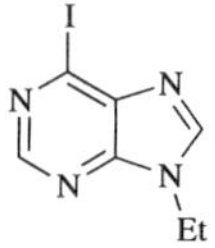	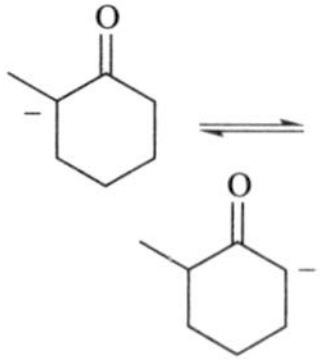	$h\nu$, NH_3, 0.5 h	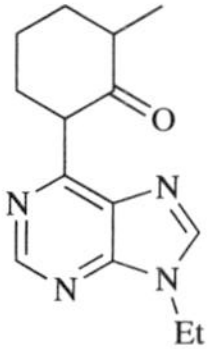 (7) + 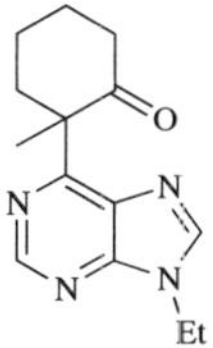 (30)	153
PhX	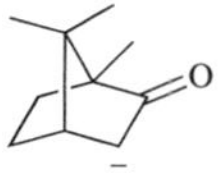	$h\nu$, NH_3	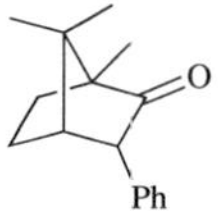 **I** (*endo*) +  **II** (*exo*)	156
X				
Cl		2.5 h	**I** (92) **I**:**II** = 99:1	
Br		2 h	**I** (95) **I**:**II** = 99:1	

TABLE II. CARBANIONS DERIVED FROM KETONES (*Continued*)

D. Enolate Ions from Cyclic Ketones (*Continued*)

Substrate	Nucleophile	Conditions	Product(s) and Yield(s) (%)	Refs.
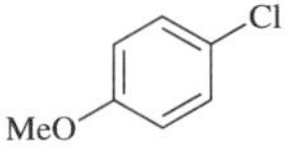	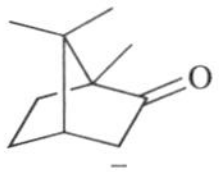	$h\nu$, NH_3, 2 h	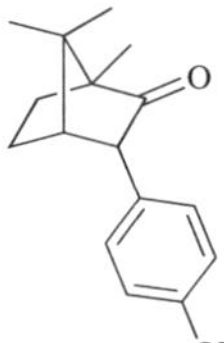 **I** (*endo*) + 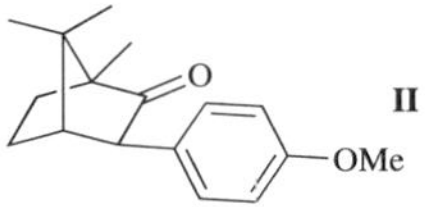 **II** (*exo*) **I** (71) **I**:**II** = 99:1	156
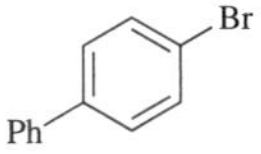		$h\nu$, NH_3, 2 h	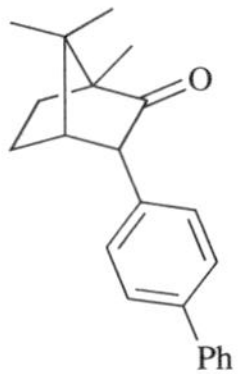 **I** (*endo*) + 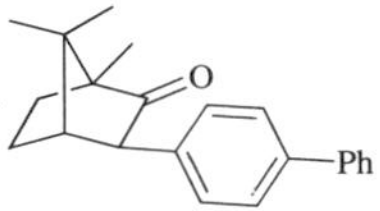 **II** (*exo*) **I** (76) **I**:**II** = 99:1	156
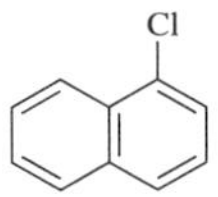		$h\nu$, NH_3, 2 h	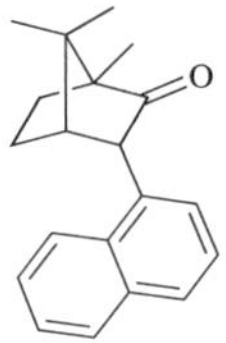 **I** (*endo*) + 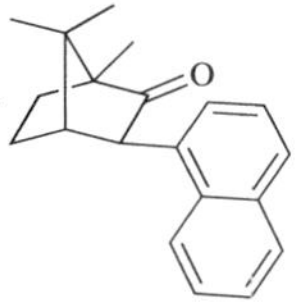 **II** (*exo*) **I** (100) **I**:**II** = 99:1	156
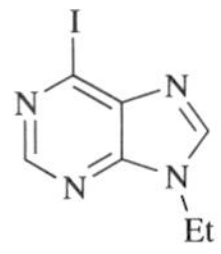	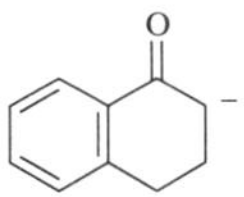	$h\nu$, NH_3, 0.5 h	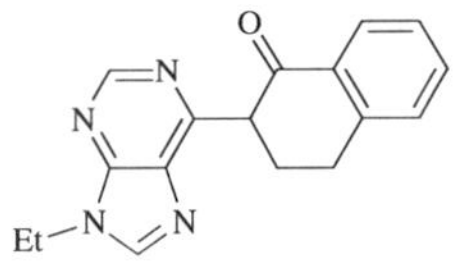 (80)	153, 154

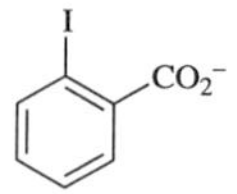	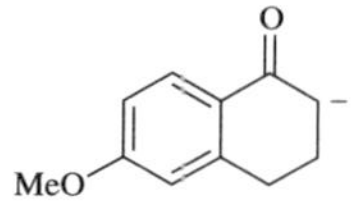	1. $h\nu$, NH_3, 1.25 h 2. H_3O^+	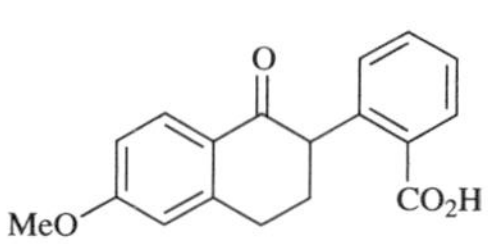	(75)	292
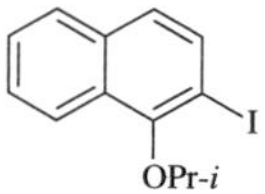		$h\nu$, NH_3, 3 h	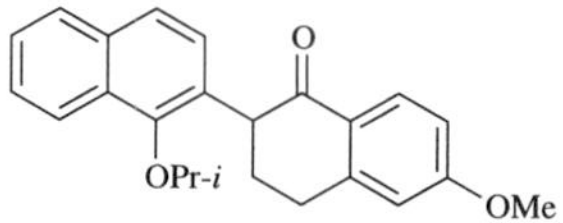	(44)	155
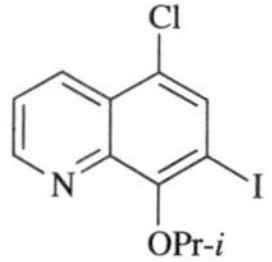		DMSO, 0.5 h	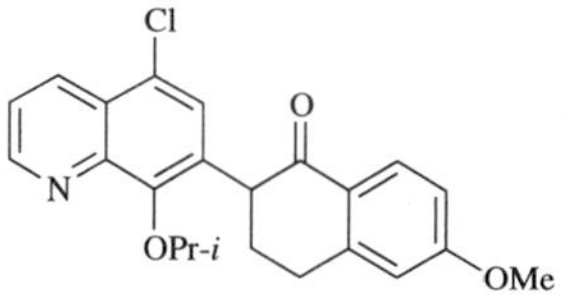	(36)	155
		$h\nu$, NH_3, 1 h	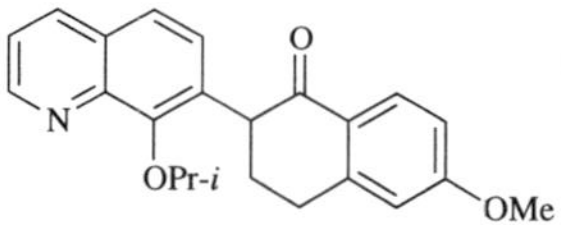	(35)	155
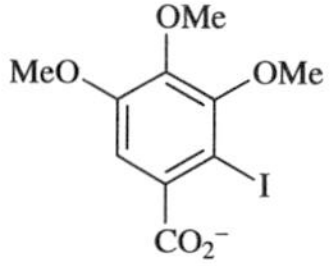	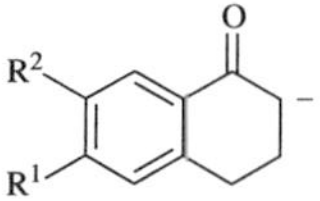	1. $h\nu$, NH_3, 1.25 h 2. H_3O^+	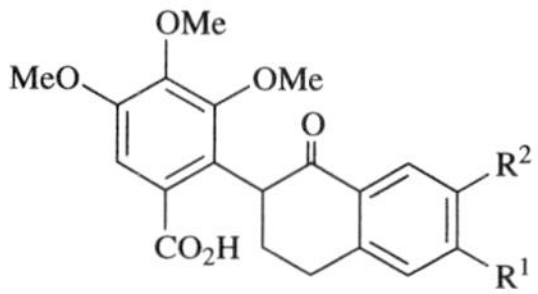	R^1 / R^2: OMe / H (60) OPr-*i* / OMe (75)	292
PhBr	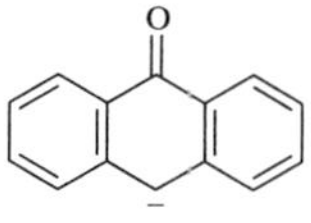	Na(Hg), NH_3, mediator	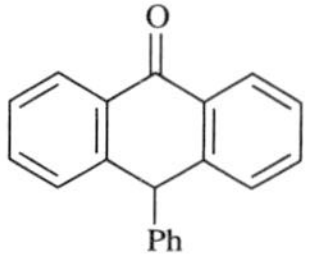 **I** (70)		101
PhI		$h\nu$, DMSO, 2 h	**I** (95)		41

TABLE II. CARBANIONS DERIVED FROM KETONES (*Continued*)

D. Enolate Ions from Cyclic Ketones (*Continued*)

Substrate	Nucleophile	Conditions	Product(s) and Yield(s) (%)	Refs.
PhBr	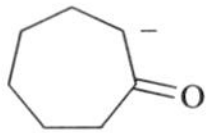	$h\nu$, NH_3, 3.5 h	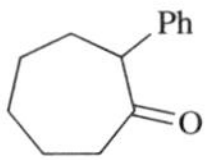Ph O (58) +  (12)	143
	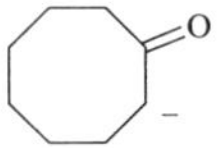	$h\nu$, NH_3, 3.5 h	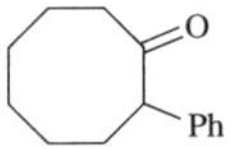O Ph (95)	143

TABLE II. CARBANIONS DERIVED FROM KETONES (*Continued*)

E. Enolate Ions from Aromatic Ketones

Substrate	Nucleophile	Conditions	Product(s) and Yield(s) (%)	Refs.
PhI	$^{-}CH_2COPh$	*h*ν, DMSO, 2 h	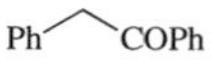 **I** (68) + 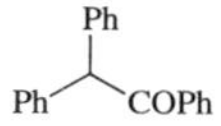 (13)	41
		*h*ν, NH_3, 2 h	**I** (0)	143
		*h*ν (quartz well), NH_3, 2 h	**I** (67)	135
		SmI_2 (THF), DMSO	**I** (47)	80
PhN_2SBu-*t*		DMSO, 1.5 h	**I** (95)	58, 161
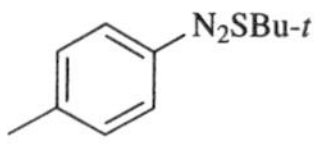		DMSO, 50°, 0.7-2 h	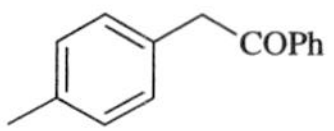 (8) + 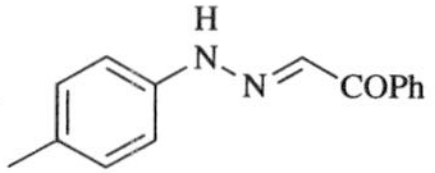 (90)	162, 57
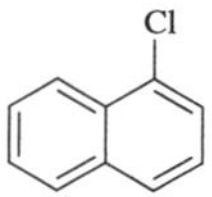		K, NH_3	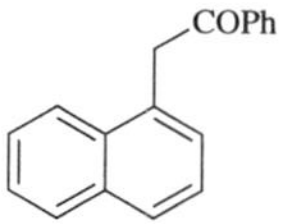 **I** (57) + 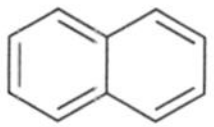 (33)	45
		*h*ν, NH_3, 3 h	**I** (8)	45
		SmI_2 (THF), DMSO	**I** (93)	80
		Na(Hg), NH_3	**I** (98)	101
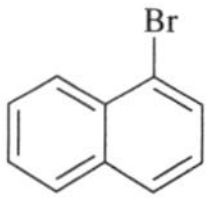		SmI_2 (THF), DMSO	**I** (97)	80
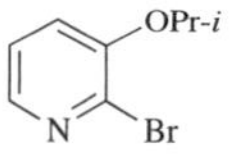		*h*ν, DMSO, 6 h	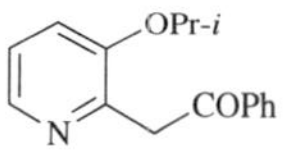 (70)	155

TABLE II. CARBANIONS DERIVED FROM KETONES (*Continued*)

E. Enolate Ions from Aromatic Ketones (*Continued*)

Substrate	Nucleophile	Conditions	Product(s) and Yield(s) (%)	Refs.
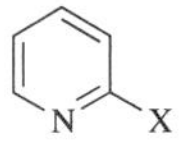	$^{-}CH_2COPh$		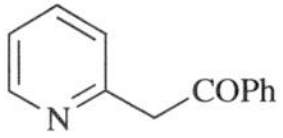	
X				
Cl		SmI_2 (THF), DMSO	(10)	80
Cl		Na(Hg), NH_3	(75)	101
Br		SmI_2 (THF), DMSO	(77)	80
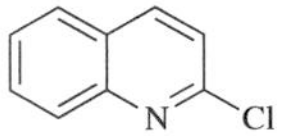		*h*ν, NH_3, 3 h[a]	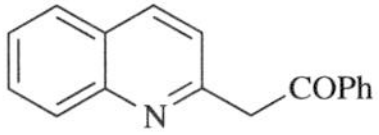 **I** (14)	142
		SmI_2 (THF), DMSO	**I** (94)	80
		Na(Hg), NH_3	**I** (98)	101
		*h*ν (near UV), NH_3, 1 h[b]	**I** (82)	157
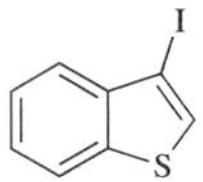		DMSO, 1 h	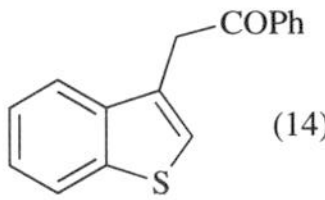 (14)	151
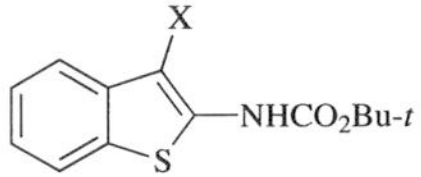		*h*ν, NH_3, –78°, 5 h	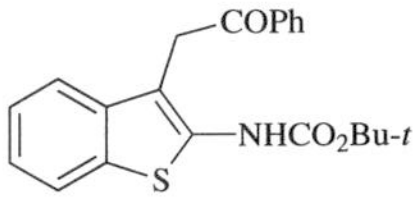 + 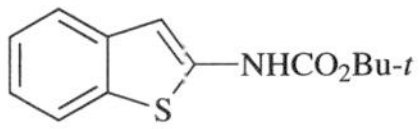	150
X				
Cl			(8) (38)	
Br			(17) (58)	
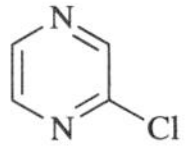		NH_3, 15 min	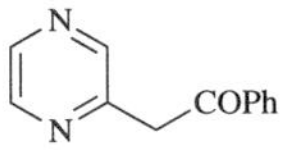 (82)	55

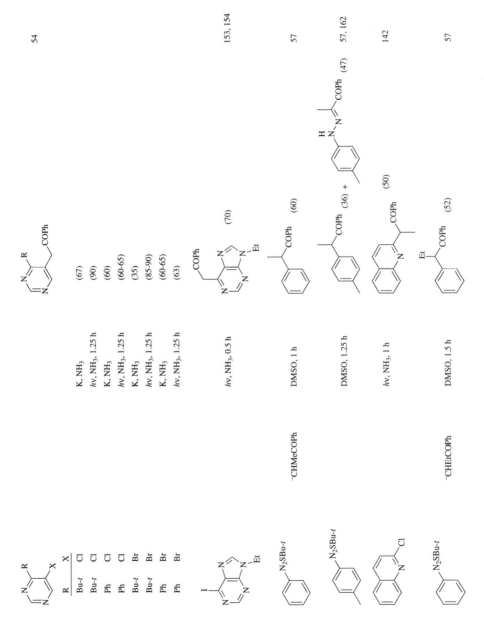

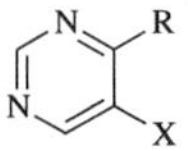 R X			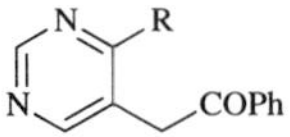	54
Bu-*t* Cl		K, NH_3	(67)	
Bu-*t* Cl		*h*ν, NH_3, 1.25 h	(90)	
Ph Cl		K, NH_3	(60)	
Ph Cl		*h*ν, NH_3, 1.25 h	(60-65)	
Bu-*t* Br		K, NH_3	(35)	
Bu-*t* Br		*h*ν, NH_3, 1.25 h	(85-90)	
Ph Br		K, NH_3	(60-65)	
Ph Br		*h*ν, NH_3, 1.25 h	(63)	
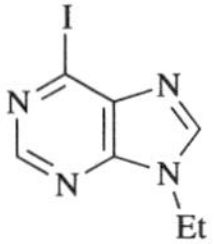		*h*ν, NH_3, 0.5 h	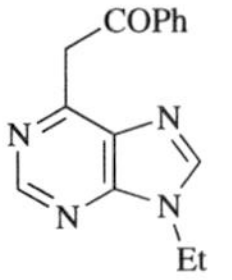 (70)	153, 154
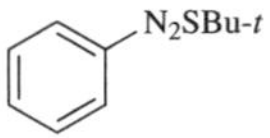	$^{-}$CHMeCOPh	DMSO, 1 h	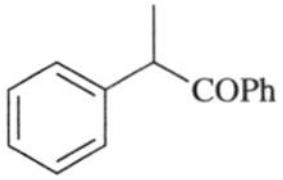 (60)	57
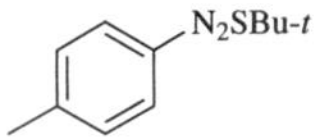		DMSO, 1.25 h	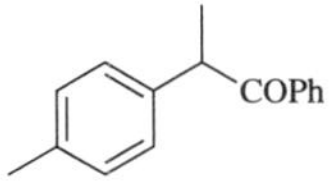 (36) + 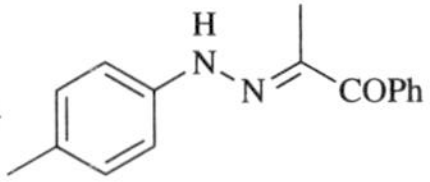 (47)	57, 162
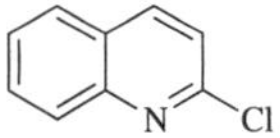		*h*ν, NH_3, 1 h	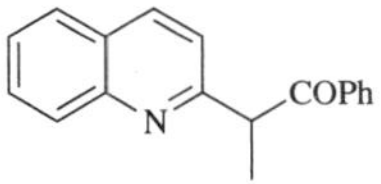 (50)	142
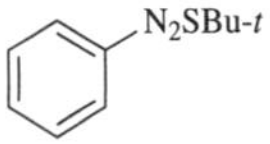	$^{-}$CHEtCOPh	DMSO, 1.5 h	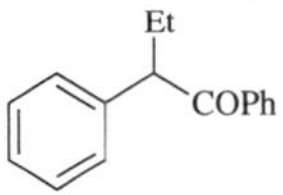 (52)	57

TABLE II. CARBANIONS DERIVED FROM KETONES (*Continued*)

E. Enolate Ions from Aromatic Ketones (*Continued*)

Substrate	Nucleophile	Conditions	Product(s) and Yield(s) (%)	Refs.
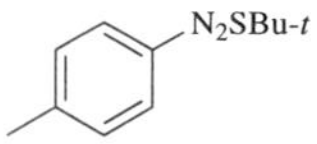	$^{-}$CHEtCOPh	DMSO, 1 h	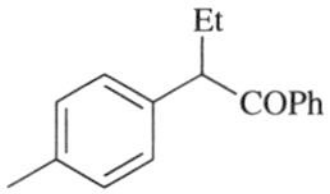 (24) + 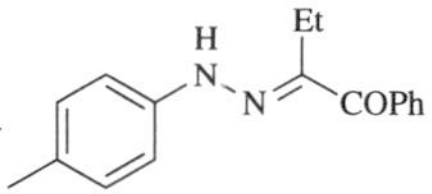 (52)	57
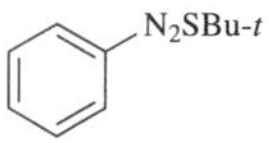	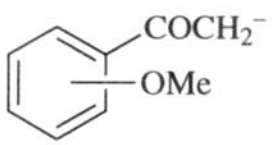	DMSO,	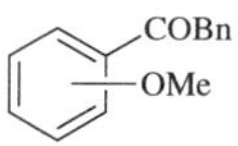	
	2-OMe	2 h	(71)	57, 161
	3-OMe	1 h	(76)	57
	4-OMe	1.7 h	(77)	57, 161
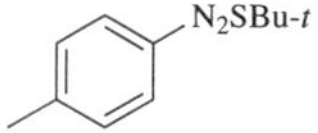	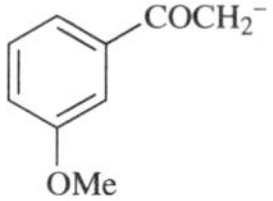	DMSO, 1 h	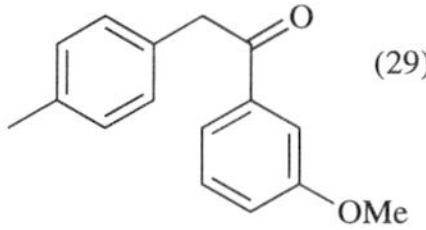 (29) + 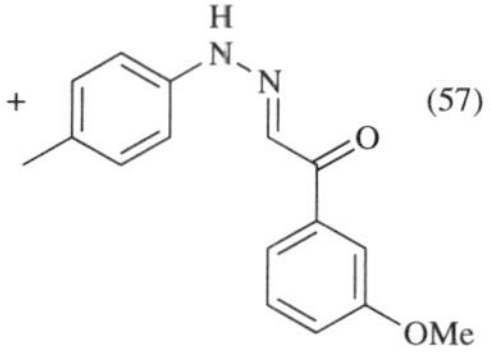 (57)	57
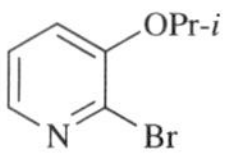	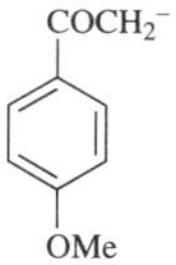	*h*ν, DMSO, 3 h	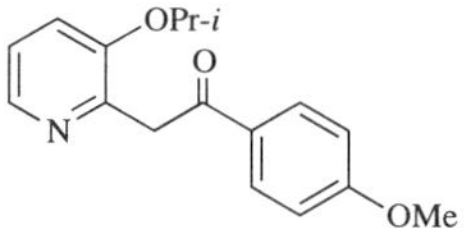 **I** (98)	155
		*h*ν, NH_3, 7 h	**I** (30)	155
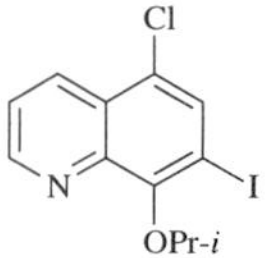		*h*ν, NH_3, 1 h	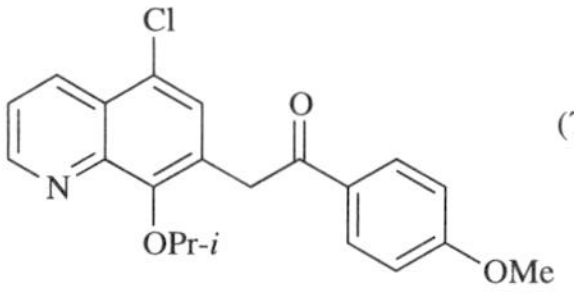 (70)	155

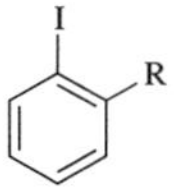	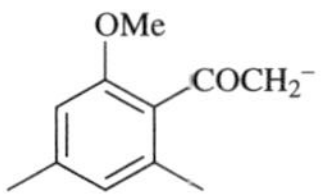		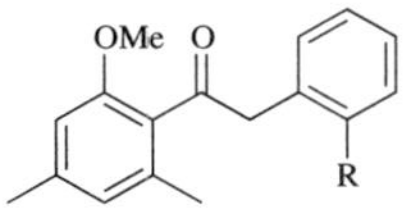	293
R				
OMe		$h\nu$, DMSO, 1 h	(18)	
COMe		$h\nu$, NH_3, 2.5 h	(18)	
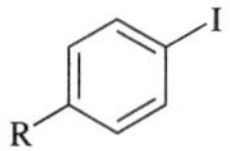		$h\nu$, DMSO, 1 h	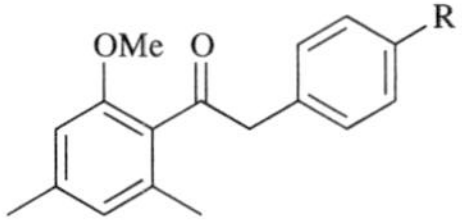	293
R				
OMe			(19)	
COMe			(13)	
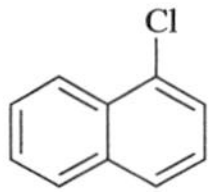	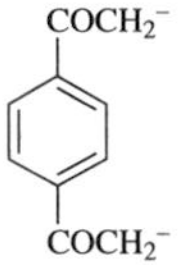	e^-, DMSO	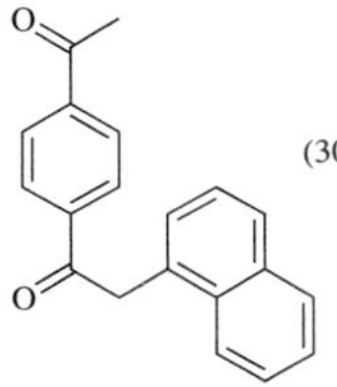 (30) + 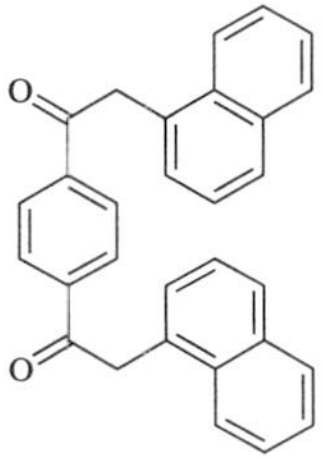 (15)	346
PhI	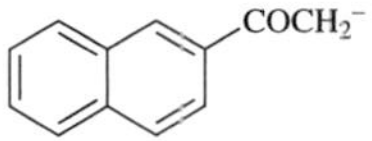	$h\nu$, DMSO, 1.5 h	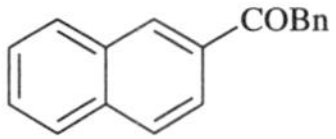 **I** (49)	158
		$FeCl_2$, DMSO, 20 min	**I** (92)	123

TABLE II. CARBANIONS DERIVED FROM KETONES (*Continued*)

E. Enolate Ions from Aromatic Ketones (*Continued*)

Substrate	Nucleophile	Conditions	Product(s) and Yield(s) (%)	Refs.
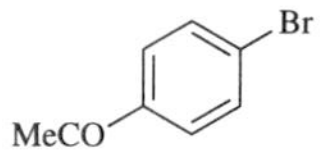	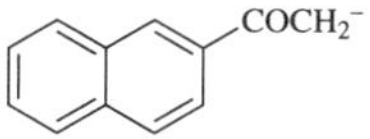		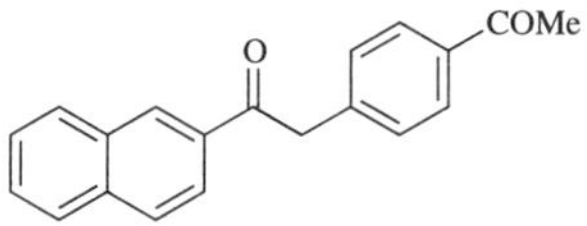 **I** + 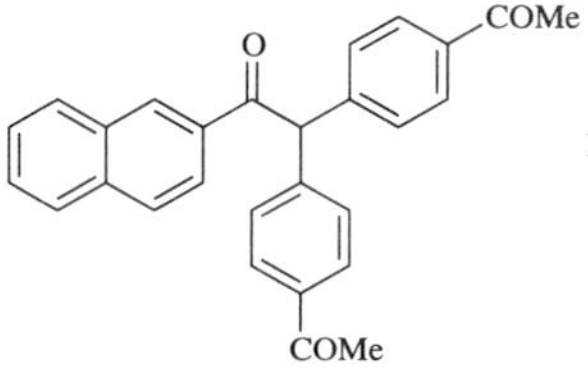 **II**	158
		hν, DMSO, 1.5 h	**I** (54) + **II** (25)	
		hν, NH_3, 1.5 h	**I** (56) + **II** (24)	
PhI	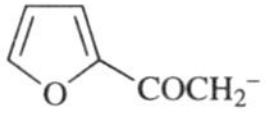	*hν*, Me_2CO, *t*-BuOK, DMSO, 3 h	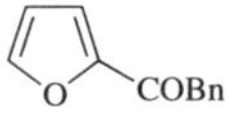 **I** (78)	159
		$FeBr_2$, DMSO	**I** (88)	159
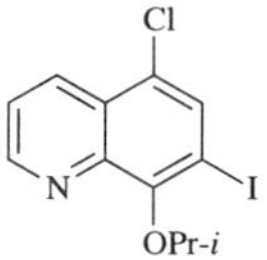		*hν*, NH_3, 1 h	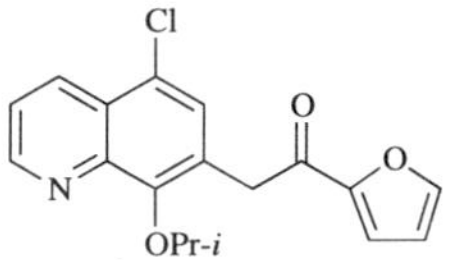 (80)	155
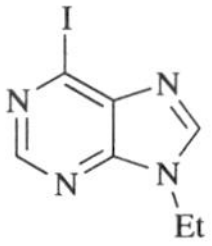		*hν*, NH_3, 0.5 h	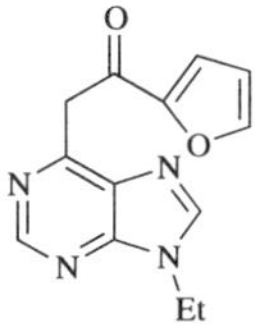 (67)	153, 154

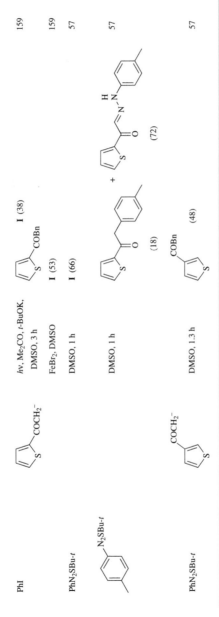

PhI	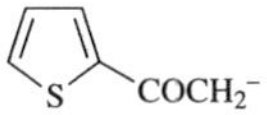	$h\nu$, Me_2CO, *t*-BuOK, DMSO, 3 h	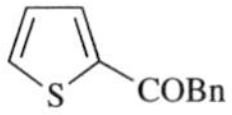 **I** (38)	159
		$FeBr_2$, DMSO	**I** (53)	159
PhN_2SBu-*t*		DMSO, 1 h	**I** (66)	57
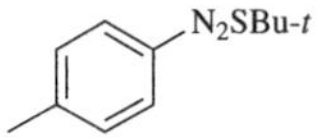		DMSO, 1 h	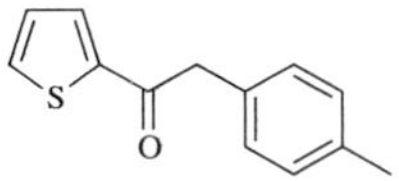 (18) + 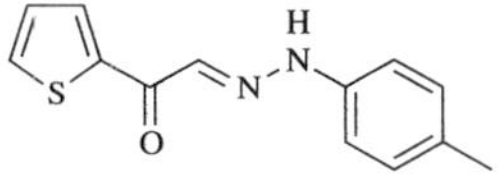 (72)	57
PhN_2SBu-*t*	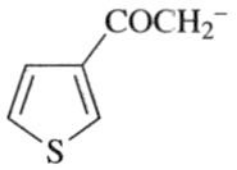	DMSO, 1.3 h	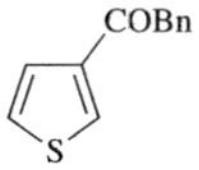 (48)	57

[a] The nucleophile was a potassium salt.

[b] The nucleophile was a lithium salt; $LiNH_2$:$LiCH_2COPh$ = 1.4.

TABLE III. CARBANIONS DERIVED FROM ESTERS OR CARBOXYLATE IONS

Substrate	Nucleophile	Conditions	Product(s) and Yield(s) (%)	Refs.
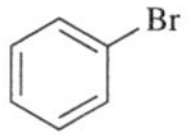	$^{-}CH_2CO_2Bu$-*t*	*h*ν, NH_3, 1.3 h	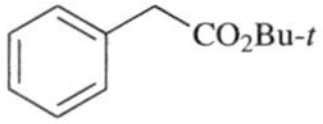 (57)	135
		1. $FeSO_4$, NH_3, 0.75 h 2. TsOH	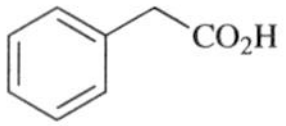 **I** (69)	76
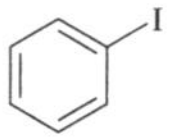		1. $FeSO_4$, NH_3, 20 min 2. TsOH	**I** (82)	76
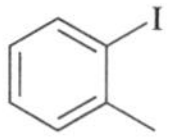		1. $FeSO_4$, NH_3, 0.5 h 2. TsOH	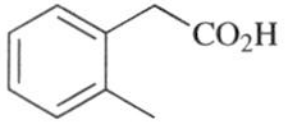 (55)	76
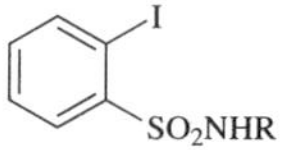		*h*ν, NH_3,	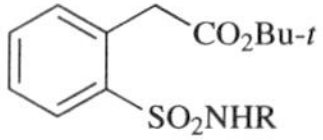	140
R				
Bu-*t*		1 h	(89)	
Bn		0.5 h	(75)	
COMe		0.5 h	(65)	
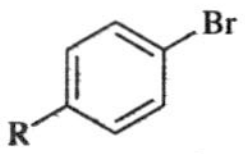			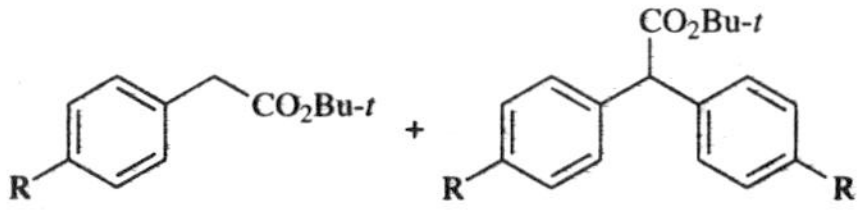	
R				
Me		**$FeSO_4$, NH_3, 3 h**	**(20)** (—)	76
OMe		***h*ν, NH_3**	**(67)** (29)	170
OMe		**$FeSO_4$, NH_3, 2 h**	**(28)** (—)	76

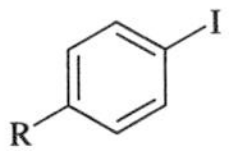	1. $FeSO_4$, NH_3, 2. TsOH	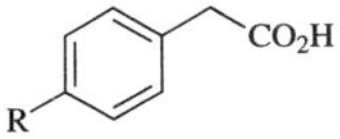	76
R: F	1. 20 min	(81)	
R: Me	1. 0.5 h	(63)	
R: OMe	1. 0.75 h	(78)	
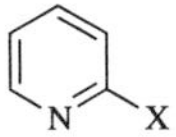	$FeSO_4$, NH_3, 0.75 h	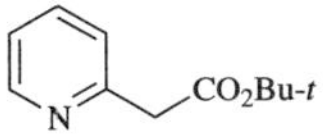 X: Cl (22); Br (89)	76
	X = Br, $h\nu$, NH_3, 5 min	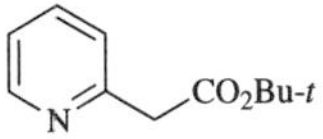 (44) + 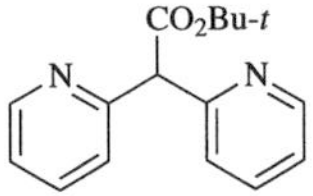 (21)	169
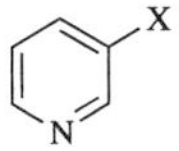	$FeSO_4$, NH_3,	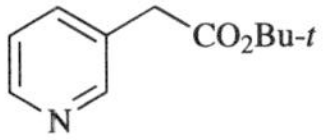	76
X: Br	0.5 h	(51)	
X: I	0.75 h	(59)	
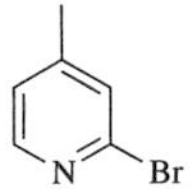	$FeSO_4$, NH_3, 0.75 h	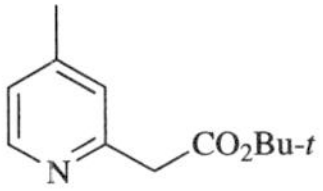 (18)	76
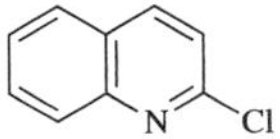	$FeSO_4$, NH_3, 0.75 h	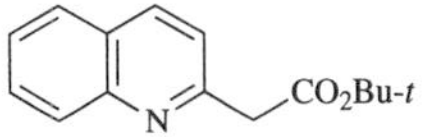 (15)	76
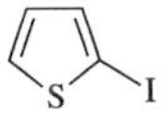	$FeSO_4$, NH_3, 1 h	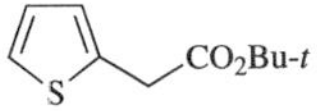 (5)	76

TABLE III. CARBANIONS DERIVED FROM ESTERS OR CARBOXYLATE IONS (*Continued*)

Substrate	Nucleophile	Conditions	Product(s) and Yield(s) (%)	Refs.
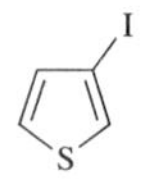	$^{-}CH_2CO_2Bu$-*t*	1. $FeSO_4$, NH_3, 0.75 h 2. TsOH	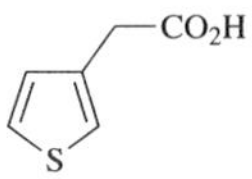 (73)	76
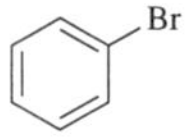	$^{-}CHMeCO_2Bu$-*t*	*h*ν, NH_3, 1.3 h	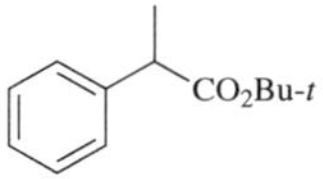 **I** (60)	135
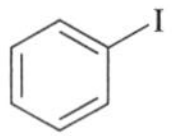		$FeSO_4$, NH_3, 1 h	**I** (51) + 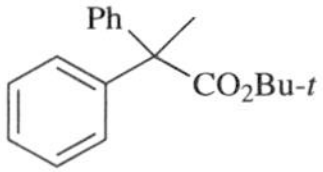 (8)	76
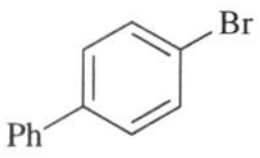	$^{-}CHMeCO_2Et$	*h*ν, NH_3, 8 h	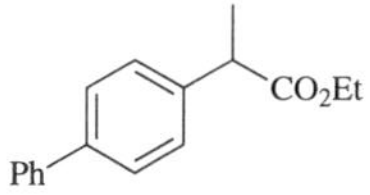 (20)	66
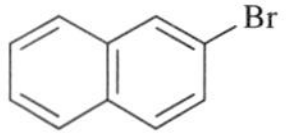		*h*ν, NH_3, 6 h	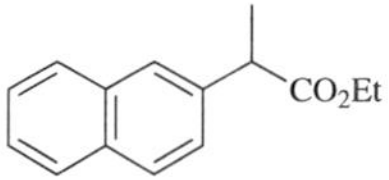 (25)	66
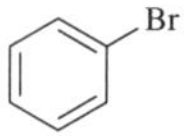	$^{-}CMe_2CO_2Bu$-*t*	*h*ν, NH_3, 1.3 h	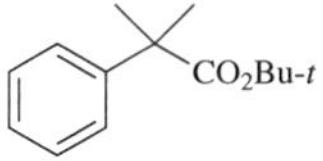 (11)	135
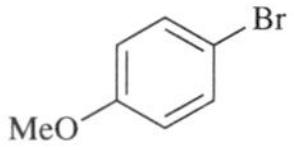		*h*ν, NH_3	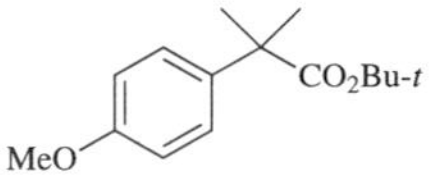 (5) + 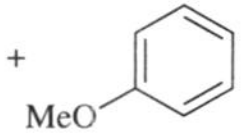 (35-50)	170
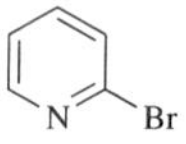	$^{-}CHPhCO_2Et$	*h*ν, NH_3, 5 min	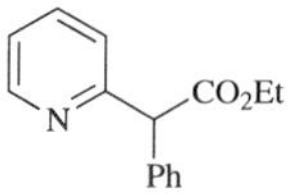 (77)	169

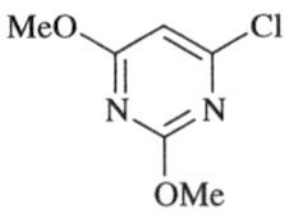		$h\nu$, NH_3, 15 min	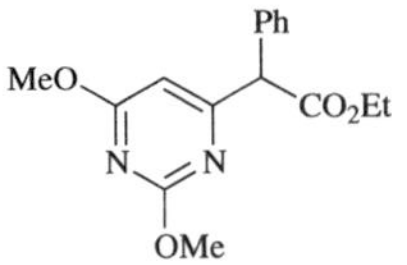 (71)	140
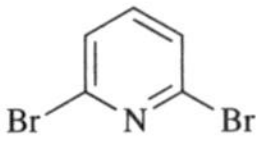		$h\nu$, NH_3, 15 min	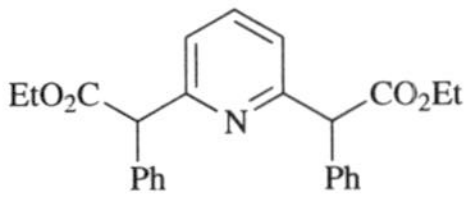 (84)	167
	$^{-}CPh_2CO_2Me$	$h\nu$, NH_3, 1 h	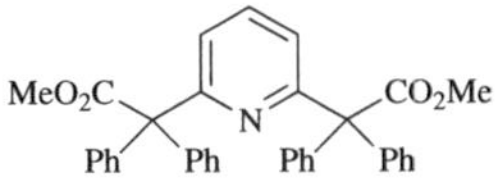 (42)	167
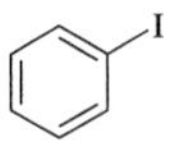	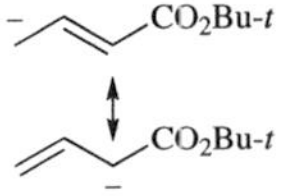	$h\nu$, NH_3, 1 h	 (42) + 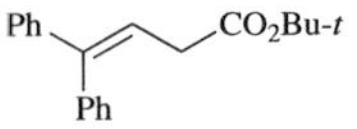 (8) + 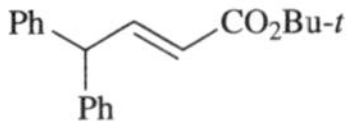 (8)	169
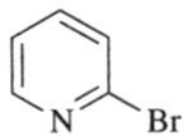		$h\nu$, NH_3, 1 h	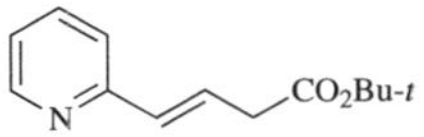 (61)	169
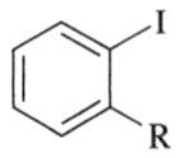	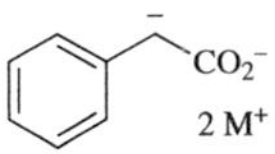	$h\nu$, NH_3, 5 h	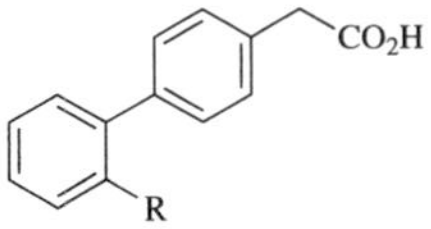 + 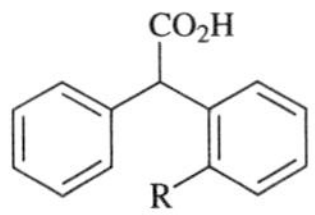	140
R	M			
H	Li		(—) (77)	
H	K		(73) (—)	
Me	K		(36) (—)	

TABLE III. CARBANIONS DERIVED FROM ESTERS OR CARBOXYLATE IONS (*Continued*)

Substrate	Nucleophile	Conditions	Product(s) and Yield(s) (%)		Refs.
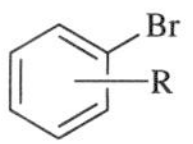	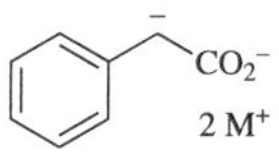	$h\nu$, NH_3, 5 h	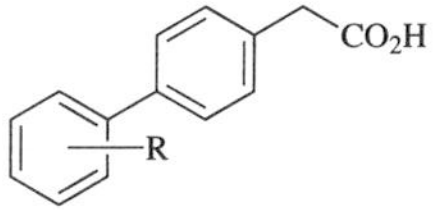 +	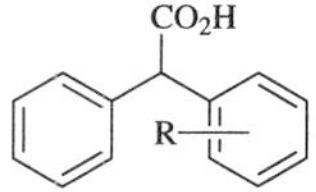	140
R	M				
2-OMe	K		(27)	(—)	
3-Me	K		(47)	(—)	
3-OMe	K		(55)	(—)	
4-Me	K		(49)	(—)	
4-CN	Li		(—)	(55)	
4-CN	K		(41)	(—)	
4-OMe	Li		(—)	(61)	
4-OMe	Na		(25)	(37)	
4-OMe	K		(64)	(—)	
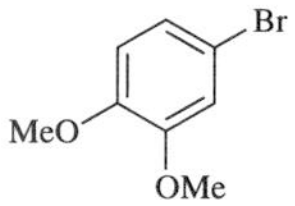	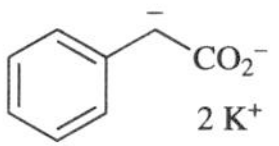	$h\nu$, NH_3, 5 h	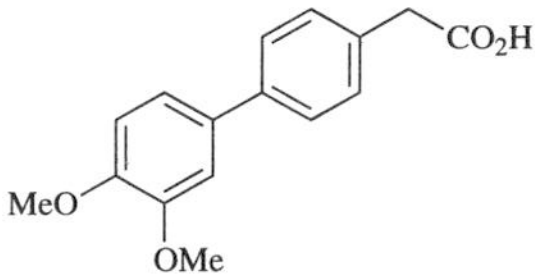 (19)		140
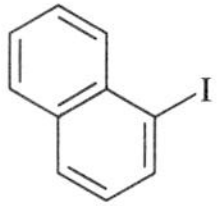		$h\nu$, NH_3, 5 h	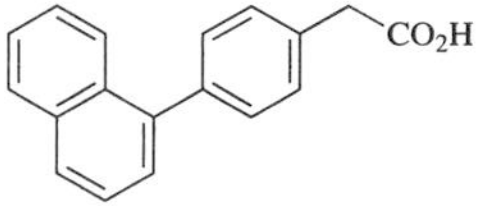 (43)		140
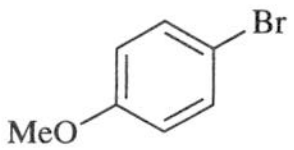	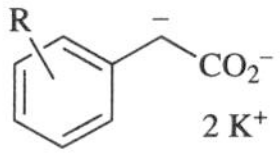	$h\nu$, NH_3, 5 h	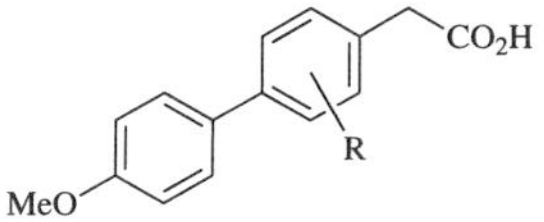		140
	R				
	2-Me		(16)		
	3-Me		(23)		
	3-OMe		(63)		

TABLE IV. CARBANIONS DERIVED FROM *N,N*-DISUBSTITUTED AMIDES OR THIOAMIDES

Substrate	Nucleophile	Conditions	Product(s) and Yield(s) (%)	Refs.
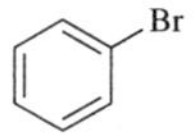	$^{-}CH_2CONMe_2$	$h\nu$, NH_3, 2 h	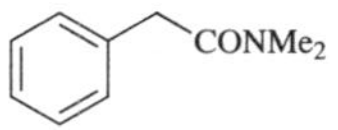 (80) + 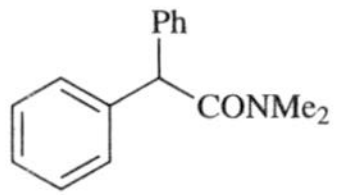 (18) **I** **II**	171
PhI	$^{-}CH_2CONMe_2$:PhI	$h\nu$, NH_3, 1.5 h		98
	15		**I** (63) + **II** (8)	
	5		**I** (47) + **II** (42)	
	2		**I** (33) + **II** (66)	
	$^{-}CH_2CONMe_2$	1. $h\nu$, NH_3, 2 h 2. MeI	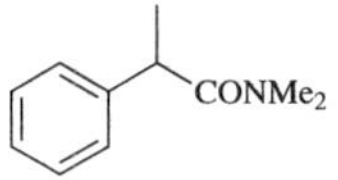 (73) + **II** (10)	164
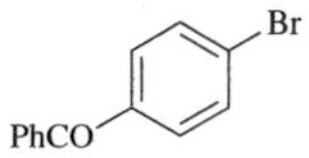		K, NH_3	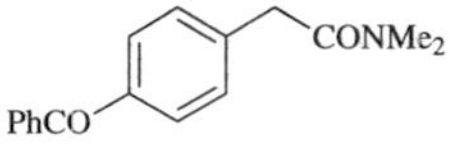 (45) + 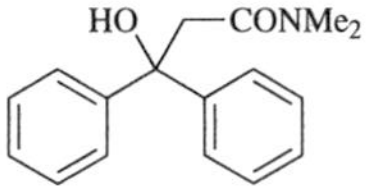 (14)	98
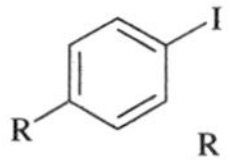 R Me OMe		$h\nu$, NH_3, 2 h	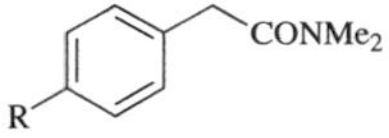 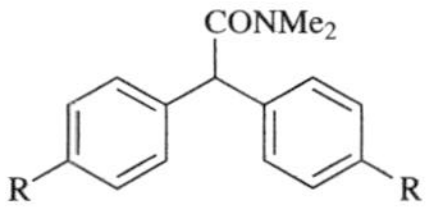 (77) (19) (57) (20)	98
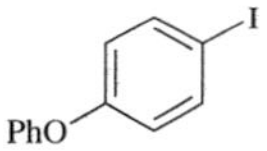		1. $h\nu$, NH_3, 2 h 2. MeI	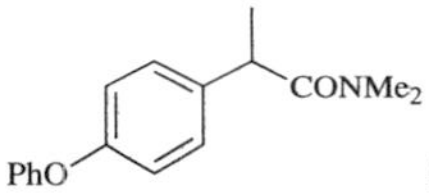 (55) + 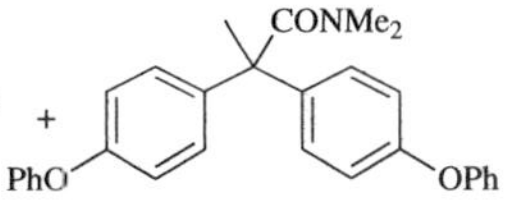 (18)	164

TABLE IV. CARBANIONS DERIVED FROM *N,N*-DISUBSTITUTED AMIDES OR THIOAMIDES (*Continued*)

Substrate	Nucleophile	Conditions	Product(s) and Yield(s) (%)	Refs.
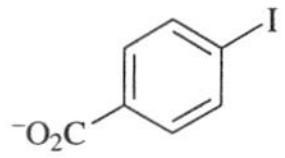	$^{-}CH_2CONMe_2$	1. $h\nu$, NH_3, 2 h 2. MeI	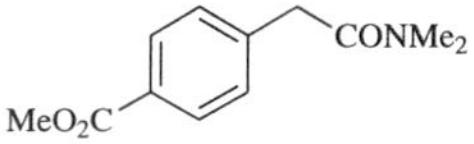 (74) + 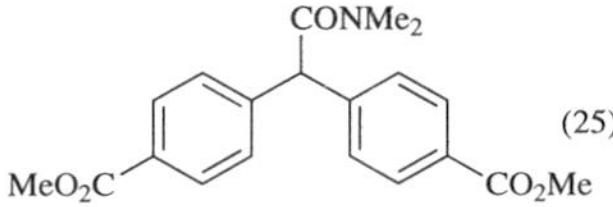 (25)	98
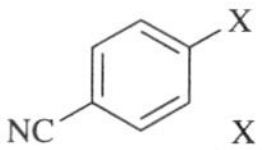			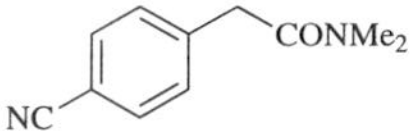 + 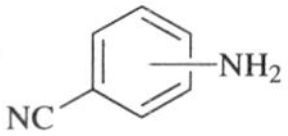	98
X = Cl		$h\nu$, NH_3, 2 h	(48) (52)	
Cl		K, NH_3	(68) (—)	
Br		K, NH_3	(98) (—)	
I		K, NH_3	(79) (—)	
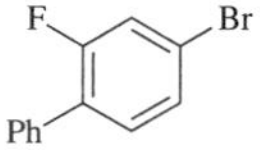		1. $h\nu$, NH_3, 2 h 2. MeI	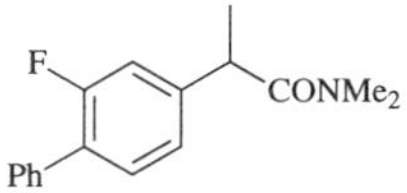 (18) + 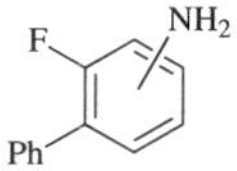 (15) + 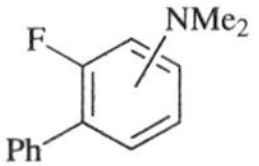 (50)	164
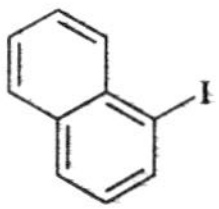		**$h\nu$, NH_3, 2 h**	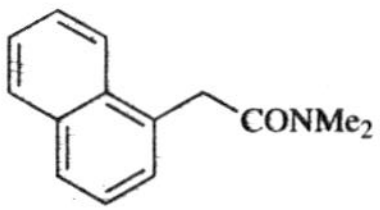 **(49)** + 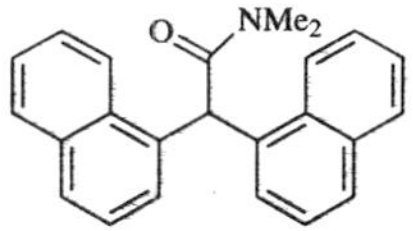 **I (22)**	**98**
		1. $h\nu$, NH_3, 2 h 2. MeI	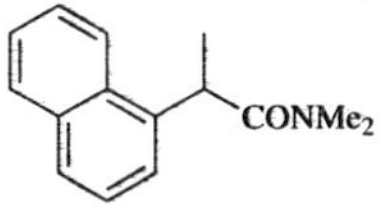 **(60) + I (18)**	**164**

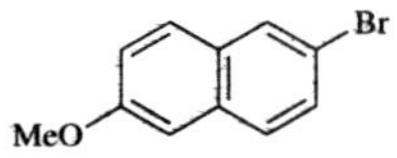	1. $h\nu$, NH_3, 2 h 2. MeI	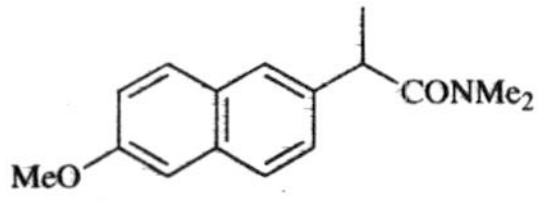 **I** + 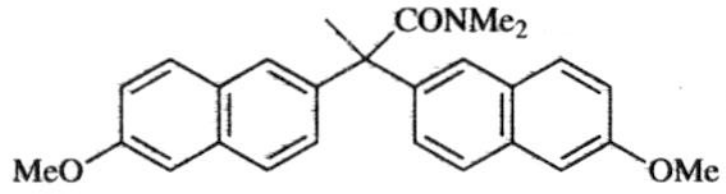 **II**	164
nucleophile:substrate			
5		**I** (52) + **II** (15)	
10		**I** (84) + **II** (10)	
15		**I** (92) + **II** (5)	
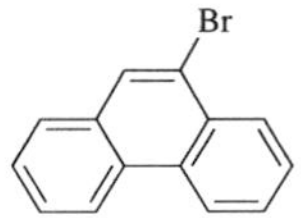	$h\nu$, NH_3, 2 h	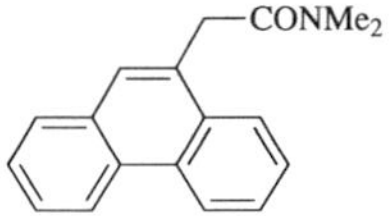 (59) + 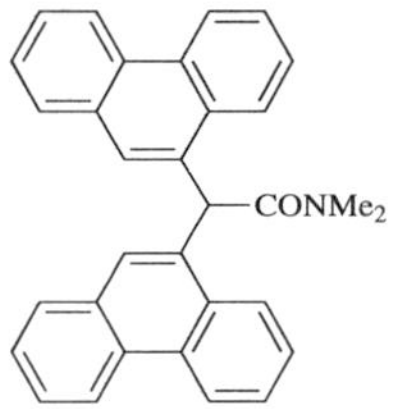 (14)	98
	1. $h\nu$, NH_3, 2 h 2. MeI	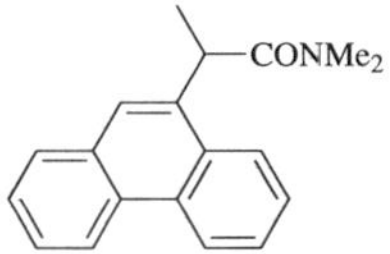 (72)	164
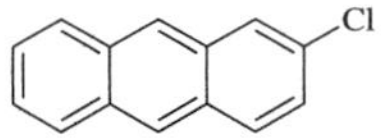	1. $h\nu$, NH_3, 2 h 2. MeI	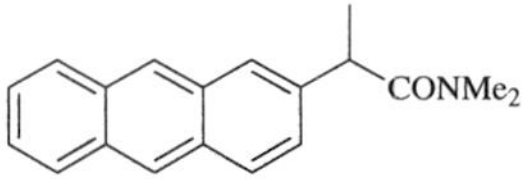 **I** (21)	164
	1. Na(Hg), NH_3, 4 h 2. MeI	**I** (77)	164

TABLE IV. CARBANIONS DERIVED FROM *N,N*-DISUBSTITUTED AMIDES OR THIOAMIDES (*Continued*)

Substrate	Nucleophile	Conditions	Product(s) and Yield(s) (%)	Refs.
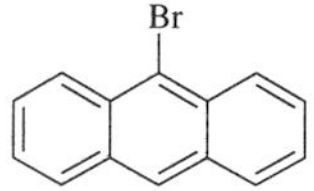	$^{-}CH_2CONMe_2$	$h\nu$, NH_3, 2 h	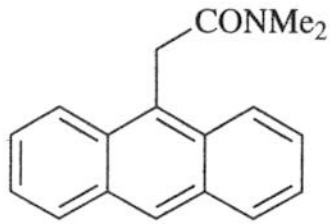 **I** (70)	164[a]
		$h\nu$, NH_3, 2 h	**I** (66) + 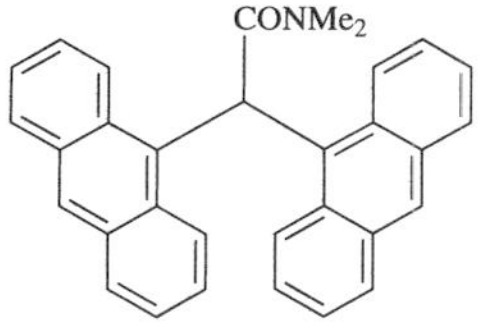 (12)	98[b]
		1. $h\nu$, NH_3, 2 h 2. MeI	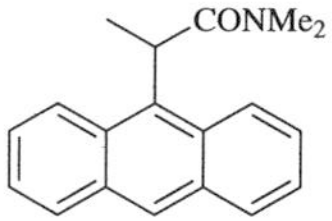 (70)	164[a]
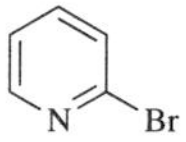		$h\nu$, NH_3, 20 min	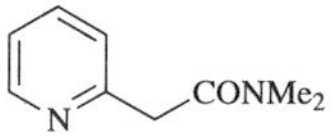 (84) + 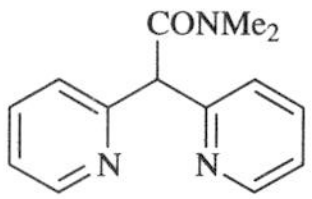 (7)	169
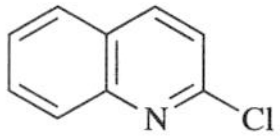		$h\nu$, NH_3, 5 min	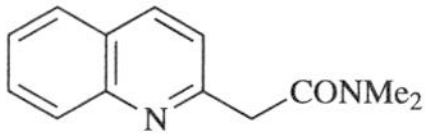 (87)	169
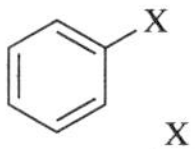	$^{-}CH_2CON(Me)Ph$	$h\nu$, NH_3,	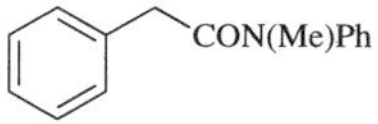	171
X				
Cl		2 h	(72)	
Br		2 h	(60)	
I		0.5 h	(80) + $Ph_2CHCON(Me)Ph$ (10)	

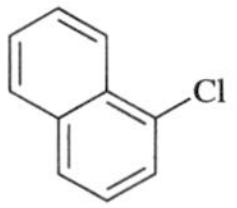		$h\nu$, NH_3, 2 h	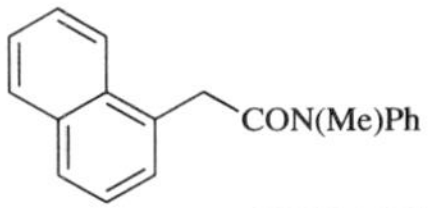 (50)	171
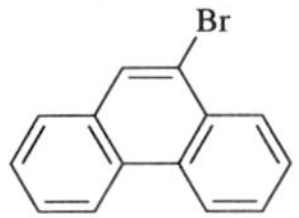		$h\nu$, NH_3, 1.5 h	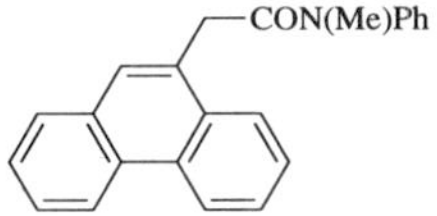 (80)	171
PhX	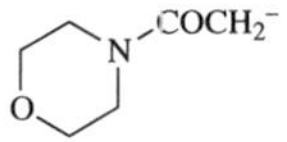		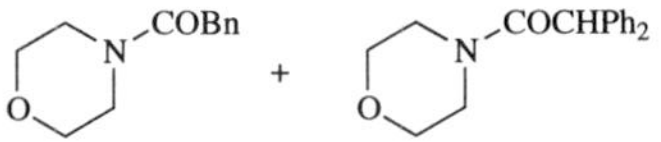	
X				
Cl		$h\nu$, NH_3, 3 h	(75) (18)	171
Br		$h\nu$, NH_3, 3 h	(56) (12)	171
Br		$FeSO_4$, NH_3	(46) (—)	76
I		$FeSO_4$, NH_3, 0.75 h	(70) (4)	76
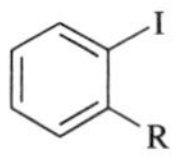		$FeSO_4$, NH_3, 1.5 h	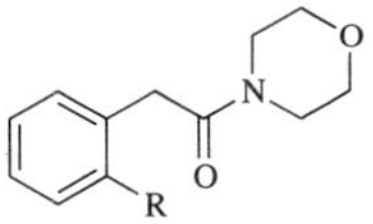	76
R				
F			(9)	
Me			(62)	
Ph			(71)	
OMe			(57)	
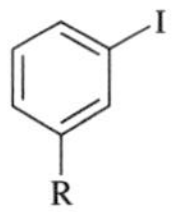		$FeSO_4$, NH_3, 1.5 h	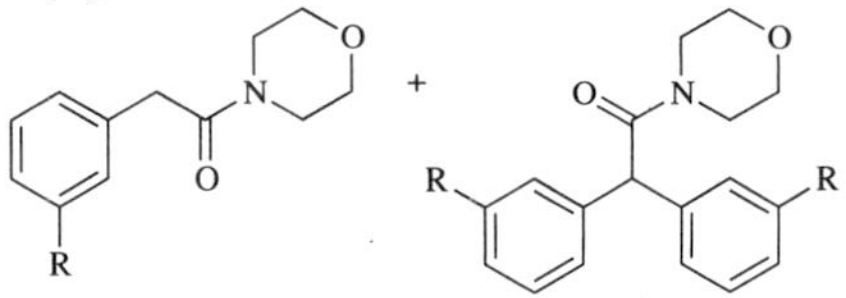	76
R				
F			(31) (15)	
Me			(73) (8)	
OMe			(31) (15)	

TABLE IV. CARBANIONS DERIVED FROM *N,N*-DISUBSTITUTED AMIDES OR THIOAMIDES (*Continued*)

Substrate	Nucleophile	Conditions	Product(s) and Yield(s) (%)		Refs.
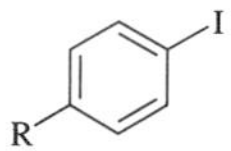	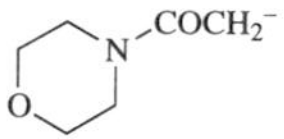	$FeSO_4$, NH_3, 1.5 h	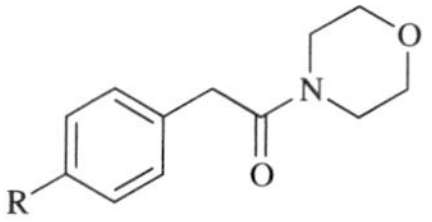 +	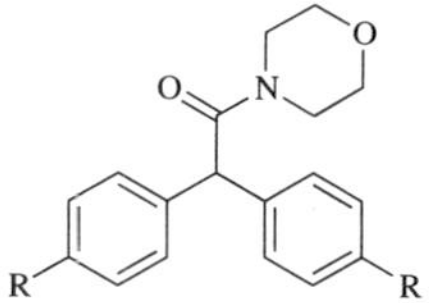	76
R					
F			(64)	(12)	
Me			(65)	(5)	
Bu-*t*			(40)	(4)	
Ph			(45)	(16)	
OMe			(60)	(6)	
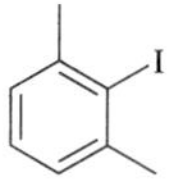		$FeSO_4$, NH_3, 1.5 h	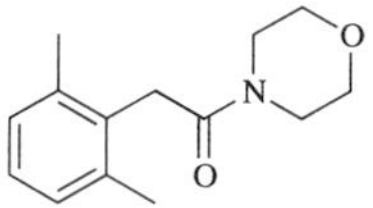 (44)		76
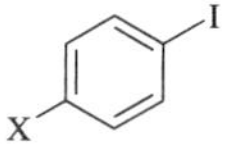		$FeSO_4$, NH_3, 1.5 h	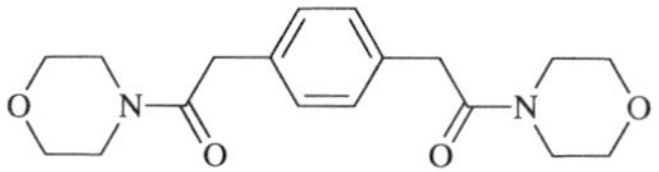	X: Cl (51); I (55)	76
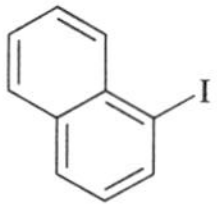		$FeSO_4$, NH_3, 1.5 h	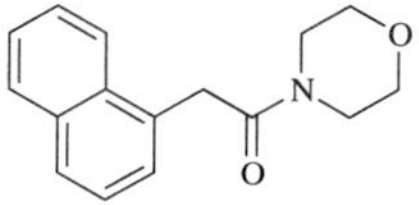 (65) +	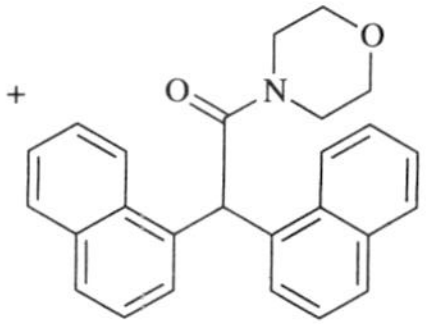 (6)	76

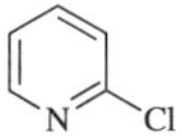		$FeSO_4$, NH_3	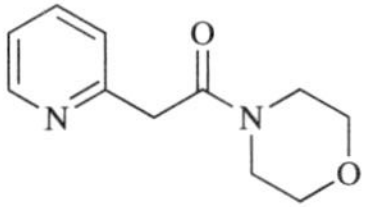 (65)	76
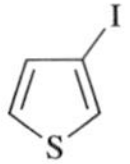		$FeSO_4$, NH_3, 1.5 h	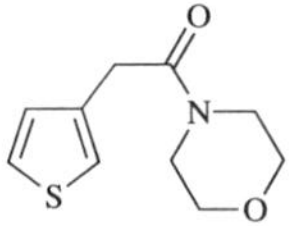 (27) + 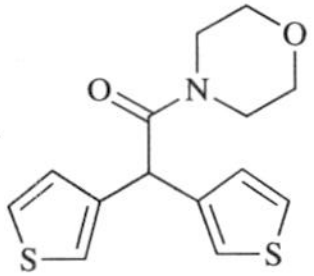 (8)	76
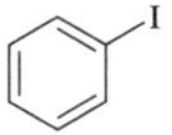	$^-$CHMeCO–N 	$FeSO_4$, NH_3, 1.5 h	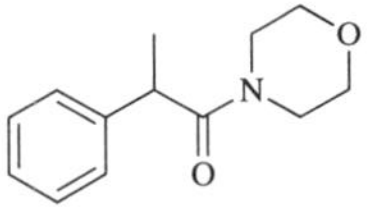 (57)	76
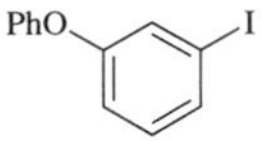		$FeSO_4$, NH_3, 1.5 h	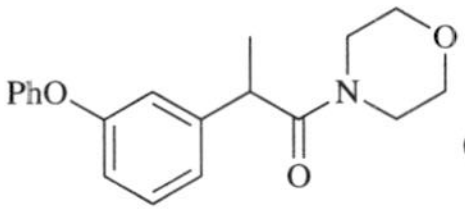 (47) + Ph_2O (33)	76
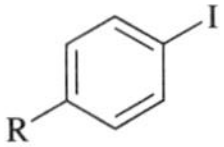 R: Me, Bu-*i*		$FeSO_4$, NH_3, 1.5 h	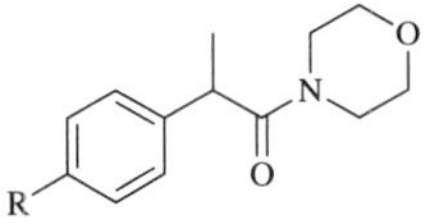 + PhR Me: (61) + (—) Bu-*i*: (42) + (13)	76
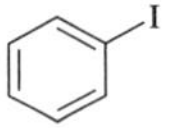	$^-$CHRCO–N 	$FeSO_4$, NH_3, 1.5 h	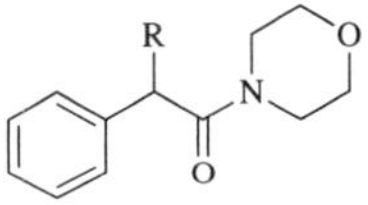 R: Et (60); Pr-*i* (63); Bu-*t* (44); *n*-C_8H_{17} (4)	76

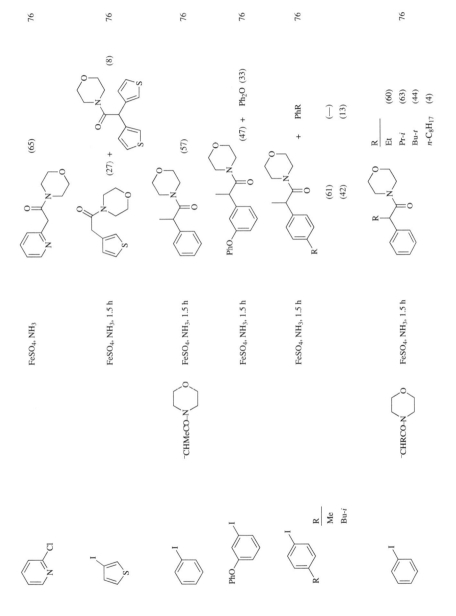

TABLE IV. CARBANIONS DERIVED FROM *N,N*-DISUBSTITUTED AMIDES OR THIOAMIDES (*Continued*)

Substrate	Nucleophile	Conditions	Product(s) and Yield(s) (%)	Refs.
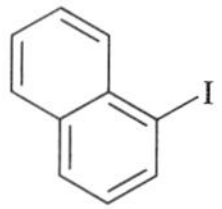	$^{-}CH(Ph)CONMe_2$	$h\nu$, NH_3, 1.5 h	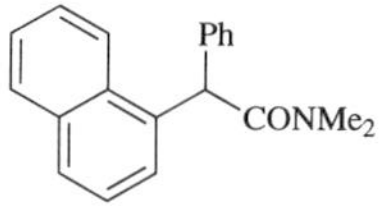 (26)	98
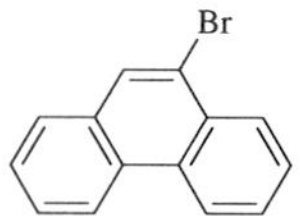		$h\nu$, NH_3, 2 h	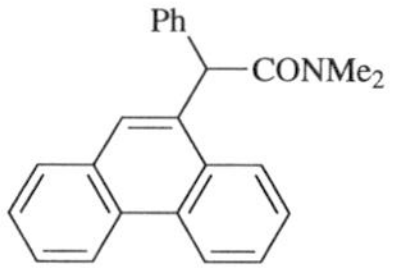 **I** (52)	98
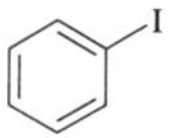	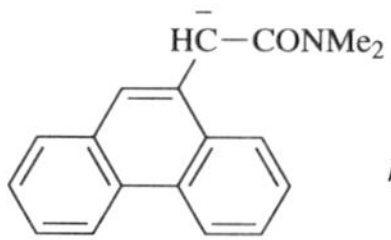	$h\nu$, NH_3, 2 h	**I** (38)	98
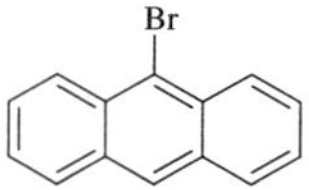	$^{-}CH(Ph)CONMe_2$	$h\nu$, NH_3, 2 h	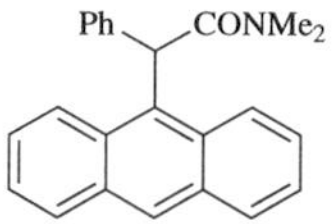 **I** (30)	98
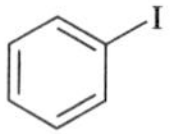	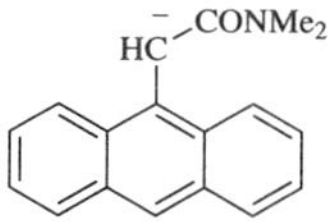	$h\nu$, NH_3, 2 h	**I** (24)	98
	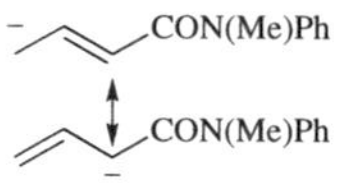	$h\nu$, NH_3, 1 h	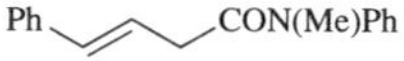 (38) + 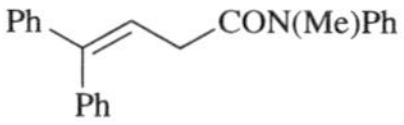 (10) + 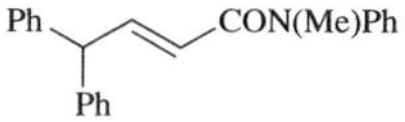 (8)	169

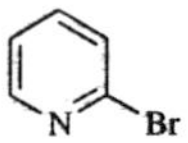		$h\nu$, NH_3, 1 h	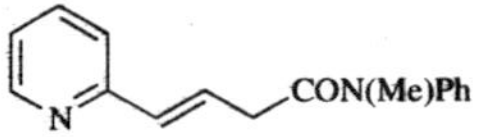 (50)	169
PhCl	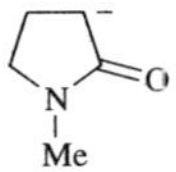	$h\nu$, NH_3, 1 h	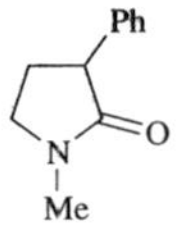 **I** (52) + 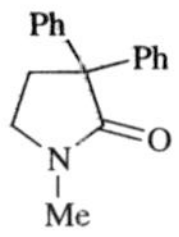 **II** (1)	172
PhBr		$h\nu$, NH_3, 1 h	**I** (51) + **II** (2)	172
PhI	nucleophile:PhI			
	15	$h\nu$, NH_3, 1 h	**I** (60) + **II** (7)	172
	2	$h\nu$, NH_3, 3 h	**II** (40)	172
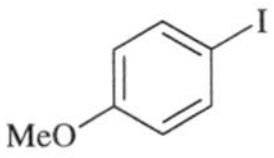		$h\nu$, NH_3, 1 h	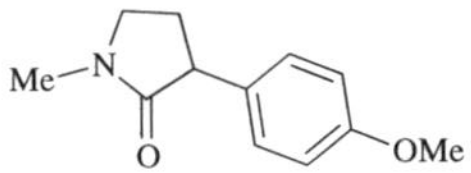 (58)	172
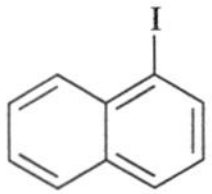		1. $h\nu$, NH_3, 1 h 2. MeI	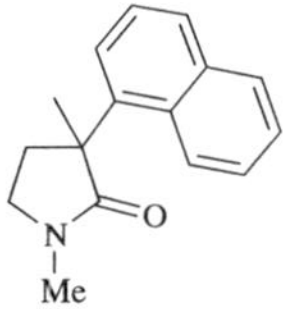 (40)	172

TABLE IV. CARBANIONS DERIVED FROM *N,N*-DISUBSTITUTED AMIDES OR THIOAMIDES (*Continued*)

Substrate	Nucleophile	Conditions	Product(s) and Yield(s) (%)	Refs.
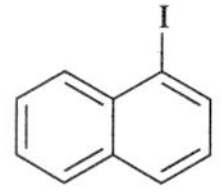	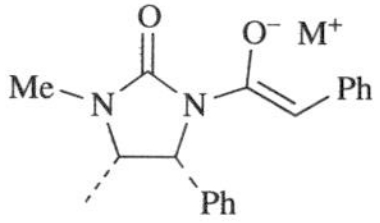	$h\nu$, NH_3, 3 h	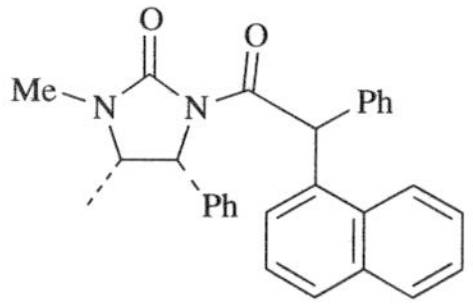	173
	M		(%) *S/R*	
	Li	–33°	(90) 6.1	
	Li	–78°	(57) >99	
	Na	–33°	(44) 2.2	
	K	–33°	(28) 1.8	
	Cs	–33°	(62) 1.7	
	Ti(IV)	–33°	(43) >99	
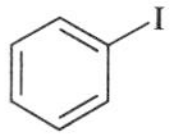		$h\nu$, DMSO, 1.5 h	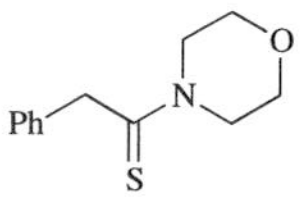 **I** (60)	78
		$FeSO_4$, DMSO, 50°, 1 h	**I** (87)	78
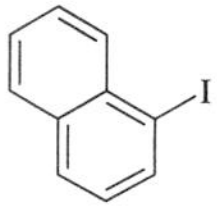		$h\nu$, DMSO, 1.5 h	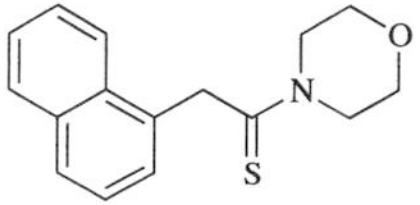 (70) + 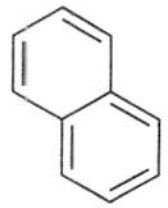 (14)	78

[a] The nucleophile:substrate ratio = 10.

[b] The nucleophile:substrate ratio = 5.

TABLE V. CARBANIONS DERIVED FROM NITRILES

Substrate	Nucleophile	Conditions	Product(s) and Yield(s) (%)	Refs.
PhF	$^{-}CH_2CN$	$h\nu$, NH_3, 2 h	PhMe **I** (4) + $PhCH_2CN$ **II** (9) + $Ph(CH_2)_2Ph$ **III** (14)	42
		K, NH_3, –78°	**I** (49) + **II** (3) + **III** (23) + C_6H_6 **IV** (10)	43
PhCl		$h\nu$, NH_3, 2 h	**I** (4) + **II** (17) + **III** (43)	42
		K, NH_3	**I** (13) + **II** (36) + **IV** (26)	4, 43
PhBr		$h\nu$, NH_3, 2 h	**I** (5) + **II** (13) + **III** (46)	42
		K, NH_3	**I** (25) + **II** (5) + **III** (14) + **IV** (43)	43
PhI		$h\nu$, NH_3, 2 h	**I** (6) + **II** (7) + **III** (63)	42
		K, NH_3	**I** (14) + **II** (5) + **III** (25) + **IV** (32)	43
$PhOP(O)(OEt)_2$		K, NH_3, –78°	**I** (43) + **II** (3) + **III** (7) + **IV** (11)	43
$Ph\overset{+}{N}Me_3I^-$		K, NH_3, –78°	**I** (26) + **II** (5) + **III** (19) + **IV** (21)	43
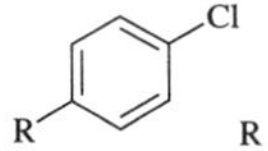		$h\nu$, NH_3	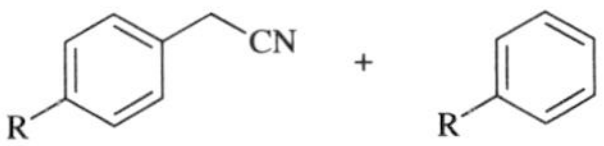	46
R: Ph		20 min	(96) + (4)	
R: PhCO		1 h	(97) + (—)	
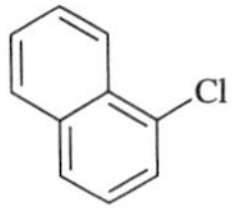		K, NH_3	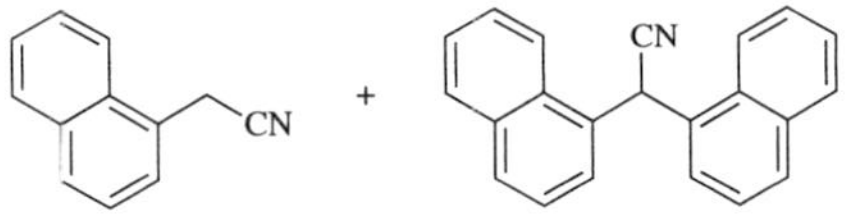 **I** (60)[a] + **II** (13)	45
		$h\nu$, NH_3, 1.7 h	**I** (39) + **II** (7)	45
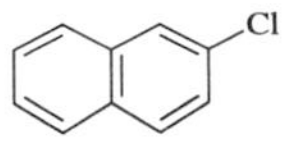		$h\nu$, NH_3, 0.75 h	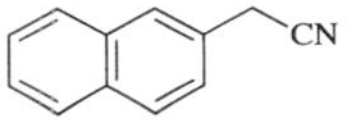 (98)	46

TABLE V. CARBANIONS DERIVED FROM NITRILES (*Continued*)

Substrate	Nucleophile	Conditions	Product(s) and Yield(s) (%)	Refs.
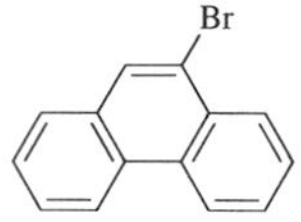	$^{-}CH_2CN$	$h\nu$, NH_3, 2 h	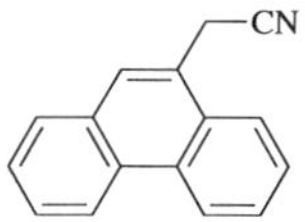 (70) + 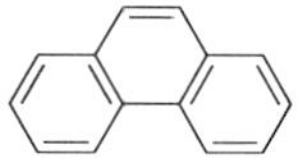 (16)	46
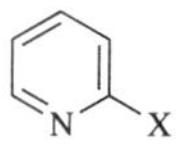 X			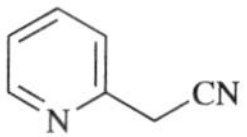	
Cl		$h\nu$, NH_3, 0.5 h	(98)	46
Br		$h\nu$, NH_3, 15 min	(75) + 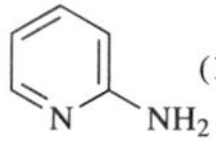 (16)	134
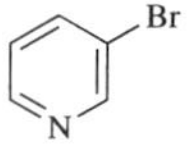		K, NH_3, –68°	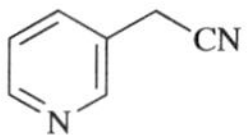 (30)	47
		$h\nu$, NH_3, –70°, 0.5 h	(80)	47
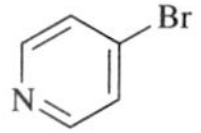		$h\nu$, NH_3, –70°, 0.7 h	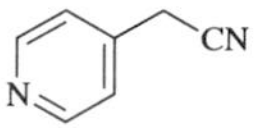 (60)	47
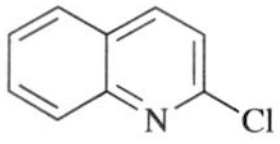		$h\nu$, NH_3, 15 min	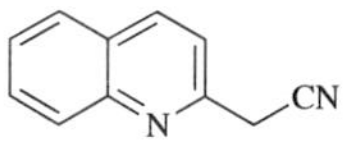 (50)	134
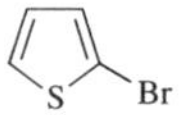		$h\nu$, NH_3, 15 min	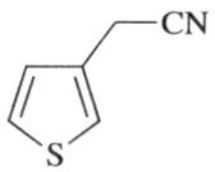 **I** (35) + 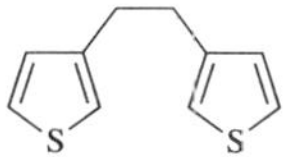 **II** (6)	174

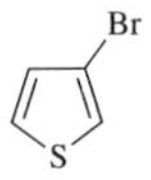		Na, NH_3, –70°	**I** (15) + **II** (5)	174
		$h\nu$, NH_3, 15 min	**I** (38) + **II** (7)	174
PhCl	$^{-}CHMeCN$	K, NH_3	PhEt **I** (34) + C_6H_6 **II** (38) + 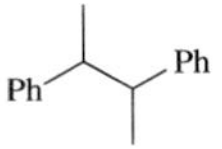 **III** (12)	43
$PhOP(O)(OEt)_2$		K, NH_3,–78°	**I** (14) + **II** (35) + **III** (14)	43
+ $PhNMe_3I^-$		K, NH_3,–78°	**I** (8) + **II** (36) + **III** (16)	43
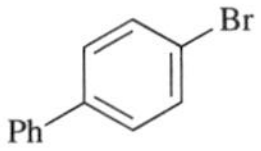		$h\nu$, NH_3, 6 h	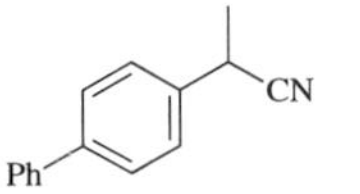 (52)	66
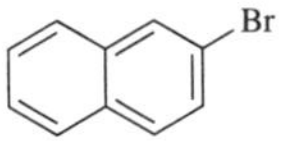		$h\nu$, NH_3, 6 h	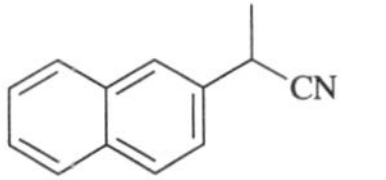 (77)	66
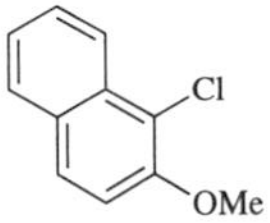		$h\nu$, NH_3, 5 h	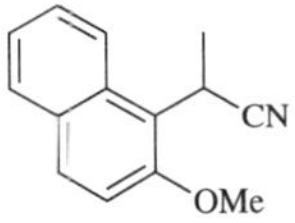 (52)	66
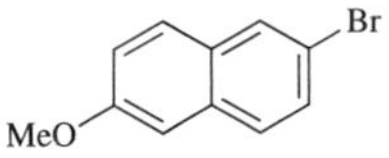		$h\nu$, NH_3, 4 h	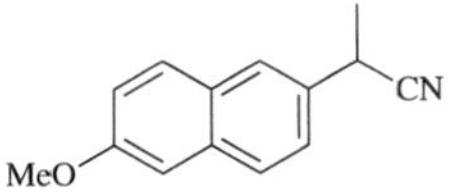 **I** (81)	66
		$h\nu$, $H_2N(CH_2)_2NH_2$, 2 h	**I** (94)	176
PhF	$^{-}CHEtCN$	K, NH_3,–78°	PhPr-*n* **I** (28) + 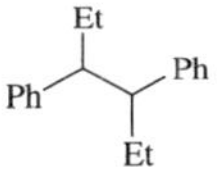 **II** (26)	43

TABLE V. CARBANIONS DERIVED FROM NITRILES (*Continued*)

Substrate	Nucleophile	Conditions	Product(s) and Yield(s) (%)	Refs.
PhI		K, NH_3	**I** (6) + **II** (24) + C_6H_6 (27) + 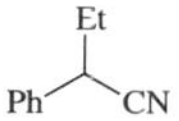 (29)	43
PhCl	^-CHEtCN	K, NH_3	**I** (37) + 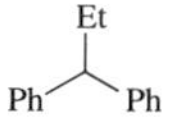 (13) + C_6H_6 (27)	43
$PhOP(O)(OEt)_2$		K, NH_3	**I** (28) + C_6H_6 (20)	43
PhCl	$^-CH(Pr\text{-}n)CN$	K, NH_3	PhBu-*n* (56) + C_6H_6 (38)	43
$PhOP(O)(OEt)_2$		K, NH_3	PhBu-*n* (38) + C_6H_6 (27) + 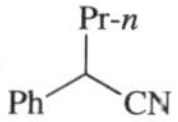 (10)	43
PhBr	$^-CMe_2CN$	K, NH_3	PhPr-*i* (27) + C_6H_6 (31)	43
PhCl	$^-CH(Pr\text{-}i)CN$	K, NH_3	PhBu-*i* (37) + C_6H_6 (31) + 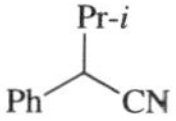 **II** (19)	43
$PhOP(O)(OEt)_2$		K, NH_3	**I** (31) + **II** (10) + C_6H_6 (32)	43
	^-CHPhCN	K, NH_3, –78°	 (12) + C_6H_6 (25)	43
$Ph\overset{+}{N}Me_3I^-$		K, NH_3, –78°	 (17) + C_6H_6 (43) Ph_3CH (12)	43
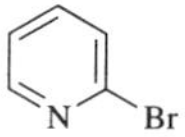		*h*ν, NH_3, 15 min	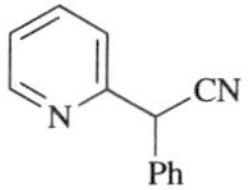 (88)	134
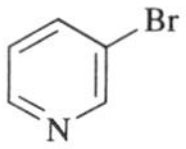		*h*ν, NH_3, 1.5 h	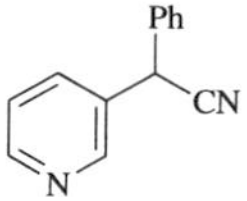 (48) + 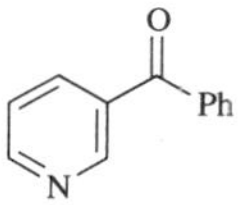 (12)	177
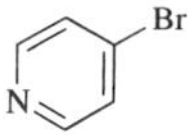		*h*ν, NH_3, 1.5 h	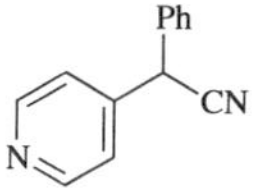 (15) + 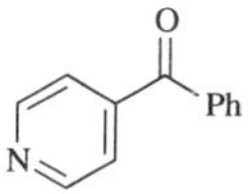 (28)	177

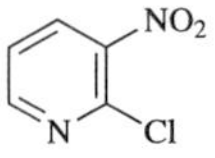	DMF, 64 h	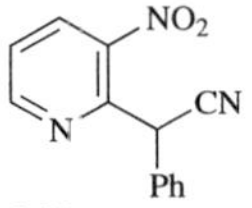 (9)	347
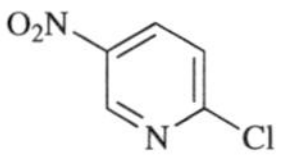	DMF, 40 h	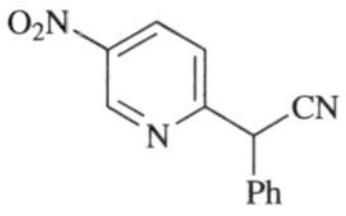 (37)	347
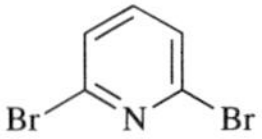	$h\nu$, NH_3, 15 min	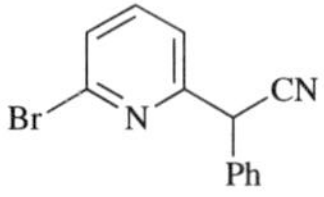 (34) + 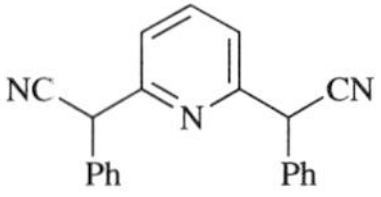 (36)	167
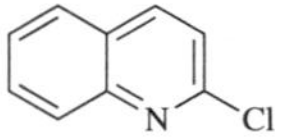	$h\nu$, NH_3, 15 min	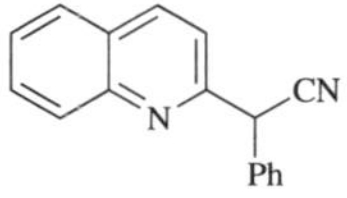 (88)	134
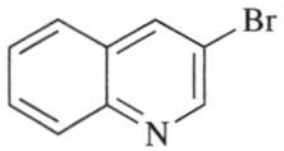	$h\nu$, NH_3, 1.5 h	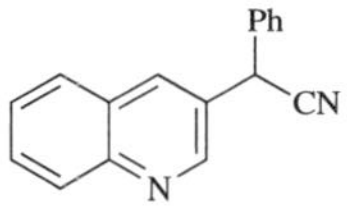 (45) + 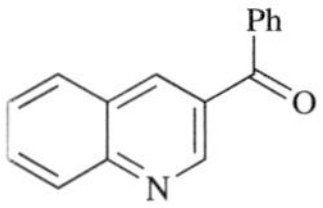 (18)	177
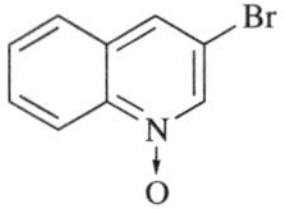	HMPA, NaOH, 1 h	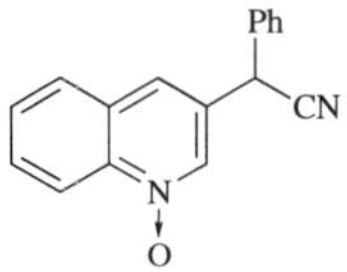 (82)	348
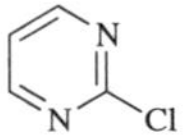	$h\nu$, NH_3, 1.5 h	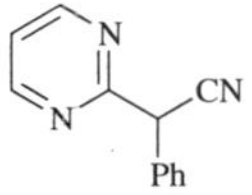 (31)	177
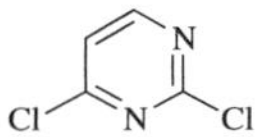	$h\nu$, NH_3, 15 min	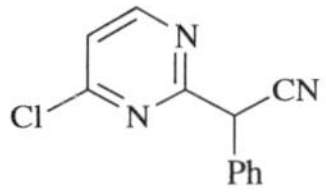 (58)	167

TABLE V. CARBANIONS DERIVED FROM NITRILES (*Continued*)

Substrate	Nucleophile	Conditions	Product(s) and Yield(s) (%)	Refs.
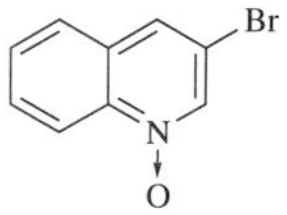Br N O	$Ph\bar{C}EtCN$	HMPA, NaOH, 1 h	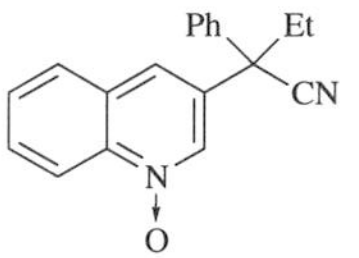Ph Et CN N O (91)	348
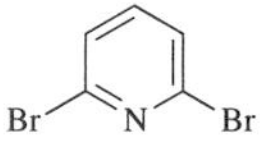Br N Br		$h\nu$, NH_3, 1.5 h	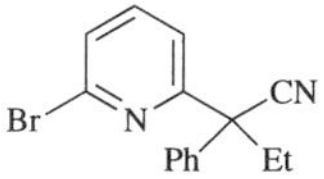Br N CN Ph Et (17) + 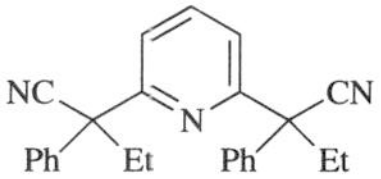NC N CN Ph Et Ph Et (25)	167
			+ 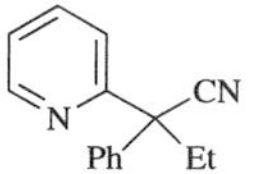N CN Ph Et (52)	
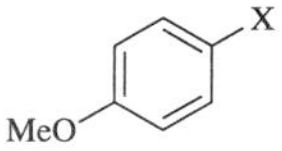X MeO	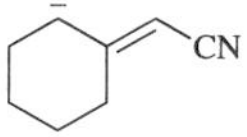CN	$h\nu$, NH_3	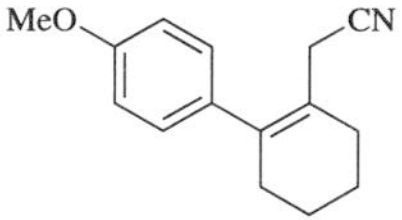MeO CN **I** + 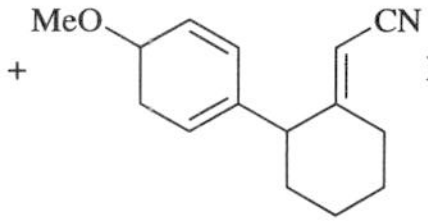MeO CN **II**	349
			+ 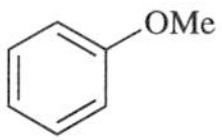OMe **III**	
X				
Br		5 h	(**I** + **II**) (69) + **III** (20)	
I		2 h	(**I** + **II**) (65) + **III** (25)	

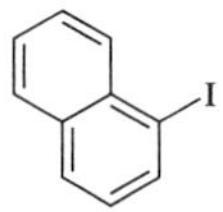	$h\nu$, NH_3, 4 h	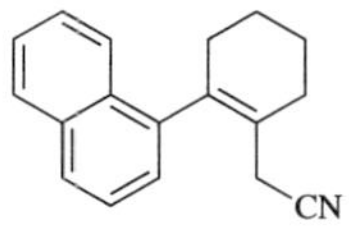 **I** + 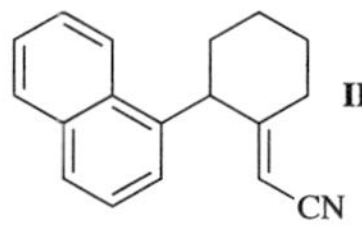 **II** + 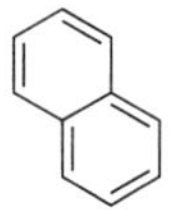 (35) (**I** + **II**) (60)	349

[a] This yield includes 1,2 and 1,4-dihydro-1-naphthylacetonitriles.

TABLE VI. β–DICARBONYL AND RELATED CARBANIONS

Substrate	Nucleophile	Conditions	Product(s) and Yield(s) (%)	Refs.
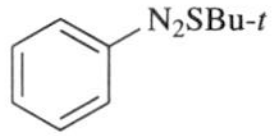	$^{-}CH(COMe)_2$	Sunlamp, DMSO, 6 h	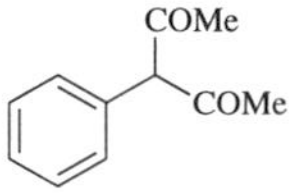 (7)	59
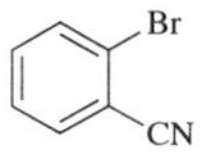		1. $h\nu$, NH_3, 1 h 2. HCl	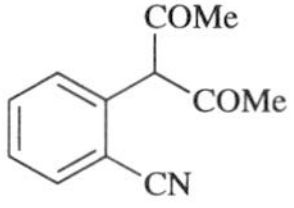 **I** (90)	107
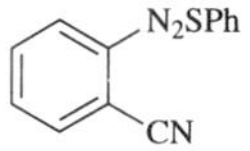		Sunlamp, DMSO, 0.75 h	**I** (67)	59
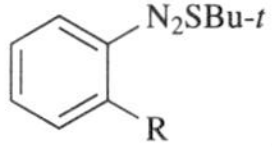		Sunlamp, DMSO,	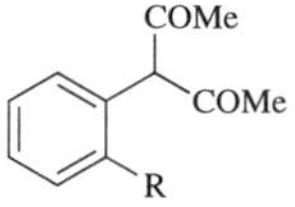	59
R: CN		22 h	(77)	
R: NO_2		3 h	(74)	
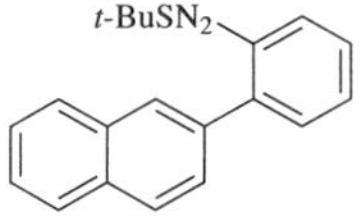		Sunlamp, DMSO, 8 h	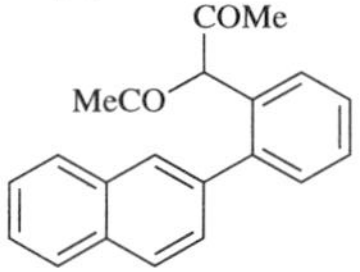 (26)	59
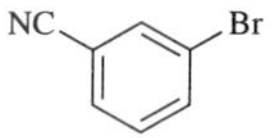		1. $h\nu$, NH_3, 1 h 2. HCl	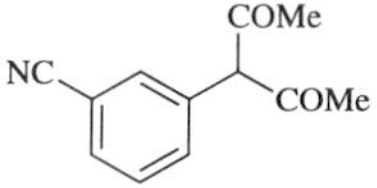 (90)	107
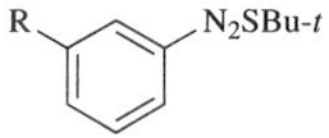			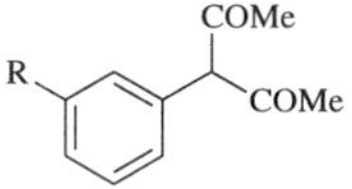	

R			
CN	Sunlamp, DMSO, 3 h	(84)	59
NO_2	Sunlamp, DMSO, 23 h	(51)	59
COPh	*h*ν, DMSO	(49)	183
COPh	Sunlamp, DMSO, 2.5 h	(58)	59
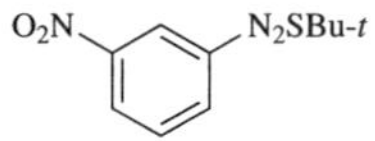	1. Lab light, DMSO, 3.5 h 2. Sunlamp, 1 h	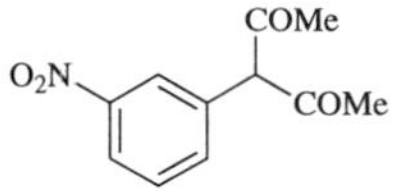 (71)	59
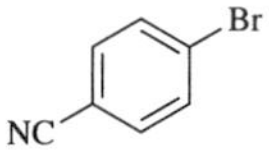	1. *h*ν, NH_3, 7 h 2. HCl	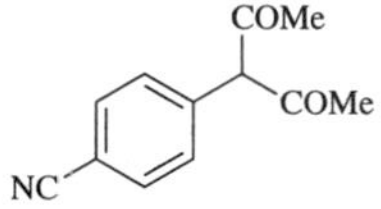 (63)	107
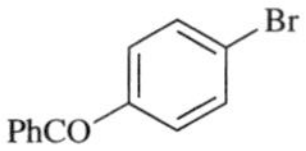	e^-, DMSO	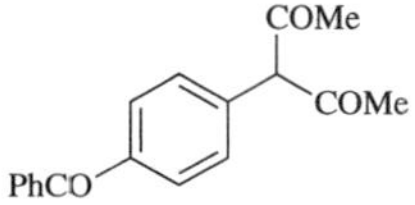 (51)	180
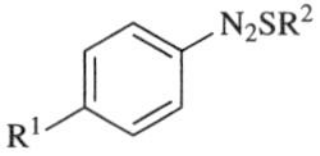		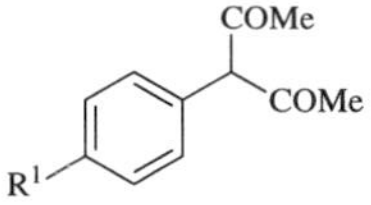	

R^1	R^2			
CN	Bu-*t*	Sunlamp, DMSO, 26 h	(74)	59
CN	Bu-*t*	e^-, DMSO	(73)	59
COMe	Bu-*t*	Sunlamp, DMSO, 27 h	(64)	59
COPr-*i*	Bu-*t*	*h*ν, DMSO	(44)	183
COPh	Bu-*t*	Sunlamp, DMSO, 28 h	(45)	59
Br	Bu-*t*	Sunlamp, DMSO, 2.5 h	(43)	59
OMe	Bu-*t*	Sunlamp, DMSO, 15 h	(9)	59
CN	Ph	Lab light, DMSO, 0.5 h	(60)	59
NO_2	Ph	Lab light, DMSO, 0.5 h	(72)	59
COPh	Ph	Sunlamp, DMSO, 26 h	(45)	59

TABLE VI. β–DICARBONYL AND RELATED CARBANIONS (*Continued*)

Substrate	Nucleophile	Conditions	Product(s) and Yield(s) (%)	Refs.
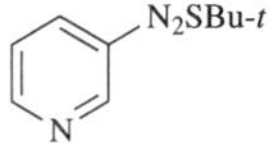	$^{-}CH(COMe)_2$	Sunlamp, DMSO, 17 h	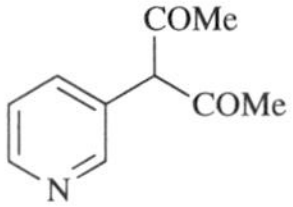 (60)	59
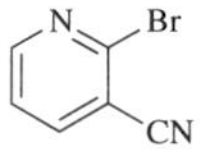		*h*ν, NH_3, 2 h	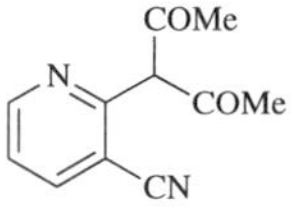 (75)	107
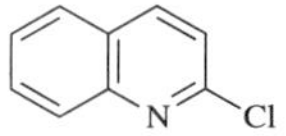		e^{-}, DMSO	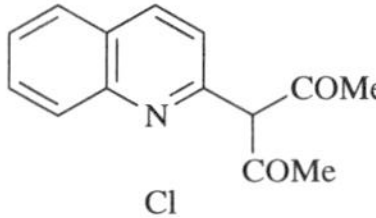 (12)	180
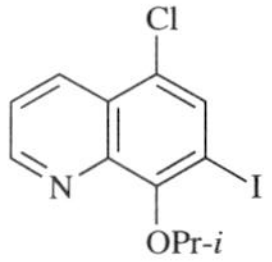		*h*ν, NH_3, 6 h	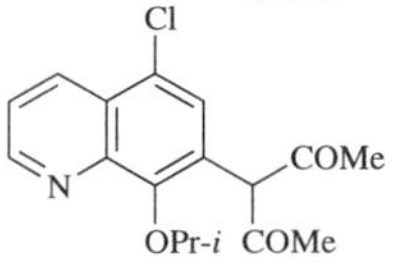 (60)	182
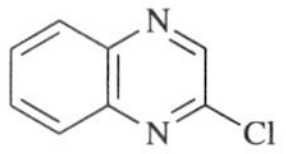	$^{-}CMe(COMe)_2$	NH_3	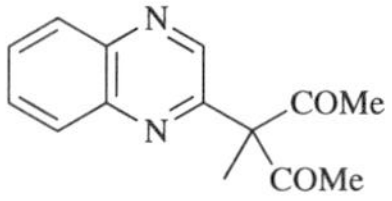 (15)	56
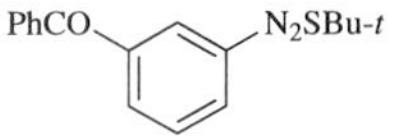	$^{-}CH(COMe)CO_2Me$	*h*ν, DMSO	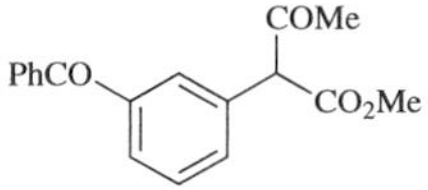 (34)	183
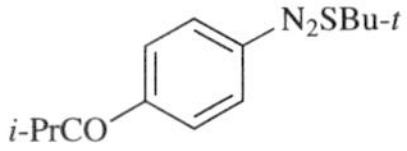		*h*ν, DMSO	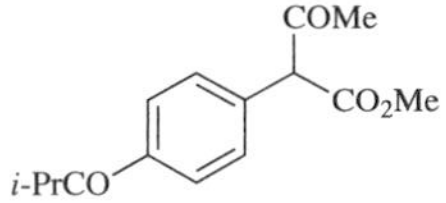 (47)	183

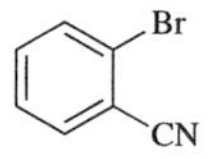		1. $h\nu$, NH_3, 1 h 2. HCl	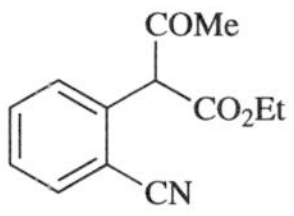 (80)	107
		1. $h\nu$, NH_3, 1 h 2. NH_4Cl	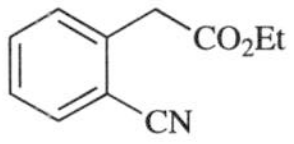 (90)	107
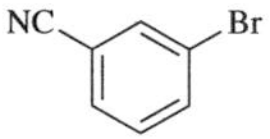		1. $h\nu$, NH_3, 1 h 2. HCl	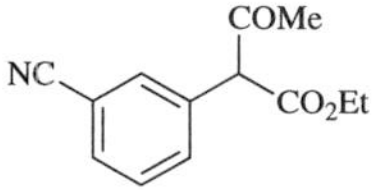 (80)	107
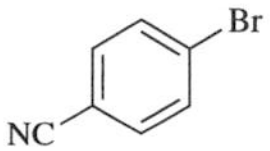		1. $h\nu$, NH_3, 1 h 2. HCl	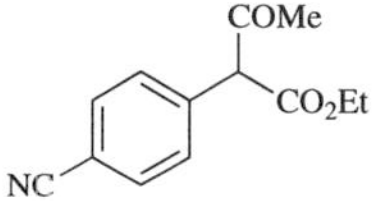 (70)	107
		1. $h\nu$, NH_3, 4 h 2. NH_4Cl	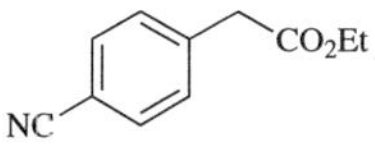 (70)	107
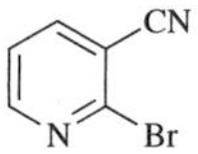		1. $h\nu$, NH_3, 4 min 2. NH_4Cl	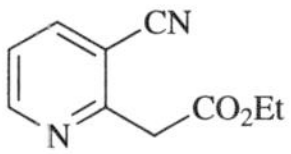 (80)	107
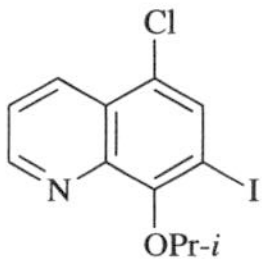		$h\nu$, NH_3, 6 h	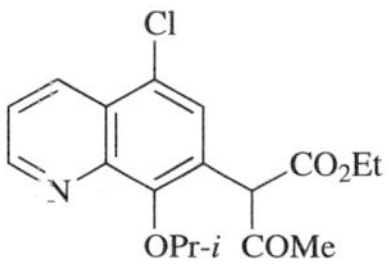 (52)	182
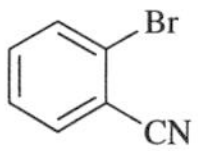	$MeCO\bar{C}MeCO_2Et$	1. $h\nu$, NH_3, 2 h 2. NH_4Cl	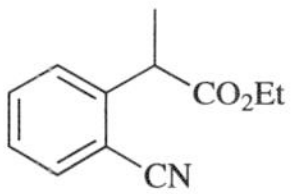 (60)	107

TABLE VI. β–DICARBONYL AND RELATED CARBANIONS (*Continued*)

Substrate	Nucleophile	Conditions	Product(s) and Yield(s) (%)	Refs.
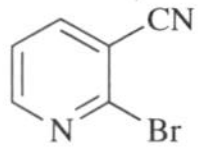	$Me\overset{-}{COC}MeCO_2Et$	1. *h*ν, NH_3, 0.5 h 2. NH_4Cl	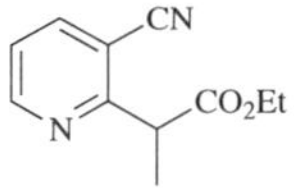 (82)	107
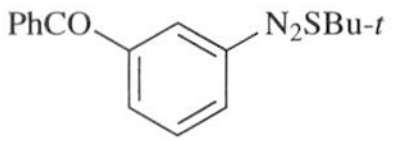	$^{-}CH(CO_2Me)_2$	*h*ν, DMSO	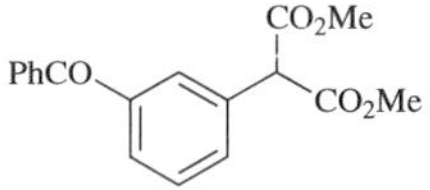 (34)	183
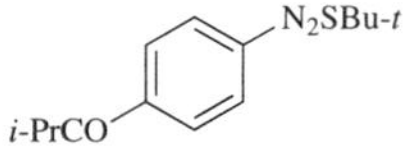		*h*ν, DMSO	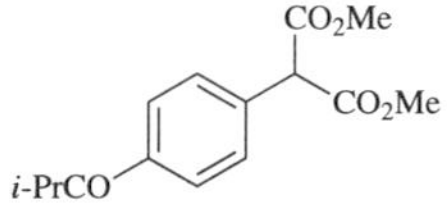 (58)	183
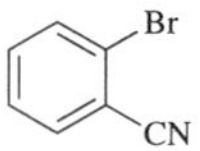	$^{-}CH(CO_2Et)_2$	*h*ν, NH_3, 6 h	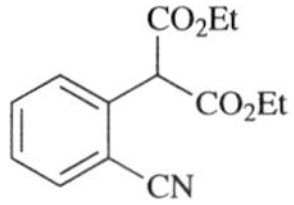 (78)	107
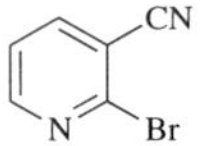		1. *h*ν, NH_3, 3 h 2. NH_4Cl	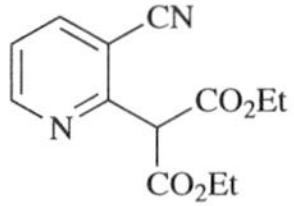 (90)	107
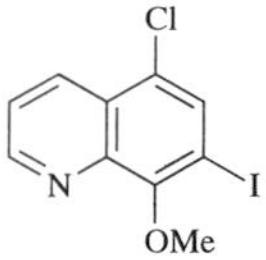		*h*ν, NH_3, 7 h	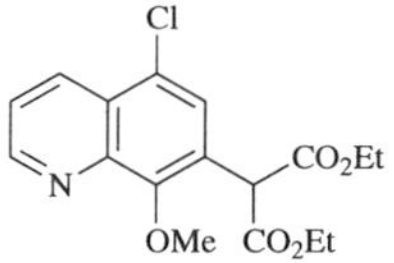 (83)	182

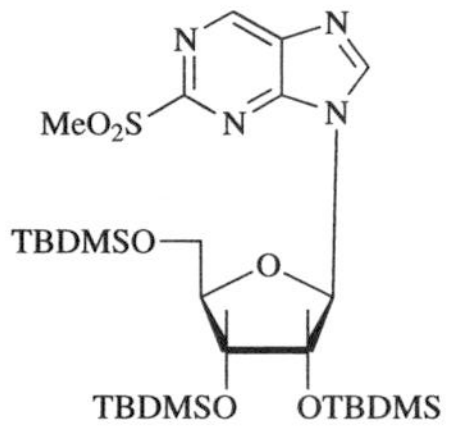		THF, 29 h	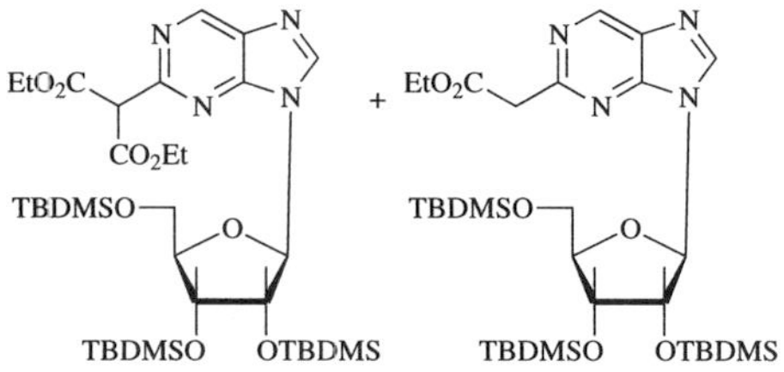 (50) + (25)	350
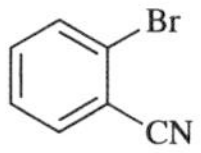	$Et\bar{C}(CO_2Et)_2$	1. $h\nu$, NH_3, 2 h 2. NH_4Cl	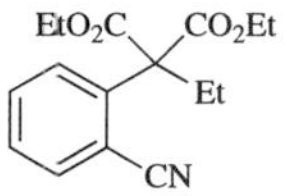 (80)	107
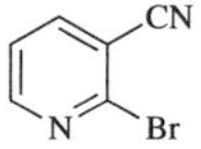		1. $h\nu$, NH_3, 10 min 2. NH_4Cl	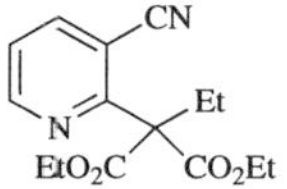 (92)	107
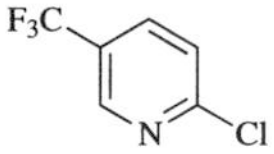	$Me\bar{C}(CO_2Et)_2$	$h\nu$, NH_3, 15 min	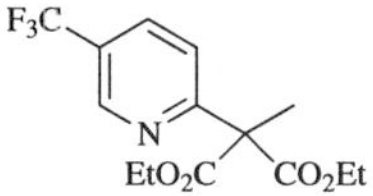 (100)	181
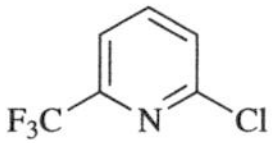		$h\nu$, NH_3, 4 h	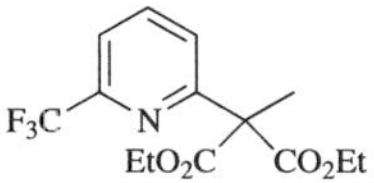 (44)	181
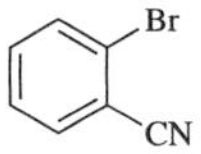	$^{-}CH(CN)CO_2Et$	$h\nu$, NH_3, 2 h	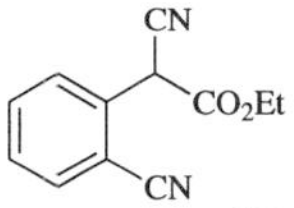 (90)	107
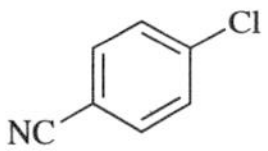		e^{-}, NH_3, mediator	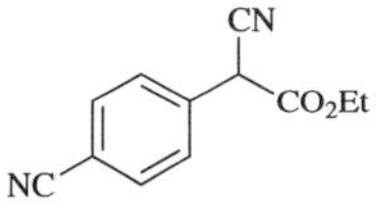 (48)	180

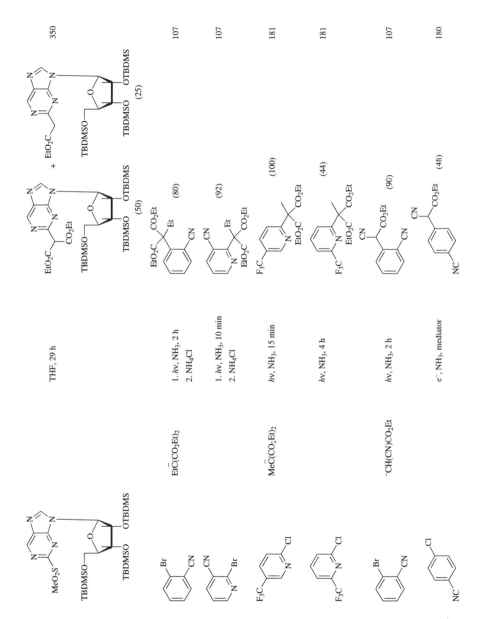

TABLE VI. β–DICARBONYL AND RELATED CARBANIONS (*Continued*)

Substrate	Nucleophile	Conditions	Product(s) and Yield(s) (%)	Refs.
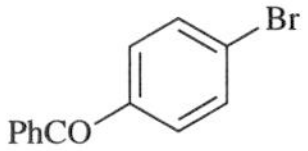	$^{-}CH(CN)CO_2Et$	e^-, DMF	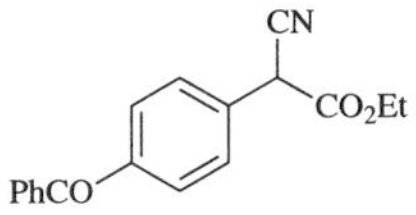 **I** (80)	180
		e^-, DMSO	**I** (94)	180
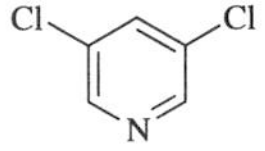		e^-, NH_3, mediator	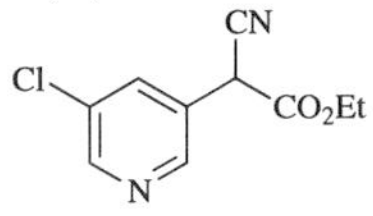 (75)	85
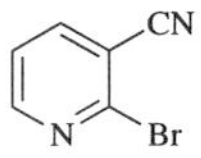		1. *hν*, NH_3, 1 h 2. NH_4Cl	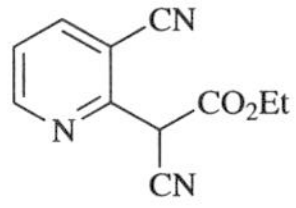 (80)	107
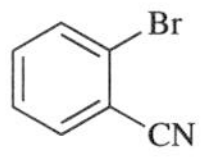	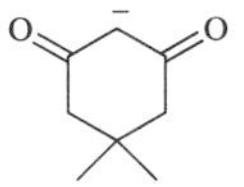	1. *hν*, NH_3, 1 h 2. NH_4Cl	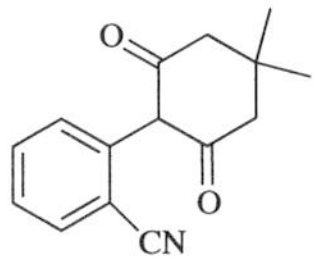 (80)	107
	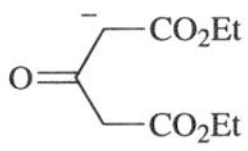	1. *hν*, NH_3, 2 h 2. NH_4Cl	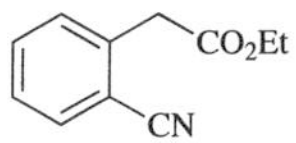 (90)	107

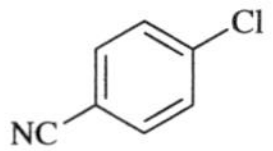	$^{-}CH(CN)_2$	1. e^-, NH_3, mediator 2. KOH	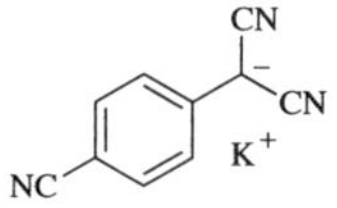 (85)	179
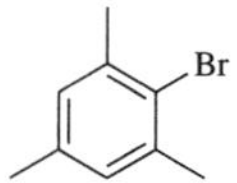	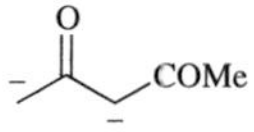	$h\nu$, NH_3	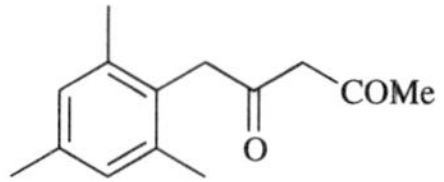 (82)	143
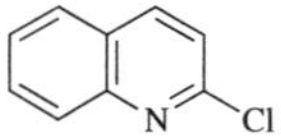	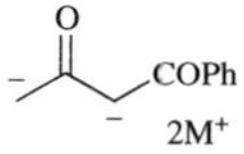	NH_3, 1 h	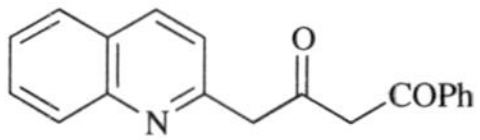	178
	M			
	Li		(71)	
	K		(17)	
	Na		(30)	

TABLE VII. OTHER C-C BOND FORMATION

A. Carbonylation Reactions

Substrate	Nucleophile	Conditions	Product(s) and Yield(s) (%)	Refs.
PhCl	$Co_2(CO)_8$, CO 1 atm	$h\nu$, NaOH, H_2O, C_6H_6, Bu_4NBr, 65°, 13 h	$PhCO_2H$ **I** (0)	188
	$Co(OAc)_2$, CO 1 atm	1. THF, t-$BuCH_2ONa$, NaH, 63°, 52 h 2. H_3O^+	**I** + $PhCO_2CH_2CMe_3$ **II** (**I** + **II**) (35-40)	186
	$Co(OAc)_2$, CO 2 atm	$h\nu$, NaOH, H_2O, EtOH, 30°, 20 h	**I** (49)	194
PhBr	$Co_2(CO)_8$, CO 1 atm	$h\nu$, NaOH, H_2O, C_6H_6, Bu_4NBr, 65°, 13 h	**I** (95)	188, 189
	$Co(OAc)_2$, CO 1 atm	1. THF, $EtCMe_2ONa$, NaH, 60-65°, 5 h 2. EtOH, H_3O^+	**I** (40) + $PhCO_2C(Me)_2Et$ (60)	187, 184
		1. THF, t-$BuCH_2ONa$, NaH, 63°, 10 h 2. H_3O^+	**I** + **II** (**I** + **II**) (80-85)	186
		Sunlamp, $EtCMe_2ONa$, THF, MeOH, NaH 40°, 6 h	**I** (9) + $PhCO_2Me$ (88)	351
	$Co(OAc)_2$, CO 1 atm	Sunlamp, $EtCMe_2ONa$, THF, ROH, NaH, 40°	**I** + $PhCO_2R$ **II**	351

R	time (h)	I	II
Me	14	(3)	(76)
Et	24	(6)	(83)
n-$C_{10}H_{21}$	79	(12)	(67)
$MeO(CH_2)_2$	20	(16)	(69)
$EtO(CH_2)_2$	15	(5)	(81)

R	time (h)	I	II
$BuO(CH_2)_2$	22	(12)	(75)
Bn	86	(16)	(70)
i-Pr	32	(2)	(82)
i-C_5H_{11}	32	(3)	(73)
c-C_5H_9	18	(13)	(79)
t-Bu	72	(6)	(72)
t-C_5H_{11}	40	(tr)	(tr)

	$Co(OAc)_2$, R^1R^2NH, CO 1 atm	1. THF, $EtCMe_2ONa$, NaH, 63°, X h 2. H_3O^+	$PhCONR^1R^2$ **I** + $PhCO_2C(Me)_2Et$ **II** + $PhCO_2H$ **III**	186

R^1	R^2	X	I	II	III
H	C_6H_{11}	20	(70-75)	(—)	(10-5)
Et	Et	45	(40-45)	(20-25)	(10-5)
i-Pr	*i*-Pr	25	(30-35)	(30-35)	(20-15)
H	*n*-Bu	48	(30-35)	(0-5)	(15-10)
n-C_8H_{17}	*n*-C_8H_{17}	40	(40-45)	(5-10)	(15-10)
H	$(CH_2)_5$	25	(45-50)	(5-10)	(20-15)

	$NaCo(CO)_4$, CO 1 atm	*hν*, $EtCMe_2ONa$, THF, 63°, 5 h	$PhCO_2C(Me)_2Et$ (70) + $PhCO_2H$ **I** (17) + C_6H_6, (3)	184
	$Co(OAc)_2$, CO 2 atm	*hν*, NaOH, H_2O, EtOH, 30°, 1 h	**I** (99)	194
PhI	$Co_2(CO)_8$, CO 1 atm	NaOH, H_2O, C_6H_6, Bu_4NBr, 65°, 23 h	**I** (25)	199
	$Co_2(CO)_8$, CO 1 atm	Sunlamp, NaOH, H_2O, C_6H_6, Bu_4NBr, 65°, 1 h	**I** (90)	189
	$Co(OAc)_2$, CO 1 atm	1. THF, *t*-$BuCH_2ONa$, NaH, 63°, 1 h 2. H_3O^+	**I** + $PhCO_2CH_2Bu$-*t* **II** + C_6H_6 (20-15) (**I** + **II**) (70-75)	186

TABLE VII. OTHER C-C BOND FORMATION (*Continued*)

A. Carbonylation Reactions (*Continued*)

Substrate	Nucleophile	Conditions	Product(s) and Yield(s) (%)	Refs.
	$Co(OAc)_2$, CO 1 atm	1. THF, $EtCMe_2ONa$, NaH, 60-65°,3 h 2. EtOH, H_3O^+	**I** (74) + $PhCO_2C(Me)_2Et$ (6)	187
	$Fe(CO)_5$, $Co_2(CO)_8$, CO 1 atm	1. EtOH, NaOH, H_2O,C_6H_6, Bu_4NBr, 60°, 23 h 2. H_3O^+	**I** (80) + PhCOPh **II** (2) + PhPh **III** (2)	199, 195, 196
		1. NaOH, H_2O, C_6H_6, Bu_4NBr, 65°, 23 h 2. H_3O^+	**I** (30) + **II** (50) + **III** (6)	199
	$Fe(CO)_5$, CO 1 atm	1. NaOH, H_2O, C_6H_6,[a] Bu_4NBr, 65°, 65 h 2. H_3O^+	**I** (18) + **II** (33) + **III** (20)	199
	$Fe(CO)_5$ aq., CO 1 atm	NaOH, H_2O, C_6H_6,[b] Bu_4NBr, 70°, 24 h	**I** (9) + **II** (85) + **III** (2)	197, 200
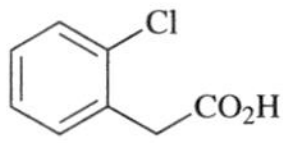	$Co_2(CO)_8$, CO 1 atm	$h\nu$, NaOH, H_2O, 65°, 24 h	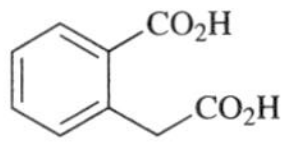 (87) + 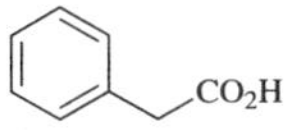 (4)	191
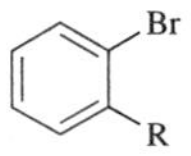	$Co(OAc)_2$, CO 1 atm	Sunlamp, THF, MeOH, NaH, $EtC(Me)_2ONa$, 40°	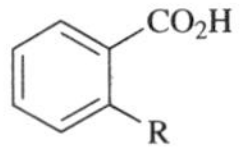 + 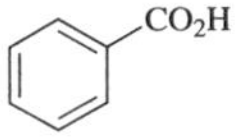	351
R				
Me		18 h	(10) (53)	
OMe		20 h	(8) (85)	
F		8 h	(3) (81)	
CF_3		18 h	(9) (84)	
CN		12 h	(78) (—)	

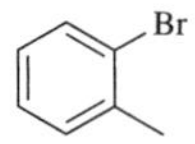	$Co_2(CO)_8$, CO 1 atm	*hν*, NaOH, H_2O, C_6H_6, Bu_4NBr, 65°, 2.25 h	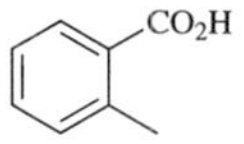 (96)	188, 189
	$Co(OAc)_2$, CO 1 atm	1. THF, *t*-$BuCH_2ONa$, NaH, 63°, 25 h 2. H_3O^+	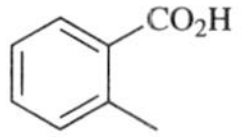 **I** + 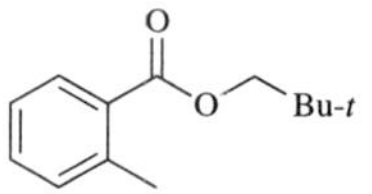 **II** (**I** + **II**) (80-85)	186
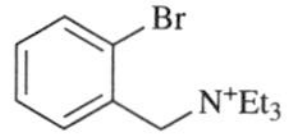	$Co_2(CO)_8$, CO 1 atm	Sunlamp, NaOH, H_2O, 65°, 12 h	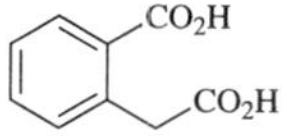 **I** (78)	190
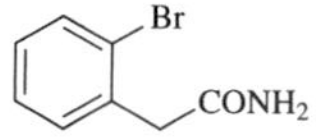		Sunlamp, NaOH, H_2O, Bu_4NBr, 65°, 12 h	**I** (90)	189
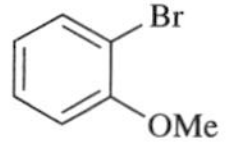		*hν*, NaOH, H_2O, C_6H_6, Bu_4NBr, 65°, 2 h	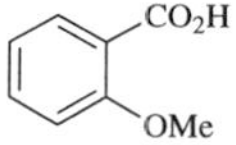 **I** (47) + 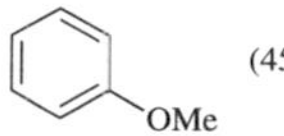 (45)	188
	$Co(OAc)_2$, CO 1 atm	1. THF, *t*-$BuCH_2ONa$, NaH, 63°, 15 h 2. H_3O^-	**I** + 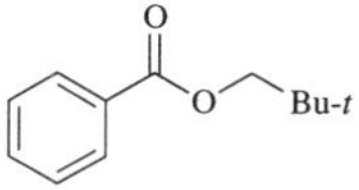 **II** (**I** + **II**) (95-100)	186
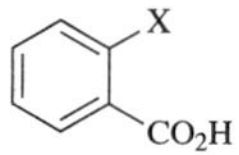	$Co_2(CO)_8$, CO 2 atm		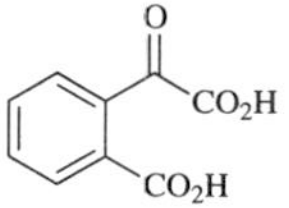 **I** + 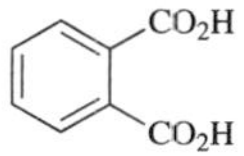 **II** + $PhCO_2H$ **III**	
X				
Br		*hν*, KOH, H_2O, 65°, 4 h	**I** (36) + **II** (38) + **III** (16)	192
I		*hν*, KOH, H_2O, 65°, 4 h	**I** (29) + **II** (33) + **III** (33)	192
Cl		*hν*, NaOH, H_2O, 65°, 18 h	**I** (29) + **II** (34) + **III** (11)	192
Cl		*hν*, NaOH, H_2O, 65°, 18 h	**I** (—) + **II** (34) + **III** (11)	191
Cl	$Co(OAc)_2$, CO 2 atm	*hν*, NaOH, H_2O, 65°, 22 h	**I** (—) + **II** (13) + **III** (26)	194

TABLE VII. OTHER C-C BOND FORMATION (*Continued*)

A. Carbonylation Reactions (*Continued*)

Substrate	Nucleophile	Conditions	Product(s) and Yield(s) (%)	Refs.
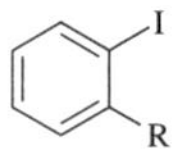	$Fe(CO)_5$, $Co_2(CO)_8$, CO 1 atm	NaOH, H_2O, Bu_4NBr, 48 h	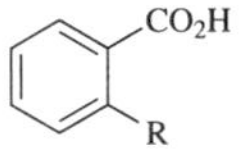 I + 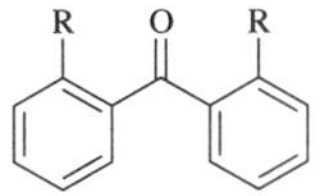 II + PhR III	
R				
Me		C_6H_6[a], 65°	I (15) + II (47) + III (15)	199
Me	$Fe(CO)_5$, CO 1 atm	C_6H_6[b], 70°	I (21) + II (31) + III (25)	200
Cl	$Fe(CO)_5$, CO 1 atm	C_6H_6[b], 70°	I (34) + II (27) + III (22)	200
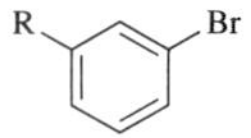	$Co(OAc)_2$, CO 1 atm	THF, *t*-$BuCH_2ONa$, NaH, 63°	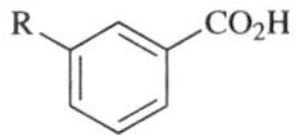 I + 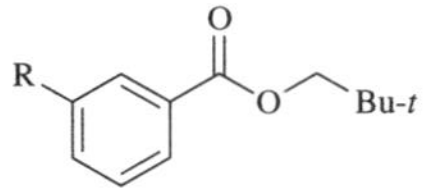 II	186
R				
Me		20 h	(I + II) (85-90)	
OMe		15 h	(I + II) (90-95)	
		Sunlamp, THF, MeOH, NaOH, $EtC(Me)_2ONa$, 40°	I + 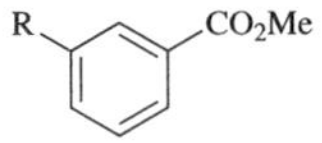 II	351
R				
Me		16 h	I (10) + II (79)	
OMe		17 h	I (8) + II (80)	
F		7 h	I (8) + II (85)	
CF_3		16 h	I (19) + II (76)	
CN		36 h	I (9) + II (72)	
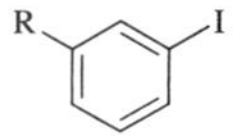	$Fe(CO)_5$, $Co_2(CO)_8$, CO 1 atm	NaOH, H_2O, C_6H_6,[a] Bu_4NBr, 65°	I + 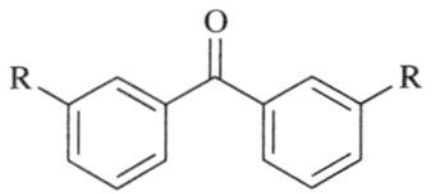 II	199
R				
Me		24 h	I (60) + II (30)	
Cl		48 h	I (27) + II (20) + PhCl (37)	

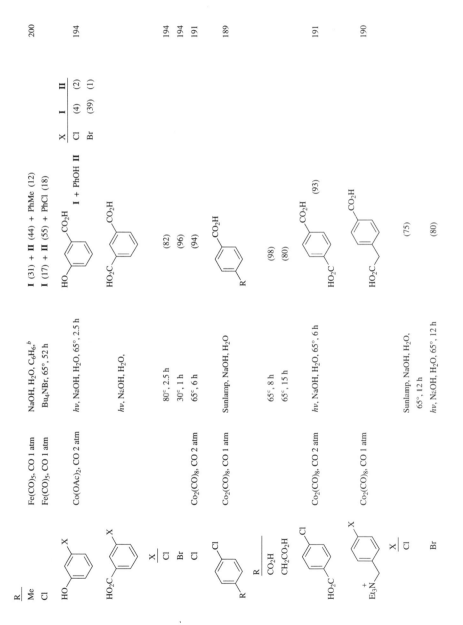

R				
Me	$Fe(CO)_5$, CO 1 atm	NaOH, H_2O, C_6H_6,[b]	**I** (31) + **II** (44) + PhMe (12)	200
Cl	$Fe(CO)_5$, CO 1 atm	Bu_4NBr, 65°, 52 h	**I** (17) + **II** (55) + PhCl (18)	
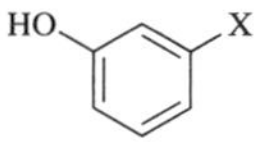	$Co(OAc)_2$, CO 2 atm	*hν*, NaOH, H_2O, 65°, 2.5 h	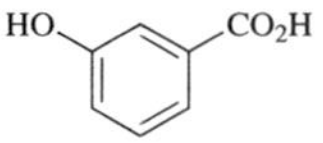 **I** + PhOH **II** X / **I** / **II** Cl / (4) / (2) Br / (39) / (1)	194
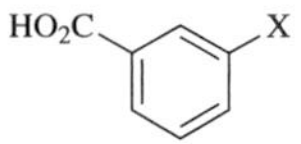		*hν*, NaOH, H_2O,	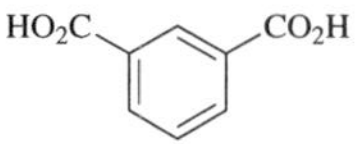	
X				
Cl		80°[c], 2.5 h	(82)	194
Br		30°[c], 1 h	(96)	194
Cl	$Co_2(CO)_8$, CO 2 atm	65°[c], 6 h	(94)	191
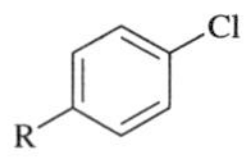	$Co_2(CO)_8$, CO 1 atm	Sunlamp, NaOH, H_2O	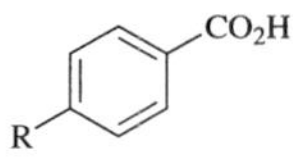	189
R				
CO_2H		65°[c], 8 h	(98)	
CH_2CO_2H		65°[c], 15 h	(80)	
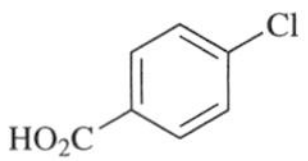	$Co_2(CO)_8$, CO 2 atm	*hν*, NaOH, H_2O, 65°, 6 h	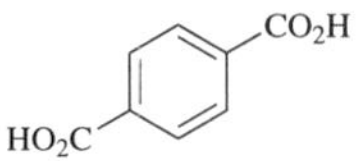 (93)	191
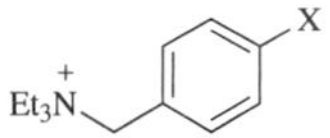	$Co_2(CO)_8$, CO 1 atm		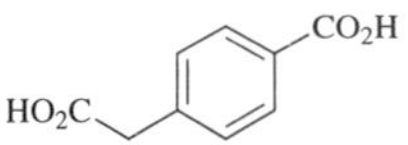	190
X				
Cl		Sunlamp, NaOH, H_2O, 65°, 12 h	(75)	
Br		*hν*, NaOH, H_2O, 65°, 12 h	(80)	

TABLE VII. OTHER C-C BOND FORMATION (*Continued*)

A. Carbonylation Reactions (*Continued*)

Substrate	Nucleophile	Conditions	Product(s) and Yield(s) (%)	Refs.
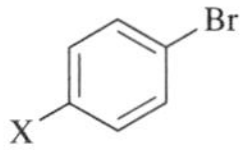	$Co_2(CO)_8$, CO 1 atm	Sunlamp, NaOH, H_2O, C_6H_6, Bu_4NBr, 65°	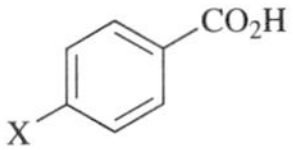	
X				
F		2 h	(97)	188
Cl		1 h	(98)	188, 189
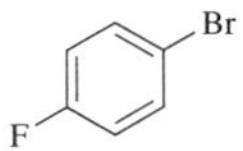	$Co(OAc)_2$, CO 1 atm	THF, *t*-$BuCH_2ONa$, NaH, 63°, 20 h	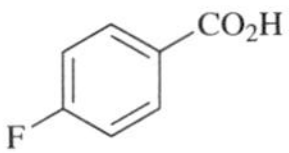 **I** + 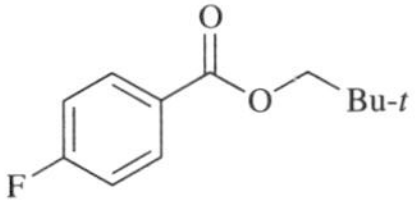 **II** (**I** + **II**) (85-90)	186
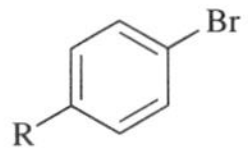	$Co_2(CO)_8$, CO 1 atm		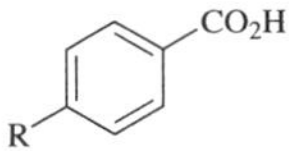 **I** + 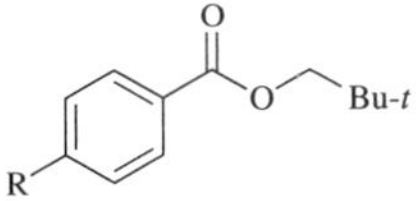 **II**	
R			**I**	
Me		*hν*, NaOH, H_2O, C_6H_6, Bu_4NBr, 65°, 1.5 h	(97)	188, 189
CO_2Et		*hν*, NaOH, H_2O, C_6H_6, Bu_4NBr, 65°, 1.5 h	(97)	189
COMe		Sunlamp, NaOH, H_2O, C_6H_6, 65°, 4.5 h	(90)	189
OH		*hν*, NaOH, H_2O, C_6H_6, Bu_4NBr, 65°, 15 h	(18)	189
OH		Sunlamp, NaOH, H_2O, 4 h	(90)	189
OMe		*hν*, NaOH, H_2O, C_6H_6, Bu_4NBr, 65°, 2 h	(94)	188, 189
NO_2		*hν*, NaOH, H_2O, C_6H_6, Bu_4NBr, 65°, 8 h	(17) + $PhNO_2$ (17)	189

R	$Co(OAc)_2$, CO 1 atm	THF, t-$BuCH_2ONa$, NaH, 63°	(**I** + **II**)	186
Me		20 h	(90-95)	
OMe		15 h	(95-100)	
NMe_2		15 h	(34-40) + $PhNMe_2$ (45-50)	
COMe		25 h	(80-85)	
	$Co(OAc)_2$, CO 1 atm	1. THF, $EtC(Me)_2ONa$, NaH, 60-65° 2. EtOH, H_3O^+	**I** + **II** + 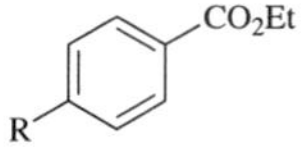 **III**	187
R				
Me		22 h	**I** (16) + **II** (21) + **III** (21)	
CO_2Et		1 h	**I** (84) + **II** (2) + **III** (10)	
	$Co(OAc)_2$, CO 1 atm	Sunlamp, $EtC(Me)_2ONa$, THF, MeOH, NaH, 40°	**I** + 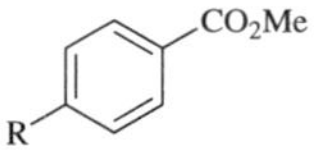 **II**	351
R			**I** / **II**	
Me		14 h	(14) / (76)	
OMe		16 h	(13) / (79)	
SMe		71 h	(12) / (71)	
F		7 h	(8) / (82)	
CF_3		15 h	(16) / (72)	
CN		35 h	(13) / (69)	
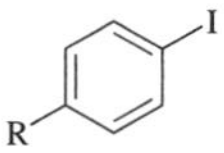	$Fe(CO)_5$, $Co_2(CO)_8$, CO 1 atm	NaOH, H_2O, C_6H_6,[a] Bu_4NBr, 65°	**I** + 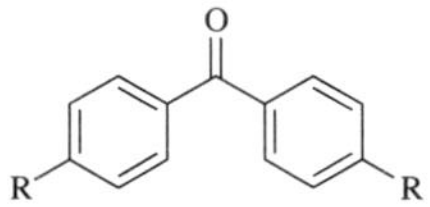 **II** + PhR **III**	199
R				
Me		48 h	**I** (20) + **II** (57) + **III** (8)	
OMe		48 h	**I** (15) + **II** (55) + **III** (10)	
Cl		60 h	**I** (35) + **II** (30) + **III** (25)	

TABLE VII. OTHER C-C BOND FORMATION (*Continued*)

A. Carbonylation Reactions (*Continued*)

Substrate	Nucleophile	Conditions	Product(s) and Yield(s) (%)	Refs.
	$Fe(CO)_5$, CO 1 atm	NaOH, H_2O, C_6H_6,[b] Bu_4NBr, 70°		200
R				
Me		52 h	**I** (19) + **II** (40) + **III** (15)	
OMe		48 h	**I** (17) + **II** (62) + **III** (8)	
Cl		52 h	**I** (20) + **II** (50) + **III** (16)	
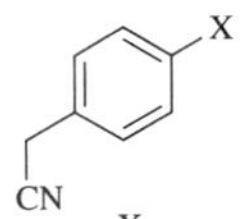	$Co_2(CO)_8$, CO, sealed tube	1. *hν*, MeONa, MeOH, 50°, Y h 2. Me_2SO_4, 60°, 0.5 h	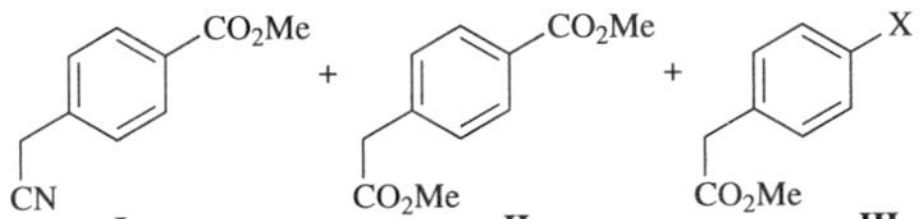 **I** + **II** + **III**	185
X		Y		
I		1 h	**I** (81) + **II** (5) + **III** (1)	
Br		1 h	**I** (44) + **II** (5) + **III** (3)	
Cl		4 h	**III** (8)	
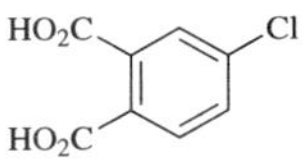	$Co_2(CO)_8$, CO 2 atm	*hν*, NaOH, H_2O, 65°, 6 h	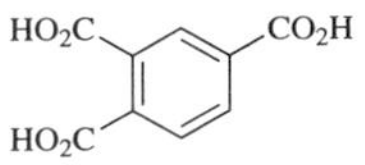 (95)	191
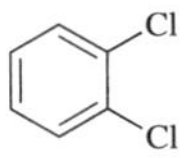		*hν*, NaOH, H_2O, 65°, 24 h	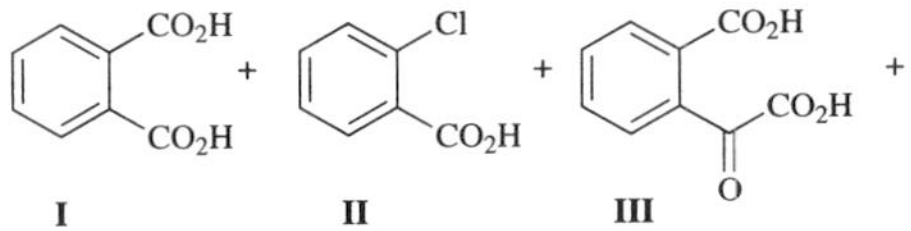 **I** + **II** + **III** + $PhCO_2H$ **IV** **I** (23) + **II** (40) + **III** (13) + **IV** (25)	192
		hν, NaOH, H_2O, EtOH, 65°, 24 h	**I** (22) + **II** (40) + **IV** (23)	191

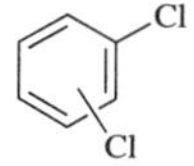	$Co(OAc)_2$, CO 1 atm	Sunlamp, $EtC(Me)_2ONa$, THF, MeOH, NaOH, 40°	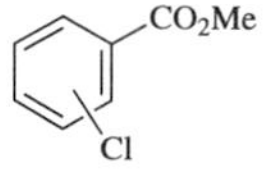 **I** + 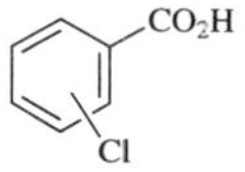 **II**	351
Cl				
o-		71 h	**I** (61) + **II** (17)	
m-		63 h	**I** (66) + **II** (17)	
p-		63 h	**I** (65) + **II** (15)	
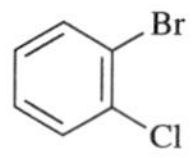	$Co_2(CO)_8$, CO 2 atm	*hν*, NaOH, H_2O, 65°, 20 h	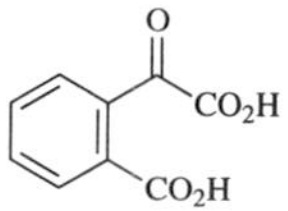 (10) + 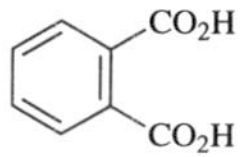 (25) + 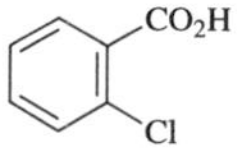 (40) + 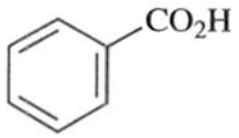 (13)	192, 191
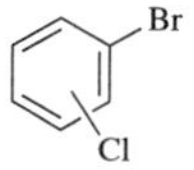	$Co(OAc)_2$, CO 1 atm	Sunlamp, $EtC(Me)_2ONa$, THF, MeOH, NaH, 40°	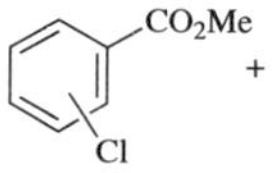 **I** + 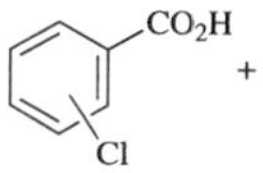 **II** + 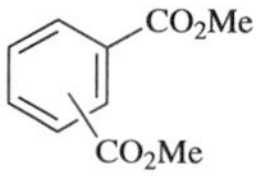 **III**	351
Cl				
o-		10 h	**I** (76) + **II** (3) + **III** (12)	
m-		8 h	**I** (86) + **II** (10) + **III** (3)	
p-		8 h	**I** (86) + **II** (3) + **III** (7)	
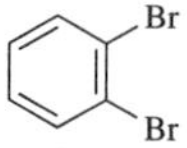	$Co_2(CO)_8$, CO 2 atm	*hν*, NaOH, H_2O, 65°, 4 h	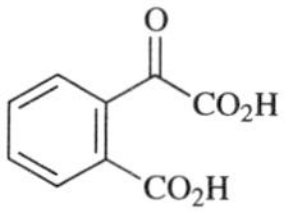 (13) + 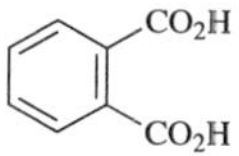 (32) + 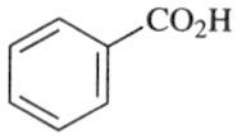 (55)	192, 191
	$Co(OAc)_2$, CO 1 atm	Sunlamp, $EtC(Me)_2ONa$, THF, MeOH, NaH, 40°, 18 h	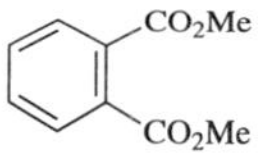 (85)	351

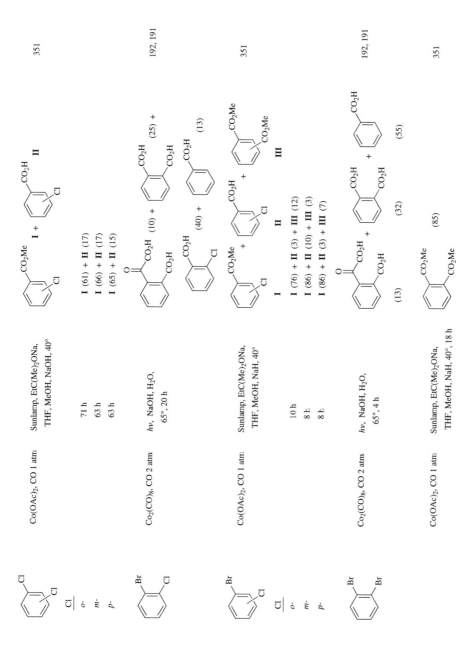

TABLE VII. OTHER C-C BOND FORMATION (*Continued*)

A. Carbonylation Reactions (*Continued*)

Substrate	Nucleophile	Conditions	Product(s) and Yield(s) (%)	Refs.
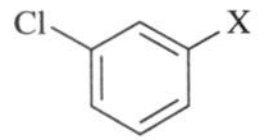	$Co_2(CO)_8$, CO 2 atm	$h\nu$, NaOH, H_2O, 65°, 20 h	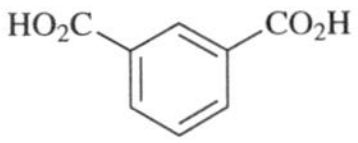 **I** X: Cl (92); Br (91)	191
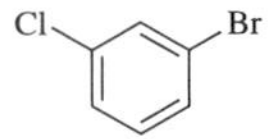	$Co(OAc)_2$, CO 1 atm	1. $EtC(Me)_2ONa$, THF, NaH, 60-65°, 10 h 2. EtOH, H_3O^+	**I** (30) + 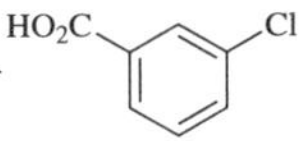 (15) + 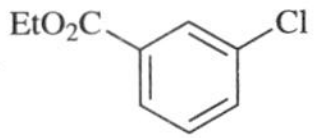 (48)	187
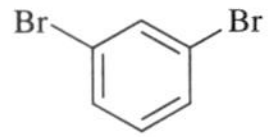	$Co_2(CO)_8$, CO 2 atm	$h\nu$, NaOH, H_2O, 65°, 2 h	**I** (94)	191
	$Co(OAc)_2$, CO 1 atm	Sunlamp, $EtC(Me)_2ONa$, THF, MeOH, NaH, 40°, 14 h	**I** (5) + 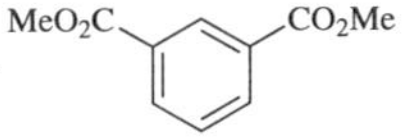 (86)	351
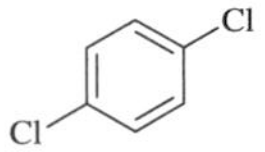	$Co(OAc)_2$, CO 2 atm	$h\nu$, NaOH, H_2O, 30°, 20 h	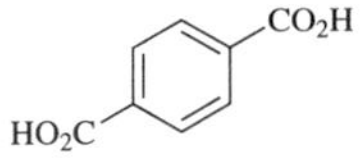 **I** (67) + 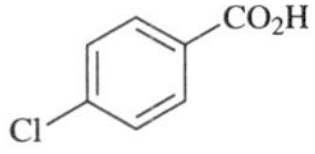 (6)	194
	$Co_2(CO)_8$, CO 2 atm	$h\nu$, NaOH, H_2O, 65°, 20 h	**I** (88)	191
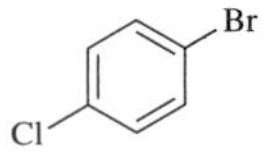	$Co_2(CO)_8$, CO 2 atm	$h\nu$, NaOH, H_2O, EtOH, 65°, 20 h	**I** (79)	191
	$Co(OAc)_2$, CO 1 atm	1. $EtC(Me)_2ONa$, THF, NaH, 60-65°, 9 h 2. EtOH, H_3O^+	**I** (20) + 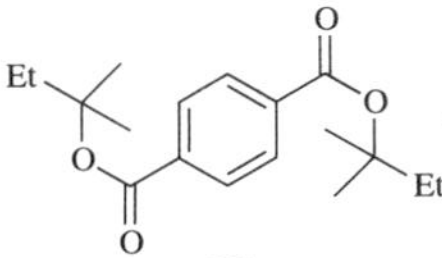 (20) + 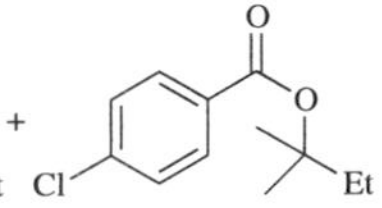 (55)	187

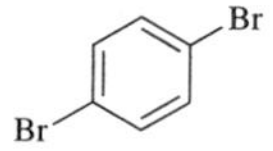	$Co_2(CO)_8$, CO 2 atm	$h\nu$, NaOH, H_2O, EtOH, 65°, 4 h	**I** (83)	191
	$Co(OAc)_2$, CO 1 atm	1. $EtC(Me)_2ONa$, THF, NaH, 60-65°, 9 h 2. EtOH, H_3O^+	**I** (60-65) + 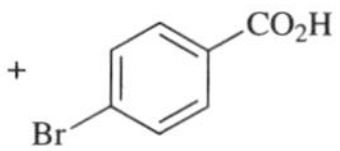 **II** (5-10) + $PhCO_2H$ (10-15)	186
	$Co(OAc)_2$, CO 2 atm	$h\nu$, NaOH, H_2O, EtOH, 30°, 4 h	**I** (97)	194
	$Co(OAc)_2$, CO 1 atm	Sunlamp, $EtC(Me)_2ONa$, THF, MeOH, NaH, 40°, 14 h	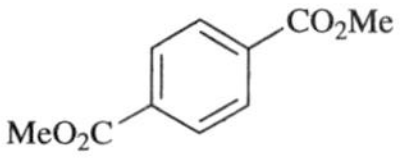 (81) + **I** (6)	351
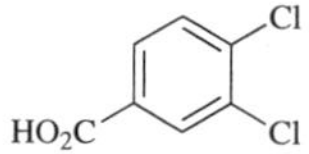	$Co_2(CO)_8$, CO 2 atm	$h\nu$, NaOH, H_2O, 65°, 6 h	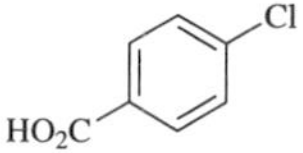 (15) + 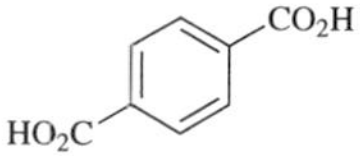 (14) + 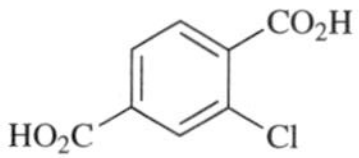 (57) + 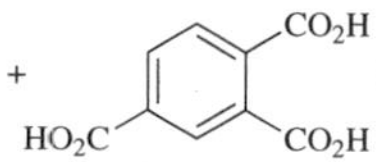 (11)	191
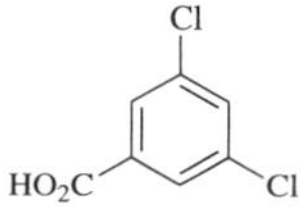		$h\nu$, NaOH, H_2O, 65°, 20 h	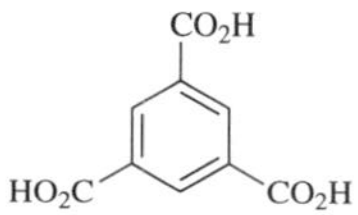 (89)	191

TABLE VII. OTHER C-C BOND FORMATION (*Continued*)

A. Carbonylation Reactions (*Continued*)

Substrate	Nucleophile	Conditions	Product(s) and Yield(s) (%)	Refs.
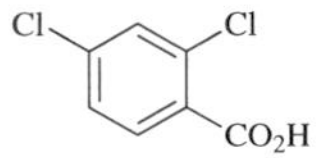	$Co_2(CO)_8$, CO 2 atm	*hν*, NaOH, H_2O, 65°, 6 h	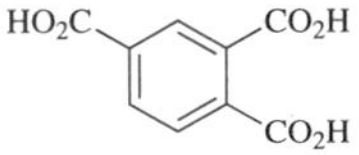 **I** (13) + 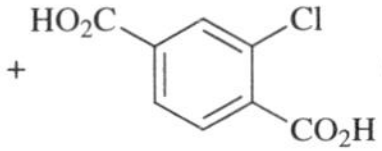 (40) + 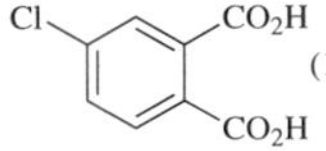 (14) + 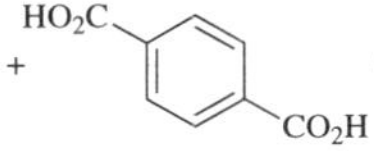 (10) + 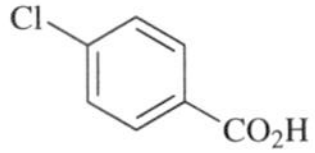 (24)	191
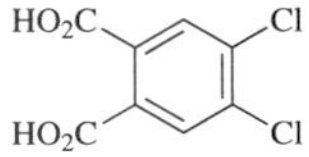		*hν*, NaOH, H_2O, 65°, 24 h	**I** (53) + 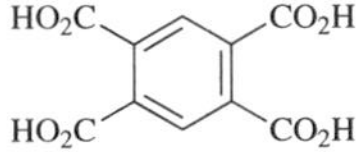 (30)	191
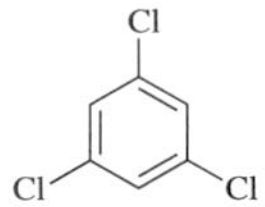	$Co(OAc)_2$, CO 2 atm	*hν*, NaOH, H_2O, EtOH, 65°, 18 h	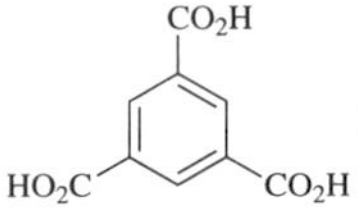 **I** (76) + 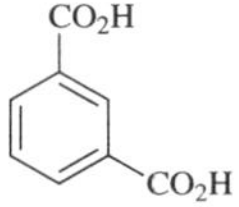 (21)	194
	$Co_2(CO)_8$, CO 2 atm	*hν*, NaOH, H_2O, EtOH, 65°, 20 h	**I** (86)	191
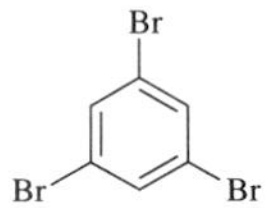		*hν*, NaOH, H_2O, EtOH, 65°, 4 h	**I** (93)	191
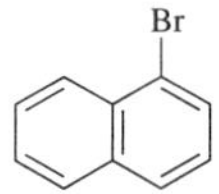	$Co_2(CO)_8$, CO 1 atm	*hν*, NaOH, H_2O, C_6H_6, Bu_4NBr, 65°, 5 h	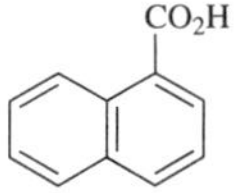 **I** (96)	188

		Sunlamp, NaOH, H_2O, EtOH, Bu_4NBr, 65°, 4.5 h	**I** (95)		189
	$Co(OAc)_2$, CO 2 atm	*hν*, NaOH, H_2O, EtOH, 65°, 4 h	**I** (99)		194
	$Co(OAc)_2$, CO 1 atm	Sunlamp, $EtC(Me)_2ONa$, THF, MeOH, NaH, 40°, 76 h	**I** (15) + 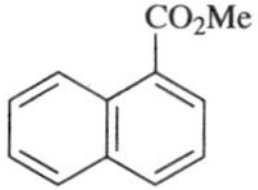 (63)		351
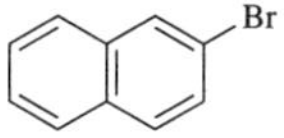	$Co_2(CO)_8$, CO 1 atm	*hν*, NaOH, H_2O, C_6H_6, Bu_4NBr, 65°, 1.5 h	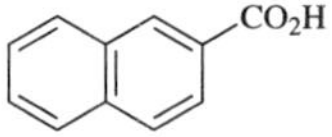 **I** (97)		189
	$Co(OAc)_2$, CO 1 atm	Sunlamp, $EtC(Me)_2ONa$, THF, MeOH, NaH, 40°, 43 h	**I** (6) + 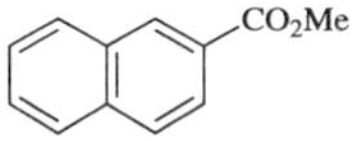 (76)		351
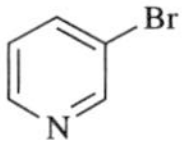		Sunlamp, $EtC(Me)_2ONa$, THF, MeOH, NaH, 40°, 61 h	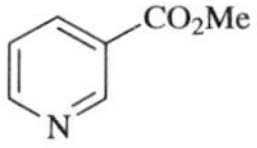 (70)		351
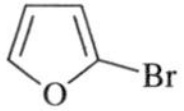		Sunlamp, $EtC(Me)_2ONa$, THF, MeOH, NaH, 40°, 42 h	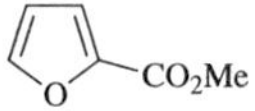 (72) +	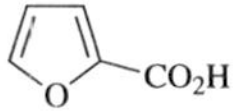 (2)	351
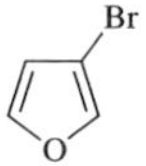		Sunlamp, $EtC(Me)_2ONa$, THF, MeOH, NaH, 40°, 36 h	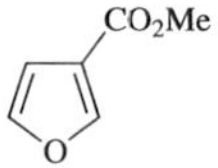 (74) +	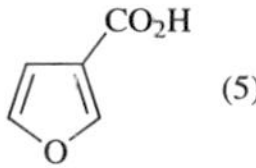 (5)	351
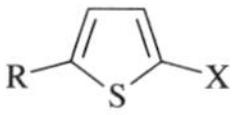		Sunlamp, $EtC(Me)_2ONa$, THF, MeOH, NaH, 40°	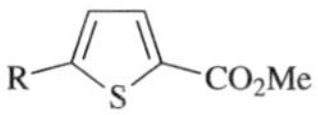 +	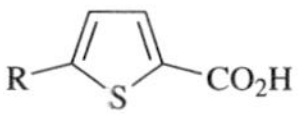	
R X					
H Cl		15 h	(81)	(7)	351
H Br		13 h	(83)	(2)	351
Me Br		19 h	(87)	(7)	351
H Br	$Co(OAc)_2$, CO 2 atm	*hν*, NaOH, H_2O, 30°, 20 h	(—)	(97)	194

TABLE VII. OTHER C-C BOND FORMATION (*Continued*)

Substrate	Nucleophile	Conditions	Product(s) and Yield(s) (%)	Refs.
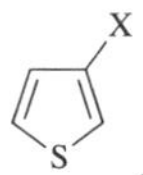	$Co(OAc)_2$, CO 1 atm	Sunlamp, $EtC(Me)_2ONa$, THF, MeOH, NaH, 40°	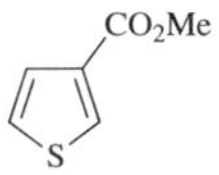 + 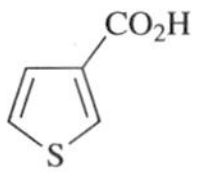	351
X				
Cl		17 h	(78) (9)	
Br		14 h	(80) (13)	
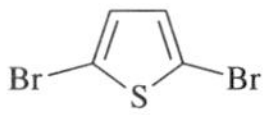		Sunlamp, $EtC(Me)_2ONa$, THF, MeOH, NaH, 40°, 54 h	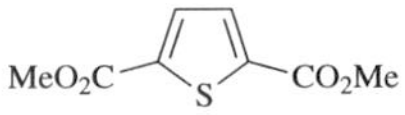 (55) + 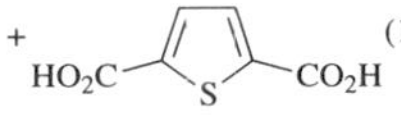 (17)	351
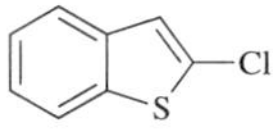		Sunlamp, $EtC(Me)_2ONa$, THF, MeOH, NaH, 40°, 48 h	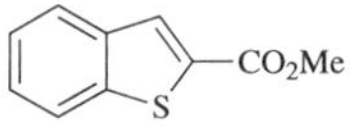 (71) + 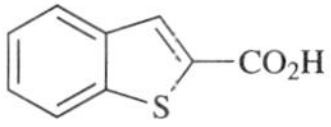 (10)	351
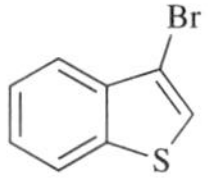		Sunlamp, $EtC(Me)_2ONa$, THF, MeOH, NaH, 40°, 48 h	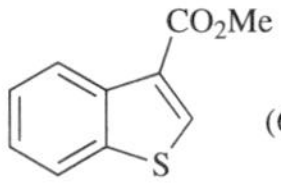 (66) + 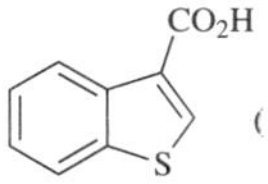 (13)	351

[a] $H_2O:C_6H_6 = 1$; substrate:$Fe(CO)_5$ = 1:0.09.

[b] $H_2O:C_6H_6 = 7$; substrate:$Fe(CO)_5$ = 1:0.29.

TABLE VII. OTHER C-C BOND FORMATION (*Continued*)

B. Reaction with Phenoxide and Related Ions

Substrate	Nucleophile	Conditions	Product(s) and Yield(s) (%)	Refs.
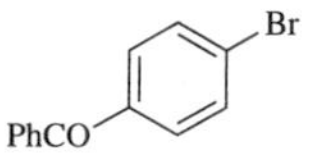	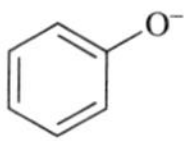	e^-, NH_3	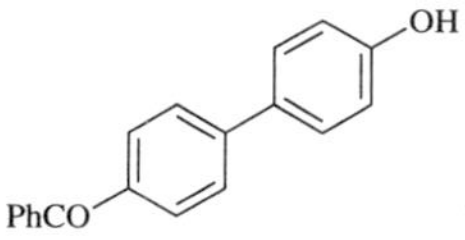 **I** (20) + 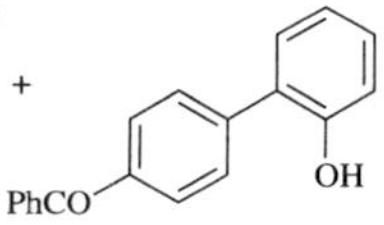 **II** (40) + 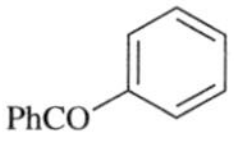 **III** (17)	205, 352
		e^-, DMSO	**I** (16) + **II** (31) + **III** (33)	205, 352
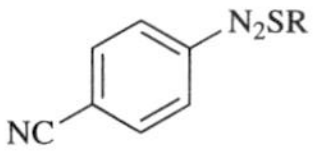		R = Bu-*t*, *hν*, DMSO, 0.7 h	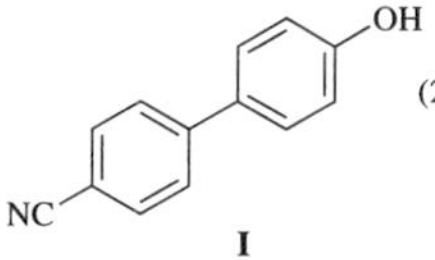 **I** (20) + 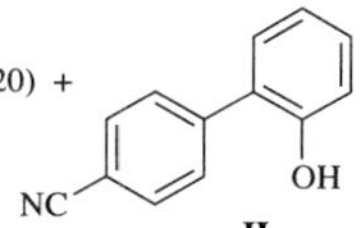 **II** (42)	217
		R = Ph, DMSO, 2 h	**I** (14) + **II** (46)	216, 217
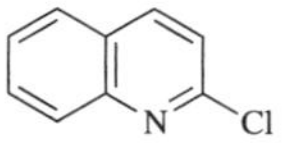		e^-, DMSO	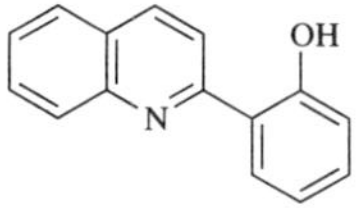 (27) + 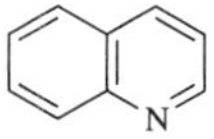 (68)	205
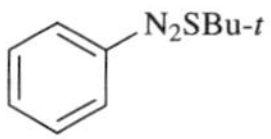	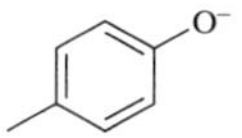	*hν*, DMSO, 1.25 h	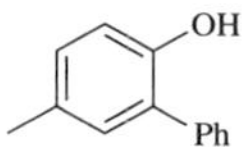 (23)	217
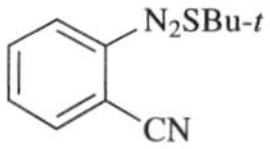		1. *hν*, DMSO, 2 h 2. MeI	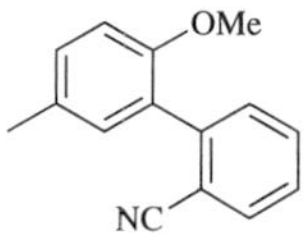 (50)	294

TABLE VII. OTHER C-C BOND FORMATION (*Continued*)

B. Reaction with Phenoxide and Related Ions (*Continued*)

Substrate	Nucleophile	Conditions	Product(s) and Yield(s) (%)	Refs.
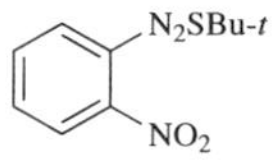 N_2SBu-*t* NO_2	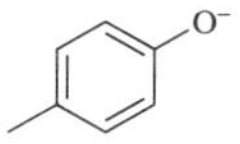 O^-	*hν*, DMSO, 0.8 h	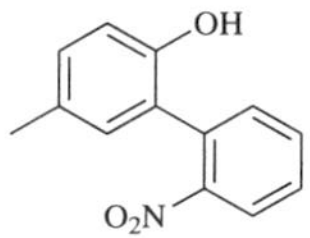 OH O_2N (52)	217
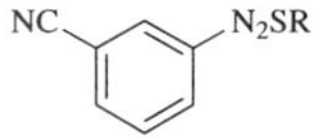 NC N_2SR		R = Ph, *hν*, DMSO, 0.75 h	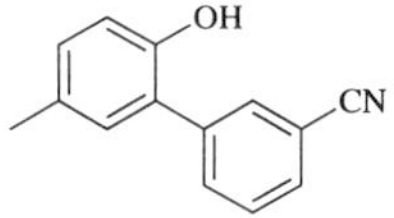 OH CN **I** (62) + 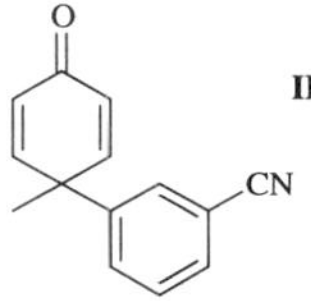 O CN **II** (11)	216, 217
		R = Bu-*t*, *hν*, DMSO, 2 h	**I** (61) + **II** (10)	217
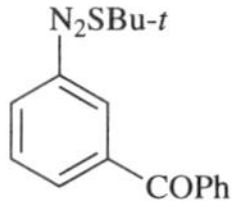 N_2SBu-*t* COPh		*hν*, DMSO, 3.5 h	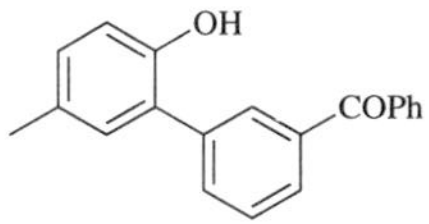 OH COPh (53)	217
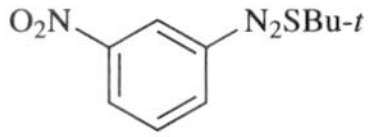 O_2N N_2SBu-*t*		*hν*, DMSO, 0.8 h	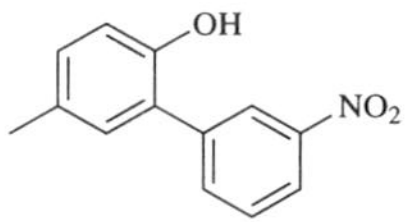 OH NO_2 (70) + 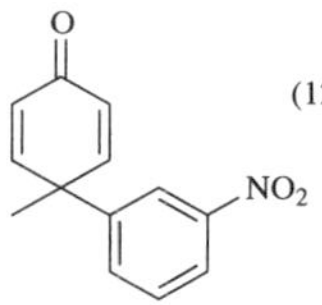 O NO_2 (12)	217
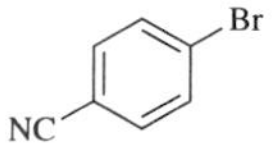 Br NC		*hν*, NH_3, 2 h	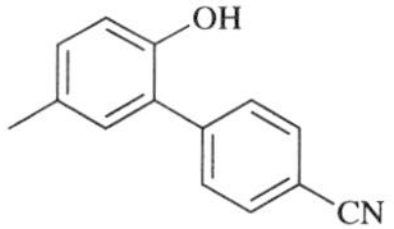 OH CN **I** (20)	105

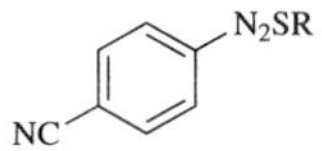	R = Bu-*t*, *h*ν, DMSO, 0.7 h	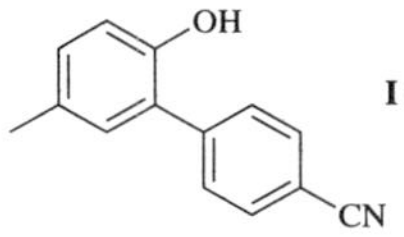 **I** (69) + 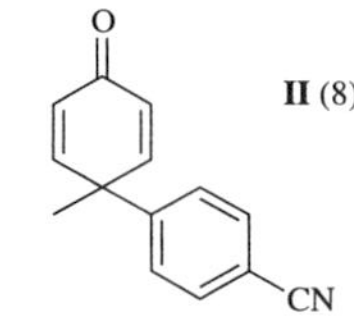 **II** (8)	217
	R = Ph, *h*ν, DMSO, 1.5 h	**I** (69) + **II** (4)	217, 216
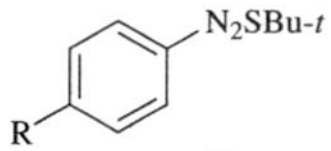	*h*ν, DMSO	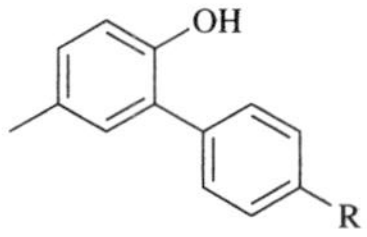	217
R			
COPh	2.5 h	(54)	
OMe	24 h	(21)	
Br	3 h	(33)	
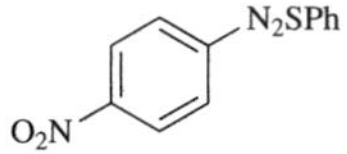	DMSO, 0.5 h	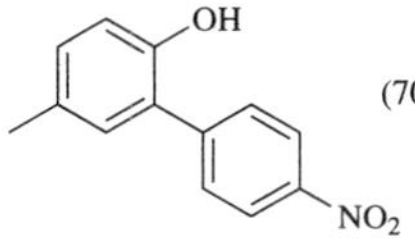 (70) + 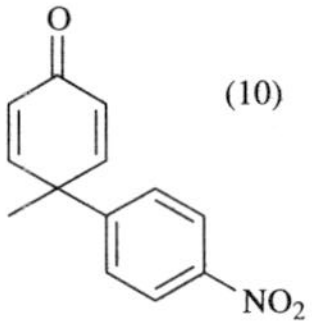 (10)	217
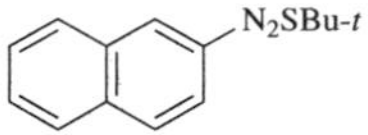	*h*ν, DMSO, 2 h	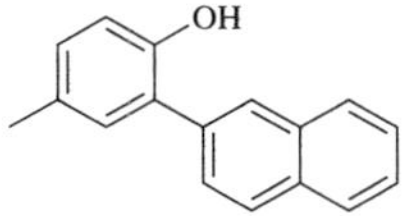 (28)	217
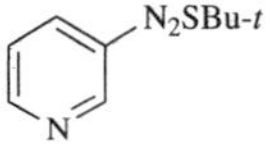	*h*ν, DMSO, 1.25 h	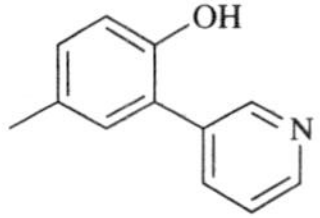 (58)	217

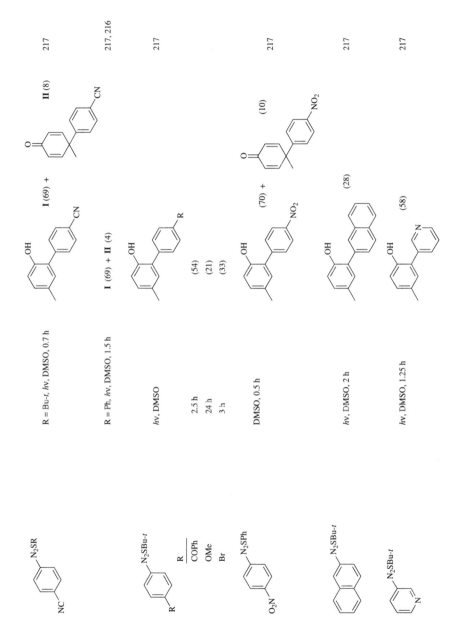

TABLE VII. OTHER C-C BOND FORMATION (*Continued*)

B. Reaction with Phenoxide and Related Ions (*Continued*)

Substrate	Nucleophile	Conditions	Product(s) and Yield(s) (%)	Refs.
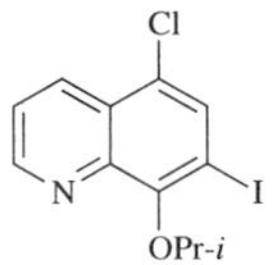	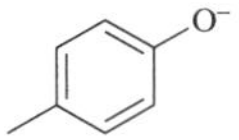	$h\nu$, NH_3, 2 h	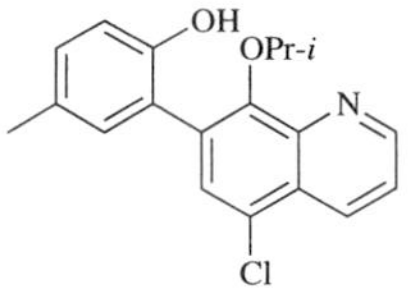 (25) + 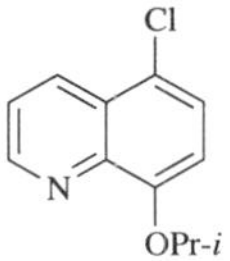 (35)	182
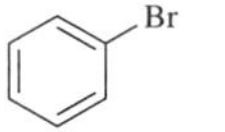	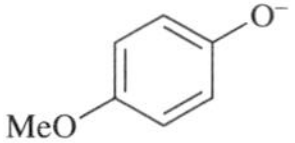	$h\nu$, NH_3, 1 h	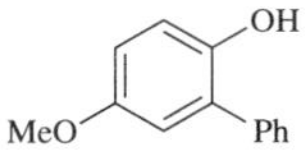 **I** (40) + 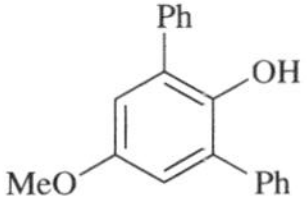 (12)	105
PhN_2SBu-*t*		$h\nu$, DMSO, 1.5 h	**I** (49)	217
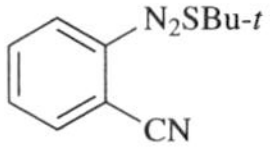		1. $h\nu$, DMSO, 2 h 2. MeI	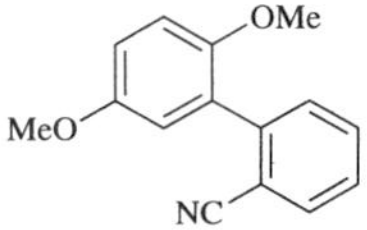 (62)	294
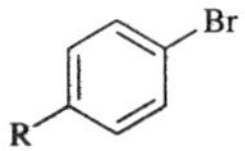		$h\nu$, NH_3	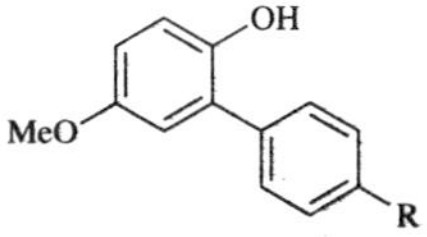	
R				
F		1-2 h	(15)	209
CN		2 h	(65)	105
CF_3		1-2 h	(20)	209

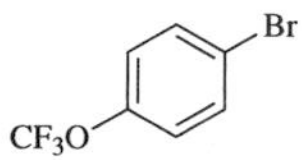	$h\nu$, NH_3. 1-2 h	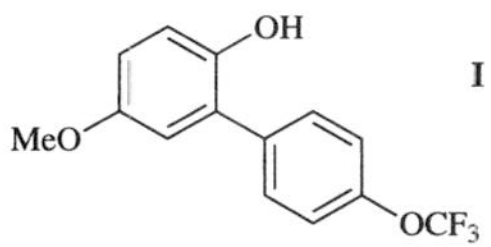 **I** + 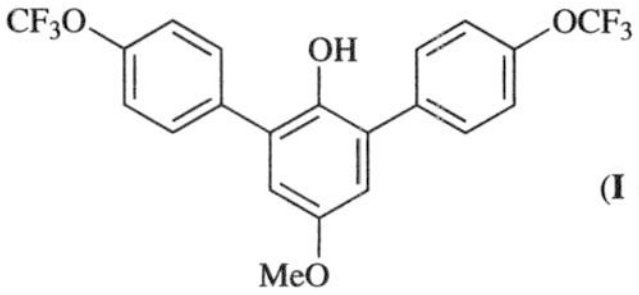 **II** (**I** + **II**) (45)	209
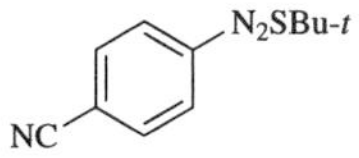	$h\nu$, DMSO, 2 h	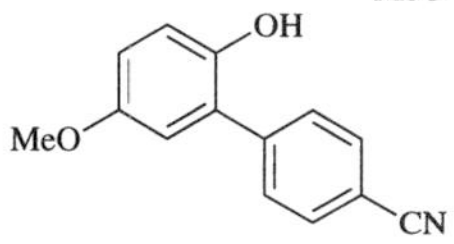 (76)	217
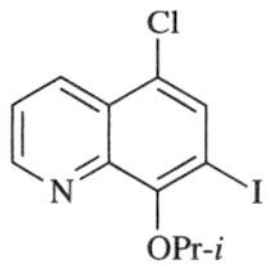	$h\nu$, NH_3, 5 h	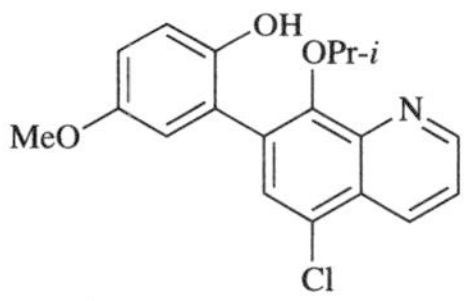 (60) + 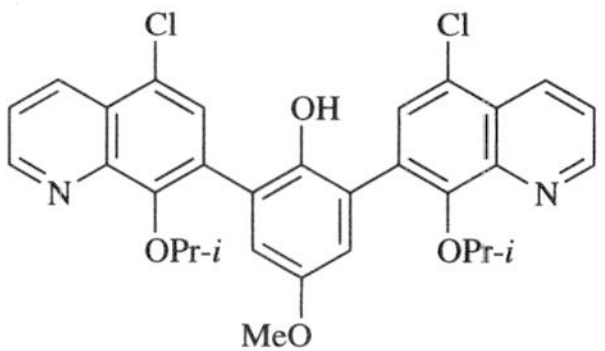 (10)	182
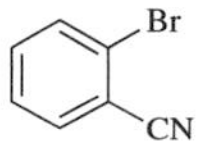 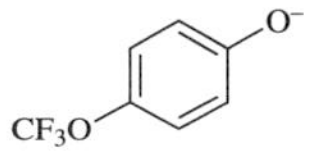	$h\nu$, NH_3. 1-2 h	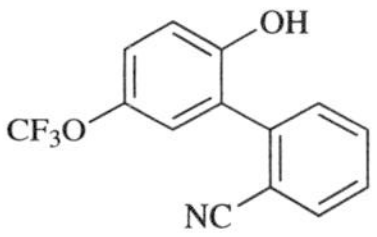 (86)	209

TABLE VII. OTHER C-C BOND FORMATION (*Continued*)

B. Reaction with Phenoxide and Related Ions (*Continued*)

Substrate	Nucleophile	Conditions	Product(s) and Yield(s) (%)	Refs.
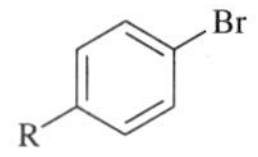	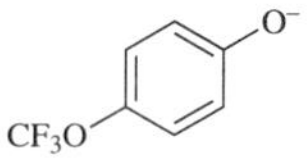	$h\nu$, NH_3. 1-2 h	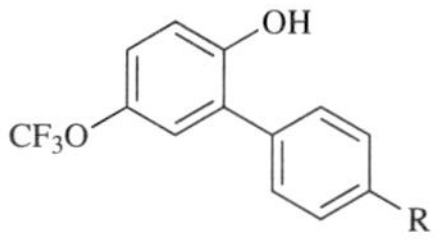 R: F (47); OCF_3 (47)	209
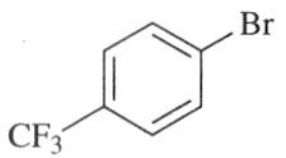		$h\nu$, NH_3. 1-2 h	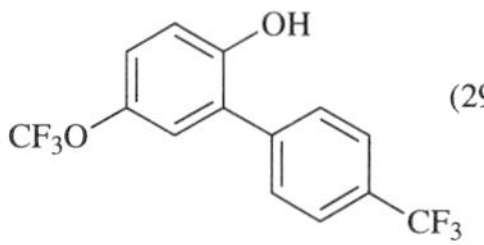 (29) + 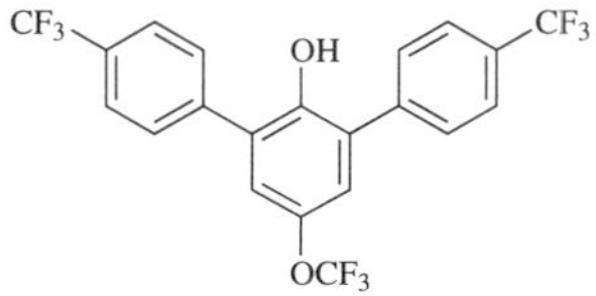 (37)	209
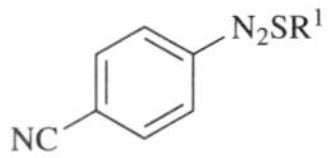	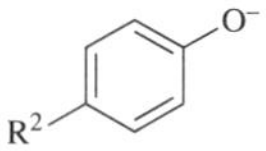	$h\nu$, DMSO	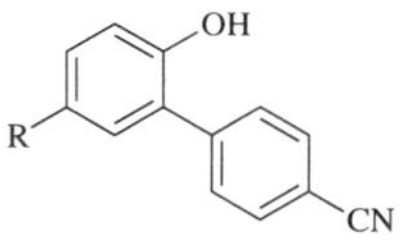	217
R^1	R^2			
Bu-*t*	CF_3	0.75 h	(66)	
Bu-*t*	Br	1 h	(56)	
Bu-*t*	NO_2	80 h	(35)	
Ph	NO_2	Sunlamp, 24 h	(51)	
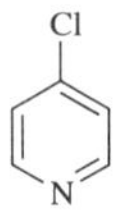	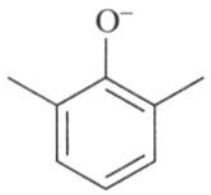	e^-, NH_3, THF, mediator	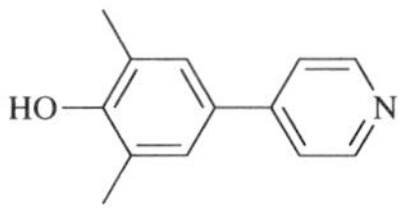 (6)	353

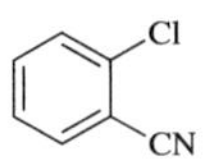	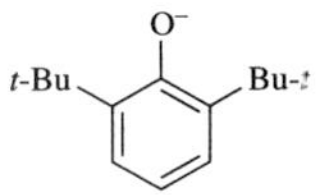	e$^-$, NH_3, mediator	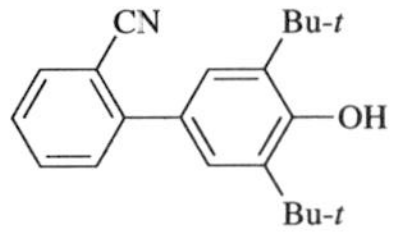 (89)	204, 84
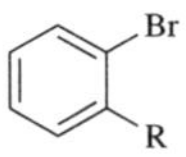		$h\nu$, NH_3. 1 h	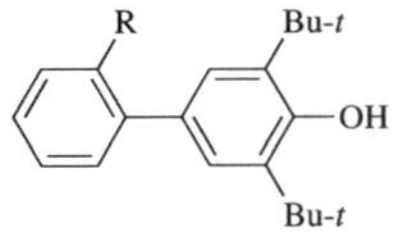	105
R: CN			(88)	
R: $CONH_2$			(85)	
R: COMe			(60)	
R: OMe			(78)	
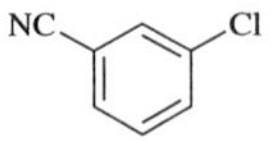		e$^-$, NH_3, mediator	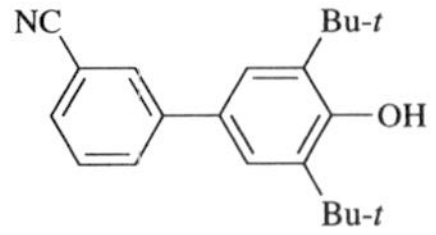 (68-80)	204, 84
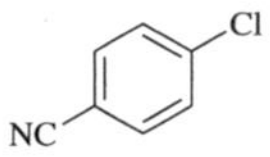		e$^-$, NH_3, mediator	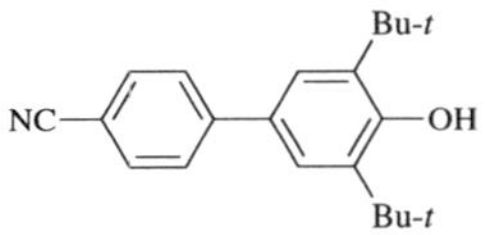 **I** (78) + 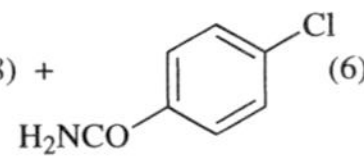 (6)	205, 84, 341
		e$^-$, DMF	**I** (60-68)	341
		$h\nu$, NH_3	**I** (70)	203
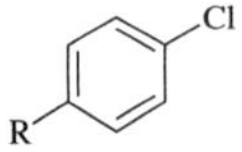		e$^-$, NH_3, mediator		
R: CO_2Me			(50)	206
R: COMe			(50)	206
R: NMe_3I			(85)	206
R: Cl			(67)	210, 85

TABLE VII. OTHER C-C BOND FORMATION (*Continued*)

B. Reaction with Phenoxide and Related Ions (*Continued*)

Substrate	Nucleophile	Conditions	Product(s) and Yield(s) (%)	Refs.
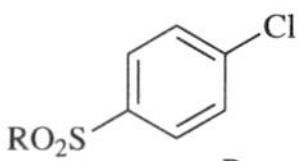Cl RO_2S R: Me, Ph, p-ClC_6H_4	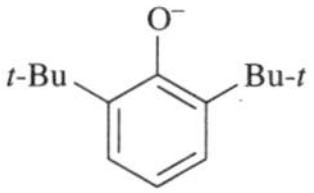O^- *t*-Bu Bu-*t*	e^-, NH_3, mediator	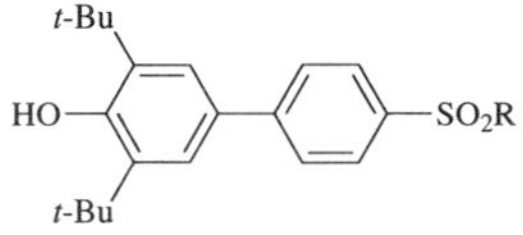*t*-Bu HO SO_2R *t*-Bu	208
Me			(90)	
Ph			(90)	
p-ClC_6H_4			(90)	
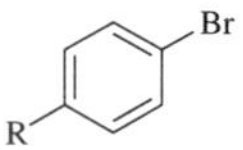Br R			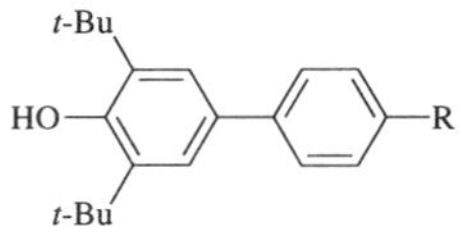*t*-Bu HO R *t*-Bu **I** + 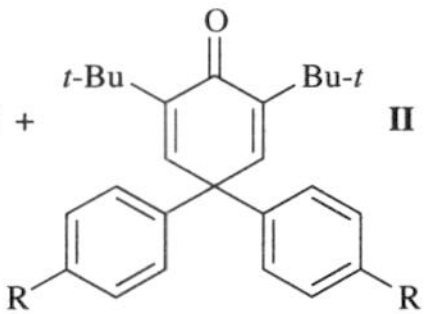O *t*-Bu Bu-*t* R R **II**	
R			**I** / **II**	
F		$h\nu$, NH_3. 1-2 h	(30) (—)	209
CN		$h\nu$, NH_3. 1 h	(96) (—)	105
SMe		e^-, NH_3, mediator	(40) (—)	206
CF_3		$h\nu$, NH_3. 1-2 h	(38) (16)	209
OCF_3		$h\nu$, NH_3. 1-2 h	(—) (10)	209
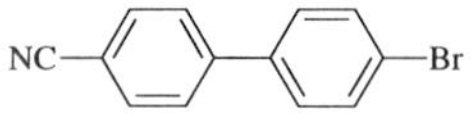NC Br		e^-, NH_3, THF, mediator	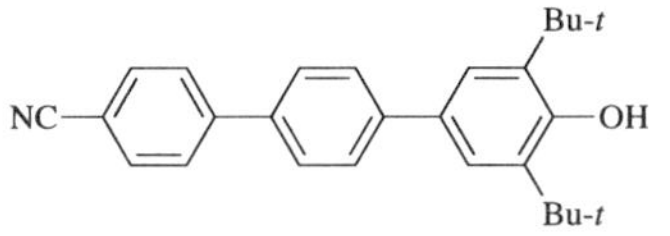NC Bu-*t* OH Bu-*t* (25)	207
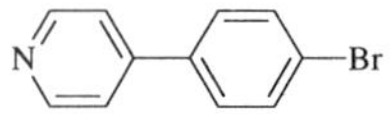N Br		e^-, NH_3, THF, mediator	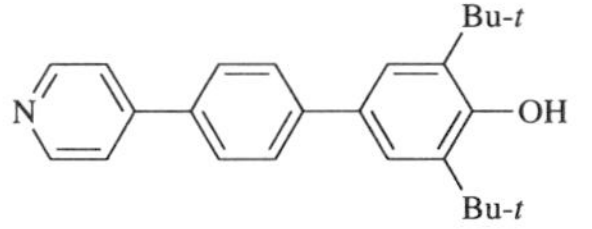N Bu-*t* OH Bu-*t* (35)	207

	e^-, NH_3, THF, mediator	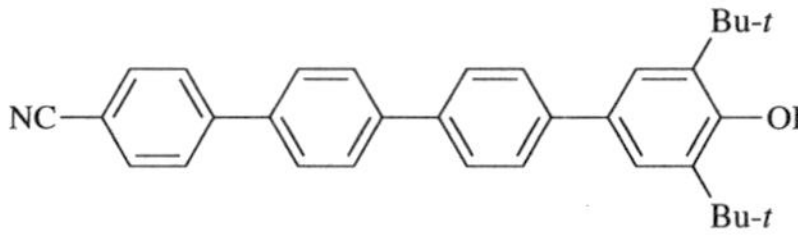 (<10)	207
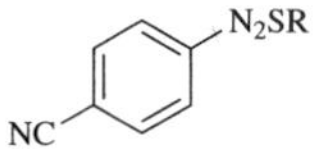	R = Bu-*t*, *h*ν, DMSO, 5 h	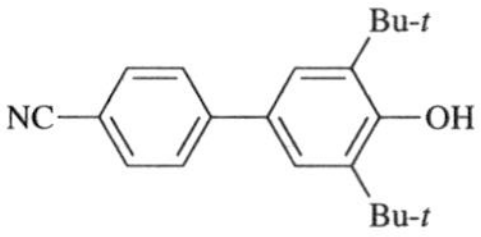 **I** (64)	217
	R = Ph, DMSO, 6 h	**I** (58)	216, 217
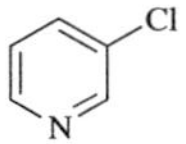	e^-, NH_3, mediator	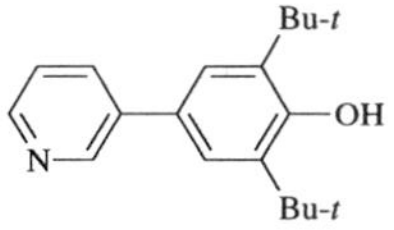 (84)	84
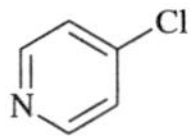	e^-, NH_3, mediator	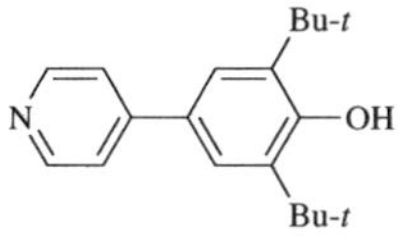 **I** (70)	354, 84
	e^-, NH_3, THF, mediator	**I** (70)	355
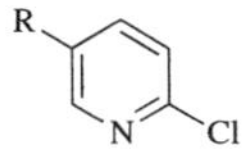		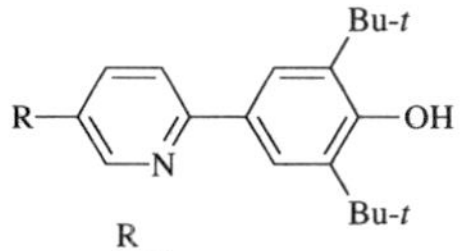	
R		R	
Cl	e^-, NH_3, mediator	H (16)	206
CN	e^-, NH_3, mediator	CN (77)	84
CN	*h*ν, NH_3	CN (70)	203

TABLE VII. OTHER C-C BOND FORMATION (*Continued*)

B. Reaction with Phenoxide and Related Ions (*Continued*)

Substrate	Nucleophile	Conditions	Product(s) and Yield(s) (%)	Refs.
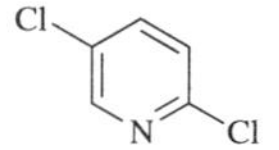	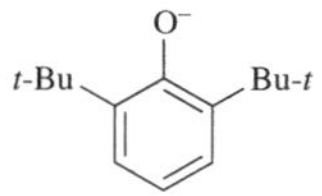	e⁻, NH_3, mediator	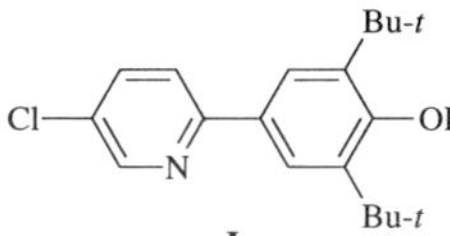 **I** + 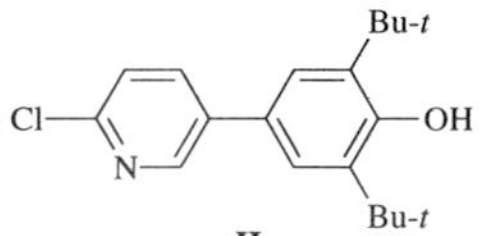 **II** **I** + **II** (35), **I**:**II** = 85:15	85
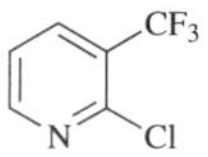		*h*ν, NH_3, 1.3 h	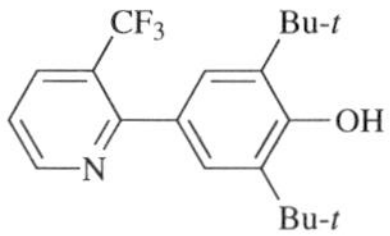 (66)	181
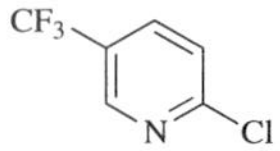		*h*ν, NH_3, 0.75 h	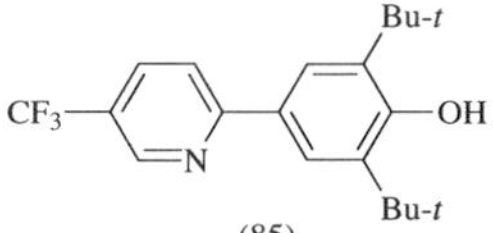 (85) **I** + 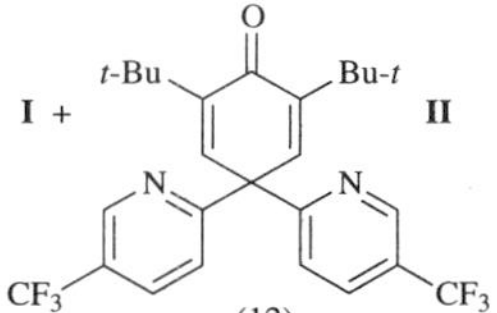 (12) **II**	181
		e⁻, NH_3, mediator	**I** (50)	202
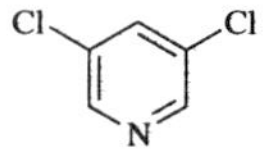		e⁻, NH_3, mediator	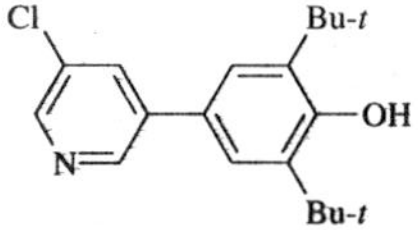 (74)	85
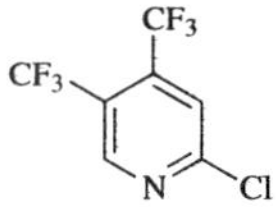		e⁻, NH_3, mediator	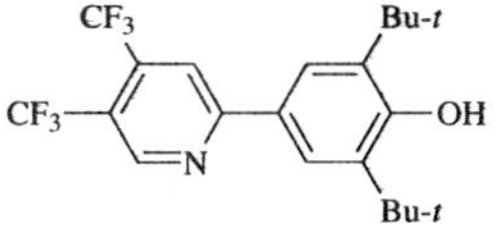 (50)	202

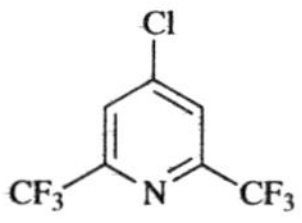		e⁻, NH_3, DMF, mediator	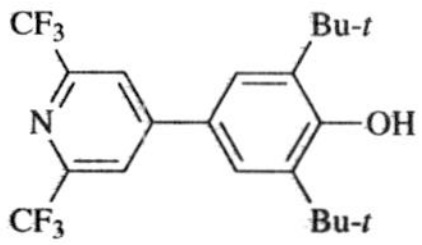	(40)	202
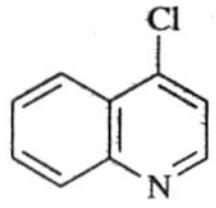		e⁻, NH_3, mediator	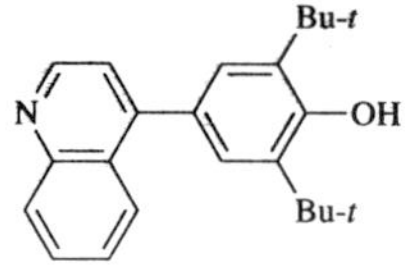	(76)	206
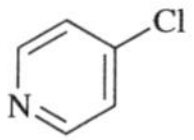	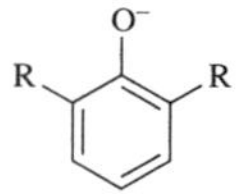	e⁻, NH_3, THF, mediator	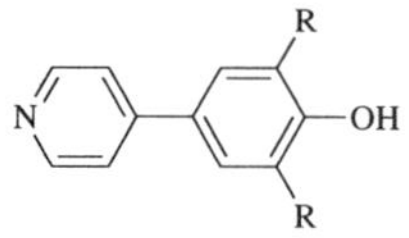	R: Pr-*i* (30); *n*-C_5H_{11} (21)	353
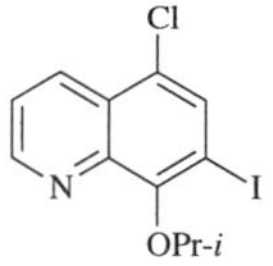	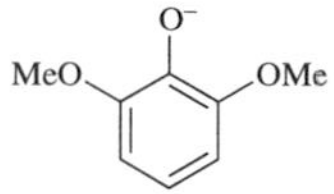	*h*ν, NH_3	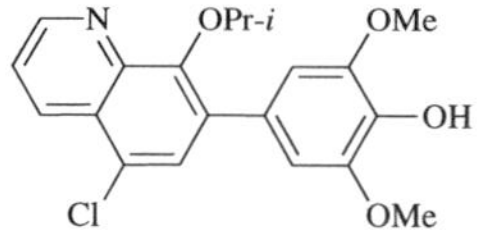	(18)	182
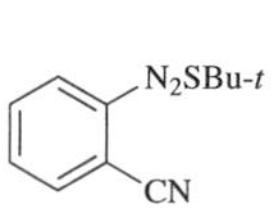	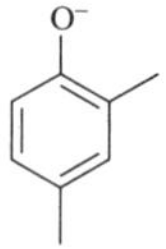	1. *h*ν, DMSO, 2 h 2. MeI	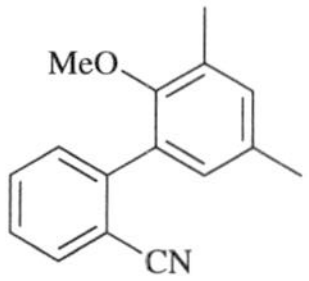	(21)	294
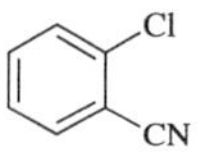	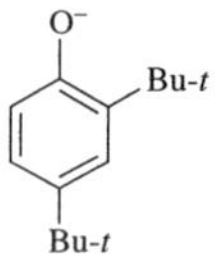	e⁻, NH_3, mediator	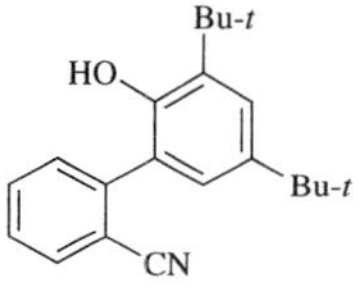	(20)[a]	204

TABLE VII. OTHER C-C BOND FORMATION (*Continued*)

B. Reaction with Phenoxide and Related Ions (*Continued*)

Substrate	Nucleophile	Conditions	Product(s) and Yield(s) (%)	Refs.
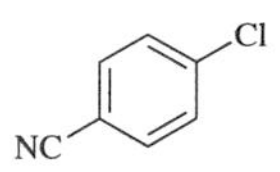	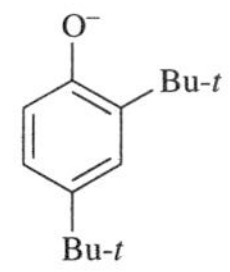	e^-, NH_3, mediator	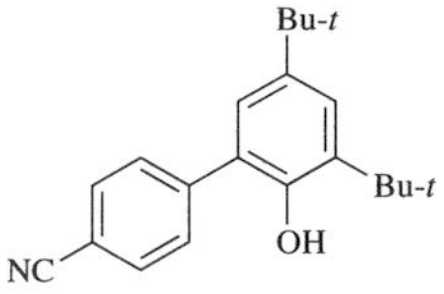 (47)	204
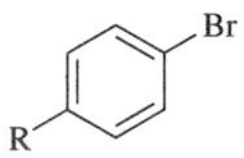		$h\nu$, NH_3, 1-2 h	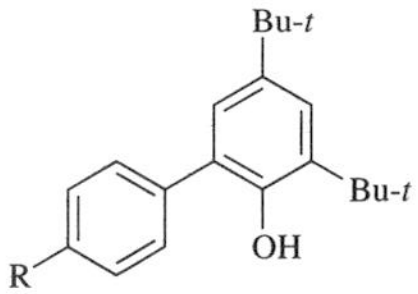	209
R: F			(48)	
R: CF_3			(27)	
R: CF_3O			(63)	
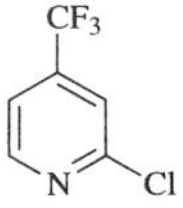		$h\nu$, NH_3, 4 h	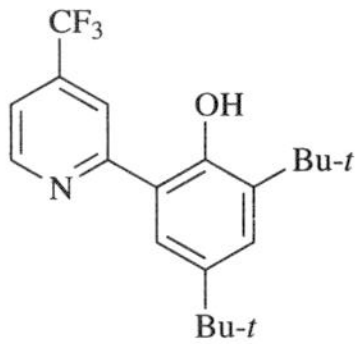 (10)	181
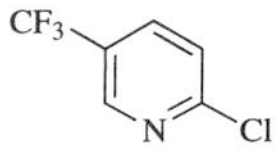		$h\nu$, NH_3, 1 h	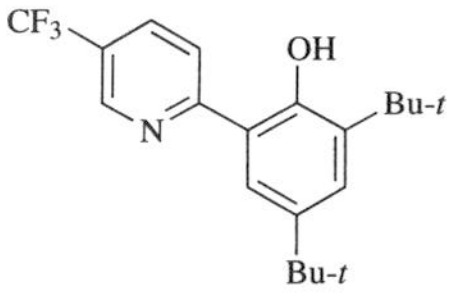 (98)	181

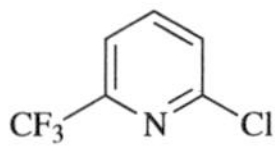	$h\nu$, NH_3, 3 h	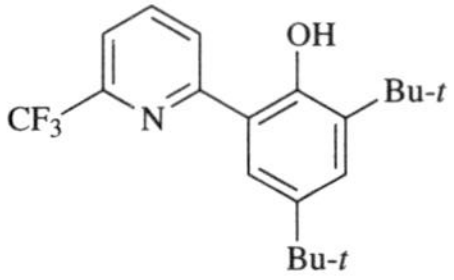 (60)	181
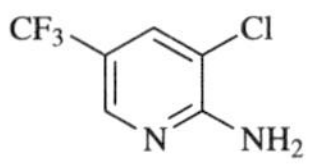	$h\nu$, NH_3, 2.5 h	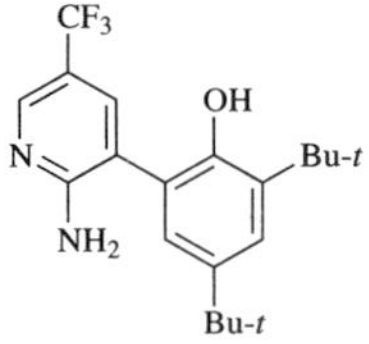 (98)	181
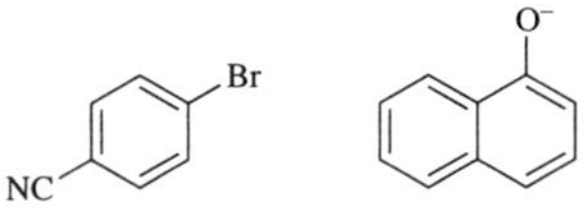	$h\nu$, NH_3, 2 h	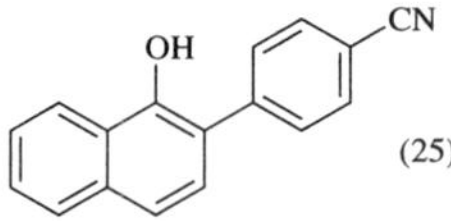 (25) + 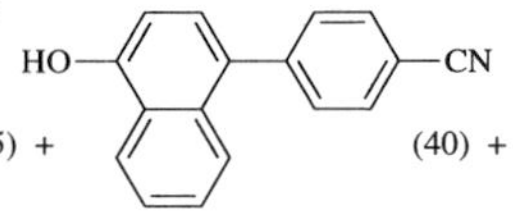 (40) + 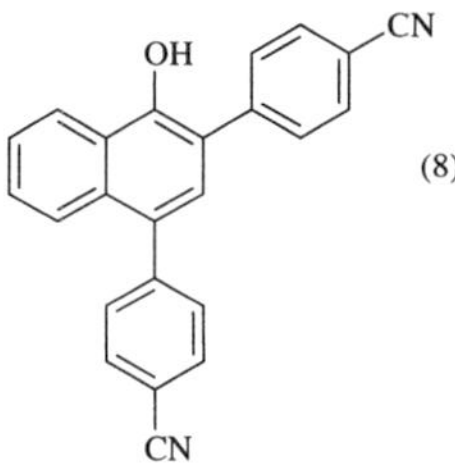 (8)	105

TABLE VII. OTHER C-C BOND FORMATION (*Continued*)

B. Reaction with Phenoxide and Related Ions (*Continued*)

Substrate	Nucleophile	Conditions	Product(s) and Yield(s) (%)	Refs.
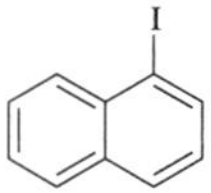I	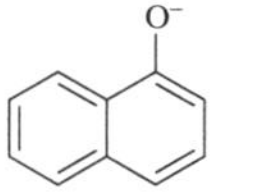O⁻	1. *h*ν, NH_3, 3 h 2. *i*-PrBr	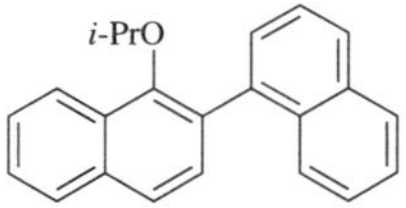*i*-PrO (22) + 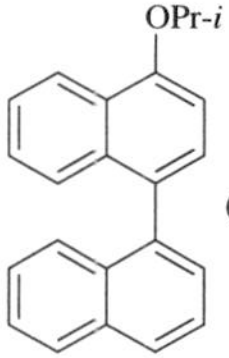OPr-*i* (37) + 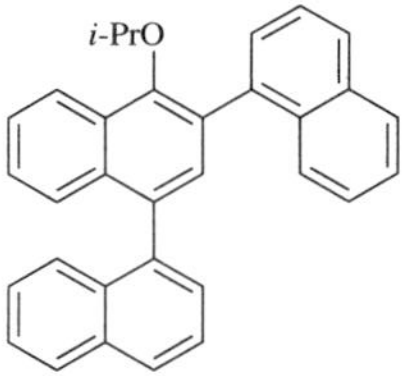*i*-PrO (20)	211
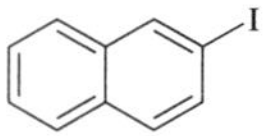I		1. *h*ν, NH_3, 2 h 2. *i*-PrBr	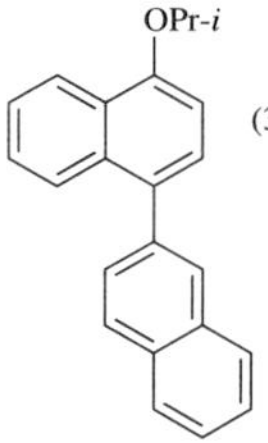OPr-*i* (36) + 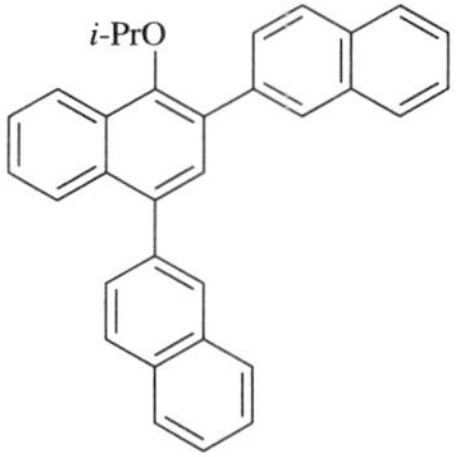*i*-PrO (20)	211
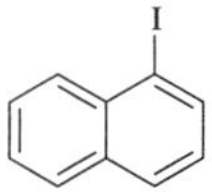I	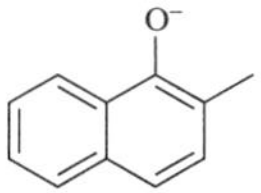O⁻	1. *h*ν, NH_3, 1 h 2. *i*-PrBr	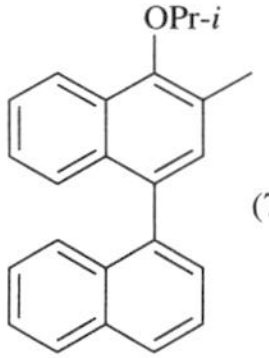OPr-*i* (70)	211

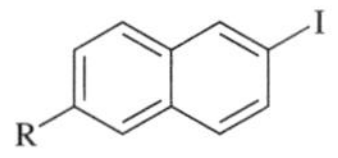	1. $h\nu$, NH_3, X h 2. *i*-PrBr	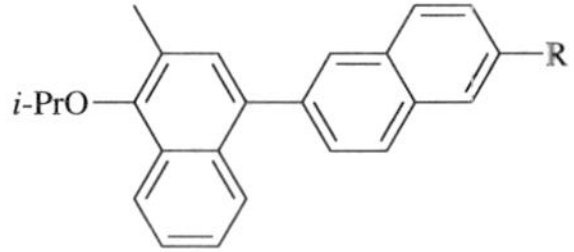 R, X: H, 5 (50); OPr-*i*, 2 (60)	211
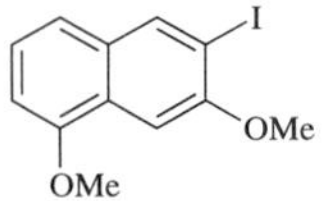	1. $h\nu$, NH_3, 1.5 h 2. *i*-PrBr	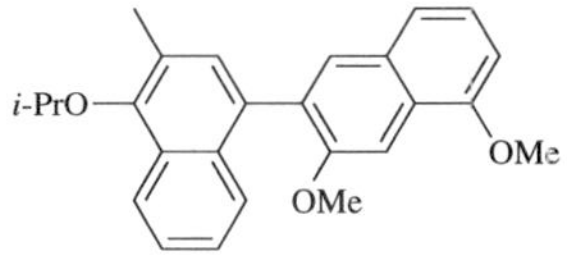 (50)	211
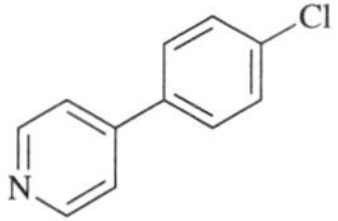 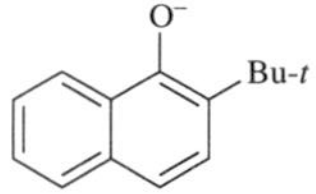	e^-, NH_3, mediator	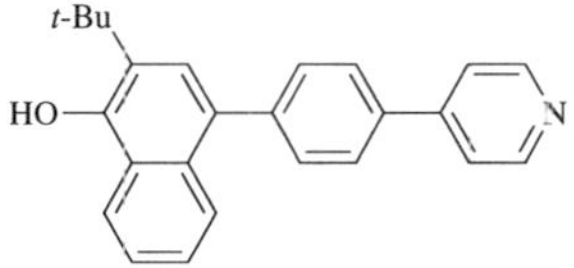 (57)	213
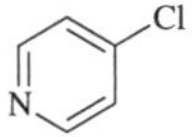	e^-, NH_3, mediator	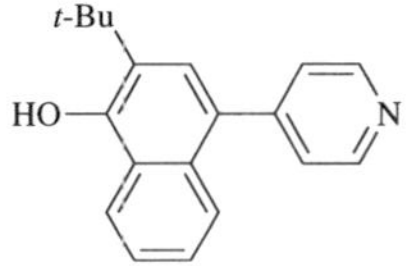 (80)	213
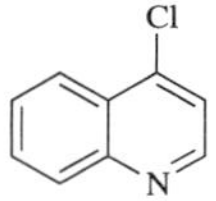	e^-, NH_3, mediator	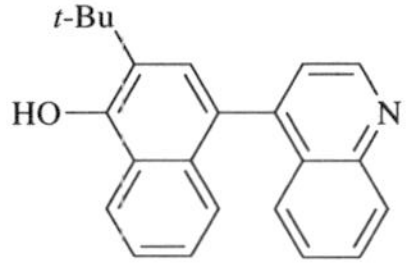 (85)	213
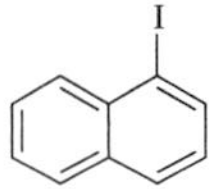 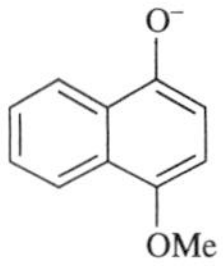	1. $h\nu$, NH_3, 3 h 2. *i*-PrBr	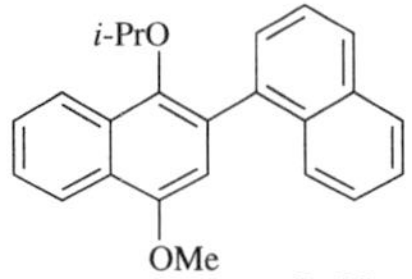 **I** (35) + 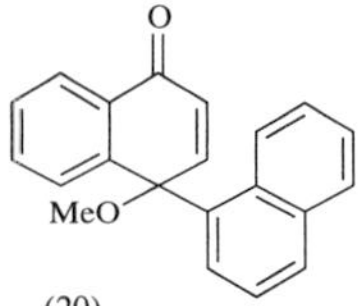 (20)	211

TABLE VII. OTHER C-C BOND FORMATION (*Continued*)

B. Reaction with Phenoxide and Related Ions (*Continued*)

Substrate	Nucleophile	Conditions	Product(s) and Yield(s) (%)	Refs.
	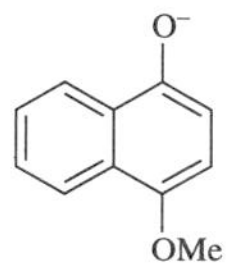	1. *h*ν, DMSO, 1 h 2. *i*-PrBr	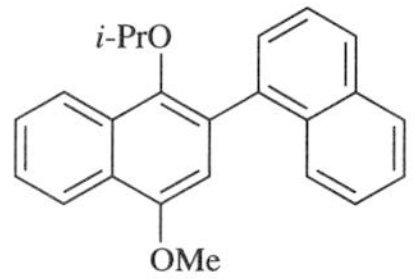 **I** (32)	211
		1. *h*ν, MeCN, 1 h 2. *i*-PrBr	**I** (47)	211
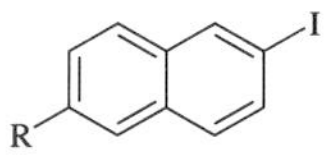			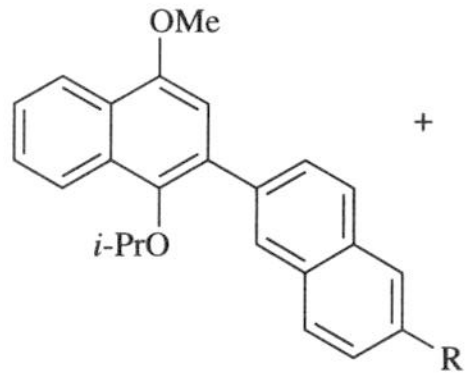 + 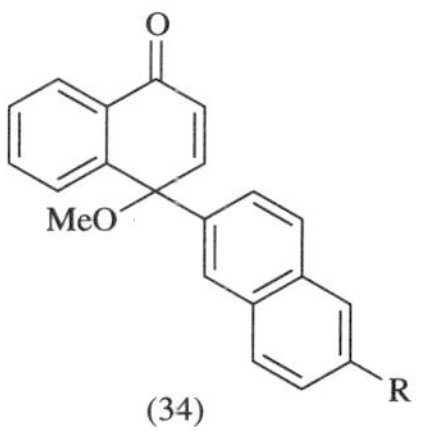	211
R / H		1. *h*ν, NH_3, 3 h 2. *i*-PrBr	(36) (34)	
OPr-*i*		1. *h*ν, MeCN, 2 h 2. *i*-PrBr	(65) (—)	
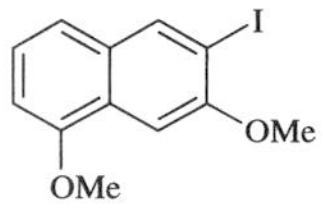		1. *h*ν, MeCN, 2 h 2. *i*-PrBr	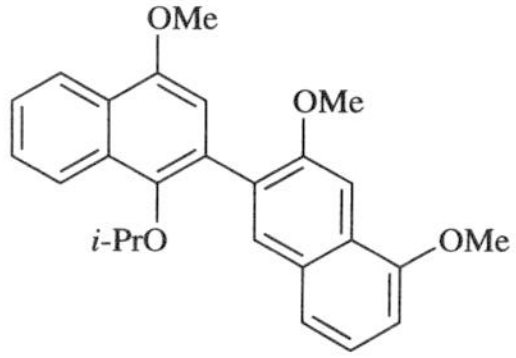 (55)	211

Substrate	Nucleophile	Conditions	Products (yield)		Refs.
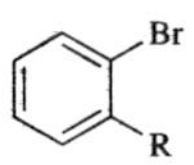	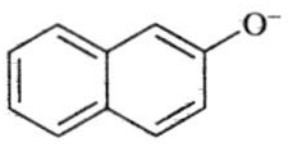	$h\nu$, NH_3	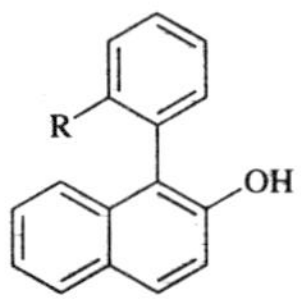 I +	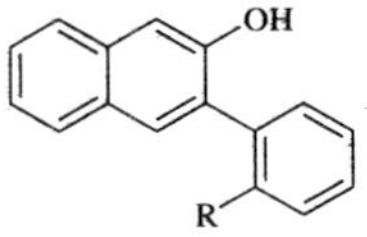 II	
R					
CN		1 h	(76)	(5)	214
NH_2		3 h	(12)	(—)	64
NH_2		KI, 3 h	(40)	(—)	64
OMe		3 h	(9)	(—)	64
OMe		KI, 3 h	(25)	(—)	64
$CONH_2$		3 h	(9)	(—)	64
$CONH_2$		KI, 3 h	(25)	(—)	64
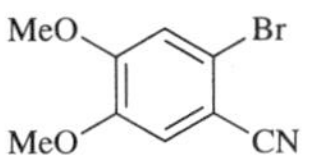		$h\nu$, NH_3, 1 h	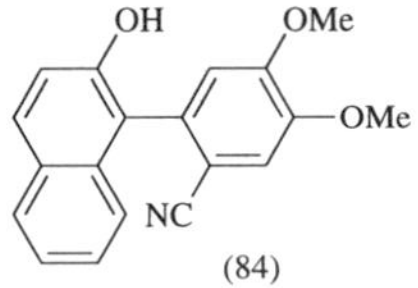 (84) +	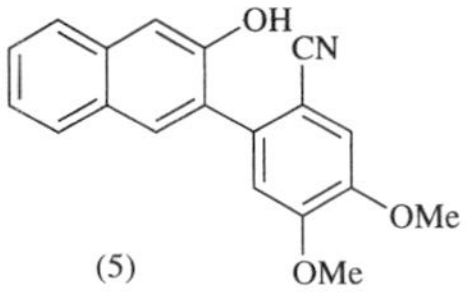 (5)	214
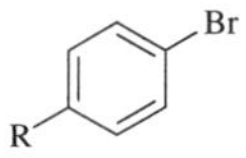		$h\nu$, NH_3, 1-2 h	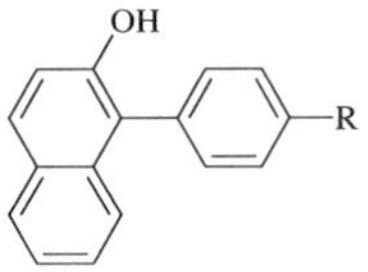		
R					
F			(31)		209
CF_3			(55)		209
OCF_3			(40)		209
CN			(85)		105
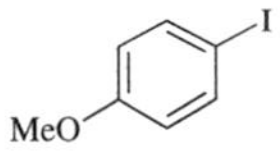		$h\nu$, NH_3, 3 h	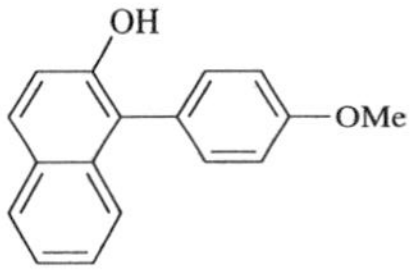	(42) + PhOMe (13)	212

TABLE VII. OTHER C-C BOND FORMATION (*Continued*)

B. Reaction with Phenoxide and Related Ions (*Continued*)

Substrate	Nucleophile	Conditions	Product(s) and Yield(s) (%)	Refs.
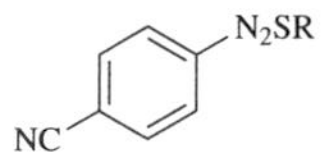	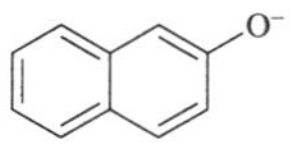	R = Bu-*t*, *h*ν, DMSO, 0.6 h	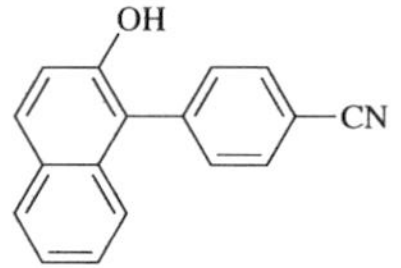 **I** (65)	217
		R = Ph, *h*ν, DMSO, 2 h	**I** (68)	217, 216
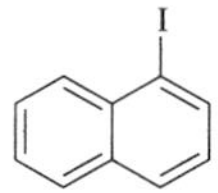		*h*ν, NH_3, 3 h	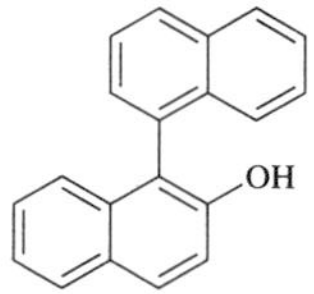 (53) + 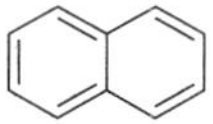 (20)	212, 211
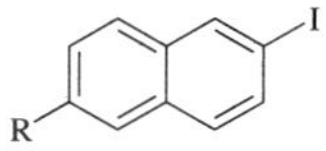		1. *h*ν, NH_3, X h 2. *i*-PrBr	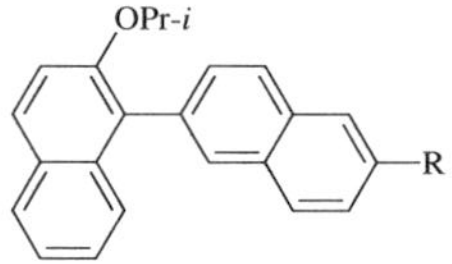 R: H, X: 3.5 (44); R: OPr-*i*, X: 1 (65)	211
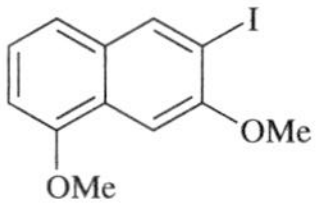		1. *h*ν, NH_3, 1 h 2. *i*-PrBr	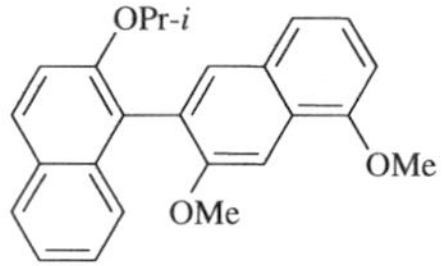 (55)	211
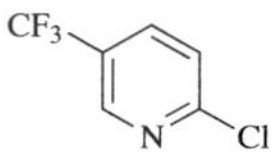		*h*ν, NH_3, 3.2 h	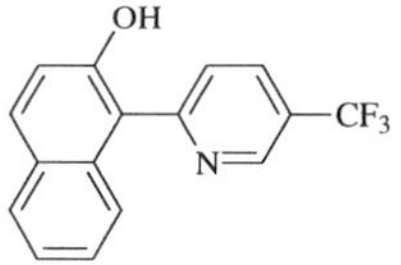 (95)	181

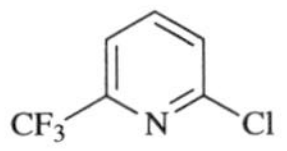	$h\nu$, NH_3, 4 h	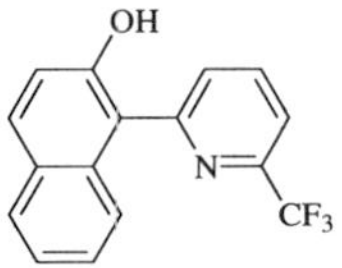 (60)	181
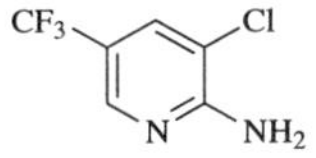	$h\nu$, NH_3, 4 h	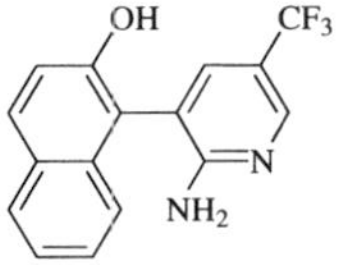 (90)	181
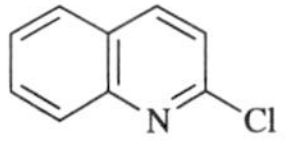	1. $h\nu$, DMSO, 2 h 2. *i*-PrBr	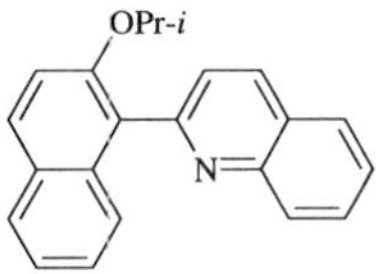 (70)	215
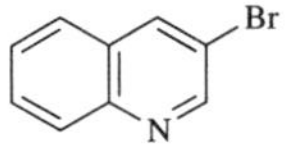	1. $h\nu$, DMSO, 0.5 h 2. *i*-PrBr	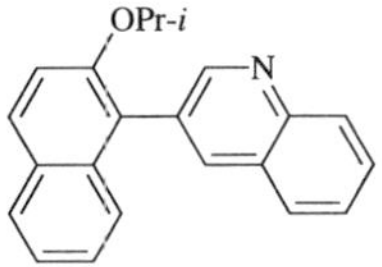 (85)	215
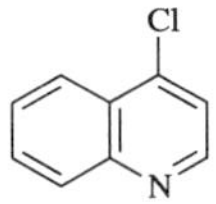	1. $h\nu$, DMSO, 2 h 2. *i*-PrBr	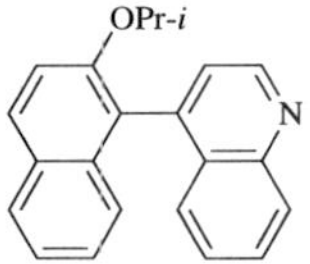 (70)	215
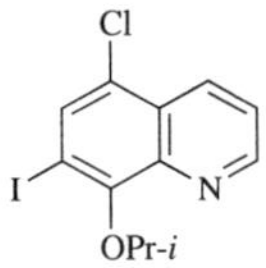	1. $h\nu$, DMSO, 2 h 2. *i*-PrBr	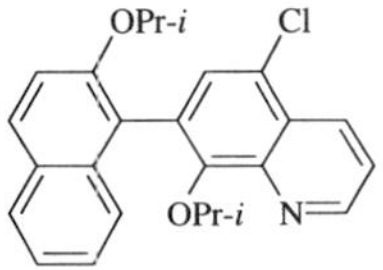 (50)	215

TABLE VII. OTHER C-C BOND FORMATION (*Continued*)

B. Reaction with Phenoxide and Related Ions (*Continued*)

Substrate	Nucleophile	Conditions	Product(s) and Yield(s) (%)	Refs.
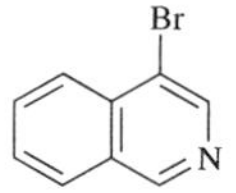Br N	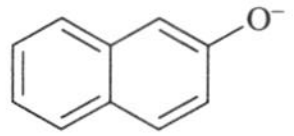O⁻	1. *hν*, DMSO, 1 h 2. *i*-PrBr	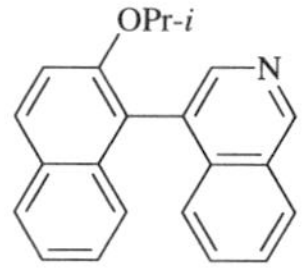OPr-*i* N (74)	215
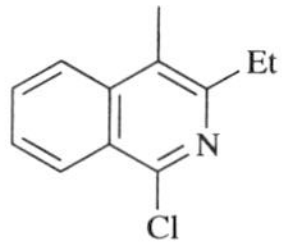Et N Cl		1. *hν*, DMSO, 1 h 2. *i*-PrBr	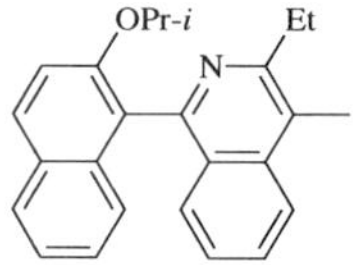OPr-*i* Et N (50)	215
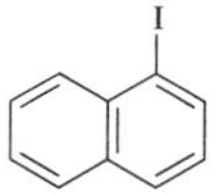I	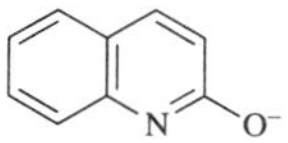N O⁻	1. *hν*, DMSO, 4 h 2. *i*-PrBr	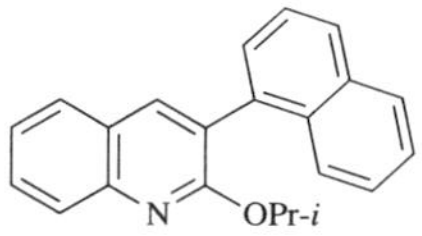N OPr-*i* (20)	215
	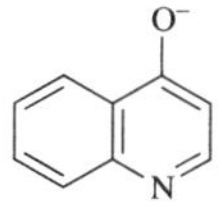O⁻ N	1. *hν*, DMSO, 4 h 2. *i*-PrBr	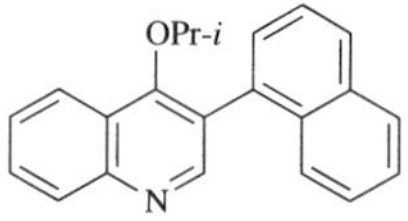OPr-*i* N (15)	215
	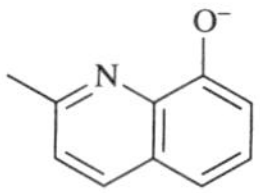O⁻ N	1. *hν*, DMSO, 4 h 2. *i*-PrBr	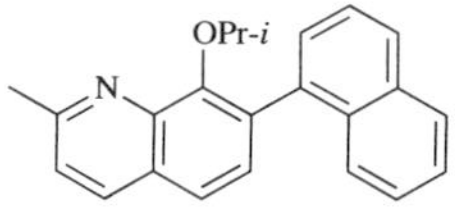OPr-*i* N **I** + 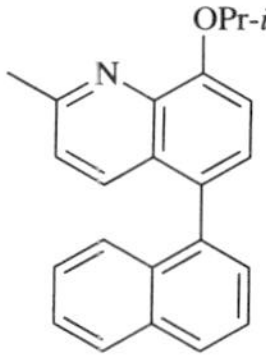OPr-*i* N **II** +	215

			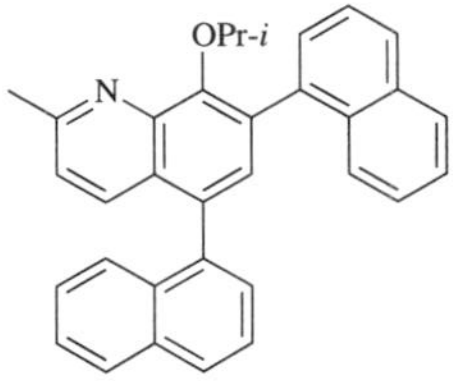 III	
			I + II + III (50)	
	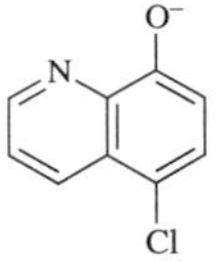	1. *hν*, DMSO, 3 h 2. *i*-PrBr	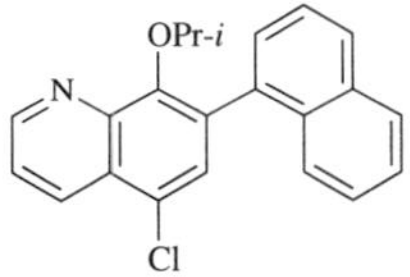 (51)	215
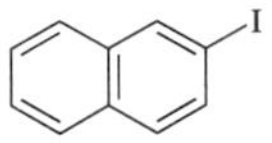		1. *hν*, DMSO, 5 h 2. *i*-PrBr	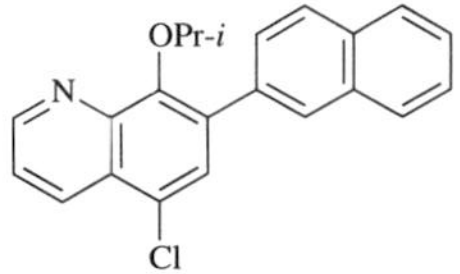 (40)	215

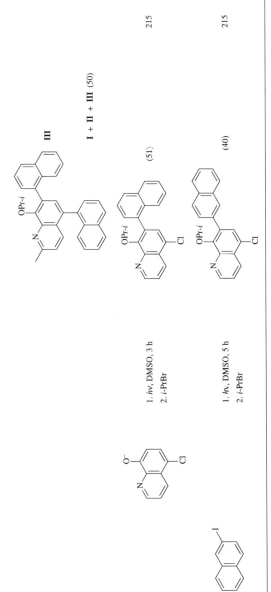

[a] The product was isolated as the lactone derivative, formed by hydrolysis of the cyanophenol during workup.

TABLE VII. OTHER C-C BOND FORMATION (*Continued*)

C. Reaction with Cyanide Ion

Substrate	Nucleophile	Conditions	Product(s) and Yield(s) (%)			Refs.
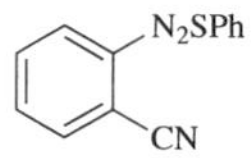	CN^-	$h\nu$, DMSO, 0.75 h	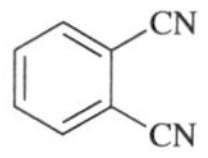 (63) +	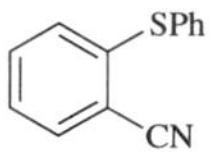 (13)		115
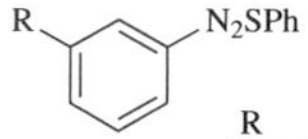		$h\nu$, DMSO, 1 h	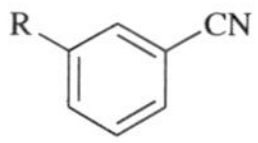 +	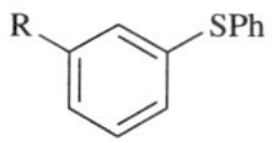		115
R						
CN			(75)	(17)		
CF_3			(61)	(13)		
OMe			(35)	(21)		
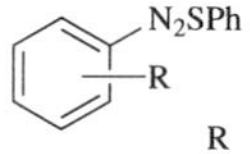			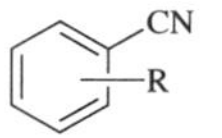 +	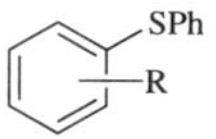 +	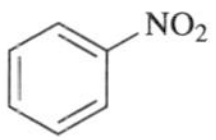	
R						
2-NO_2		Daylight, DMSO, 1 h	(14)	(22)	(31)	115
3-NO_2		Daylight, DMSO, 1 h	(4)	(15)	(31)	115
4-NO_2		$h\nu$, DMSO, 1 h	(36)	(40)	(—)	115
4-NO_2		Daylight, DMSO, 1 h	(48)	(28)	(—)	114, 115
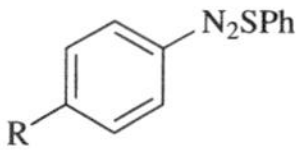			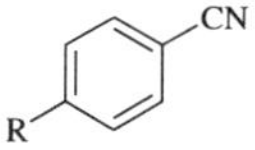 +	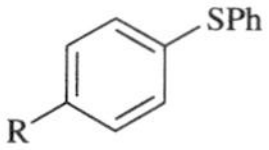		
R						
F		$h\nu$, DMSO, 1 h	(37)	(20)		115
CN		$h\nu$, DMSO, 1 h	(71)	(19)		114, 115
CN		e^-, DMSO	(71)	(20)		114, 115
CN		e^-, MeCN	(80)	(9)		219
COMe		daylight, DMSO, 184 h	(19)	(3)		115
COMe		$h\nu$, DMSO, 1 h	(50)	(19)		114, 115

R				
COPh	daylight, DMSO, 66 h	(27)	(21)	115
COPh	$h\nu$, DMSO, 1 h	(69)	(21)	114, 115
COPh	e^-, DMSO	(40)	(36)	114, 115
CF_3	$h\nu$, DMSO, 1 h	(60)	(22)	114, 115
SO_2Ph	$h\nu$, DMSO, 1 h	(74)	(16)	114, 115
SO_2Ph	e^-, DMSO	(72)	(15)	114, 115
OMe	$h\nu$, DMSO, 1 h	(9)	(20)	115
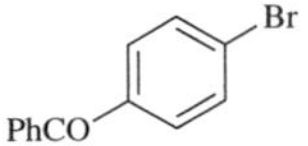	e^-, MeCN	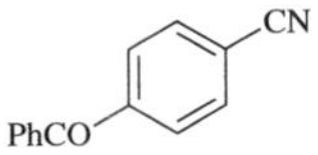 (95)		88, 89, 86
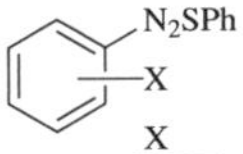	$h\nu$, DMSO, 1 h	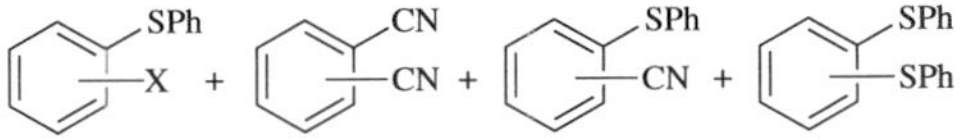		115
X				
2-Cl		(—) (53) (10) (3)		
3-Br		(5) (50) (20) (3)		
4-Br		(8) (54) (15) (3)		
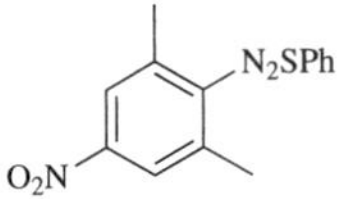	Daylight, DMSO, 1 h	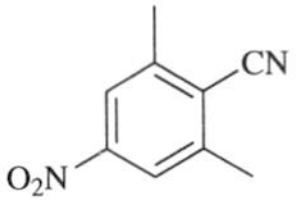 (46) +	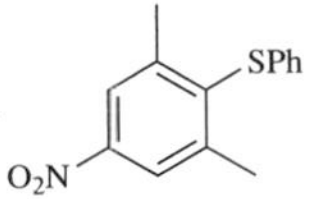 (27)	115
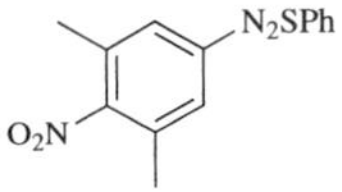	$h\nu$, DMSO, 1 h	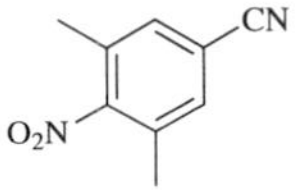 (60) +	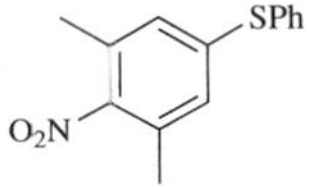 (19)	115
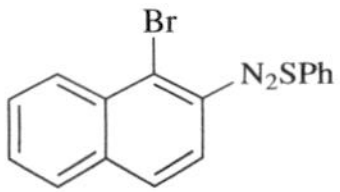	Sunlamp, DMSO, 1 h	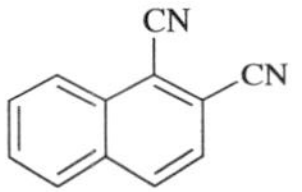 (27)		127

TABLE VII. OTHER C-C BOND FORMATION (*Continued*)

C. Reaction with Cyanide Ion (*Continued*)

Substrate	Nucleophile	Conditions	Product(s) and Yield(s) (%)	Refs.
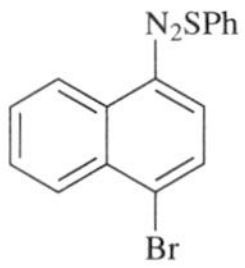	CN^-	Sunlamp, DMSO, 1 h	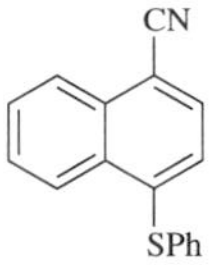 (21) + 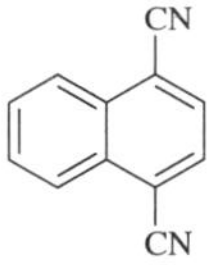 (48) + 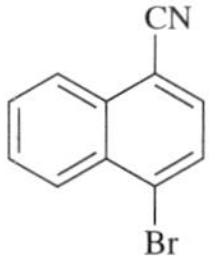 (3)	127
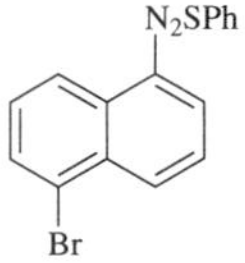		*hν*, DMSO, 1 h	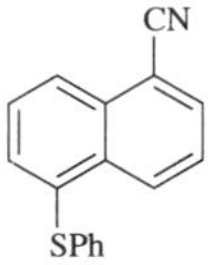 (22) + 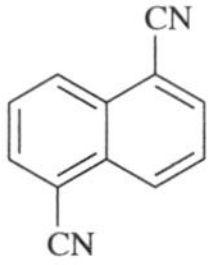 (46) + 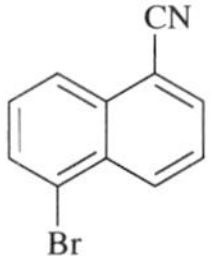 (8)	127
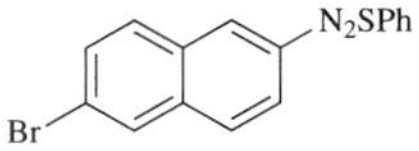		*hν*, DMSO, 1 h	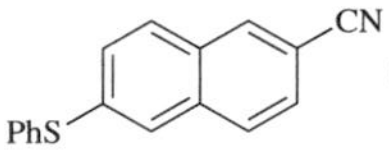 (17) + 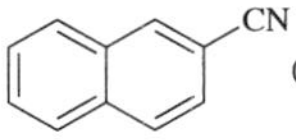 (11) + 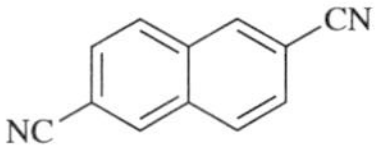 (30)	127
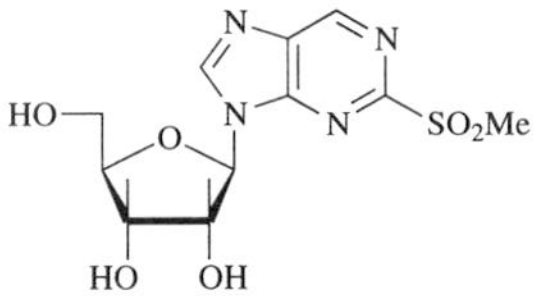		DMF, 0°, 15 h	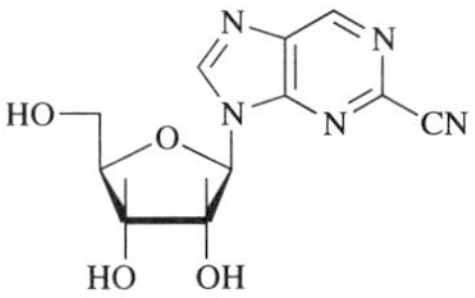 (40)	350

TABLE VII. OTHER C-C BOND FORMATION (*Continued*)

D. Reaction with Nitronate Ions

Substrate	Nucleophile	Conditions	Product(s) and Yield(s) (%)	Refs.
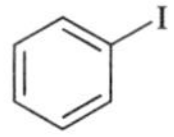	$^{-}CH_2NO_2$	$h\nu$, *t*-BuOK, Me_2CO DMSO, 2 h	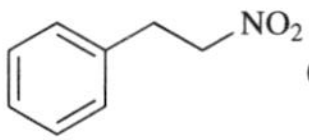 (11) + 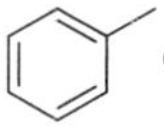 (1) +  (2)	41
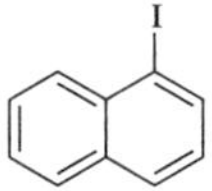		$h\nu$, *t*-BuOK, Me_2CO DMSO, 2 h	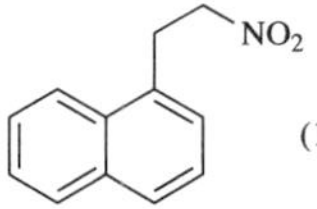 (10) + 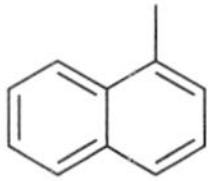 (10) + 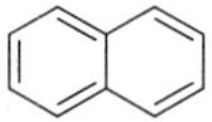 (6)	41
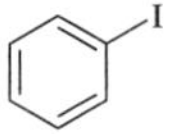	$^{-}C(Me)_2NO_2$	e^-, DMSO	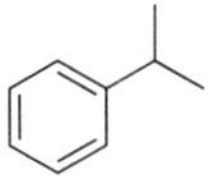 (29) +  (71)	40
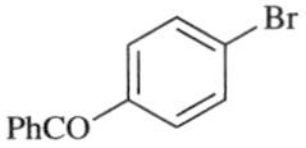		e^-, DMSO, 18-crown-6	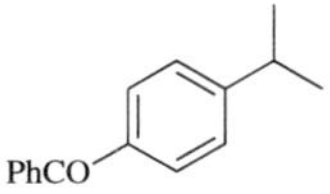 (50) + 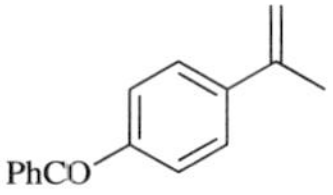 (7) + 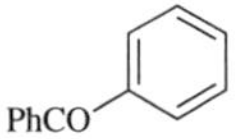 (26)	40

TABLE VII. OTHER C-C BOND FORMATION (*Continued*)

E. Reaction with Other Carbanions

Substrate	Nucleophile	Conditions	Product(s) and Yield(s) (%)	Refs.
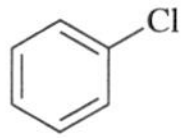	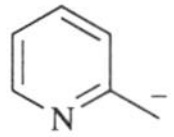	K, NH_3, –77°	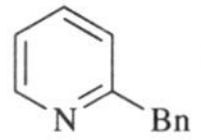 **I** (48) +  (14)	220
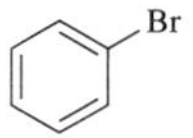		NH_3, 1 h	**I** (57)	220
		*h*ν, NH_3, 1 h	**I** (73)	220
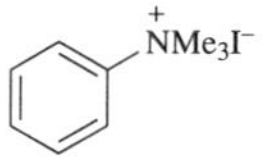		*h*ν, NH_3, 1 h	**I** (66) + 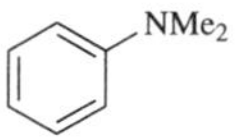 (9)	220
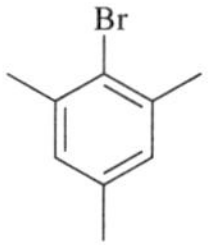		*h*ν, NH_3, 1 h	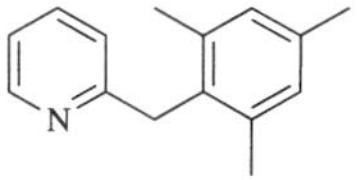 (87) + 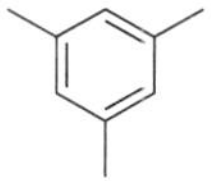 (4)	220
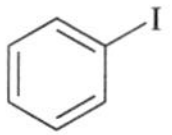	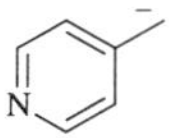	K, NH_3, –77°	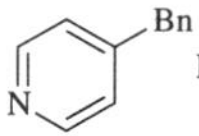 **I** (51) + 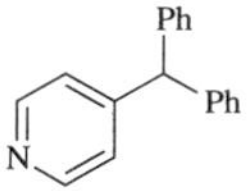 **II** (12) +  (12)	220
		NH_3, –77°	**I** (14)	220
$Ph\overset{+}{N}Me_3I^-$		*h*ν, NH_3, 1 h	**I** (88) + **II** (3)	220
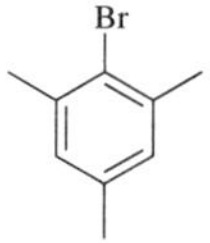		*h*ν, NH_3, 2 h	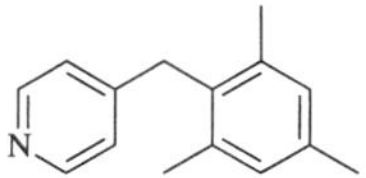 (55) + 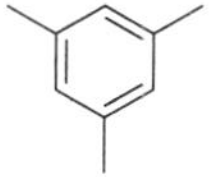 (3)	220

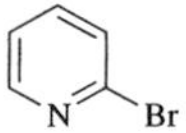		$h\nu$, NH_3, 15 min	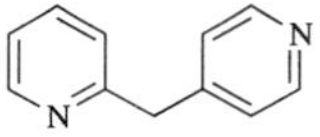 (47) + 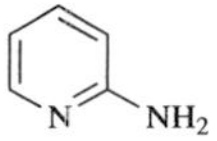 (46)	134
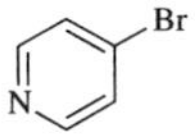		$h\nu$, NH_3, 15 min	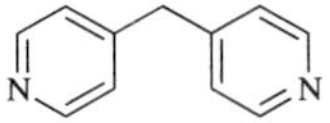 (78)	174
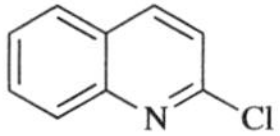		$h\nu$, NH_3, 15 min	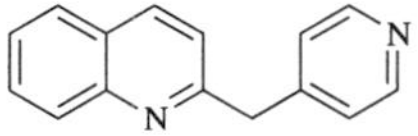 (24) + 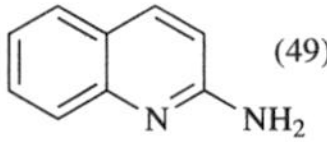 (49)	174
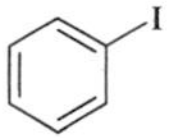	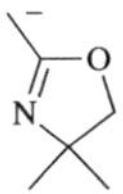	$h\nu$, NH_3, 5 min	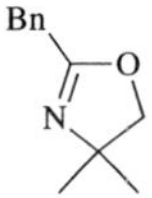 (56) + 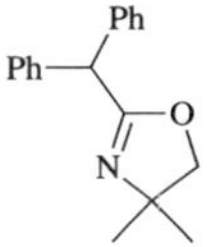 (28)	169
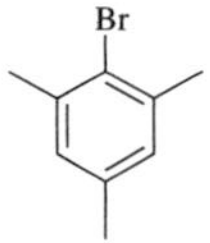		$h\nu$, NH_3, 1 h	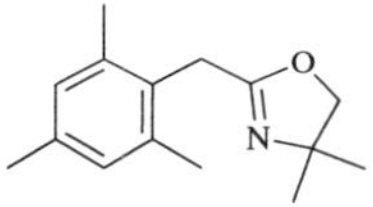 (94)	169
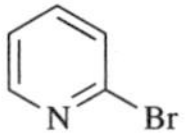		$h\nu$, NH_3, 5 min	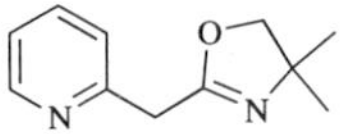 (26) + 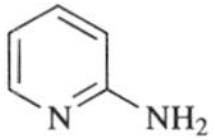 (50)	169
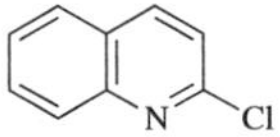		NH_3, 5 min	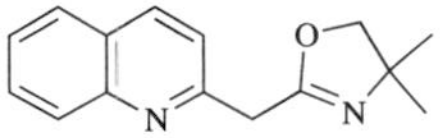 (45)	169
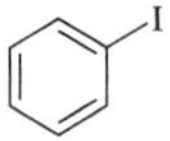	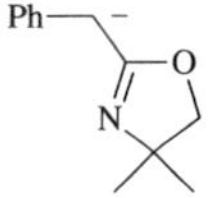	$h\nu$, NH_3, 1 h	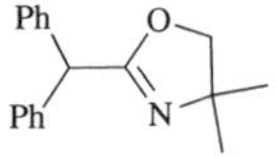 (62)	169
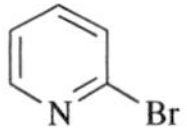		$h\nu$, NH_3, 5 min	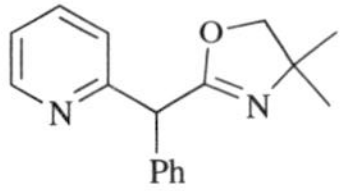 (70)	169

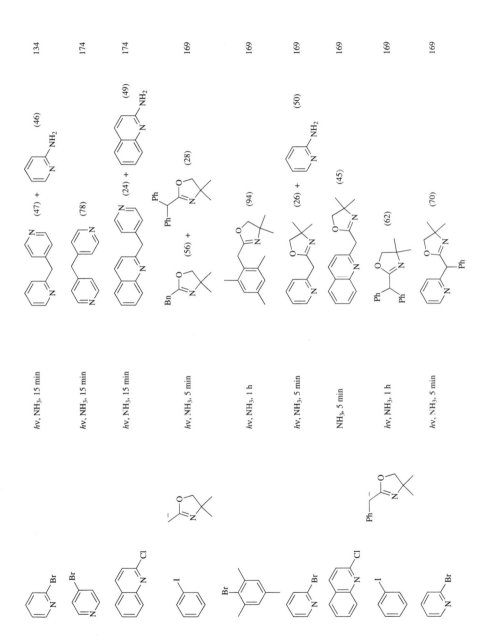

TABLE VII. OTHER C-C BOND FORMATION (*Continued*)

E. Reaction with Other Carbanions (*Continued*)

Substrate	Nucleophile	Conditions	Product(s) and Yield(s) (%)	Refs.
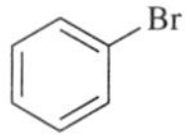	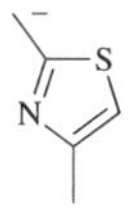	$h\nu$, NH_3, 1.7 h	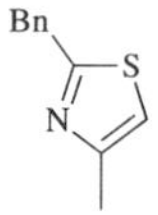 (59) + 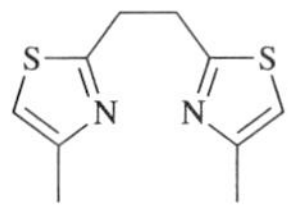 **I** (14)	169
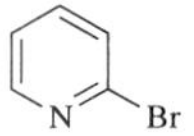		$h\nu$, NH_3, 1.7 h	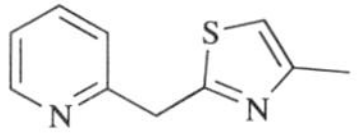 (37) + **I** (12) + 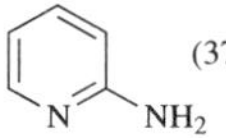 (37)	169
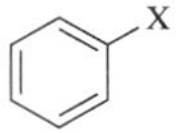	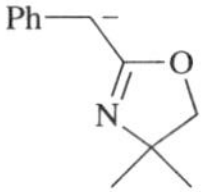	$h\nu$, NH_3, 1.7 h	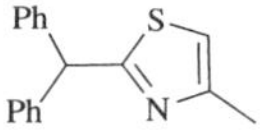 X: I (39); Br (57)	169
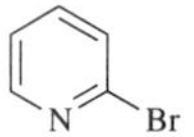		$h\nu$, NH_3, 1.7 h	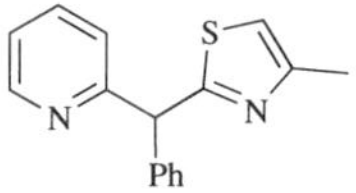 (94)	169
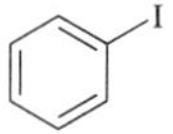	$^{-}CH_2P(O)(OMe)_2$	$h\nu$, NH_3, 1 h	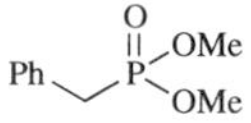 (47) + 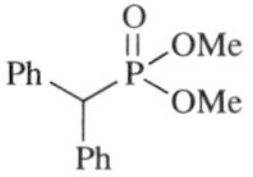 (18)	169
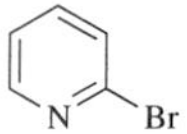		$h\nu$, NH_3, 1 h	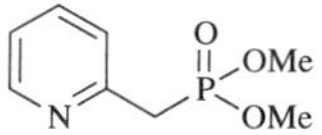 (68) + 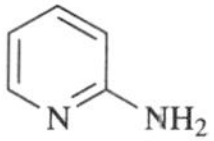 (10)	169

TABLE VIII. OTHER NUCLEOPHILES

A. Nucleophiles Derived from Tin

Substrate	Nucleophile	Conditions	Product(s) and Yield(s) (%)	Refs.
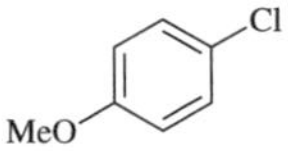	Me_3Sn^-	$h\nu$, NH_3, 1 h	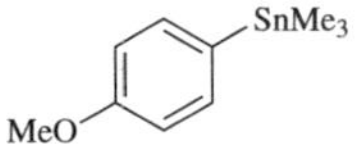 (100)	223
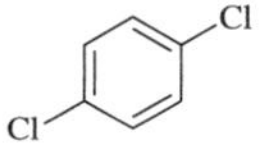		$h\nu$, NH_3, 1 h	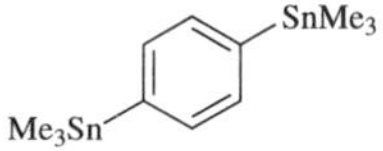 (88)	223
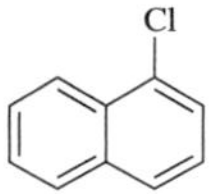		$h\nu$, NH_3, 1 h	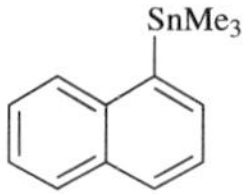 (90)	223
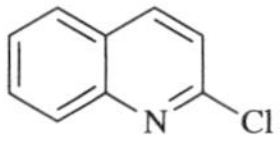		$h\nu$, NH_3, 1 h	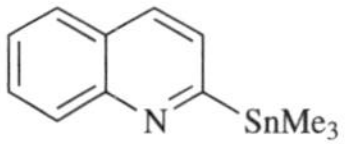 (96)	223
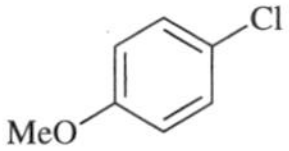		1. $h\nu$, NH_3, 1 h 2. Na, *t*-BuOH 3. 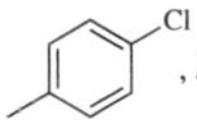 , $h\nu$, 1 h	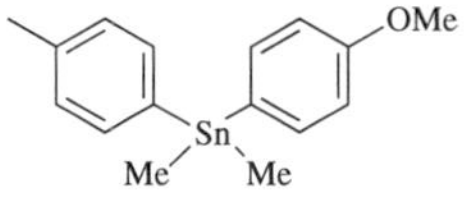 (89)	224
		1. $h\nu$, NH_3, 1 h 2. Na, *t*-BuOH 3. 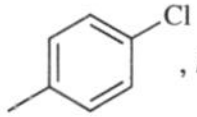 , $h\nu$, 1 h 4. Na, *t*-BuOH 5. PhCl, $h\nu$, 2 h	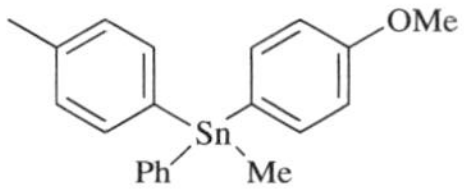 (47)[a]	224

TABLE VIII. OTHER NUCLEOPHILES (*Continued*)

A. Nucleophiles Derived from Tin (*Continued*)

Substrate	Nucleophile	Conditions	Product(s) and Yield(s) (%)	Refs.
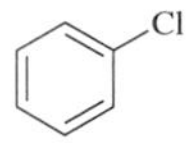	Me_3Sn^-	1. *hν*, NH_3, 0.7 h 2. Na, *t*-BuOH 3. 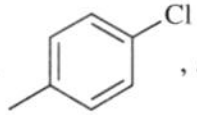, *hν*, 1 h 4. Na, *t*-BuOH 5. 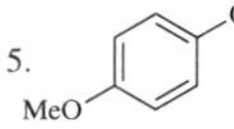, *hν*, 1.5 h	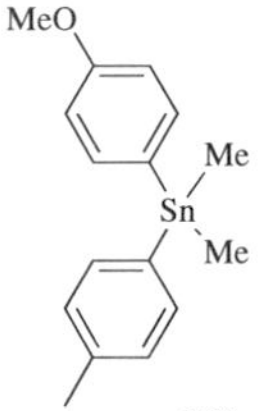 (50) **I** + 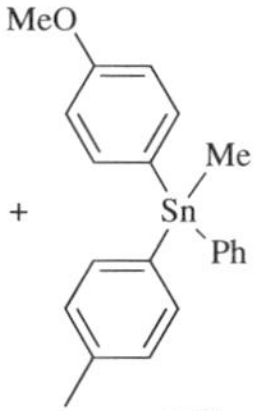 (30) + 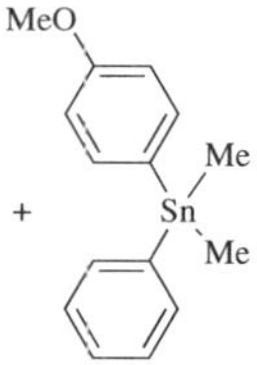 (15)	224
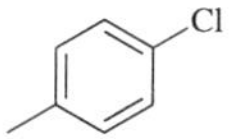	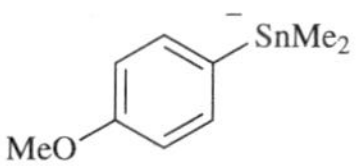	*hν*, NH_3, 1 h	**I** (100)	224
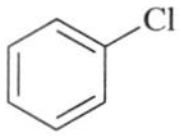	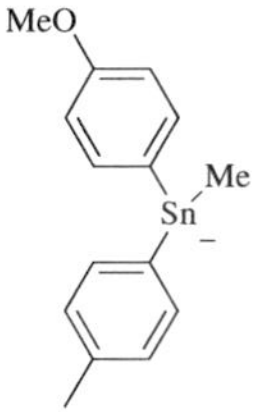	*hν*, NH_3, 1 h	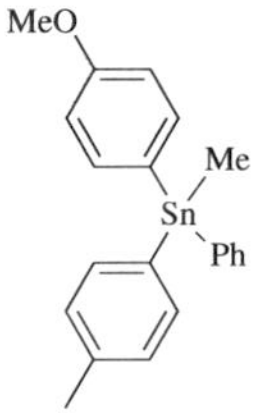 (31)[a]	224
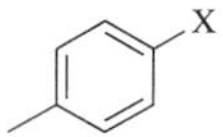	Ph_3Sn^-	*hν*, NH_3	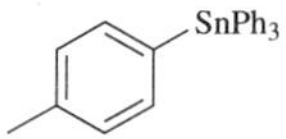	223
X				
Cl		2 h	(75)	
Br		1 h	(62)	
I		1 h	(38)	

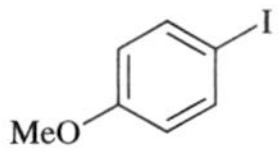		$h\nu$, NH_3, 1 h	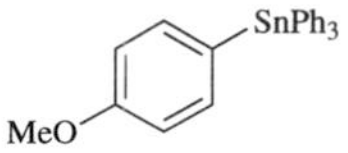 (20)	223
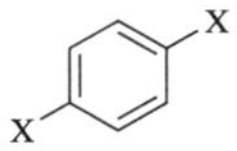		$h\nu$, NH_3, Y h	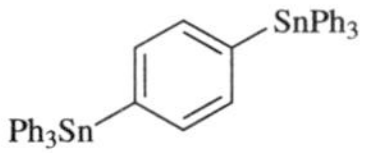	223
			X Y	
			Cl 3 (69)	
			Br 1 (22) + Ph_4Sn (40)	
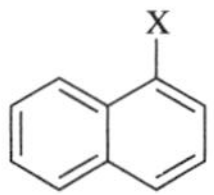		$h\nu$, NH_3	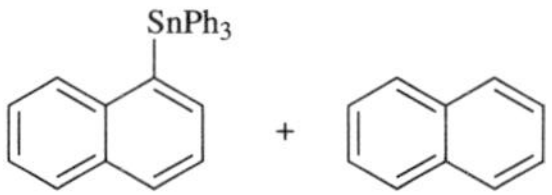	223
X				
Cl		3 h	(80) (9)	
Br		1 h	(75) (32)	
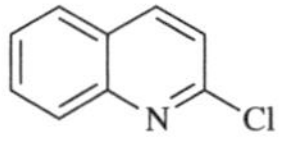		$h\nu$, NH_3, 1 h	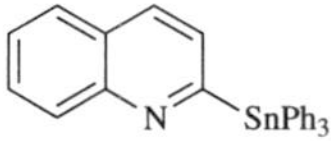 (80)	223
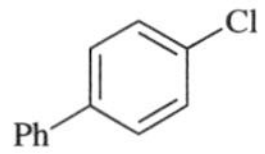	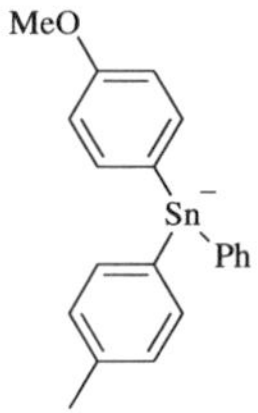	$h\nu$, NH_3, 2.3 h[b]	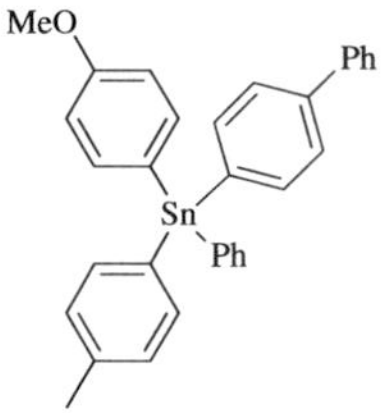 (25)[a]	224

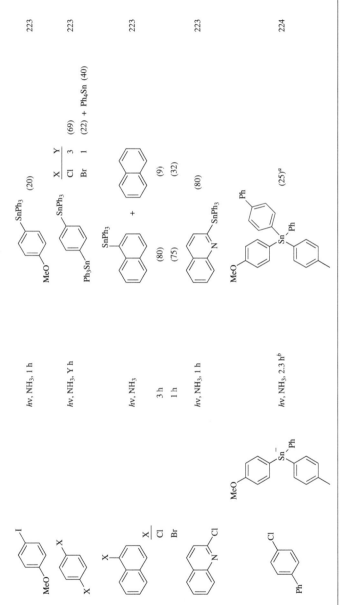

[a] Other stannyl products were also formed.

[b] The nucleophile was formed by reductive cleavage of *p*-anisylmethylphenyl(tolyl-*p*)stannane with sodium metal. Minor amounts of anions derived from Sn-Ar bond fragmentation were also formed.

TABLE VIII. OTHER NUCLEOPHILES (*Continued*)

B. Nucleophiles Derived from Nitrogen

Substrate	Nucleophile	Conditions	Product(s) and Yield(s) (%)	Refs.
PhOPh	NH_2^-	K, NH_3	$PhNH_2$ **I** (53)	4
$PhOP(O)(OEt)_2$		K, NH_3	**I** (73)	225
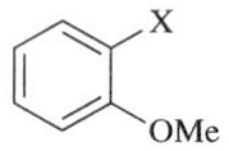		K, NH_3	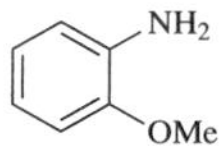 + 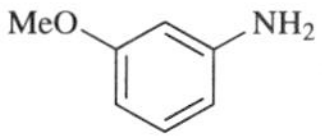 + 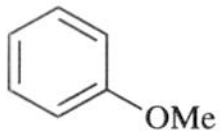	4
X = Cl			(57) (26) (5)	
X = Br			(64) (—) (16)	
X = I			(67) (—) (11)	
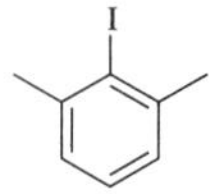		K, NH_3	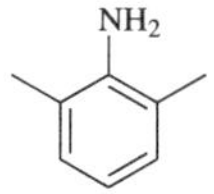 **I** (64) + 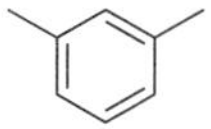 (25)	4
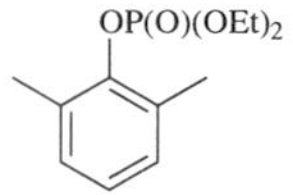		K, NH_3	**I** (78)	225
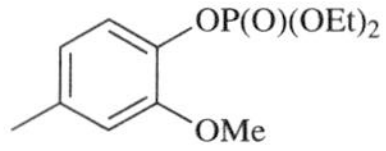		K, NH_3	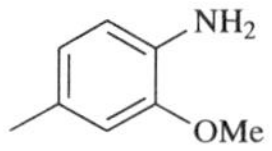 (56)	225
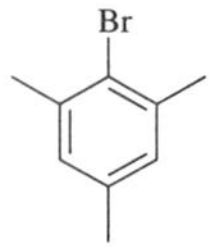		*h*ν, NH_3, 2 h	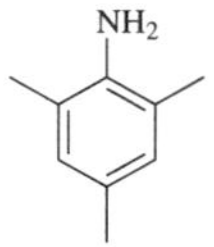 (70) + 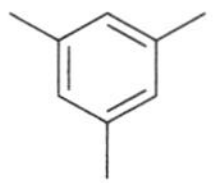 (6)	126

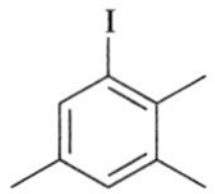		NH_3	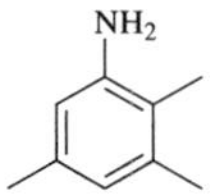 **I** (65) + 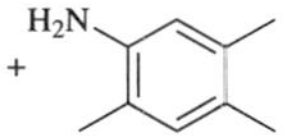 **II** (11) + 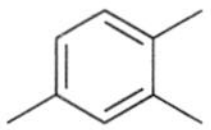 **III** (10)	3
		K, NH_3	**I** (54) + **III** (30)	4
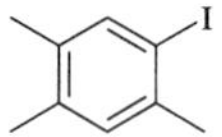		NH_3	**I** (32) + **II** (51)	3
		K, NH_3	**I** (50) + **III** (40)	4
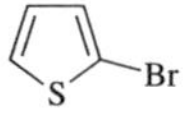		NH_3, 15 min	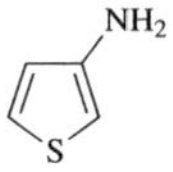 **I** (48) +  **II** (9)	152
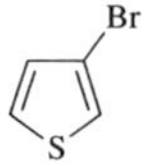		NH_3, 1 h	**I** (79) + **II** (4)	152
		$h\nu$, NH_3, 1 h	**I** (63) + **II** (1)	152
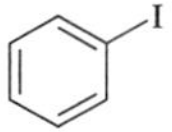	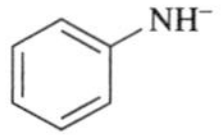	K, NH_3	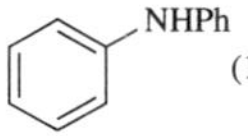 (10) + 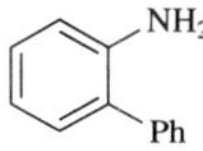 (6) + 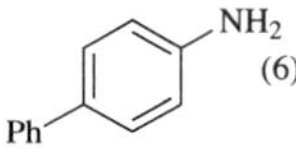 (6)	4
	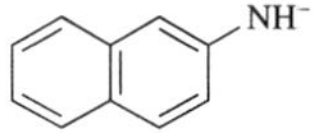	$h\nu$, NH_3, 1 h	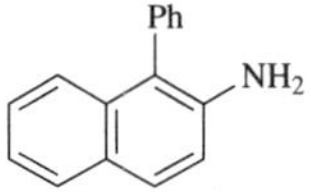 (47) + 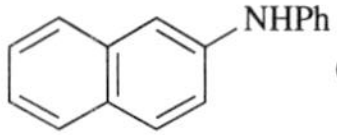 (1)	226

TABLE VIII. OTHER NUCLEOPHILES (*Continued*)

B. Nucleophiles Derived from Nitrogen (*Continued*)

Substrate	Nucleophile	Conditions	Product(s) and Yield(s) (%)	Refs.
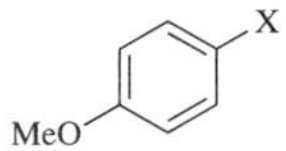 X: Br, I	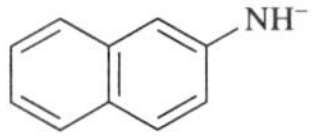	$h\nu$, NH_3, 3 h	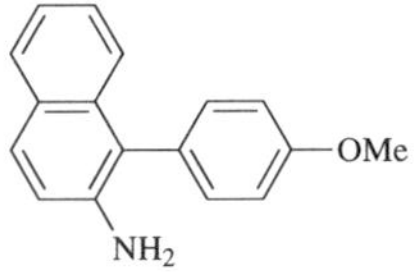 + 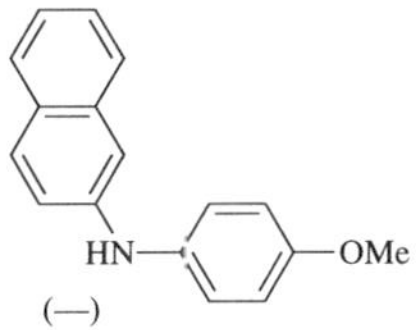 X = Br: (12), (—) X = I: (63), (6)	226
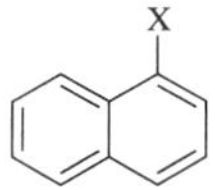		$h\nu$, NH_3, 3 h	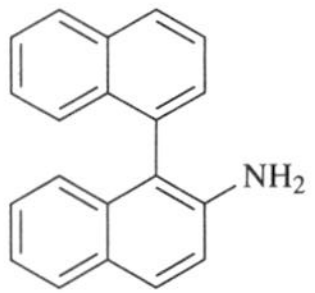 X = Br (25) X = I (45)	226
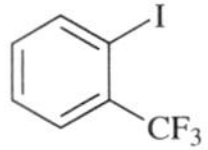	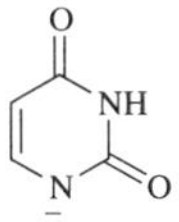	e^-, DMSO, mediator	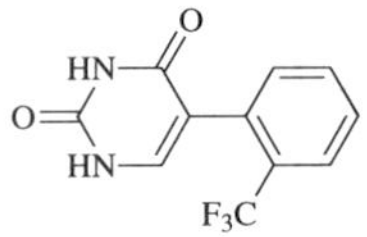 (38)	356, 230
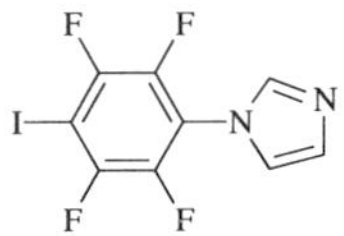		e^-, DMSO, mediator	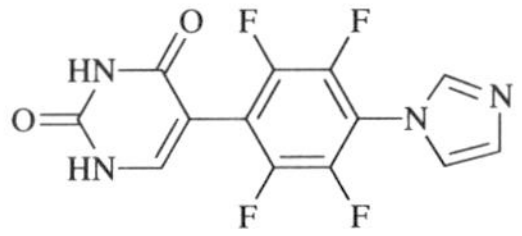 (50)	356
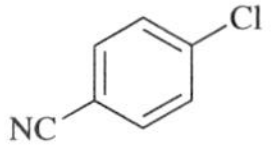		e^-, DMSO	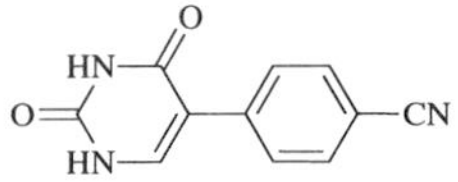 (50)	230

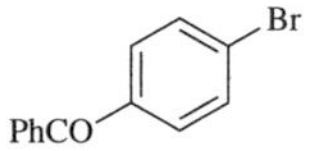	e⁻, DMSO	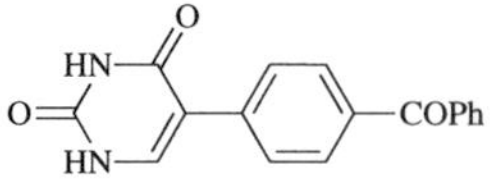 (55)	230
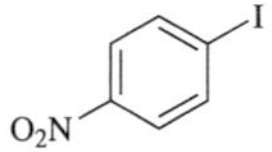	e⁻, DMSO	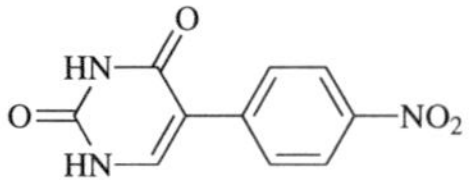 (55)	230
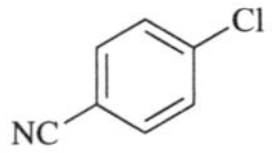 	e⁻, NH_3, mediator	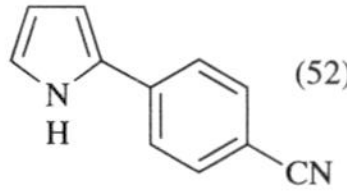 (52) + 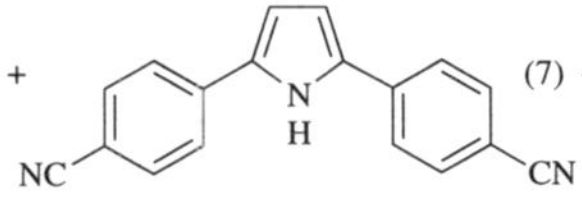 (7) + 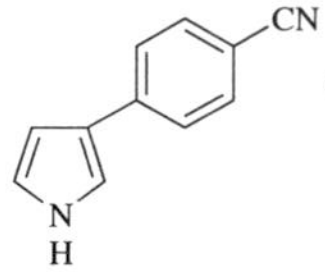 (7) + 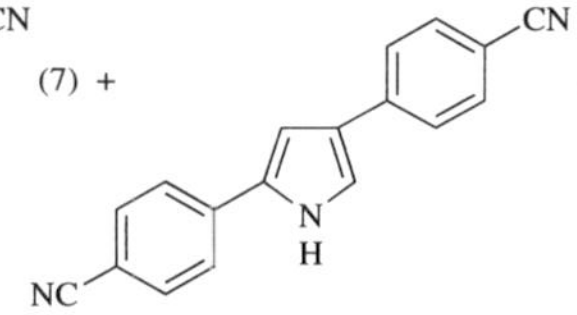 (7)	227
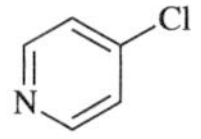	e⁻, NH_3, mediator, 1.5 h	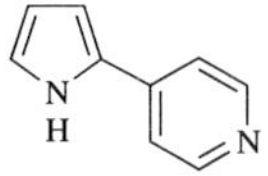 (60) + 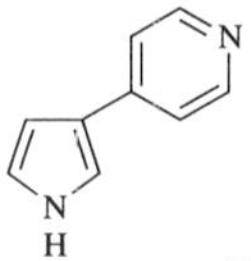 (3)	227
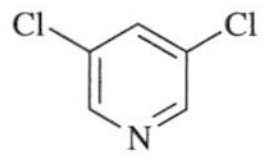	e⁻, NH_3, mediator, 1.5 h	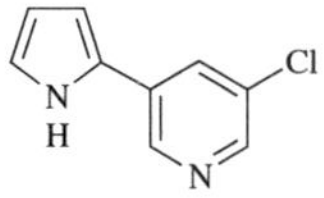 (67) + 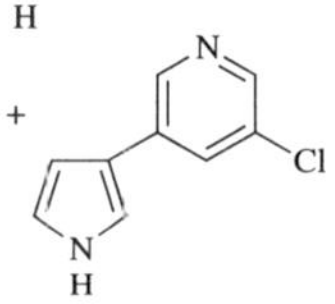 (6)	227

TABLE VIII. OTHER NUCLEOPHILES (*Continued*)

B. Nucleophiles Derived from Nitrogen (*Continued*)

Substrate	Nucleophile	Conditions	Product(s) and Yield(s) (%)	Refs.
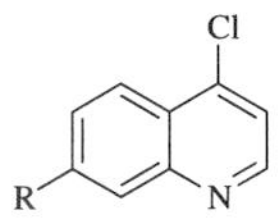 R: H R: CF_3		e^-, NH_3, mediator, 1.5 h	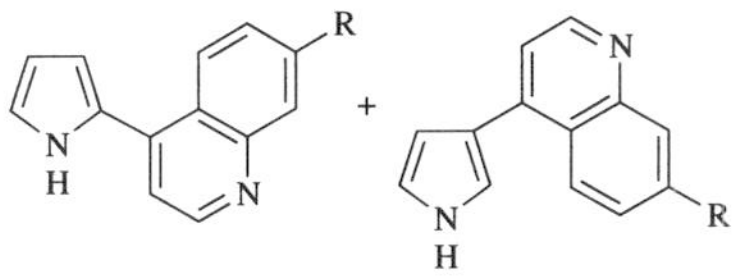 (53) + (8) (65) + (14)	227
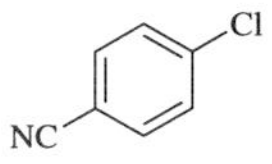	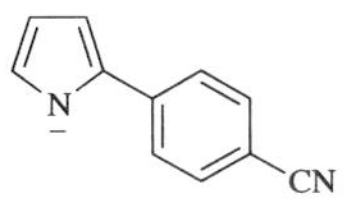	e^-, NH_3, mediator	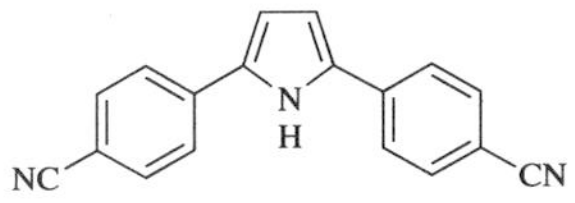 (60) + 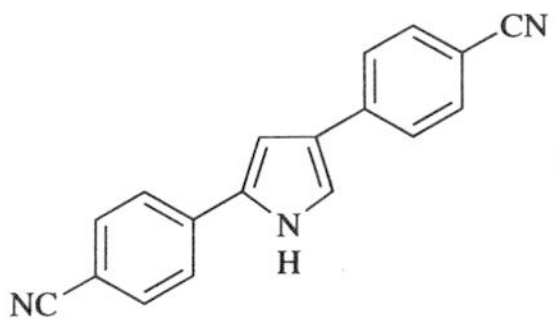 (20)	229
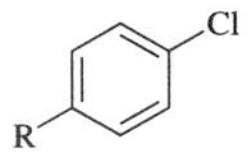 R: CN R: SO_2Ph	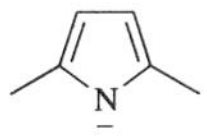	e^-, NH_3, mediator	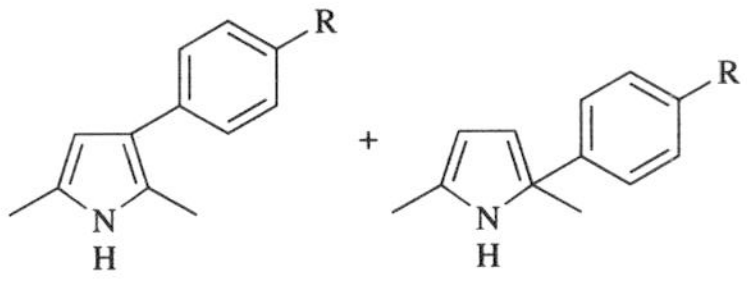 (40) + (20) (40) + (11)	229, 228

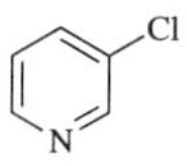		e⁻, NH_3, mediator	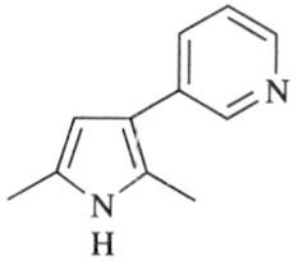 (40) + 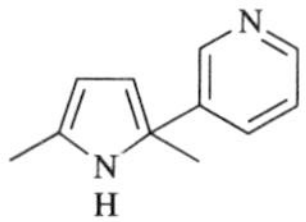 (25)	229, 228
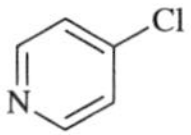		e⁻, NH_3, mediator	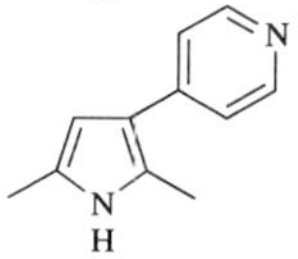 (40) + 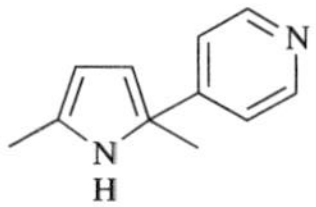 (25)	229, 228
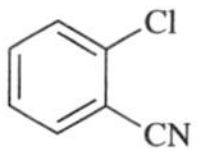	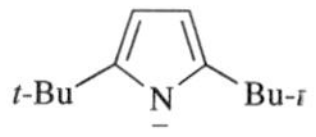	e⁻, NH_3, mediator	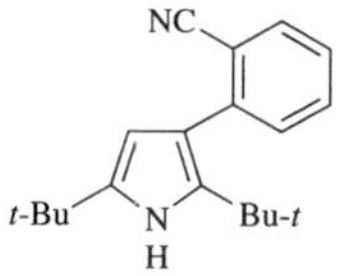 (30) + 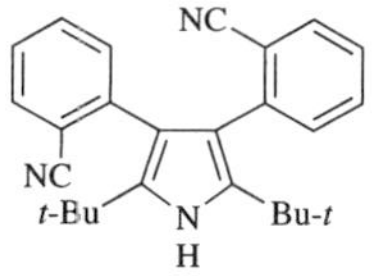 (25)	229
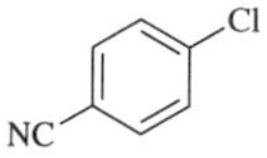		e⁻, NH_3, mediator	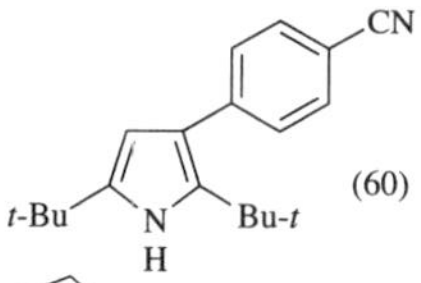 (60) + 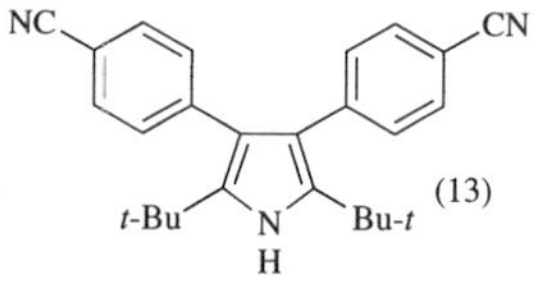 (13)	229
	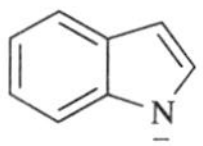	e⁻, NH_3, mediator, 1.5 h	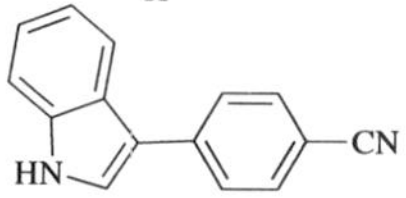 (60)	227
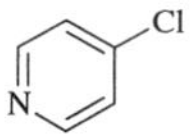		e⁻, NH_3, mediator, 1.5 h	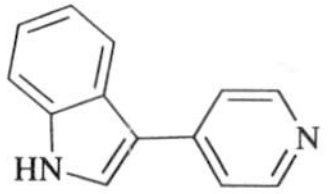 (60)	227
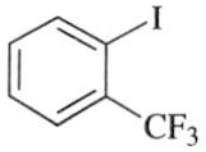	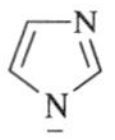	e⁻, DMSO, mediator	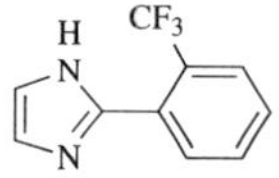 **I** + 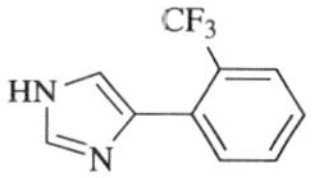 **II** **I** + **II** (35); **I**:**II** = 1	356

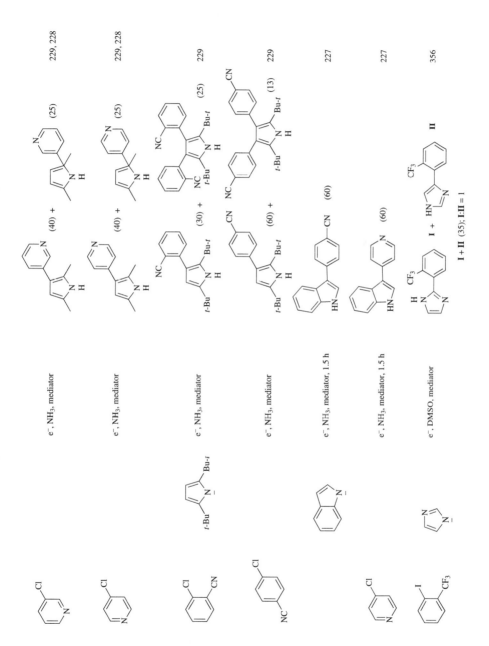

TABLE VIII. OTHER NUCLEOPHILES (*Continued*)

B. Nucleophiles Derived from Nitrogen (*Continued*)

Substrate	Nucleophile	Conditions	Product(s) and Yield(s) (%)	Refs.
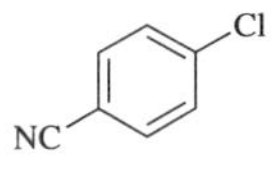	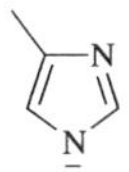	e^-, NH_3, mediator	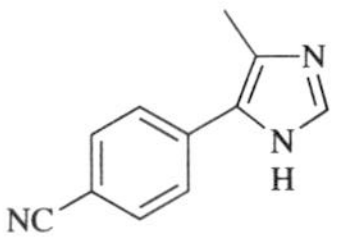 (40) + 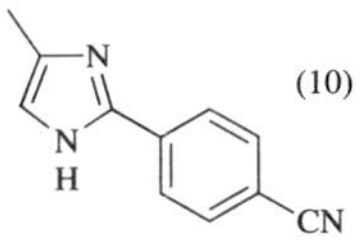 (10)	229
	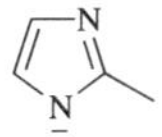	e^-, NH_3, mediator	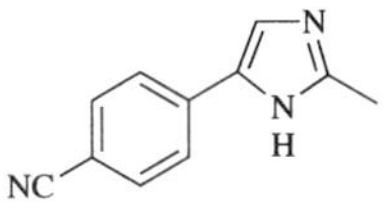 (40)	229
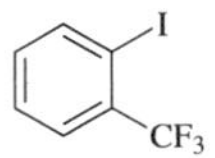	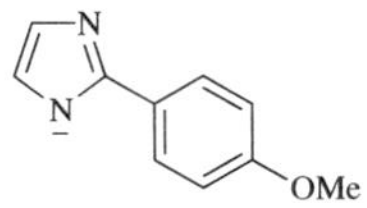	e^-, DMSO, mediator	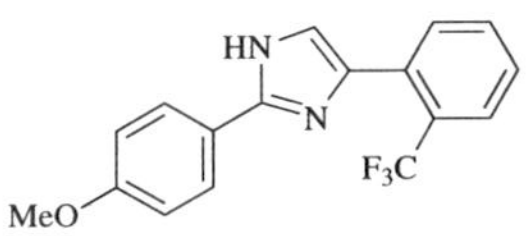 (55)	356
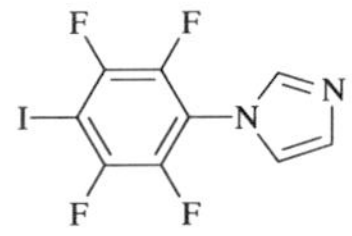	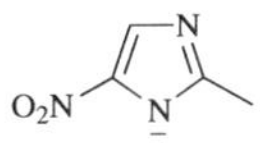	e^-, DMSO, mediator	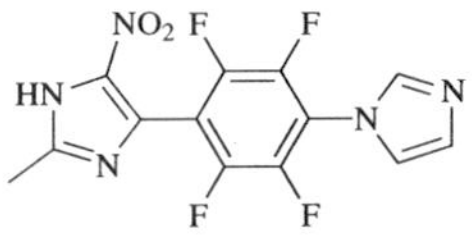 (35)	356

TABLE VIII. OTHER NUCLEOPHILES (*Continued*)

C. Nucleophiles Derived from Phosphorus

Substrate	Nucleophile	Conditions	Product(s) and Yield(s) (%)	Refs.
PhI	P^{3-}	1. $h\nu$, NH_3, 2 h 2. [O]	Ph_3PO (75)	242
PhCl		1. $h\nu$, NH_3, 1 h 2. Na, 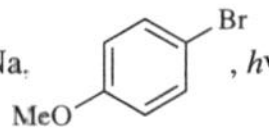Br MeO, $h\nu$, 1 h 3. [O]	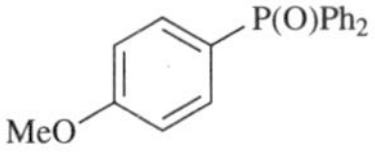P(O)Ph2 MeO (55)	242
PhSPh	Ph_2P^-	1. Visible light, DMSO, 15 h 2. [O]	Ph_3PO **I** (70) + PhSH **II** (83)	67
PhSOPh		Visible light, DMSO, 15 h	**I** (80) + **II** (16)	67
$PhSO_2Ph$		1. $h\nu$, DMSO, 15 h 2. [O]	**I** (80)	67
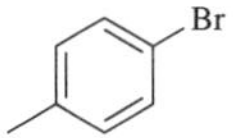Br		1. $h\nu$, NH_3, 1.5 h 2. [O]	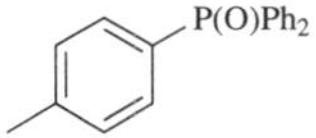P(O)Ph2 (57)	71
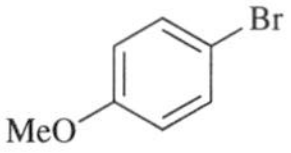Br MeO		1. $h\nu$, NH_3, 1.5 h 2. [O]	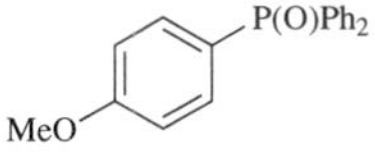P(O)Ph2 MeO **I** (82)	240
		1. Na(Hg), NH_3, mediator 2. [O]	**I** (85) + PhOMe (17)	102
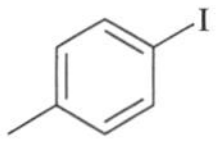I		1. NH_3, 5.5 h 2. [O]	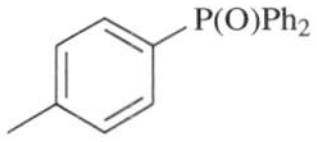P(O)Ph2 **I** (89)	71
		1. $h\nu$, NH_3, 1 h 2. [O]	**I** (90)	71
		1. DMSO, 2 h 2. [O]	**I** (78)	71

TABLE VIII. OTHER NUCLEOPHILES (*Continued*)

C. Nucleophiles Derived from Phosphorus (*Continued*)

Substrate	Nucleophile	Conditions	Product(s) and Yield(s) (%)	Refs.
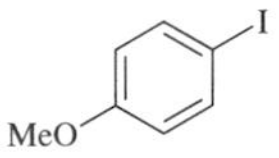	Ph_2P^-	NH_3, 0.75 h	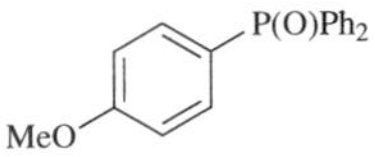 **I** (26) + PhOMe **II** (5)	241
		1. Na(Hg), NH_3, mediator 2. [O]	**I** (40) + **II** (63)	102
		1.))) , NH_3, rt, 0.75 h 2. [O]	**I** (75) + **II** (7)	241
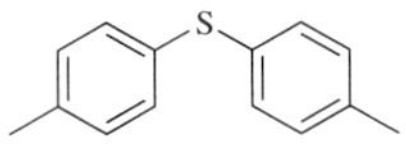		1. DMSO, 2.5 h 2. [O]	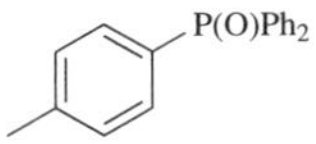 (72) + 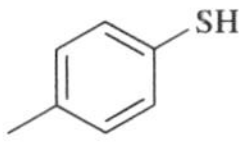 (100)	67
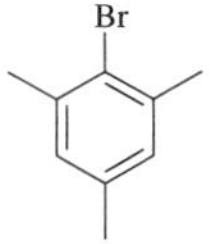		1. *h*ν, NH_3, 1 h 2. [O]	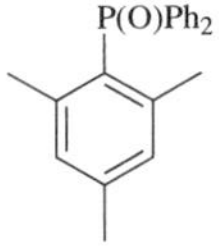 (45)	126
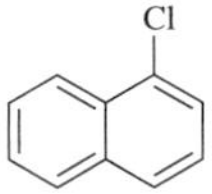		1. Na(Hg), NH_3, mediator 2. [O]	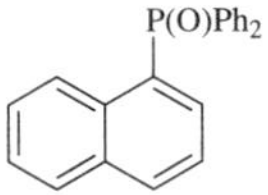 **I** (83) + 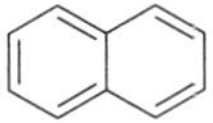 **II** (19)	102
		1. Na(Hg), NH_3, rt, 2 h 2. [O]	**I** (66)	357
		1.))) , NH_3, rt, 1 h 2. [O]	**I** (30) + **II** (4)	241
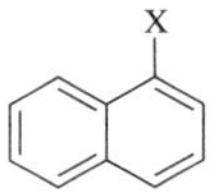		1.))) , NH_3, rt, 1 h 2. [O]	X = Br, **I** (94) + **II** (6)	241, 357

		1.))) , NH_3, rt, 15 min 2. [O]	X = I, **I** (70) + **II** (30)	241
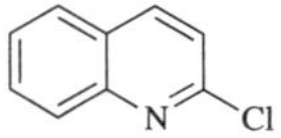		1. Na(Hg), NH_3 2. [O]	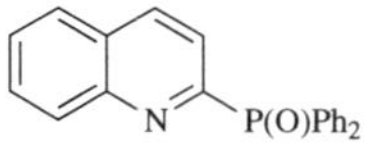 (96)	102
PhI	Ph_2PO^-	$h\nu$, NH_3, 20 min	Ph_3PO (95)	72
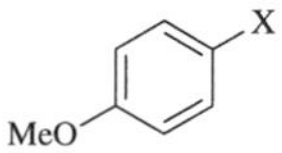		1. $h\nu$, NH_3, Y h 2. [O]	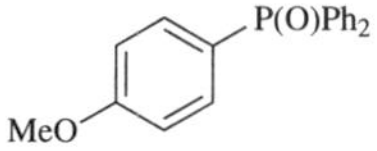 X / Y: Br / 2 (81); I / 1 (92)[a]	240
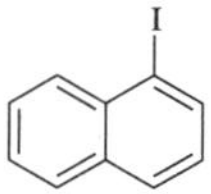		1. $h\nu$, NH_3, 1 h 2. [O]	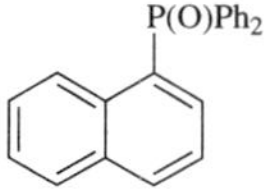 (79)	240
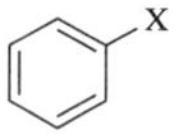	Bn_2PO^-	1. $h\nu$, NH_3, 1 h 2. [O]	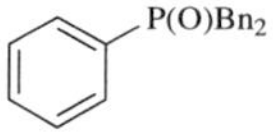 X: Br (80); I (85)	240
PhBr	$PhP(OBu)O^-$	$h\nu$, NH_3, 1.2 h	$Ph_2P(O)OBu$ **I** (72)	72
PhI		$h\nu$, NH_3, 15 min	**I** (95)	72
		$h\nu$, DMSO, 1 h	**I** (42)	72
PhBr	$(Me_2N)_2PO^-$	$h\nu$, NH_3, 1.5 h	$PhP(O)(NMe_2)_2$ **I** (65) + C_6H_6 **II** (4)	72
PhI		$h\nu$, NH_3, 0.5 h	**I** (64) + **II** (8)	72
	$(EtO)_2PS^-$	$h\nu$, NH_3, 15 min	$PhP(S)(OEt)_2$ (95)	72
	$(MeO)_2PO^-$	$h\nu$, NH_3, 0.8 h	$PhP(O)(OMe)_2$ (93)	232
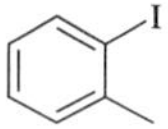		$h\nu$, NH_3, 2 h	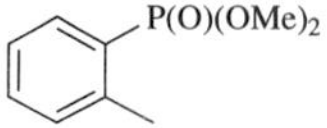 (82)	231

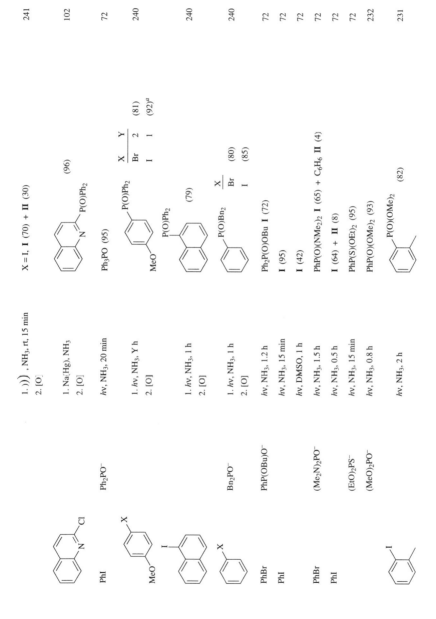

TABLE VIII. OTHER NUCLEOPHILES (*Continued*)

C. Nucleophiles Derived from Phosphorus (*Continued*)

Substrate	Nucleophile	Conditions	Product(s) and Yield(s) (%)	Refs.
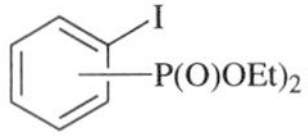	$(MeO)_2PO^-$	*hν*, NH_3	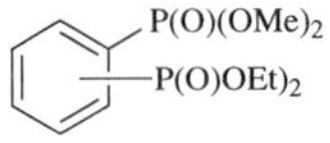	234
		1.5 h	2-$P(O)(OEt)_2$ (60)	
		1 h	3-$P(O)(OEt)_2$ (70)	
		20 min	4-$P(O)(OEt)_2$ (99)	
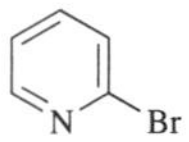		*hν*, NH_3, 2 h	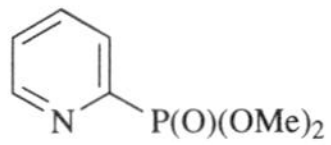 (44)	140
PhI	$(BuO)_2PO^-$	*hν*, NH_3, 1 h	$PhP(O)(OBu)_2$ (88)	232
	$(EtO)_2PO^-$	*hν*, NH_3, 0.75 h	$PhP(O)(OEt)_2$ **I** (96)	232, 340, 344
		hν, *t*-BuOH, 4.5 h	**I** (74)	340
		hν, MeCN, 4 h	**I** (94)	340
		hν, MeCN, THF, 2.5 h	**I** (96)	235
		$FeSO_4$, NH_3, 0.3 h	**I** (98)	75, 345
PhSPh		*hν*, DMSO, 5 h	$PhP(O)(OEt)_2$ **I** (59) + Ph_2S_2 (23) + C_6H_6 (8)	67
		hν, DMF, 2.5 h	**I** (72) + PhSH **II** (21)	67
		hν, NH_3, –76°, 2.5 h	**I** (70) + **II** (74) + PhSEt **III** (14)	67
PhSOPh		*hν*, DMF, 14 h	**I** (64) + **II** (15) + **III** (27) + C_6H_6 (6)	67
$PhSO_2Ph$		*hν*, DMSO, 4 h	**I** (63) + C_6H_6 (7)	67
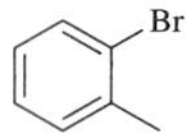		*hν*, MeCN, THF, 10 h	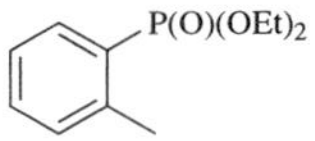 **I** (40)	65

	$h\nu$, MeCN, THF, NaI, 10 h	**I** (95)	65
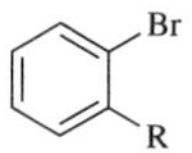	$h\nu$, NH_3	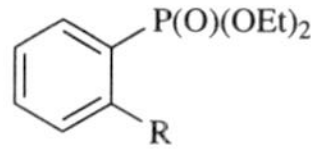	64
R			
NH_2	0.5 h	(60)	
NH_2	KI, 0.5 h	(100)	
OMe	3 h	(8)	
OMe	KI, 3 h	(90)	
$CONH_2$	3 h	(9)	
$CONH_2$	KI, 3 h	(63)	
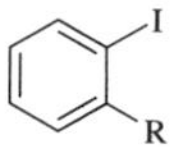	$h\nu$, NH_3	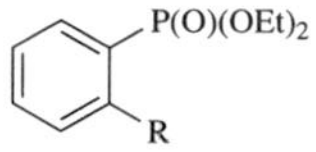	
R			
F	0.8 h	(80)	238
Me	1 h	(71)	231
OMe	1.25 h	(85)	231
NH_2	3 h	(98)	234
NH_2	$h\nu$, THF, MeCN, 2.5 h	(87)	235
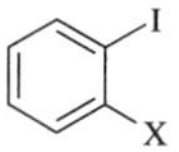	$h\nu$, NH_3, 0.8 h	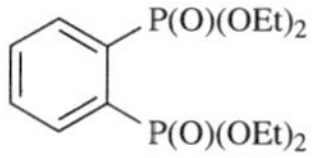	238
X			
Cl		(82)	
Br		(72)	
I		(70)	
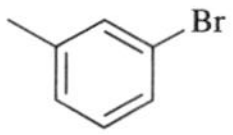	$h\nu$, MeCN, THF, 10 h	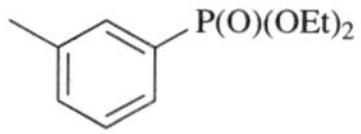 **I** (49)	65
	$h\nu$, MeCN, THF, NaI, 10 h	**I** (99)	65

TABLE VIII. OTHER NUCLEOPHILES (*Continued*)

C. Nucleophiles Derived from Phosphorus (*Continued*)

Substrate	Nucleophile	Conditions	Product(s) and Yield(s) (%)	Refs.
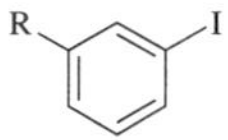	$(EtO)_2PO^-$		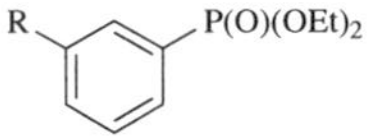	
R				
F		$h\nu$, NH_3, 0.8 h	(96)	232
Cl		$h\nu$, NH_3, 0.7 h	(89)	232
Cl		$h\nu$, NH_3, 1 h	(91)	231
CF_3		$h\nu$, NH_3, 0.9 h	(95)	232
NH_2		$h\nu$, NH_3, 3 h	(75)	234
NH_2		$h\nu$, MeCN, THF, 2.5 h	(90)	235
Me		$h\nu$, NH_3, 1.25 h	(91)	231
OMe		$h\nu$, NH_3, 1.25 h	(86)	231
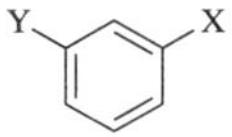		$h\nu$, NH_3, T min	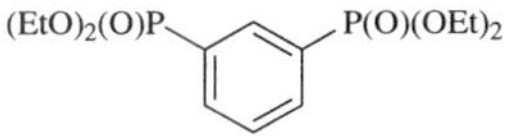	
Y X		T		
Cl Br		10	(96)	239
Br Br		120	(69)	116
I I		7	(63)	116
I I		60	(87)	232
I I		90	(94)	232
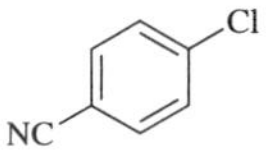		e^-, NH_3	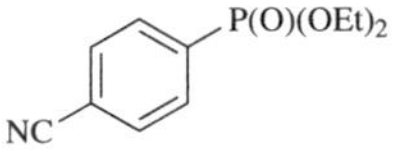 (100)	86, 358

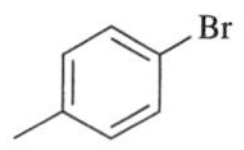	$h\nu$, MeCN, THF, 10 h	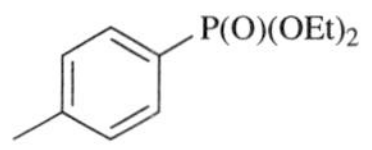 **I** (47)	65
	$h\nu$, MeCN, THF, NaI, 10 h	**I** (84)	65
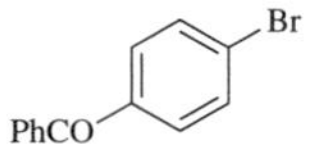	e^-, DMSO	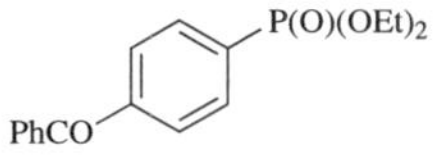 (10) + 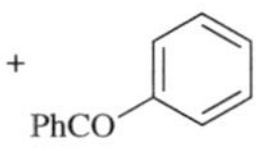 (42)	40
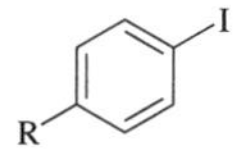		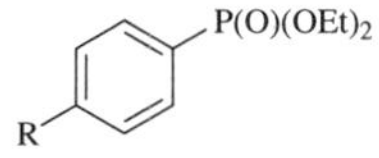	
R			
Me	$h\nu$, NH_3, 1.25 h	(95)	232
NH_2	$h\nu$, MeCN, THF, 2.5 h	(90)	235
NEt_2	$h\nu$, MeCN, THF, 2.5 h	(0)	235
NH_2	$h\nu$, NH_3, 3 h	(95)	234
OMe	$h\nu$, NH_3, 1.1 h	(95)	232
F	$h\nu$, NH_3, 1.5 h	(91)	231
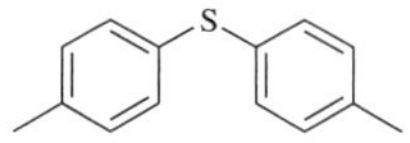	$h\nu$, DMF, 3 h	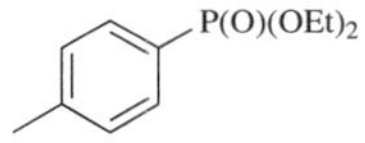 (75) + 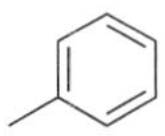 (21)	67
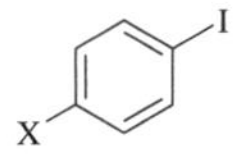	$h\nu$, NH_3	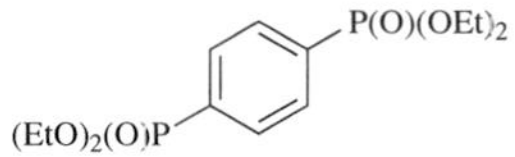	
X			
Cl	1.5 h	(59)	231
Br	4 h	(56)	231
I	3.4 h	(87)	232

TABLE VIII. OTHER NUCLEOPHILES (*Continued*)

C. Nucleophiles Derived from Phosphorus (*Continued*)

Substrate	Nucleophile	Conditions	Product(s) and Yield(s) (%)	Refs.
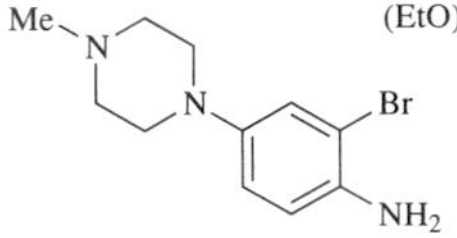	$(EtO)_2PO^-$	$h\nu$, NH_3, 3 h	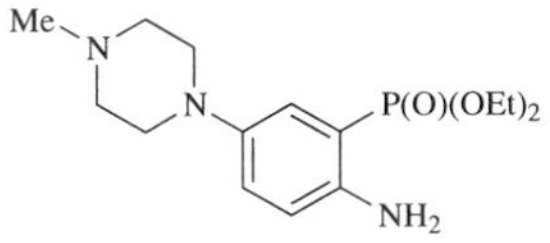 (60)	233
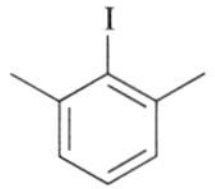		$h\nu$, NH_3, 1 h	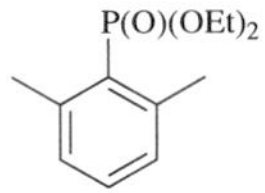 (87)	232
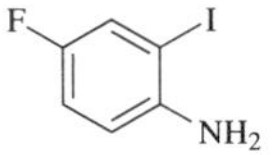		$h\nu$, NH_3, 1.5 h	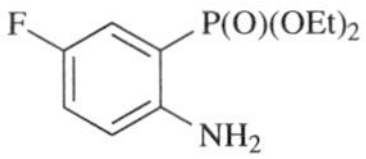 (90)	233
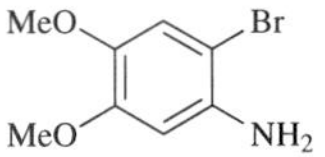		$h\nu$, NH_3, 3 h	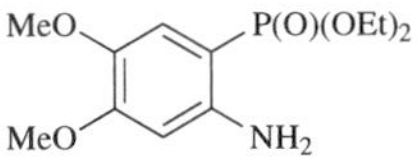 **I** (20)	64
		$h\nu$, NH_3, KI, 3 h	**I** (75)	64
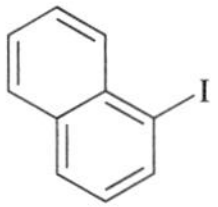		$h\nu$, NH_3, 2.2 h	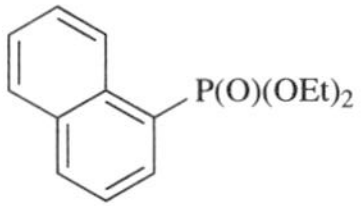 **I** (93)	232
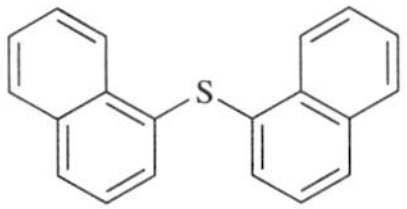		$h\nu$, DMF, 20 h	**I** (28) + 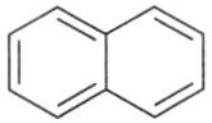 (33)	67

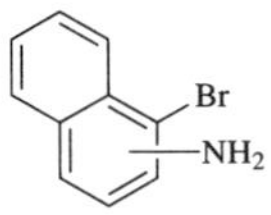Br NH$_2$	$h\nu$, NH_3	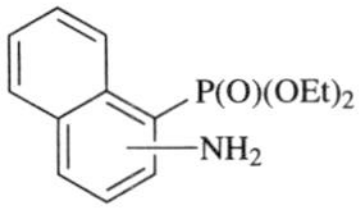P(O)(OEt)$_2$ NH$_2$ 2-NH_2 3 h (85) 4-NH_2 1 h (60)	237, 234 234
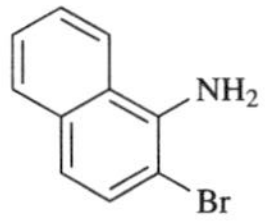NH$_2$ Br	$h\nu$, NH_3, 3 h	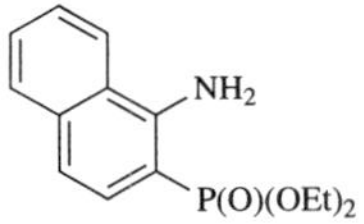NH$_2$ P(O)(OEt)$_2$ (75)	237, 234
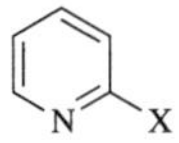N X	$h\nu$, MeCN, THF	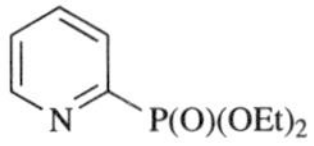N P(O)(OEt)$_2$	
X			
Br	2.5 h	(78)	235
Br	NaI, 3 h	(81)	65
I	2.5 h	(78)	235
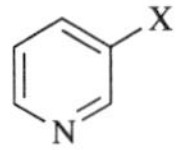X N		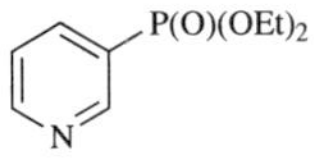P(O)(OEt)$_2$ N	
X			
Br	$h\nu$, MeCN, THF, 2.5 h	(40)	235
Br	$h\nu$, MeCN, THF, NaI, 5 h	(80)	65
I	Sunlight, MeCN, THF, 7 h	(70)	235
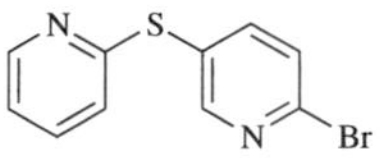N S N Br	$h\nu$, NH_3, 3 h	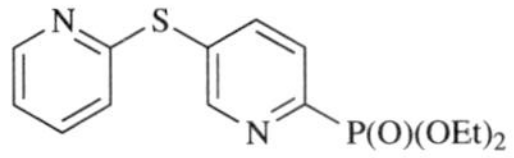N S N P(O)(OEt)$_2$ (76)	234
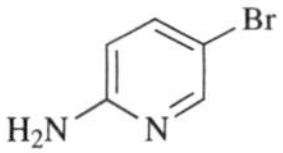Br H$_2$N N	$h\nu$, NH_3, 3 h	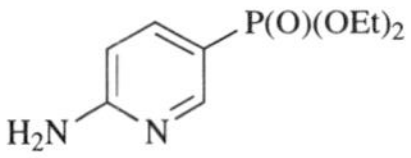P(O)(OEt)$_2$ H$_2$N N (60)	234

TABLE VIII. OTHER NUCLEOPHILES (*Continued*)

C. Nucleophiles Derived from Phosphorus (*Continued*)

Substrate	Nucleophile	Conditions	Product(s) and Yield(s) (%)	Refs.
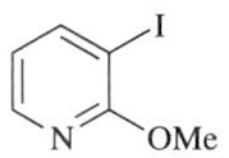	$(EtO)_2PO^-$	$h\nu$, NH_3, 1 h	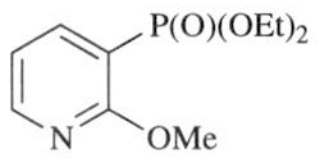 (78)	139
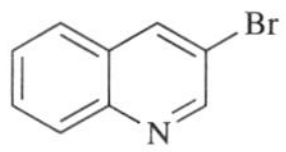		$h\nu$, MeCN, THF, 2.5 h	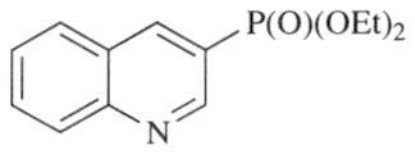 (76)	235
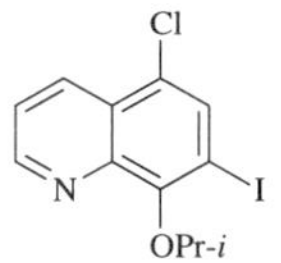		$h\nu$, NH_3, 0.5 h	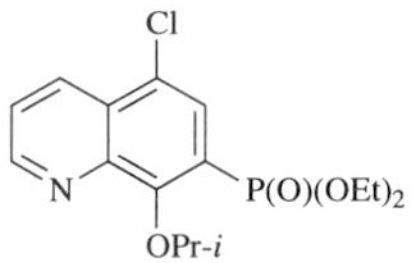 (70)	244
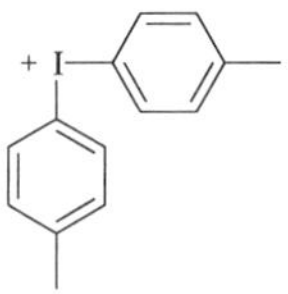	PPh_3	$h\nu$, $(CD_3)_2CO$, 23 min	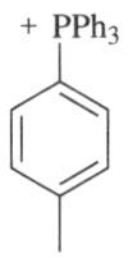 (60) + 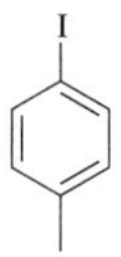 **I** (40)	359
	$P(OMe)_3$	$h\nu$, $(CD_3)_2CO$, PAIBN[b], 23 min	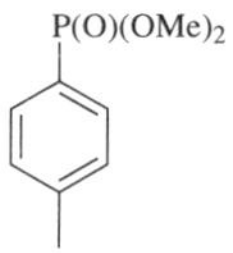 (50) + **I** (50)	359

[a] The sodium salt of the nucleophile was used.

[b] Phenylazoisobutyronitrile was added as a photoinitiator.

TABLE VIII. OTHER NUCLEOPHILES (*Continued*)

D. Nucleophiles Derived from As and Sb

Substrate	Nucleophile	Conditions	Product(s) and Yield(s) (%)	Refs.
PhBr	As^{3-}	$h\nu$, NH_3, 1 h	Ph_3As (75)	242
PhCl		1. $h\nu$, NH_3, 1 h 2. Na 3. 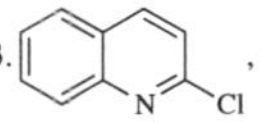 , $h\nu$, .1 h	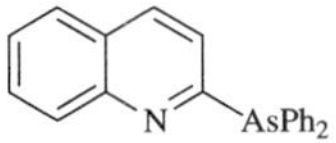 (90)	242
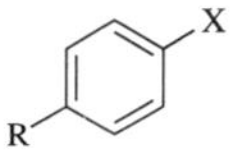	Ph_2As^-	$h\nu$, NH_3	Ph_3As + 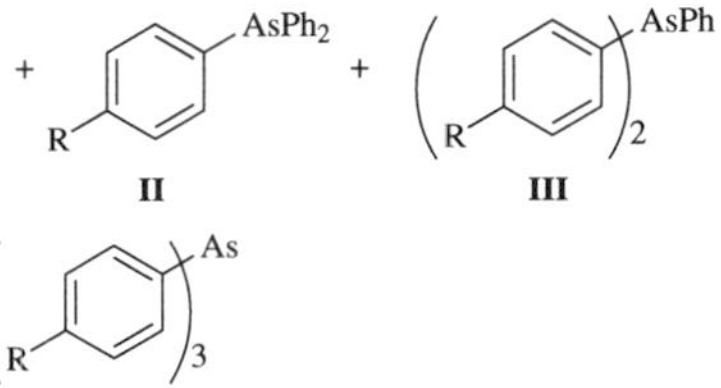 I, II, III, IV	

X	R	Conditions	I	II	III	IV	Refs.
Cl	Me	1 h	(15)	(61)	(15)	(2)	118
Cl	OMe	1 h	(30)	(48)	(19)	(3)	118
Br	Me	1 h	(21)	(62)	(15)	(1)	118
Br	OMe	5 min	(20)	(60)	(22)	(—)	118
I	Me	1 h	(20)	(50)	(25)	(4)	118
I	OMe	5 min	(18)	(78)	(5)	(—)	118
I	OMe	1 h	(12)	(56)	(27)	(5)	240

Substrate	Nucleophile	Conditions	Product(s) and Yield(s) (%)	Refs.
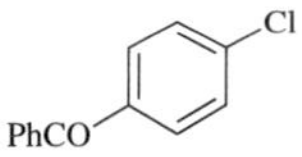		$h\nu$, NH_3, 0.5 h	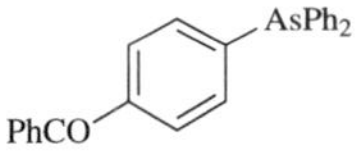 (100)	118

TABLE VIII. OTHER NUCLEOPHILES (*Continued*)

D. Nucleophiles Derived from As and Sb (*Continued*)

Substrate	Nucleophile	Conditions	Product(s) and Yield(s) (%)	Refs.
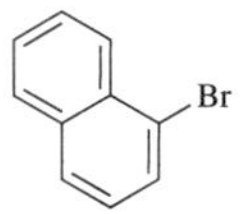	Ph_2As^-	$h\nu$, NH_3, 10 min	Ph_3As (33) + 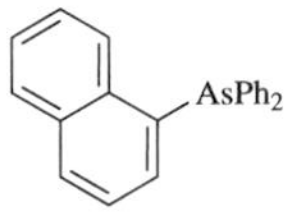 (40) + 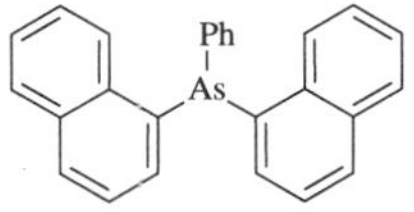 (21)	119
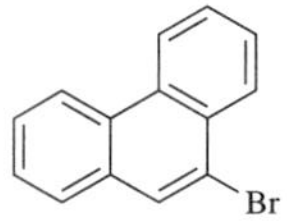		$h\nu$, NH_3, 1 h	Ph_3As (20) + 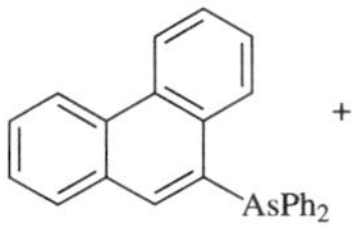 (60) + 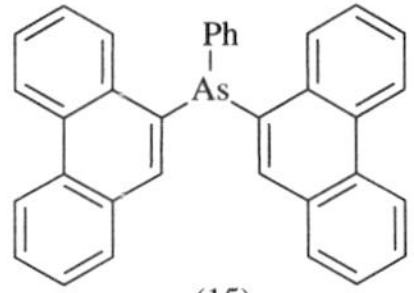 (15)	119
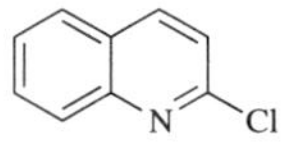		$h\nu$, NH_3, 5 min	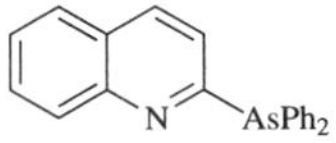 **I** (60)	119
		$h\nu$, NH_3, 1 h	**I** (76)	240
PhCl	Sb^{3-}	$h\nu$, NH_3, 1 h	Ph_3Sb (45)	242
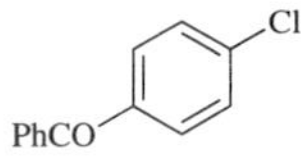	Ph_2Sb^-	$h\nu$, NH_3, 0.5 h	Ph_3Sb (30) + 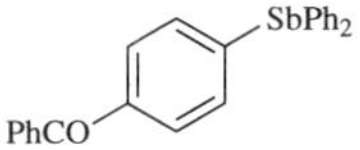 (33) + 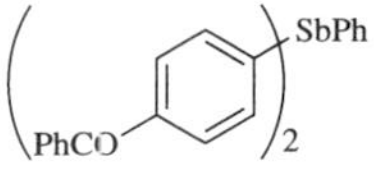 (15)	119
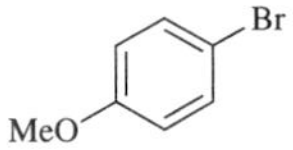		$h\nu$, NH_3, 0.5 h	Ph_3Sb (30) + 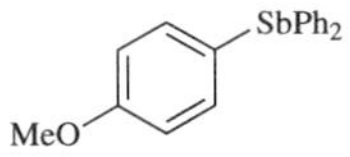 (45) + 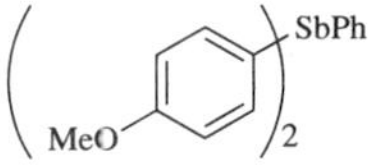 (19)	119

TABLE VIII. OTHER NUCLEOPHILES (*Continued*)

E. Nucleophiles Derived from Sulfur

Substrate	Nucleophile	Conditions	Product(s) and Yield(s) (%)	Refs.
PhI	MeS^-	1. $h\nu$, NH_3, 1 h 2. BnCl	PhSMe (12) + PhSBn (19)	48
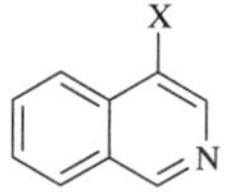		NH_3, 4 h	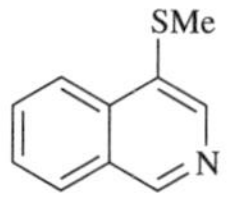 X: Cl (66); Br (80)	243
PhI	EtS^-	1. $h\nu$, NH_3, 3.3 h 2. BnCl	PhSEt (30) + PhSBn (44) + Ph_2S (3)	160
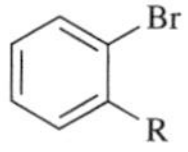		$h\nu$, NH_3	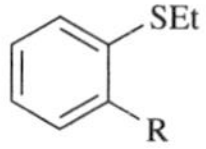	49
R: CN		0.5 h	(85)	
CHO		1 h	(70)	
COMe		1 h	(90)	
$CONH_2$		1.5 h	(85)	
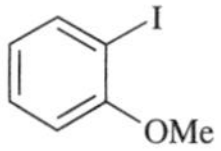		1. $h\nu$, NH_3, 0.75 h 2. MeI	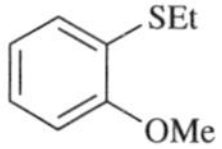 (20) + 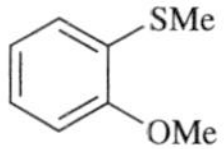 (45)	49
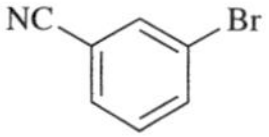		$h\nu$, NH_3, 0.5 h	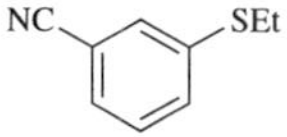 (90)	49
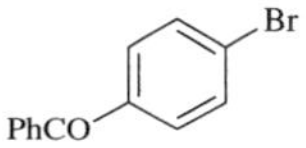		$h\nu$, NH_3, 0.5 h	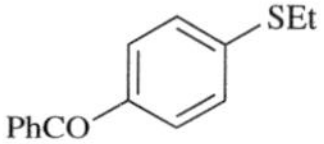 (95)	49
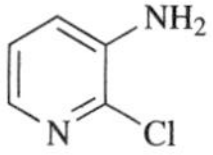		$h\nu$, NH_3, 6 h	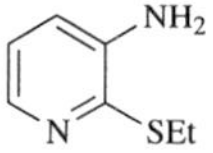 (35)	49

TABLE VIII. OTHER NUCLEOPHILES (*Continued*)

E. Nucleophiles Derived from Sulfur (*Continued*)

Substrate	Nucleophile	Conditions	Product(s) and Yield(s) (%)	Refs.
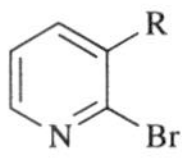	EtS^-	$h\nu$, NH_3	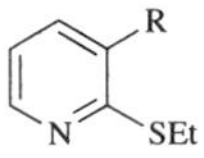	49
R				
H		2 h	(74)	
CN		0.5 h	(87)	
OMe		1.5 h	(61)	
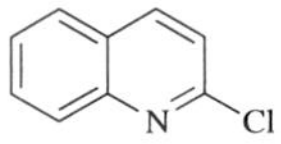		$h\nu$, NH_3, 0.75 h	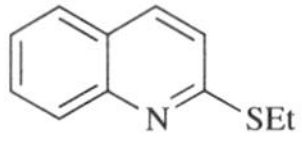 (85)	49
PhI	n-BuS^-	1. $h\nu$, NH_3, 3 h 2. BnCl	PhSBu-*n* (14) + PhSBn (16)	48
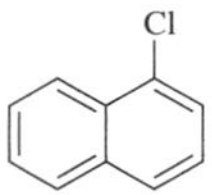		$h\nu$, NH_3, 2.8 h	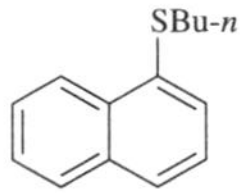 (81) + 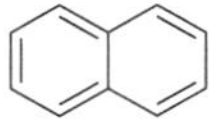 (18)	45
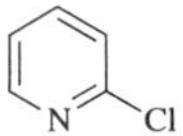		1. $h\nu$, NH_3, 3 h 2. BnCl	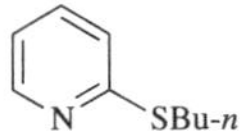 (72-85) + 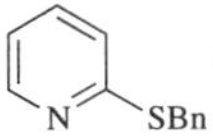 (14-15)	48
PhI	t-BuS^-	1. $h\nu$, NH_3, 3 h 2. BnCl	PhSBu-*t* (36) + PhSBn (33)	48
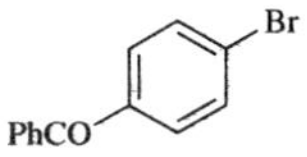		e^-, DMSO	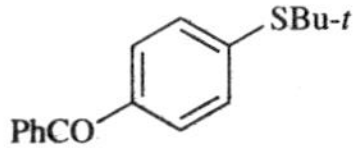 (60)	88, 89
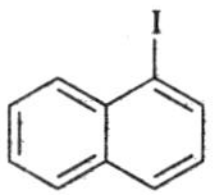		$h\nu$, NH_3, 3 h	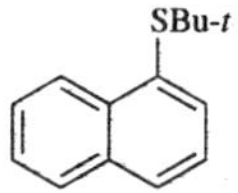 (88)	48

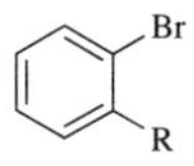 R: CN; COMe	BnS^-	1. hν, NH_3, 6 h 2. MeI	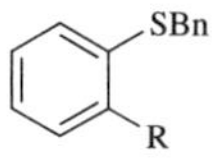 + 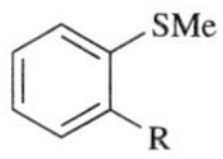 CN: (15) (85) COMe: (66) (6)	49
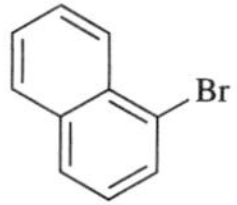		1. hν, NH_3, 3 h 2. MeI	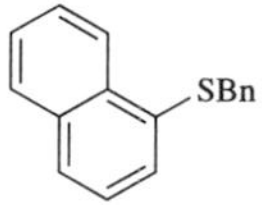 (2-4) + 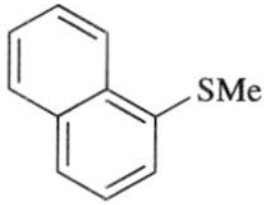 (15-20)	48
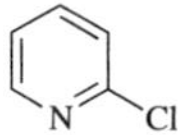		1. hν, NH_3, 3 h 2. MeI	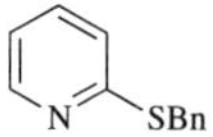 (11) + 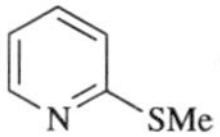 (23)	48
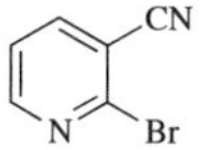		hν, NH_3, 15 min	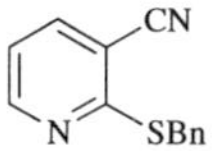 (82)	49
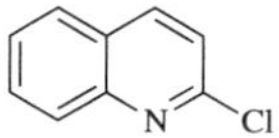		hν, NH_3, 1.5 h	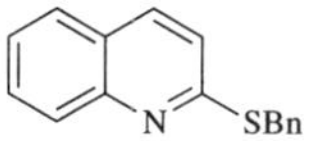 (69)	48
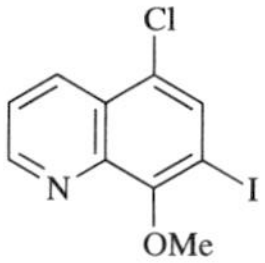		1. hν, NH_3, 2 h 2. MeI	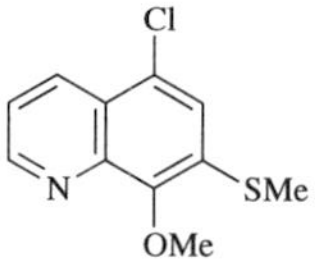 (54) + 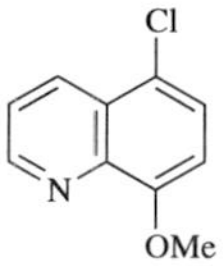 (41)	244
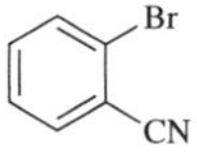	$HO(CH_2)_2S^-$	hν, NH_3, 2 h	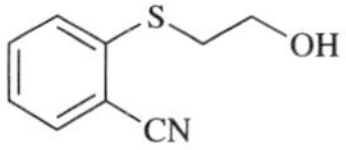 (85)	49
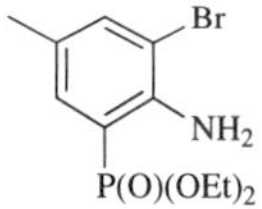		hν, NH_3, 1.5 h	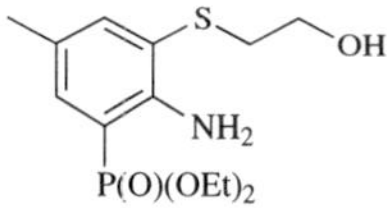 (65)	233

TABLE VIII. OTHER NUCLEOPHILES (*Continued*)

E. Nucleophiles Derived from Sulfur (*Continued*)

Substrate	Nucleophile	Conditions	Product(s) and Yield(s) (%)		Refs.
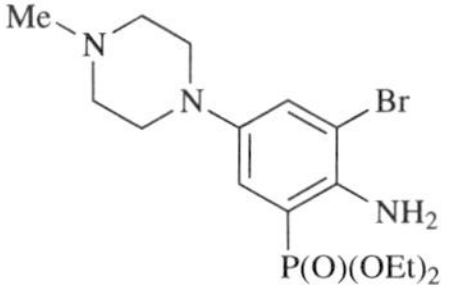	$HO(CH_2)_2S^-$	*h*ν, NH_3, 1 h	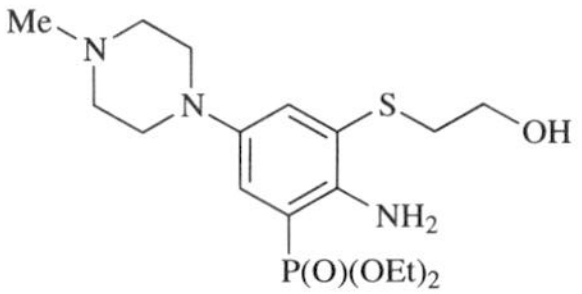 (80)		233
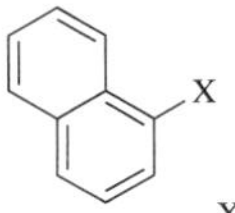		*h*ν, NH_3	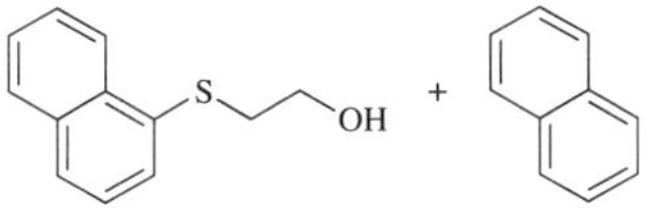		45
X					
Cl		2.75 h	(71)	(19)	
I		2.5 h	(74)	(8)	
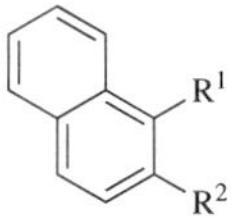		*h*ν, NH_3	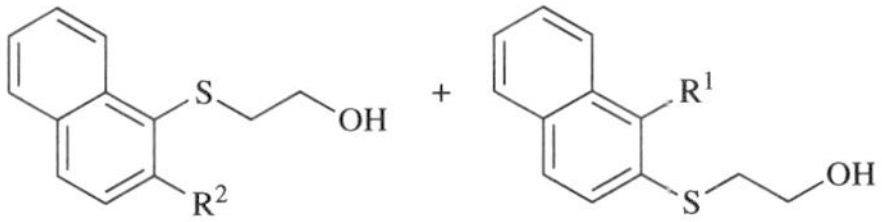		237
R^1 R^2					
Br NH_2		3 h	(88)	(—)	
NH_2 Br		1 h	(—)	(85)	
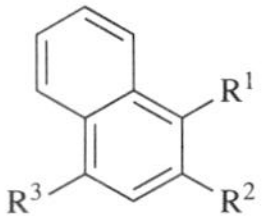		*h*ν, NH_3	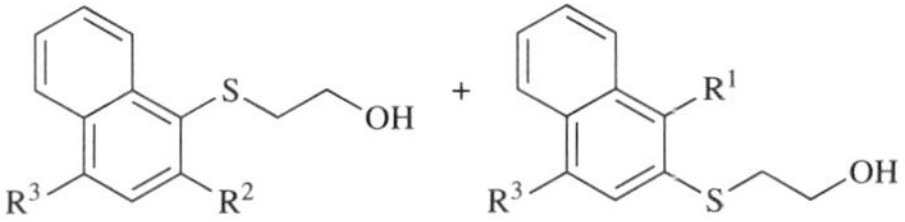		234
R^1 R^2 R^3					
I $P(O)(OEt)_2$ H		1.5 h	(88)	(—)	
I H $P(O)(OEt)_2$		1.5 h	(75)	(—)	
$P(O)(OEt)_2$ I H		1 h	(—)	(75)	

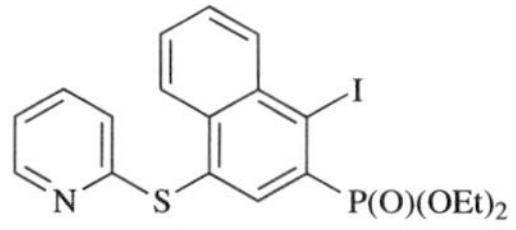		$h\nu$, NH_3, 0.5 h	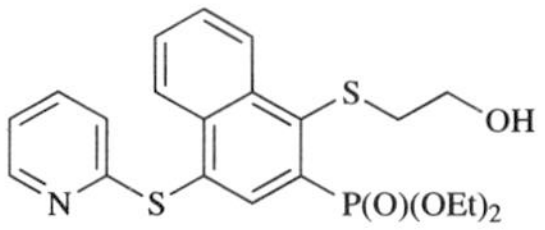 (74)	234
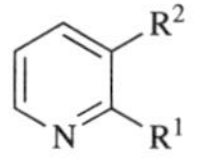		$h\nu$, NH_3	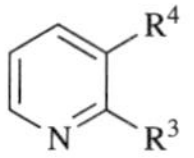	
R^1 R^2			R^3 R^4	
Cl H		3 h	$S(CH_2)_2OH$ H (85)	245
Br H		0.5 h	$S(CH_2)_2OH$ H (100)	49
Cl CF_3		0.75 h	$S(CH_2)_2OH$ CF_3 (97)	245
Cl NH_2		6 h	$S(CH_2)_2OH$ NH_2 (45)	49
Br CN		1 h	$S(CH_2)_2OH$ CN (100)	49
Br OMe		1 h	$S(CH_2)_2OH$ OMe (60)	49
OMe I		1 h	OMe $S(CH_2)_2OH$ (80)	139
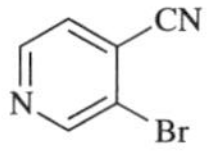		$h\nu$, NH_3, 1 h	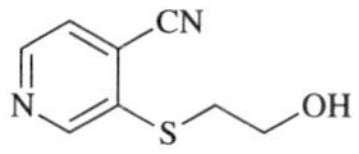 (100)	49
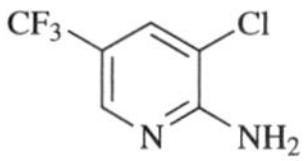		$h\nu$, NH_3, 15 min	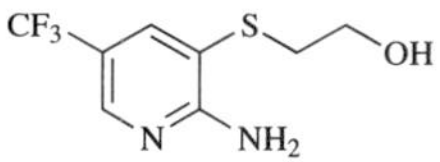 (85)	245
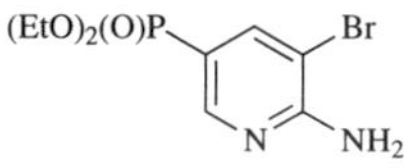		$h\nu$, NH_3, 1 h	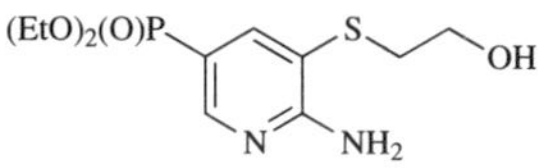 (60)	234
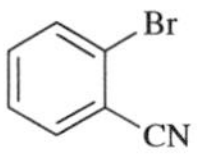	$EtO_2CCH_2S^-$	1. $h\nu$, NH_3, 1 h 2. MeI	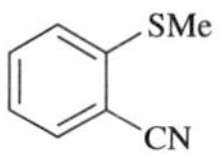 (100)	49
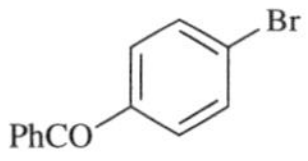		$h\nu$, NH_3, 1 h	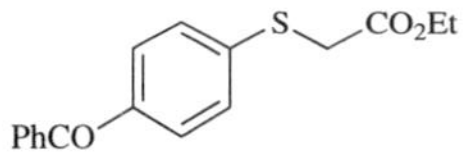 (68)	49

TABLE VIII. OTHER NUCLEOPHILES (*Continued*)

E. Nucleophiles Derived from Sulfur (*Continued*)

Substrate	Nucleophile	Conditions	Product(s) and Yield(s) (%)	Refs.
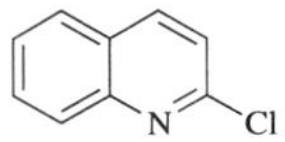	$EtO_2CCH_2S^-$	1. *hν*, NH_3, 6 h 2. MeI	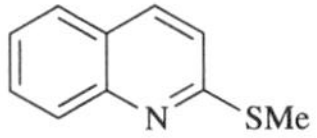 (75)	49
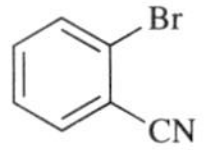	$EtO_2C(CH_2)_2S^-$	1. *hν*, NH_3, 3 h 2. MeI	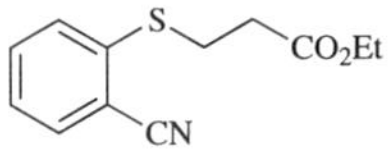 (70) + 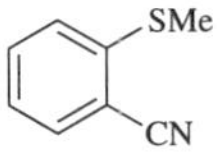 (15)	49
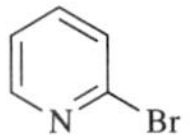		*hν*, NH_3, 3 h	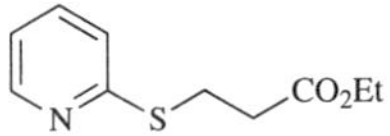 (32)	49
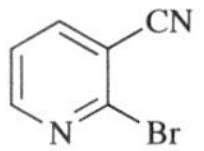		*hν*, NH_3, 0.5 h	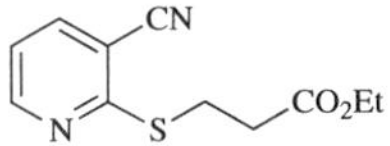 (90)	49
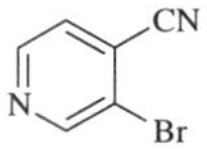		*hν*, NH_3, 0.5 h	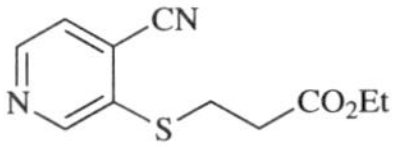 (68)	49
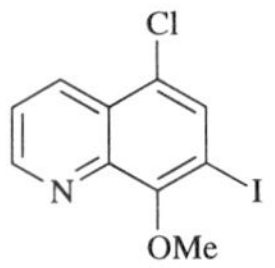		1. *hν*, NH_3, 0.5 h 2. MeI	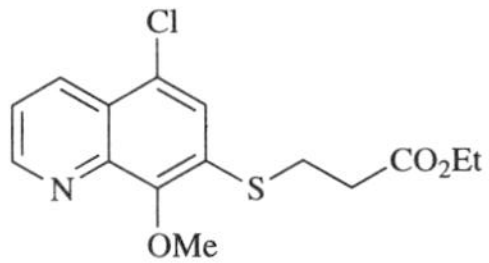 (65) + 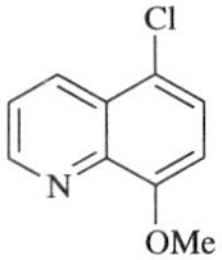 (30) + 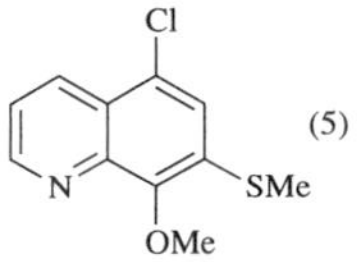 (5)	244
$PhN_2^+BF_4^-$	$MeCOS^-$	DMSO, 0.5 h	PhSCOMe (60) + Ph_2S (2) + Ph_2S_2 (5)	34

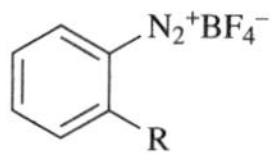	DMSO	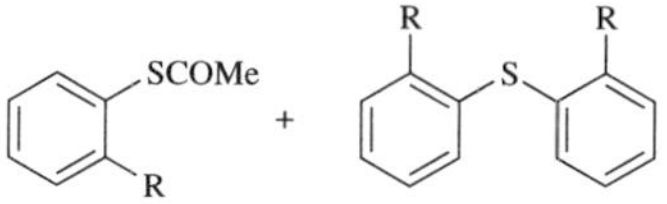		
R				
Me	0.5 h	(60)	(5)	34
Ph	0.6 h	(56)	(—)	267
SPh	0.7 h	(55)	(—)	267
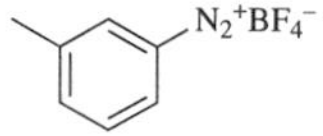	DMSO, 0.5-0.8 h	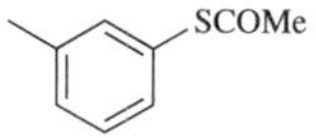 (47)		266
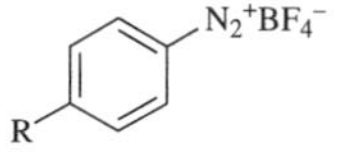	DMSO	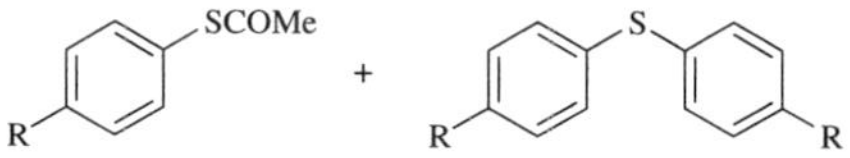		34
R				
Me	0.7 h	(55)	(4)	
OMe	0.5 h	(60)	(—)	
NO_2	0.75 h	(47)	(—)	
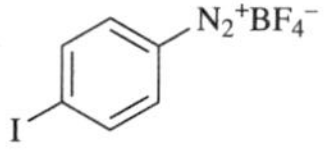	DMSO, 1.5 h	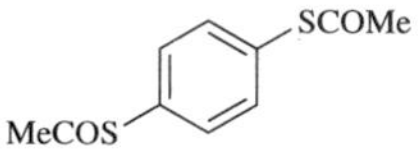 **I** (3) + 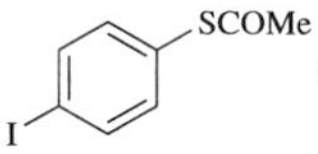 (41)		34
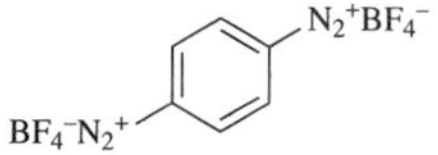	DMSO, 1 h	**I** (34) + 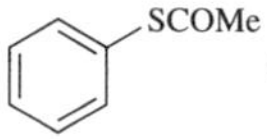 (8)		34
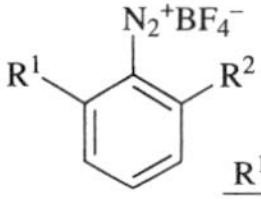	DMSO	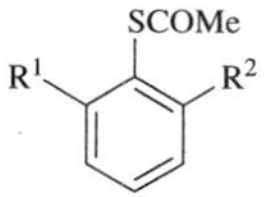		
R^1 / R^2				
Me / Me	0.5 h	(48)		34
Et / Et	0.75 h	(36)		267
Me / Pr-*i*	1 h	(45)		267
Pr-*i* / Pr-*i*	1 h	(36)		267

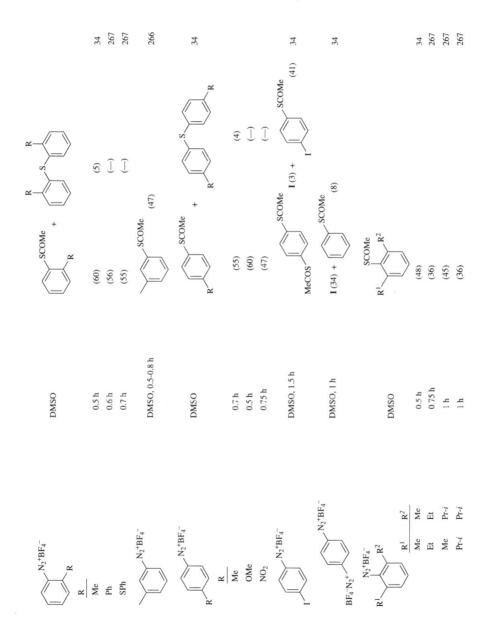

TABLE VIII. OTHER NUCLEOPHILES (*Continued*)

E. Nucleophiles Derived from Sulfur (*Continued*)

Substrate	Nucleophile	Conditions	Product(s) and Yield(s) (%)	Refs.
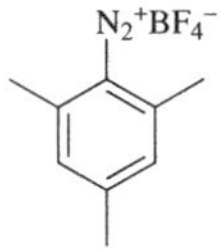	$MeCOS^-$	DMSO, 0.5 h	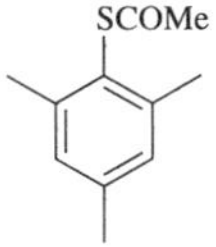 (39)	34
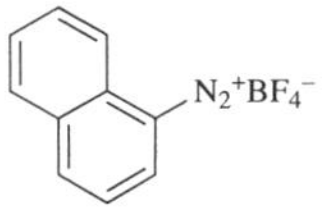		DMSO, 0.8 h	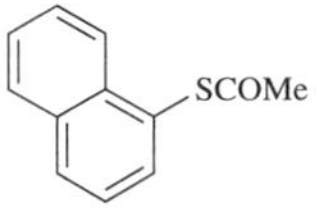 (60) + 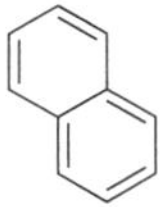 (8)	34
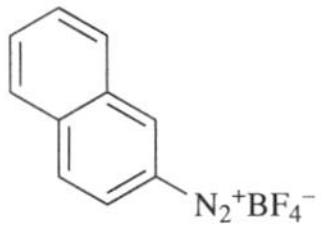		DMSO, 0.5 h	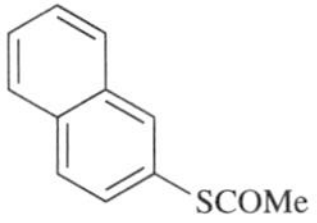 (60)	34
$PhN_2^+BF_4^-$	$PhCOS^-$	DMSO, 1 h	PhSCOPh (62)	266
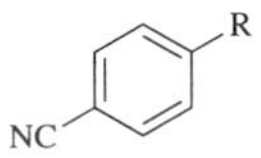		DMSO	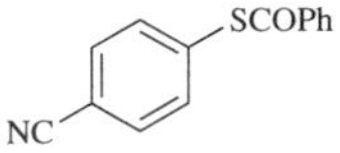 R: $N_2^+BF_4^-$ 0.5 h (87); N_2SOPh 22 min (76)	266
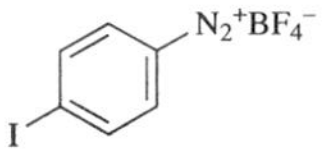		DMSO, 1 h	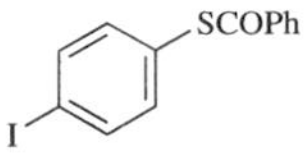 (74) + 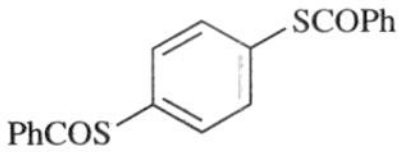 (6)	266
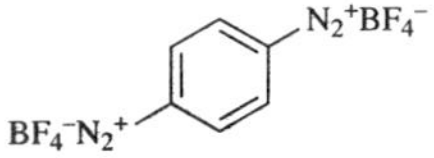		DMSO, 2 h	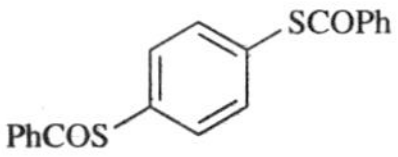 (30)	266
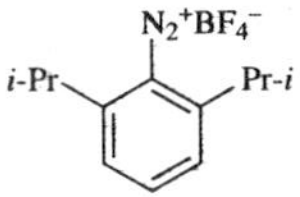		DMSO, 1 h	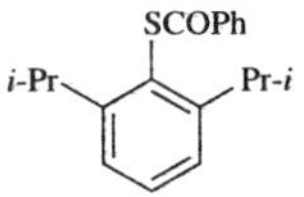 (57)	266

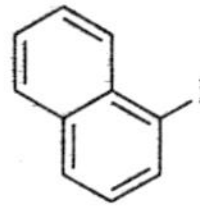 $N_2^+BF_4^-$		DMSO, 4 h	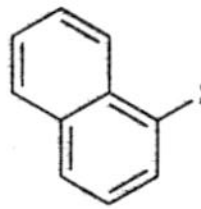 SCOPh (70)	266
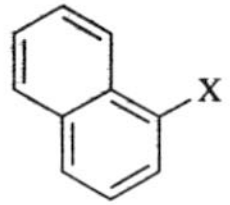	$^-S(CH_2)_2S^-$	$h\nu$, NH_3	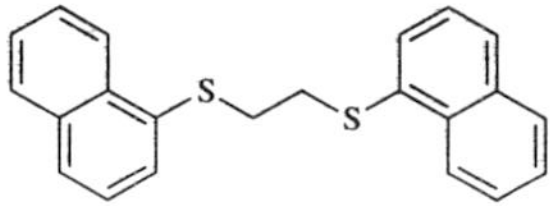 **I** + 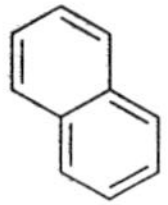 **II** + 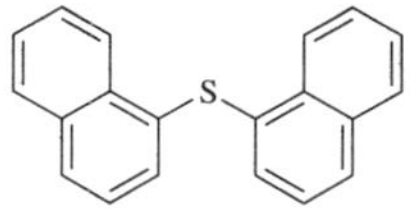 **III**	246
X: Br		3.3 h	**I** (50), **II** (—), **III** (—)	
X: I		1.7 h	**I** (50), **II** (17), **III** (18)	
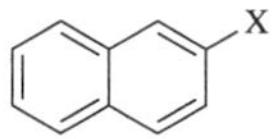		$h\nu$, NH_3, 2 h	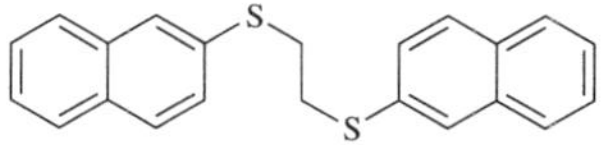 **I** + 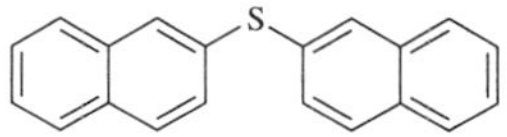 **II** + 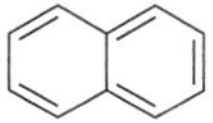 **III**	246
X: Br			**I** (27), **II** (18), **III** (—)	
X: I			**I** (40), **II** (14), **III** (10)	
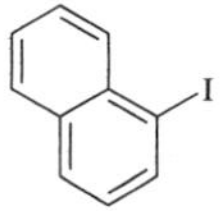	$^-S(CH_2)_3S^-$	$h\nu$, NH_3, 1 h	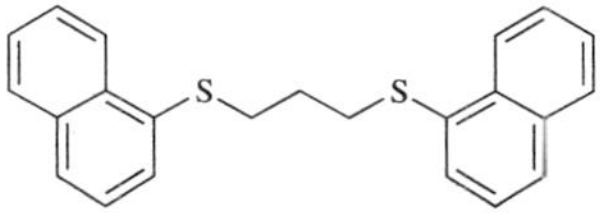 (50) + 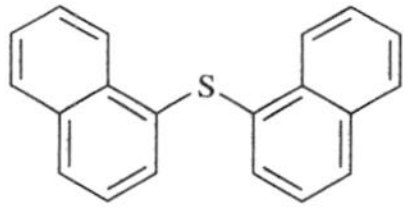 (10) + 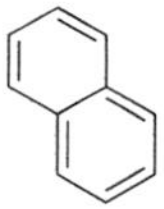 (24)	246

TABLE VIII. OTHER NUCLEOPHILES (*Continued*)

E. Nucleophiles Derived from Sulfur (*Continued*)

Substrate	Nucleophile	Conditions	Product(s) and Yield(s) (%)	Refs.
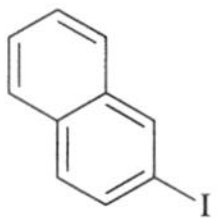		$h\nu$, NH_3, 3 h	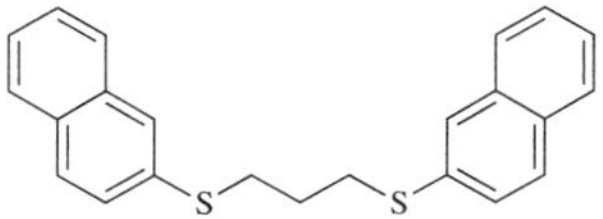 **I** (13)	246
		$h\nu$, DMSO, 1 h	**I** (27) + $C_{10}H_8$ (10)	246
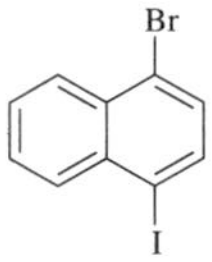		$h\nu$, NH_3, 3 h	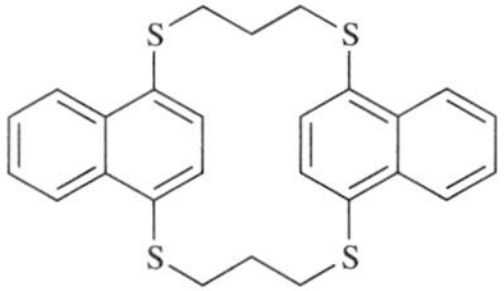 (13)[a] + 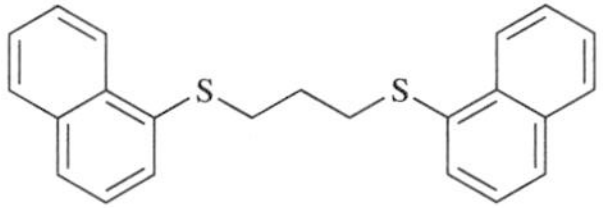 (42)	360
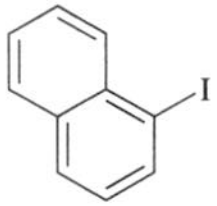	$^-S(CH_2)_4S^-$	$h\nu$, NH_3, 3 h	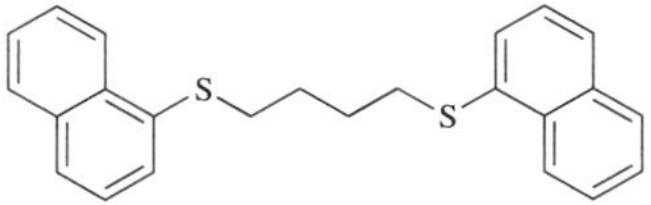 (25) + 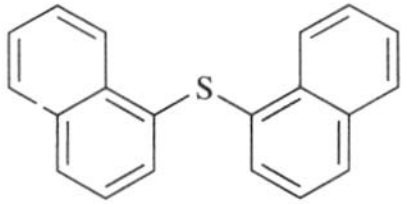 (9) + 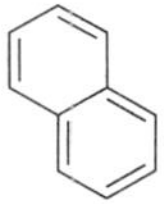 (25)	246

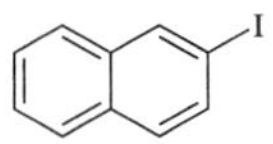		$h\nu$, NH_3, 2 h	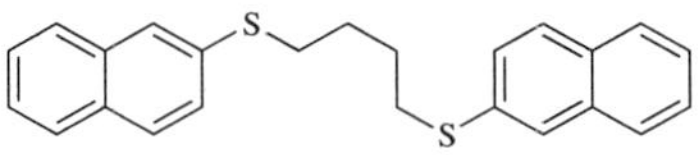 (39) + 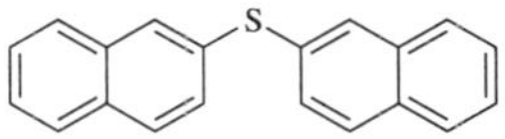 (14)	246
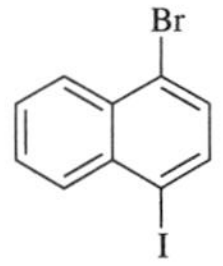		$h\nu$, NH_3, 3 h	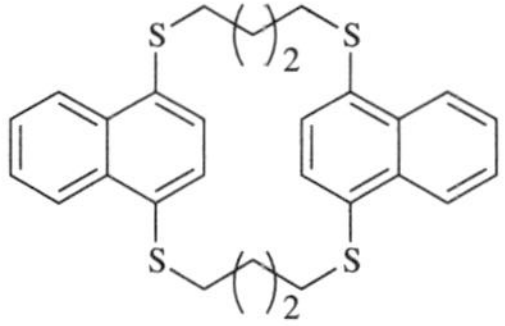 (16) + 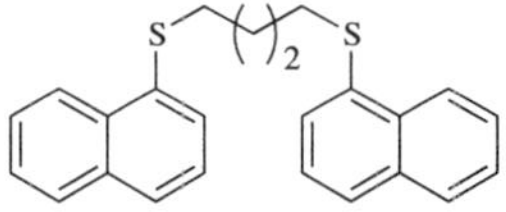 (48)	360
	$^{-}S(CH_2)_2XS(CH_2)_2^{-}$ ($^{-}S(CH_2)_2$–X–$(CH_2)_2S^{-}$)	$h\nu$, NH_3, 3 h	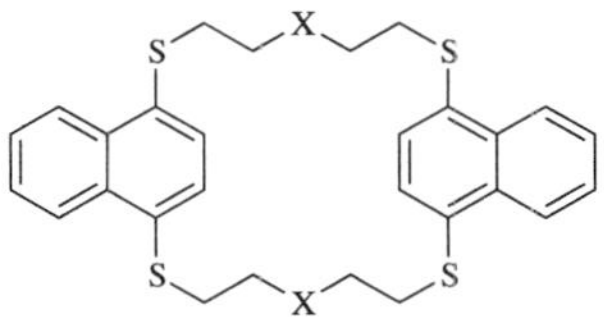 **I** + 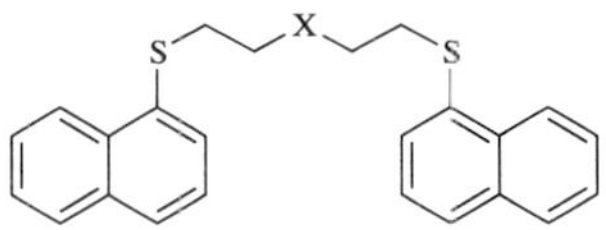 **II**	360
X: O			**I** (16), **II** (45)	
X: S			**I** (20), **II** (50)	

TABLE VIII. OTHER NUCLEOPHILES (*Continued*)

E. Nucleophiles Derived from Sulfur (*Continued*)

Substrate	Nucleophile	Conditions	Product(s) and Yield(s) (%)	Refs.
PhBr	PhS^-	$h\nu$, NH_3, 2 h	Ph_2S **I** (23)	247
		e^-, DMSO, mediator	**I** (56) + C_6H_6 (37)	82
PhI		$h\nu$, NH_3, 1.2 h	**I** (94)	247
$PhN_2^+BF_4^-$		DMSO, 3.5 h	**I** (97)	261
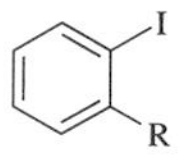		$h\nu$, NH_3	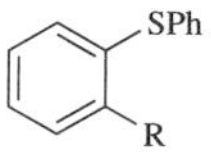 **I**	
R				
Me		2.8 h	(68)	247
OMe		1.5 h	(91)	247
NH_2		3 h	(10)	234
$P(O)(OEt)_2$		3 h	(15)	234
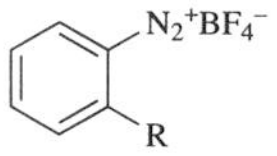	R			
	Me	DMSO, 5.5 h	**I** (82)	261
	F	DMSO, 3 h	**I** (80)	262
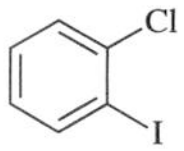		$h\nu$, NH_3	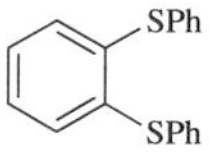 1 h (89)	263
			1.5 h (77)	264
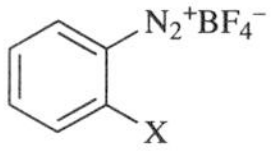		DMSO	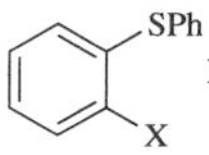 **I** + 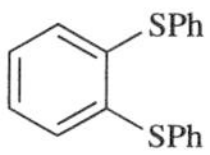 **II** + 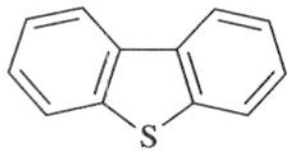 **III**	262
X				
Cl		5.3 h	**I** (13) + **II** (48) + **III** (4)	
I		6 h	**I** (8) + **II** (28) + **III** (4)	

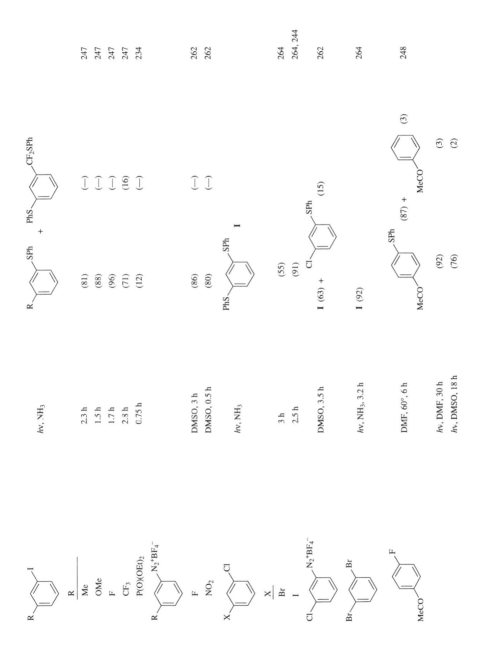

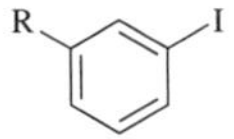	$h\nu$, NH_3	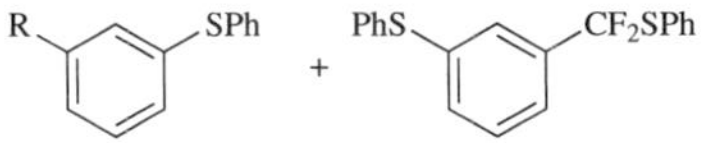		
R				
Me	2.3 h	(81)	(—)	247
OMe	1.5 h	(88)	(—)	247
F	1.7 h	(96)	(—)	247
CF_3	2.3 h	(71)	(16)	247
$P(O)(OEt)_2$	0.75 h	(12)	(—)	234
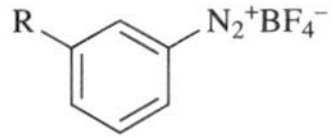				
F	DMSO, 3 h	(86)	(—)	262
NO_2	DMSO, 0.5 h	(80)	(—)	262
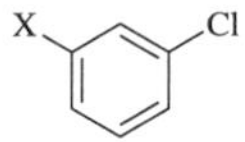	$h\nu$, NH_3	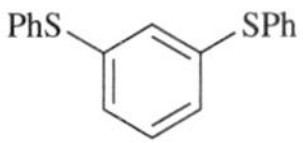 **I**		
X				
Br	3 h	(55)		264
I	2.5 h	(91)		264, 244
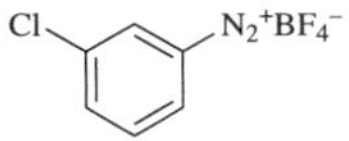	DMSO, 3.5 h	**I** (63) + 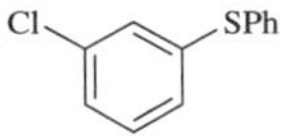 (15)		262
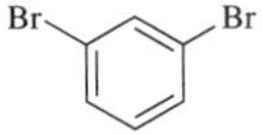	$h\nu$, NH_3, 3.2 h	**I** (92)		264
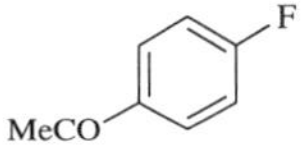	DMF, 60°, 6 h	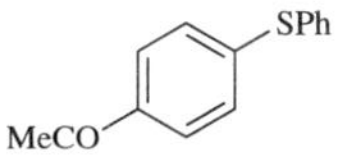 (87) +	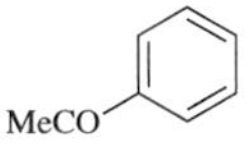 (3)	248
	$h\nu$, DMF, 30 h	(92)	(3)	
	$h\nu$, DMSO, 18 h	(76)	(2)	

TABLE VIII. OTHER NUCLEOPHILES (*Continued*)

E. Nucleophiles Derived from Sulfur (*Continued*)

Substrate	Nucleophile	Conditions	Product(s) and Yield(s) (%)	Refs.
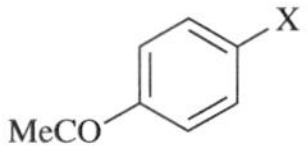	PhS^-		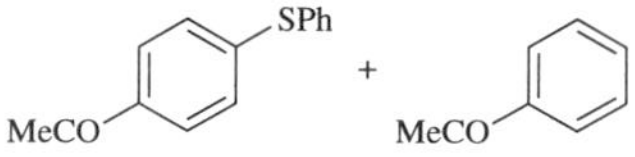	
X				
Cl		DMF, 60°, 4 h	(87), (3)	248
Cl		*h*ν, DMF, 20 h	(84), (3)	248
Cl		*h*ν, DMSO, 12 h	(74), (3)	248
Br		DMF, 60°, 1 h	(97), (3)	248
Br		*h*ν, DMF, 7 h	(96), (4)	248
Br		*h*ν, DMSO, 5 h	(74), (3)	248
Br		*h*ν, MeCN, 4 h	(72), (3)	248
Br		e^-, MeCN	(95), (—)	255, 254
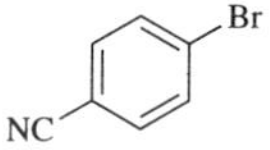		e^-, MeCN	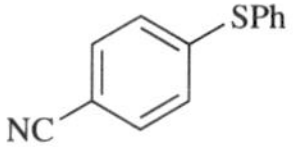 (80)	88
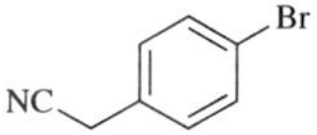		*h*ν, DMSO, H_2O, 50°, 10 h	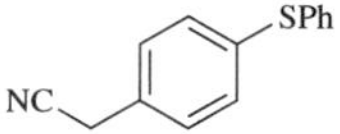 (91) + 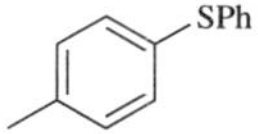 (7)	185
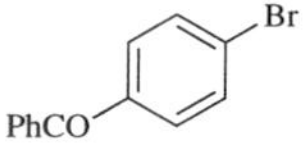			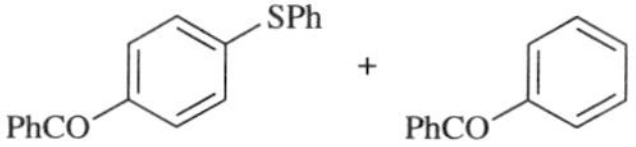	
		DMF, 60°, 0.5 h	(98), (3)	248
		*h*ν, DMF, 1 h	(96), (5)	248
		*h*ν, DMSO, 3 h	(97), (3)	248
		*h*ν, MeCN, 1 h	(87), (4)	248
		e^-, MeCN	(95), (—)	89, 256
		e^-, DMSO	(80), (10)	40

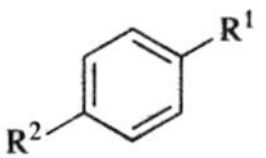

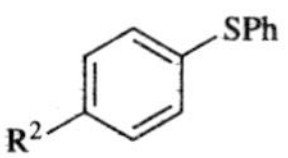 + PhR^2

R^1	R^2	Conditions	Product	PhR^2	Ref.
I	Me	$h\nu$, NH_3, 6 h	(72)	(—)	247
I	OMe	$h\nu$, NH_3, 0.5 h	(76)	(—)	247
I	OPh	$h\nu$, NH_3, 2 h	(92)	(—)	247
I	CN	e^-, MeCN	(20)	(80)	88
I	$P(O)(OEt)_2$	$h\nu$, NH_3, 1 h	(10)	(—)	234
$N_2^+BF_4^-$	F	DMSO, 4 h	(75)	(—)	262
$N_2^+BF_4^-$	Me	DMSO, 7.5 h	(79)	(—)	261
$N_2^+BF_4^-$	Me	$h\nu$, DMSO, 0.8 h	(43)	(—)	262
$N_2^+BF_4^-$	NO_2	DMSO, 15 min	(79)	(—)	261
$N_2^+BF_4^-$	OMe	DMSO, 3.5 h	(72)	(—)	261

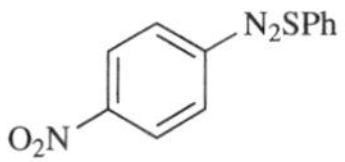

DMSO, 15 min

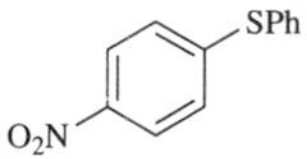 (93) 262

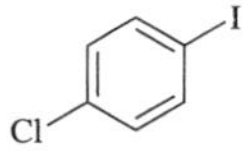

$h\nu$, NH_3

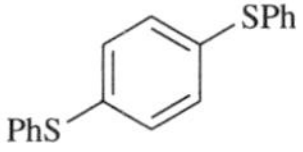

Time	Yield	Ref.
1 h	(77)	263
2 h	(89)	264

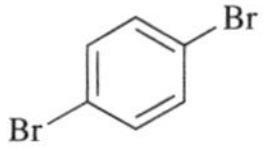

$h\nu$, NH_3, 5 h

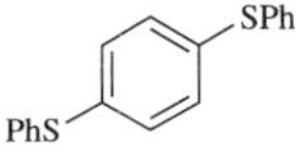 (64) 264

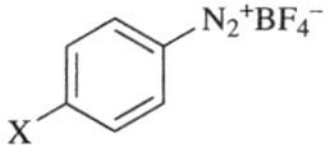

DMSO

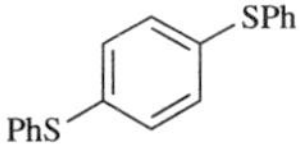 + 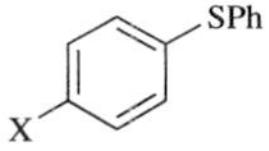 262

X	Time	Yield (first product)	Yield (second product)
Cl	3 h	(66)	(12)
Br	5.3 h	(69)	(8)
I	6.5 h	(63)	(13)

TABLE VIII. OTHER NUCLEOPHILES (*Continued*)

E. Nucleophiles Derived from Sulfur (*Continued*)

Substrate	Nucleophile	Conditions	Product(s) and Yield(s) (%)	Refs.
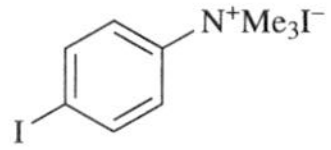N+Me3I− I	PhS^-	$h\nu$, NH_3, 3 h	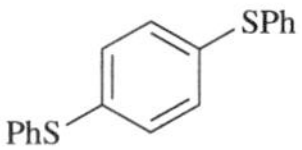SPh PhS (95)	264
		$h\nu$, H_2O, 1.7 h	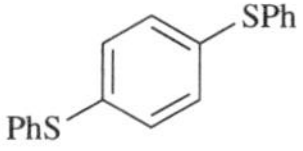SPh PhS (12) + 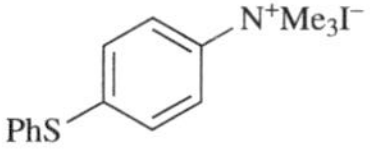N+Me3I− PhS (38)	340
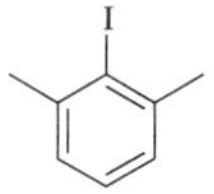I		$h\nu$, NH_3, 2.3 h	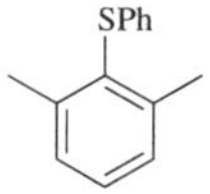SPh (19) + 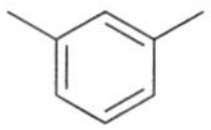 (28) + Ph_2S_2 (36)	247
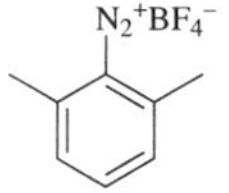N2+BF4−		DMSO, 2 h	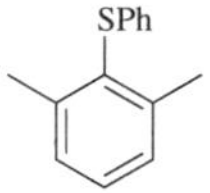SPh (76)	261
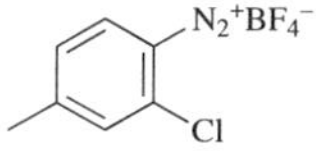N2+BF4− Cl		DMSO, 5.3 h	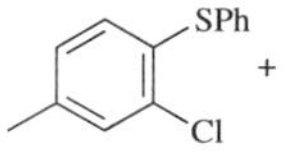SPh Cl (9) + 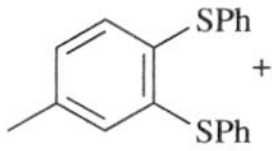SPh SPh (34) + 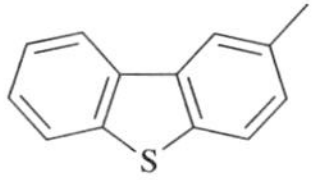S (3)	262
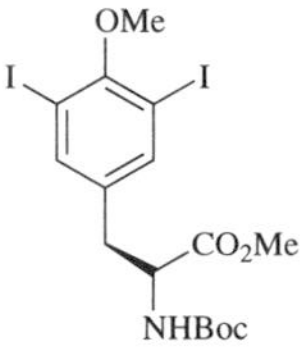OMe I I CO2Me NHBoc		$h\nu$, NH_3, 1 h	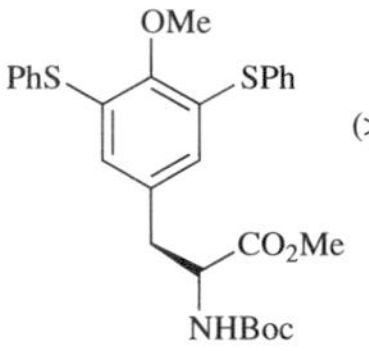OMe PhS SPh CO2Me NHBoc (>90)	265

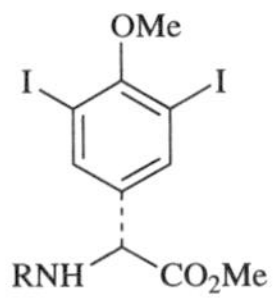	$h\nu$, NH_3, 1 h	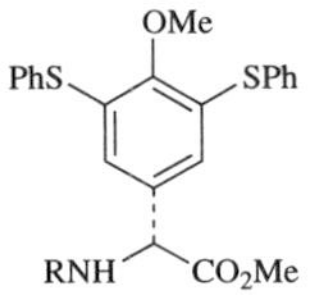 R: H (96)[b]; Boc (81)[c]	265
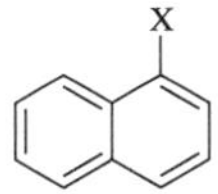 X		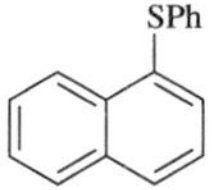	
Br	e^-, MeCN	(32)	88
Br	e^-, MeCN, mediator	(87)	258, 257
Br	e^-, DMSO	(100)	88
I	$h\nu$, NH_3, 1.25 h	(85)	247
$N_2^+BF_4^-$	DMSO, 3.5 h	(82)	261
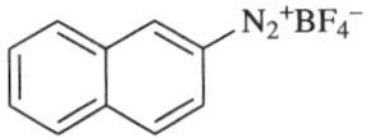	DMSO, 5.5 h	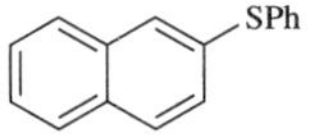 (78)	261
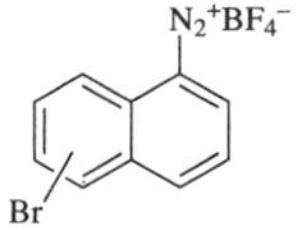	DMSO	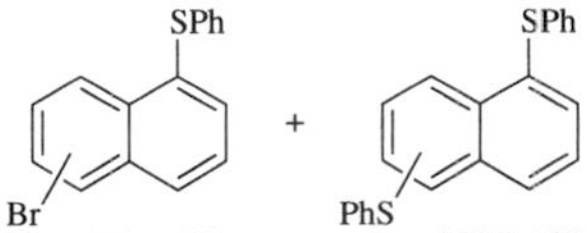	127
	6 h	4-Br (20) + 4-PhS (62)	
	2 h	5-Br (22) + 5-PhS (56)	
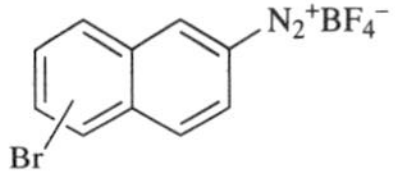	DMSO	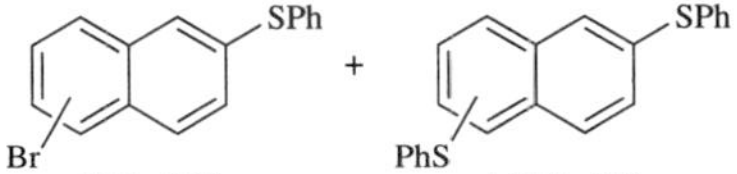	127
	1. 25°, 1 h 2. 60°, 3 h	6-Br (19) + 4-PhS (45)	
	25° overnight	1-Br (0) + 1-PhS (57)	

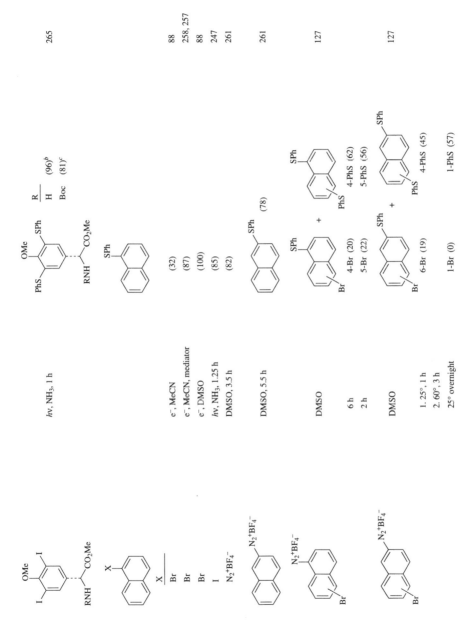

TABLE VIII. OTHER NUCLEOPHILES (*Continued*)

E. Nucleophiles Derived from Sulfur (*Continued*)

Substrate	Nucleophile	Conditions	Product(s) and Yield(s) (%)	Refs.
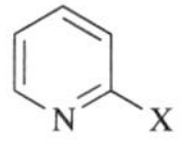	PhS^-		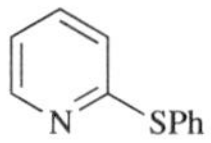	
X				
Br		DMF, 80°, 4 h	(52)	249
Br		*h*ν, NH_3, 1.5 h	(21)	134
Br		HMPA, 80°, 4 h	(65)	249
I		DMF, 80°, 4 h	(58)	249
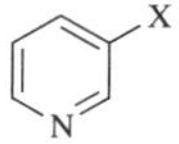			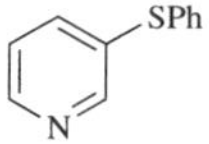	
X				
Br		DMF, 80°, 4 h	(24)	249
Br		*h*ν, NH_3, –65°, 0.75 h	(23)	47
I		*h*ν, MeONa, MeOH, 4.3 h	(63) + C_5H_5N (17)	252
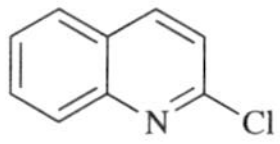		*h*ν, MeCN, 4 h	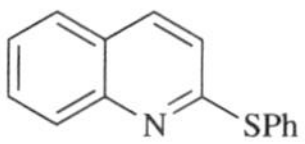 **I** (94)	250
		e^-, NH_3	**I** (96)	260, 259
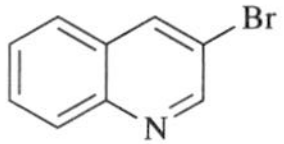		*h*ν, MeONa, MeOH, 3.9 h	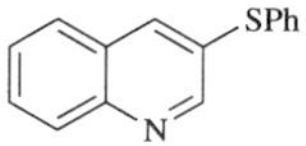 (78) + 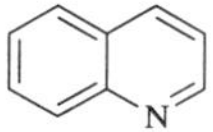 (21)	252
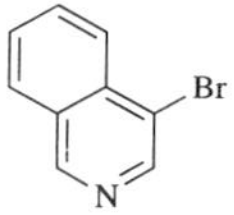		*h*ν, MeONa, MeOH, 1.5 h	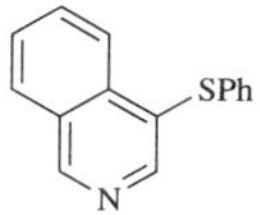 (65) + 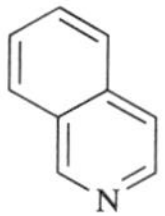 (35)	251, 252

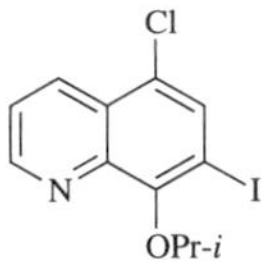	$h\nu$, NH_3, 2 h	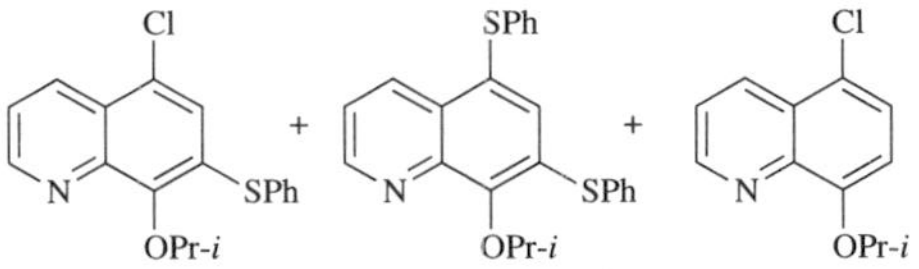 (70) + (10) + (9)	244
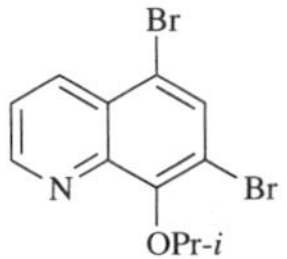	$h\nu$, NH_3, 5 h	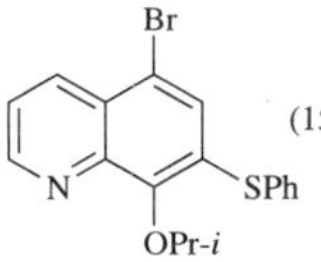 (15) + 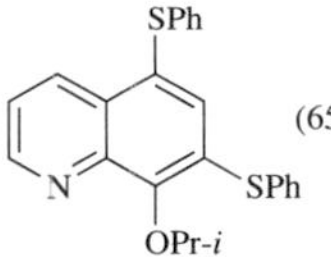 (65)	244
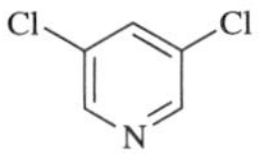	e^-, NH_3, mediator	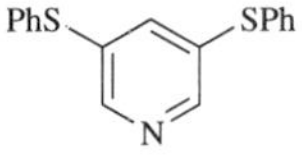 (60)	85
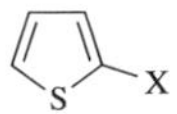		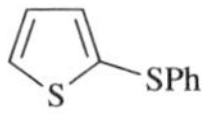 **I** + 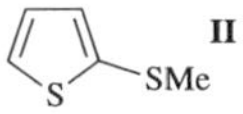 **II** +  **III**	253
X			
Cl	1. *hν*, MeCN, 0.5 h 2. MeI	**I** (10) + **II** (14) + **III** (6)	
Br	1. *hν*, MeCN, 0.5 h 2. MeI	**I** (26) + **II** (17) + **III** (9)	
Br	1. *hν*, MeCN, PhCN, 2 h 2. MeI	**I** (46) + **II** (7) + **III** (7)	
I	1. *hν*, MeCN, 0.5 h 2. MeI	**I** (52) + **II** (10) + **III** (8)	
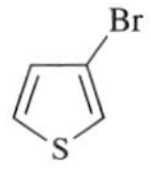	1. *hν*, MeCN, 1.5 h 2. MeI	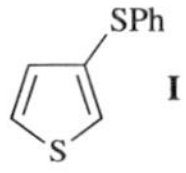 **I** (38) + 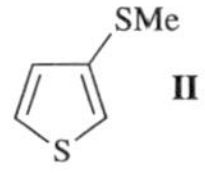 **II** (18) +  **III** (5)	253
	1. *hν*, MeCN, PhCN, 5 h 2. MeI	**I** (64) + **II** (6) + **III** (7)	253

TABLE VIII. OTHER NUCLEOPHILES (*Continued*)

E. Nucleophiles Derived from Sulfur (*Continued*)

Substrate	Nucleophile	Conditions	Product(s) and Yield(s) (%)	Refs.
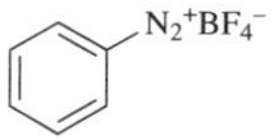	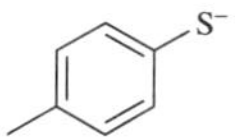	DMSO, 3.5 h	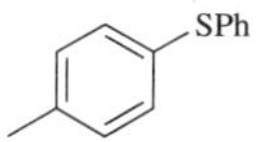 (82)	261
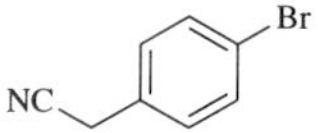		$h\nu$, DMSO, H_2O, 50°, 10 h	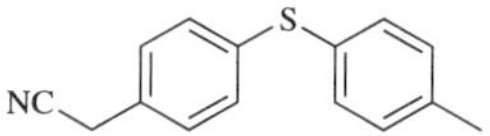 (90) + 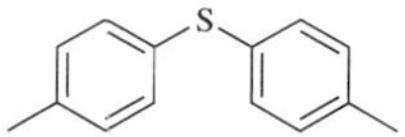 (7)	185
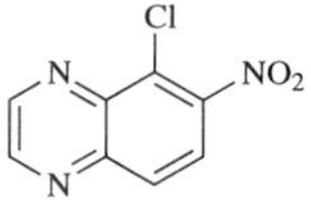		1. MeOH, 0°, 4 h 2. rt, 98 h 3. 50°, 17 h	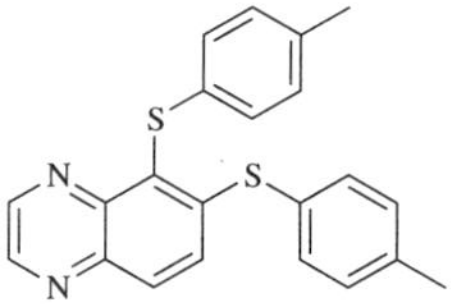 (56)	361
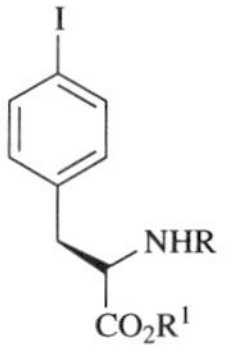	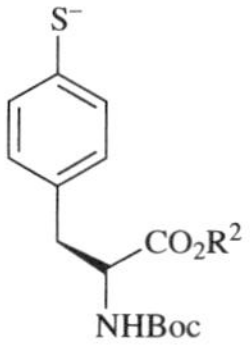	$h\nu$, NH_3, 1 h	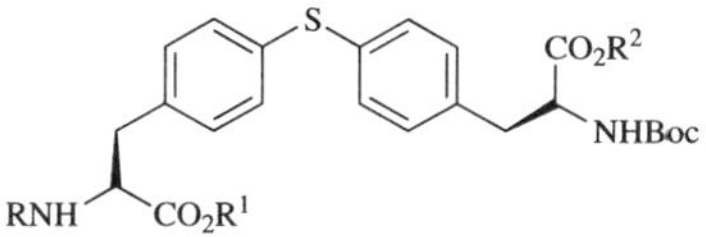	265
R R[1]	R[2]			
Boc Me	Me		(84)	
Boc H	Me		(88)	
H Me	Me		(81)	
Boc Me	H		(76)	

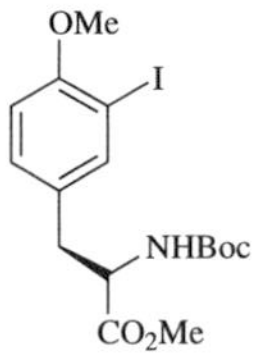	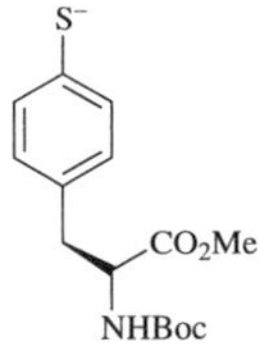	$h\nu$, NH_3, 1 h	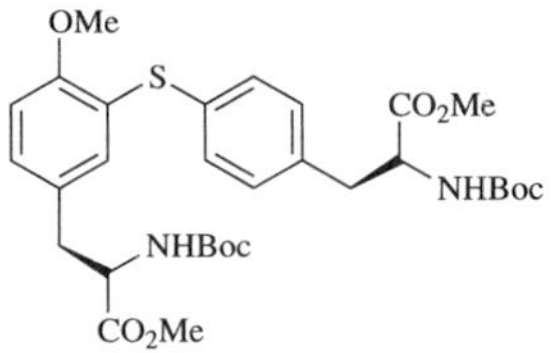 (91)	265
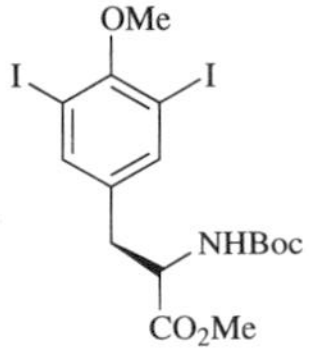		$h\nu$, NH_3, 1 h	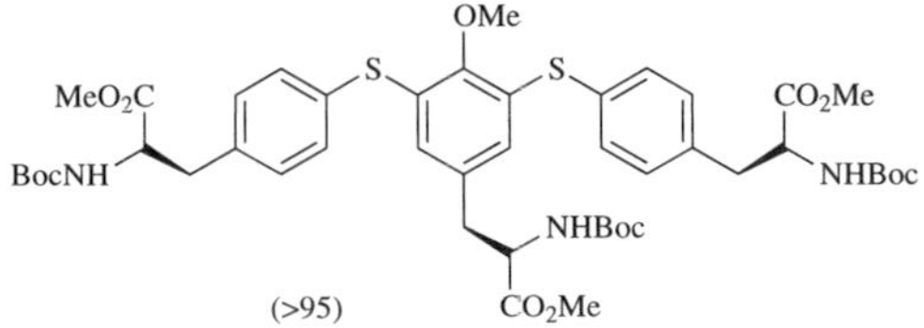 (>95)	265
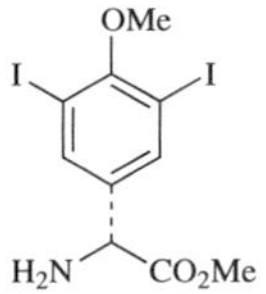		$h\nu$, NH_3, 1 h	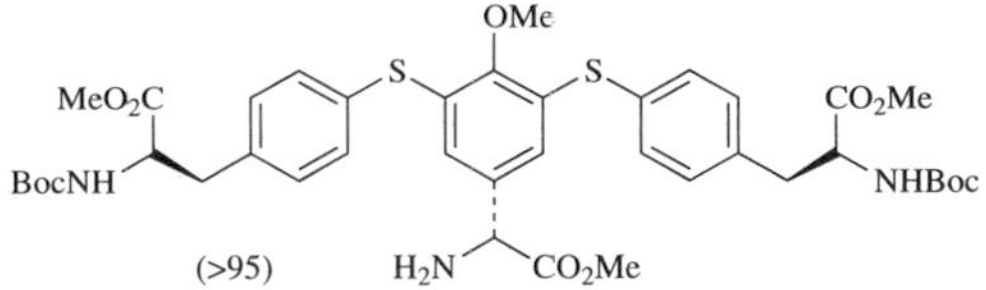 (>95)	265
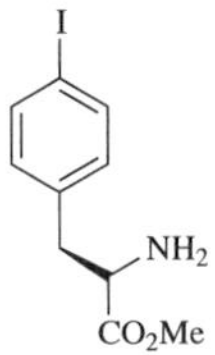	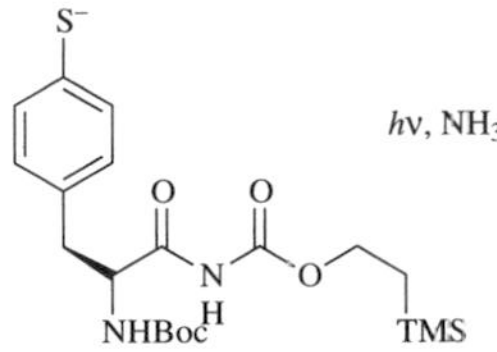	$h\nu$, NH_3, 1 h	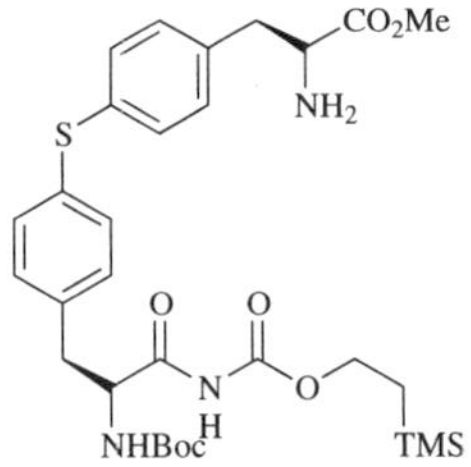 (73)	265
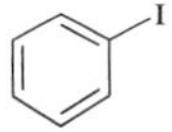	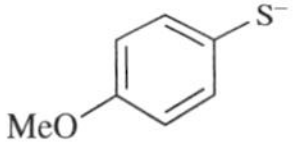	$h\nu$, NH_3, 2 h	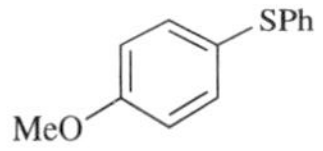 (71)	160

TABLE VIII. OTHER NUCLEOPHILES (*Continued*)

E. Nucleophiles Derived from Sulfur (*Continued*)

Substrate	Nucleophile	Conditions	Product(s) and Yield(s) (%)	Refs.
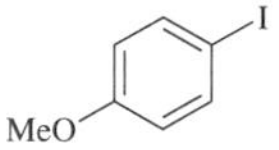	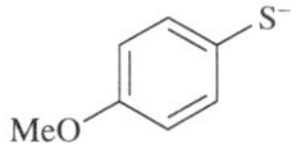	$h\nu$, NH_3, 1.7 h	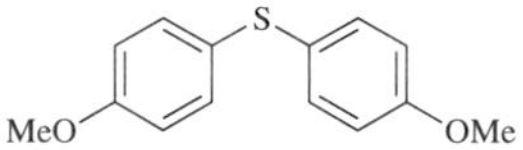 (73)	160
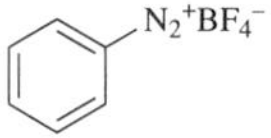	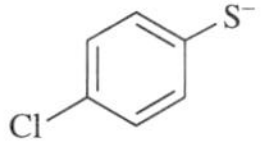	DMSO, 3 h	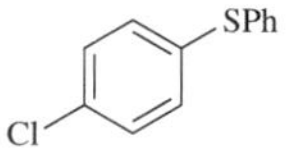 (34) + 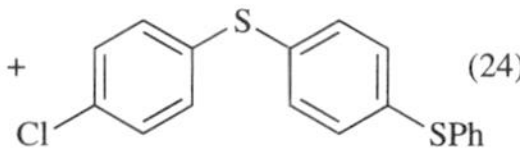 (24) + 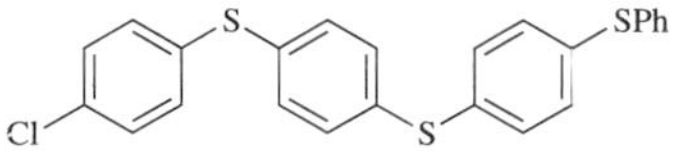 (9)	262
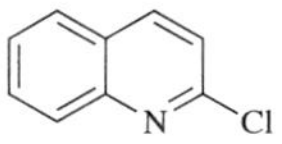		e^-, NH_3	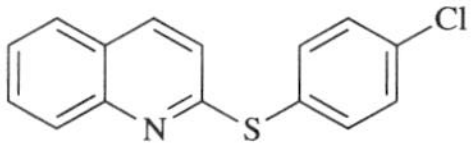 (80)	260
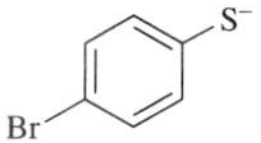	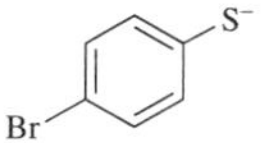	DMSO, 24 h[d]	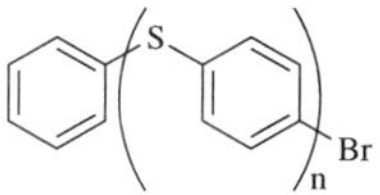 n = 6.2 (16)	362
		DMSO, 24 h[e]	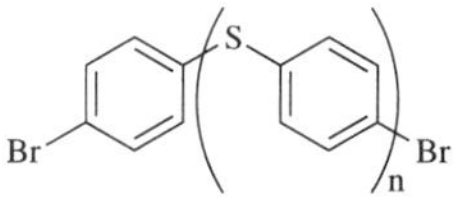 n = 9 (25)	362
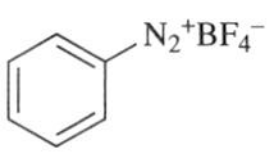	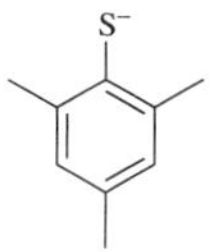	DMSO, 6.5 h	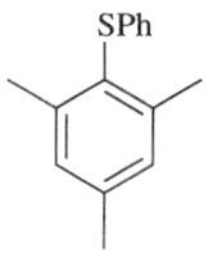 (67)	261

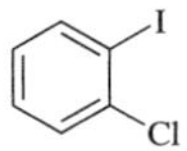	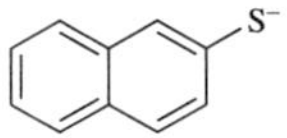	*h*ν, NH_3, 3 h	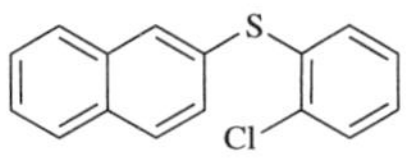 (81)	308
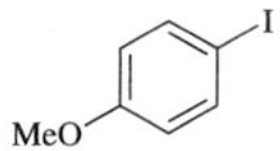		1. *h*ν, NH_3, 3 h 2. MeI	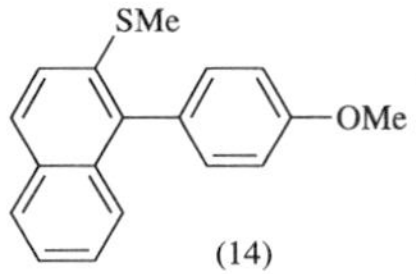 (14) + 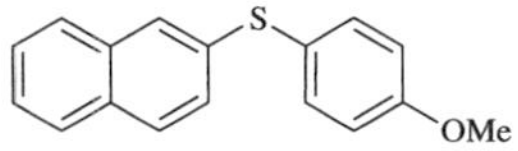 (65)	125
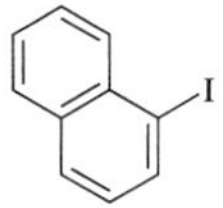		*h*ν, NH_3, 3 h	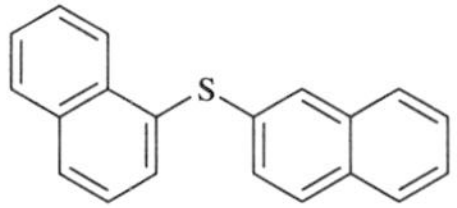 (95)	125
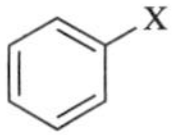	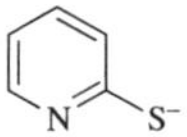	*h*ν, NH_3, TBAH[f]	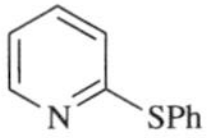 X: Br, 6 h (9); I, 3 h (68)	363
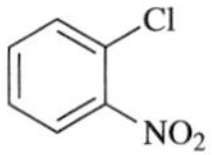		*h*ν, NH_3, TBAH,[f] 3 h	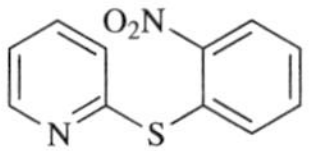 (25)	363
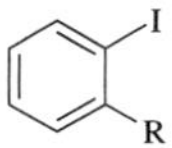		*h*ν, NH_3	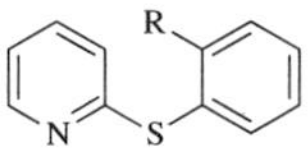 R: NH_2, 1 h (74); $P(O)(OEt)_2$, 1.25 h (81)	234
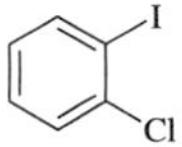		*h*ν, NH_3, 1 h	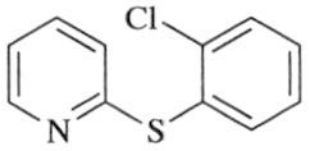 (70) + 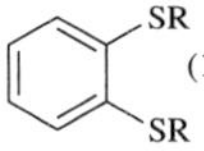 (10) R = 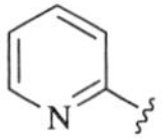	263
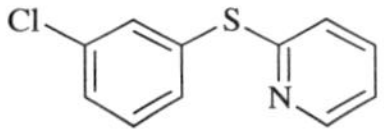		*h*ν, NH_3, 2 h	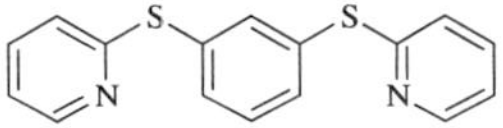 (95)	263

TABLE VIII. OTHER NUCLEOPHILES (*Continued*)

E. Nucleophiles Derived from Sulfur (*Continued*)

Substrate	Nucleophile	Conditions	Product(s) and Yield(s) (%)	Refs.
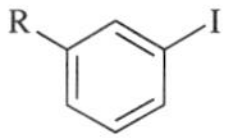	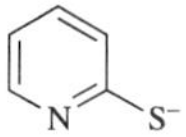		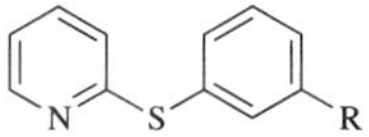	
R				
NO_2		$h\nu$, NH_3, TBAH,[f] 0.5 h	(44)	363
NH_2		$h\nu$, NH_3, 3 h	(56)	234
$P(O)(OEt)_2$		$h\nu$, NH_3, 15 min	(50)	234
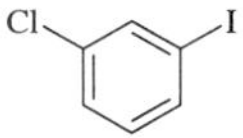		$h\nu$, NH_3, 1 h	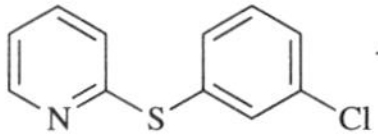 (83) + 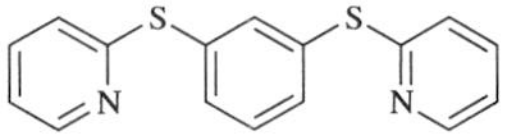 (12)	244
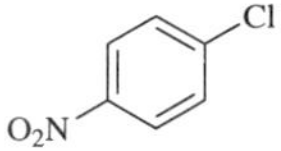		$h\nu$, NH_3, TBAH,[f] 6 h	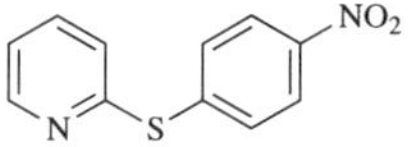 (12)	363
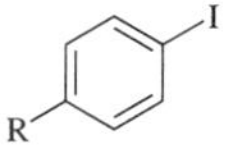			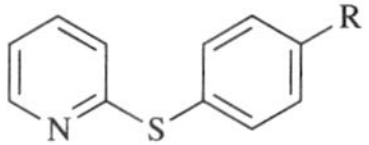	
R				
$P(O)(OEt)_2$		$h\nu$, NH_3, 1.5 h	(60)	234
SPh		$h\nu$, NH_3	(92)	234
Cl		$h\nu$, NH_3, 1 h	(87) + 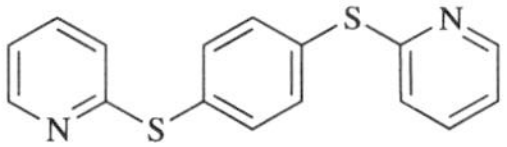 (12)	263
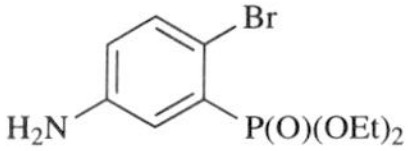		$h\nu$, NH_3, 3 h	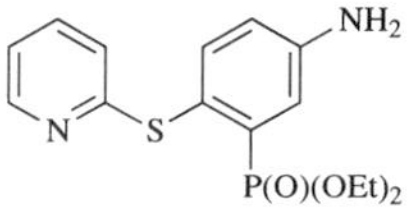 (60)	234

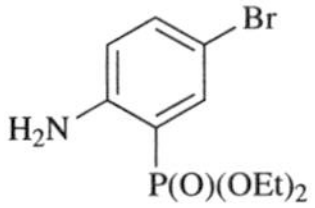	$h\nu$, NH_3, 3 h	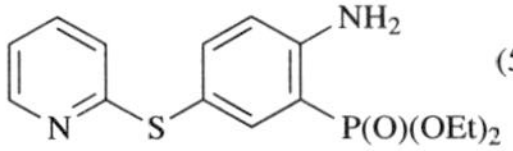	(56)	234
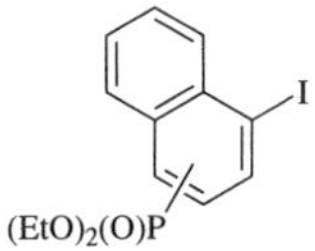	$h\nu$, NH_3, 0.5 h	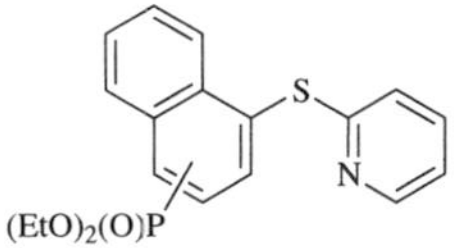	2-$P(O)(OEt)_2$ (75) 4-$P(O)(OEt)_2$ (68)	234
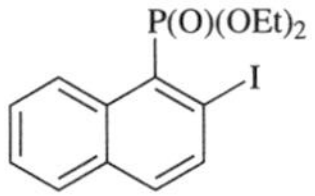	$h\nu$, NH_3, 1 h	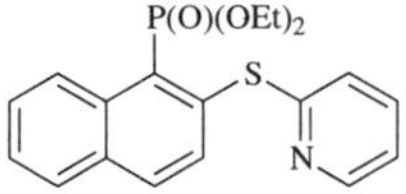	(65)	234
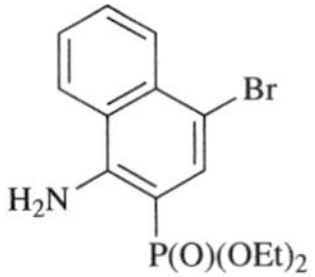	$h\nu$, NH_3, 0.5 h	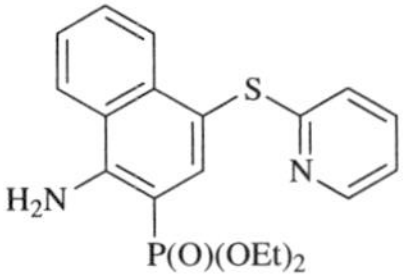	(85)	234
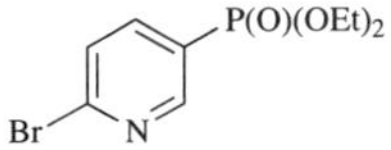	$h\nu$, NH_3, 3 h	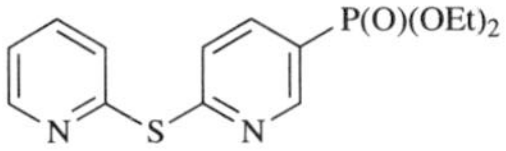	(65)	234
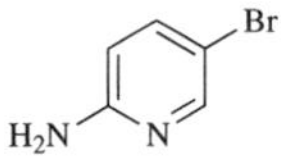	$h\nu$, NH_3, 3 h	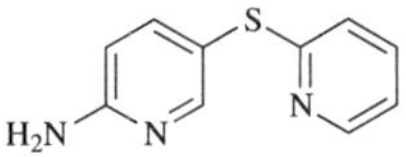	(60)	234
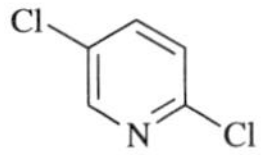	e^-, NH_3, mediator	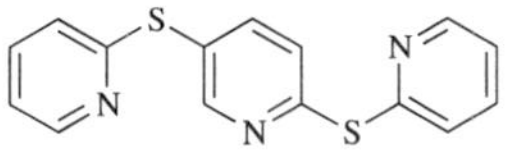	(40)	85
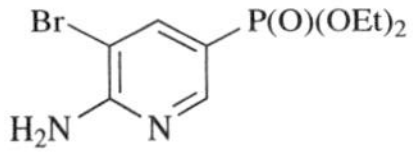	$h\nu$, NH_3, 1.5 h	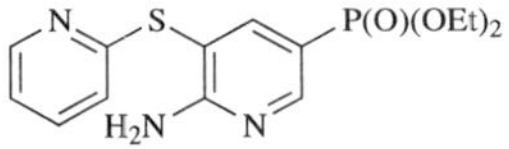	(98)	234

TABLE VIII. OTHER NUCLEOPHILES (*Continued*)

E. Nucleophiles Derived from Sulfur (*Continued*)

Substrate	Nucleophile	Conditions	Product(s) and Yield(s) (%)	Refs.
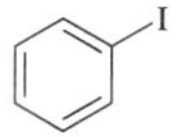	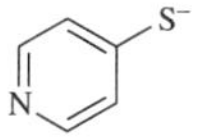	*hν*, NH_3, TBAH,[f] 6 h	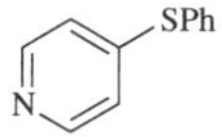 (47)	363
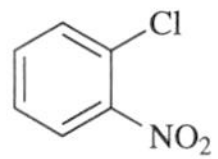		*hν*, NH_3, TBAH,[f] 6 h	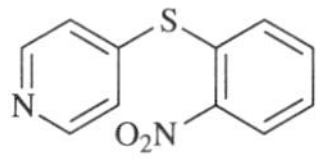 (32)	363
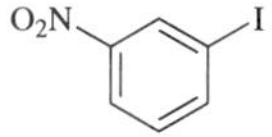		*hν*, NH_3, TBAH,[f] 0.5 h	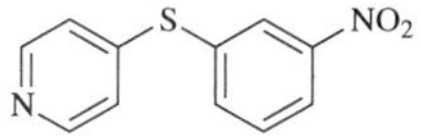 (32)	363
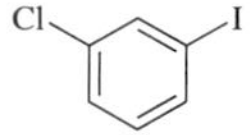	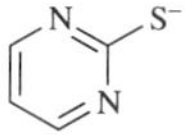	*hν*, NH_3, 1 h	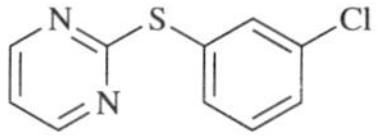 (100)	244
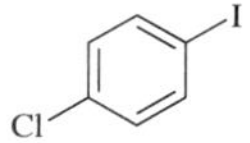		*hν*, NH_3, 1 h	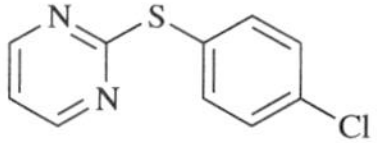 (100)	263
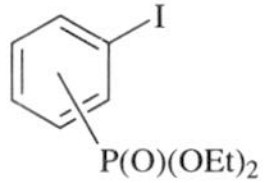		*hν*, NH_3	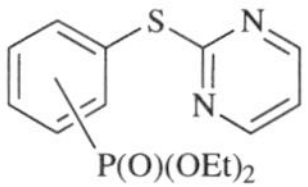	234
		0.75 h	2-$P(O)(OEt)_2$ (60)	
		15 min	3-$P(O)(OEt)_2$ (50)	
		0.75 h	4-$P(O)(OEt)_2$ (50)	
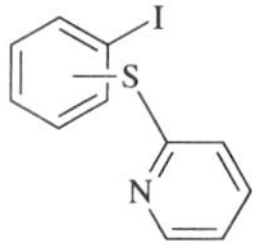		*hν*, NH_3	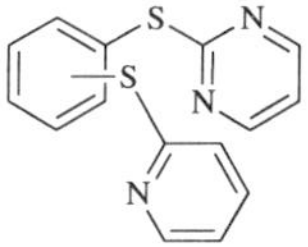	234
		0.75 h	2-SC_5H_4N (90)	
		1 h	3-SC_5H_4N (68)	

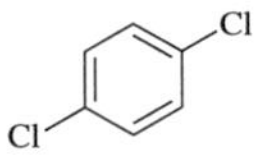	e⁻, NH_3, mediator	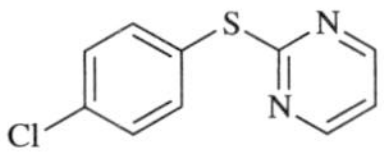 (10) + 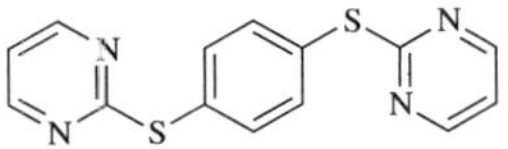 (52)	85
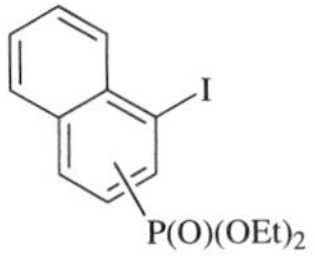	$h\nu$, NH_3	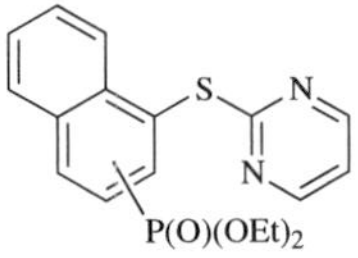	234
	0.5 h	2-$P(O)(OEt)_2$ (75)	
	2 h	4-$P(O)(OEt)_2$ (65)	
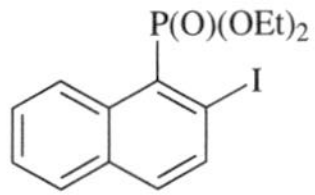	$h\nu$, NH_3, 2 h	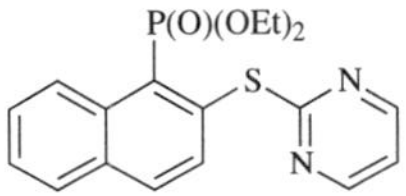 (73)	234
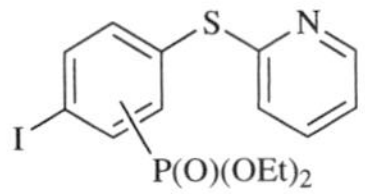	$h\nu$, NH_3	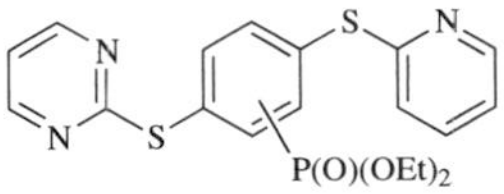	234
	3 h	2-$P(O)(OEt)_2$ (54)	
	2 h	3-$P(O)(OEt)_2$ (78)	
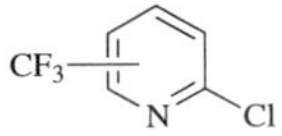	$h\nu$, NH_3	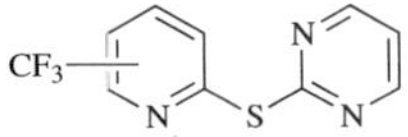	245
	10 min	3-CF_3 (96)	
	4 h	4-CF_3 (11)	
	0.5 h	5-CF_3 (89)	
	1.5 h	6-CF_3 (62)	

TABLE VIII. OTHER NUCLEOPHILES (*Continued*)

E. Nucleophiles Derived from Sulfur (*Continued*)

Substrate	Nucleophile	Conditions	Product(s) and Yield(s) (%)	Refs.
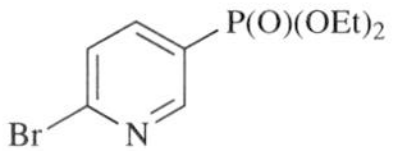	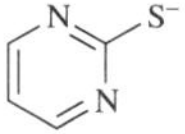	$h\nu$, NH_3	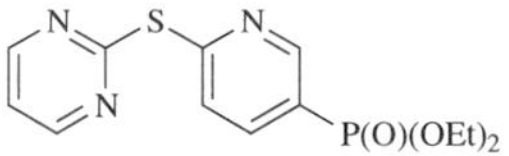 (77)	234
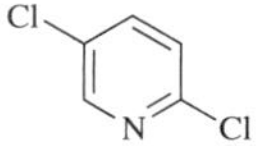		e^-, NH_3, mediator	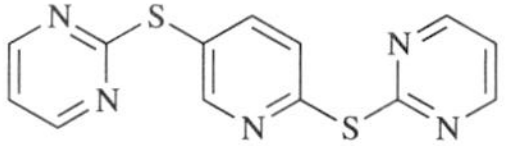 (45)	85
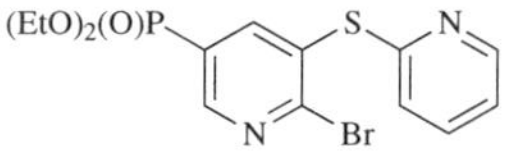		$h\nu$, NH_3, 3 h	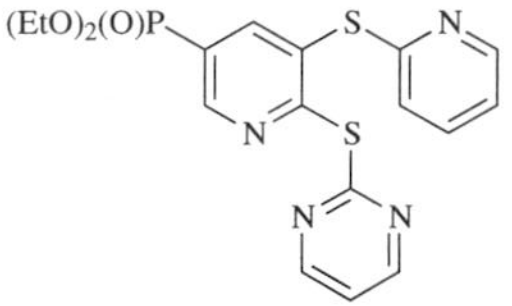 (60)	234
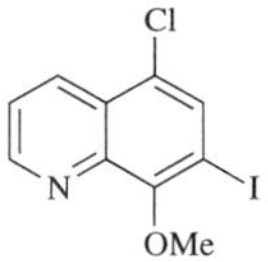		$h\nu$, NH_3, 1 h	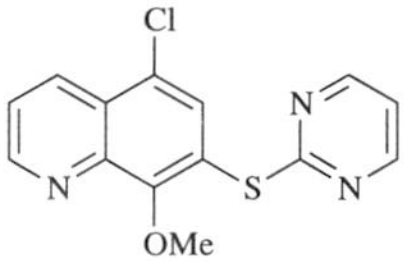 (90)	244
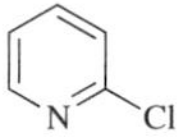	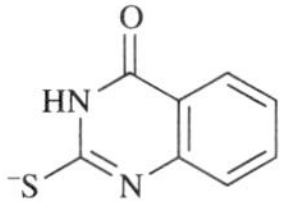	$h\nu$, NH_3, 4 h	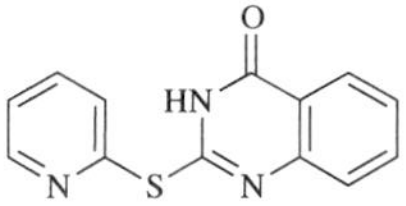 (15)	245
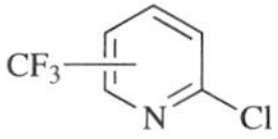		$h\nu$, NH_3	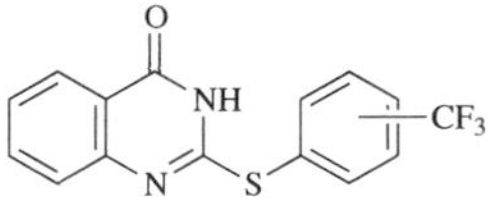	245

		20 min	3-CF_3 (98)	
		4 h	4-CF_3 (1)	
		20 min	5-CF_3 (97)	
		1.5 h	6-CF_3 (45)	
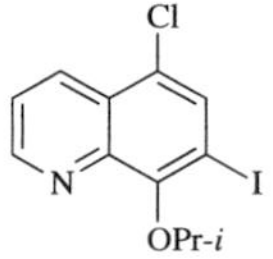		$h\nu$, NH_3, 1.5 h	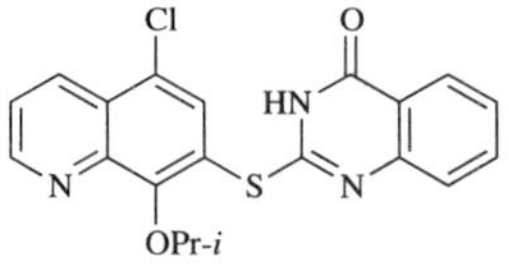 (100)	244
	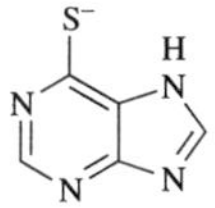	$h\nu$, NH_3, 2 h	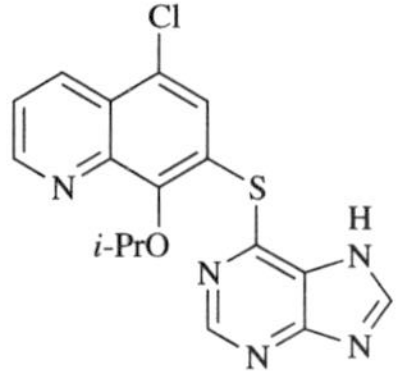 (88)	244
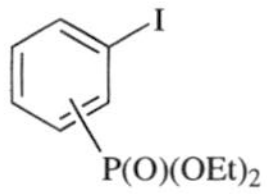	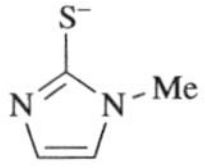	$h\nu$, NH_3	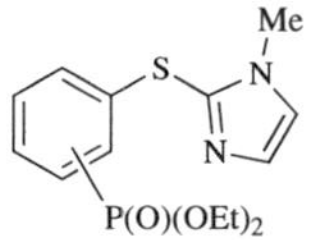	234
		1.5 h	2-$P(O)(OEt)_2$ (49)	
		3 h	4-$P(O)(OEt)_2$ (39)	
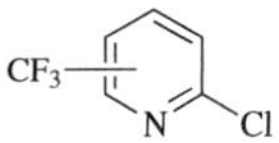		$h\nu$, NH_3	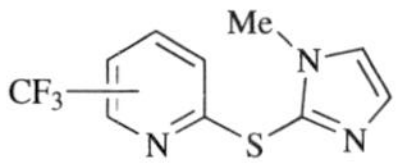	245
		15 min	3-CF_3 (92)	
		4 h	4-CF_3 (4)	
		0.5 h	5-CF_3 (80)	

TABLE VIII. OTHER NUCLEOPHILES (*Continued*)

E. Nucleophiles Derived from Sulfur (*Continued*)

Substrate	Nucleophile	Conditions	Product(s) and Yield(s) (%)	Refs.
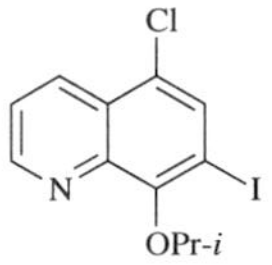	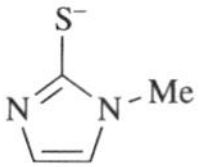	$h\nu$, NH_3, 1 h	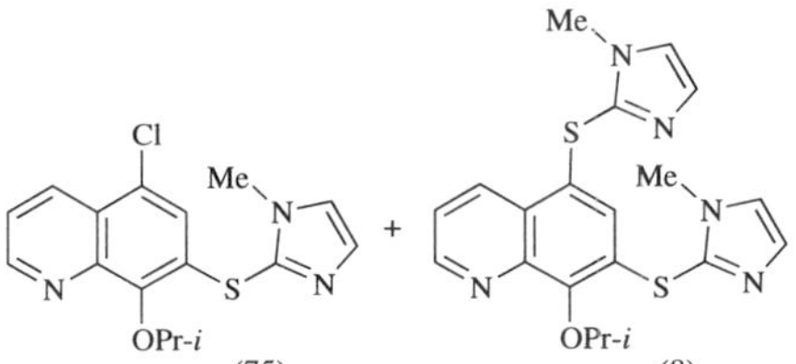 (75) + (8)	244
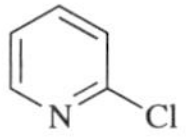	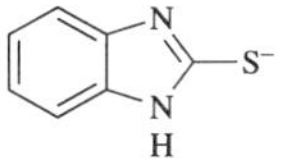	$h\nu$, NH_3, 4 h	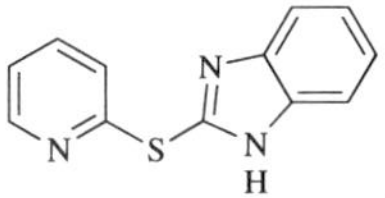 (21)	245
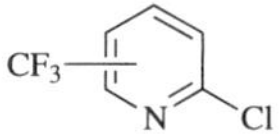		$h\nu$, NH_3	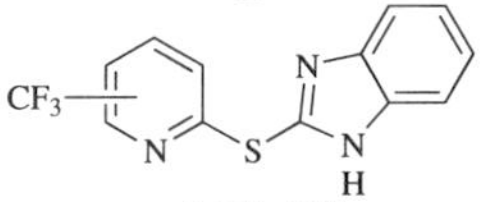	245
		0.5 h	3-CF_3 (98)	
		4 h	4-CF_3 (6)	
		0.5 h	5-CF_3 (53)	
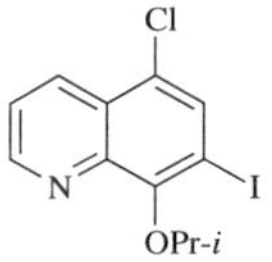		$h\nu$, NH_3, 1 h	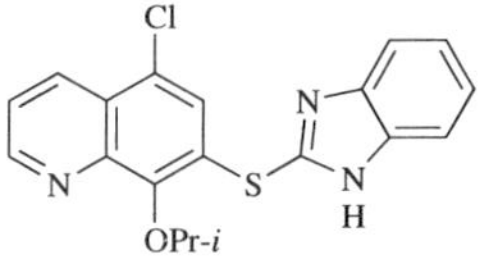 (100)	244
	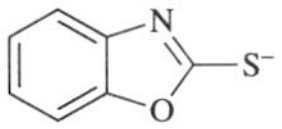	$h\nu$, NH_3, 1 h	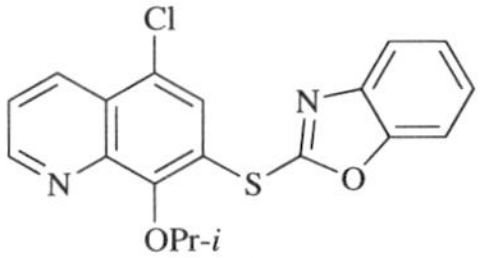 (80)	244

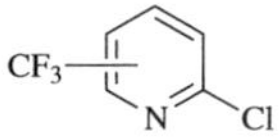	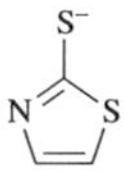	$h\nu$, NH_3, 4 h	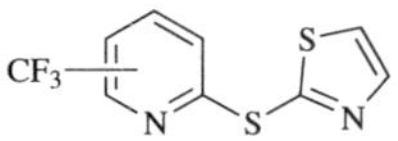 3-CF_3 (78) 4-CF_3 (1) 5-CF_3 (95)	245
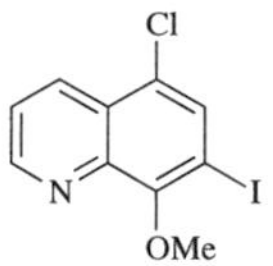		$h\nu$, NH_3, 4 h	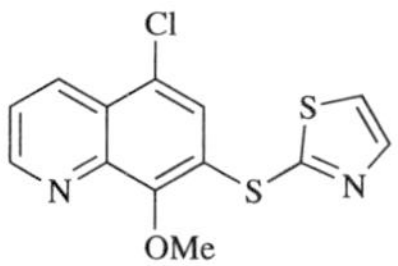 (77)	244
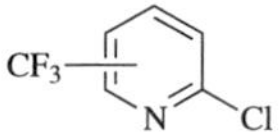	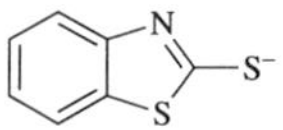	$h\nu$, NH_3	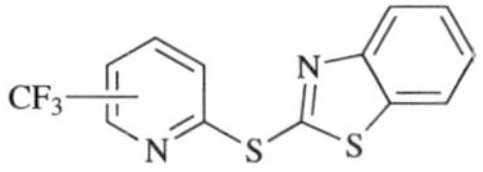	245
		4 h	3-CF_3 (96)	
		4.5 h	4-CF_3 (7)	
		2.5 h	5-CF_3 (95)	
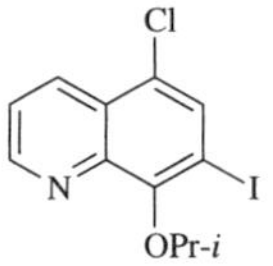		$h\nu$, NH_3, 1 h	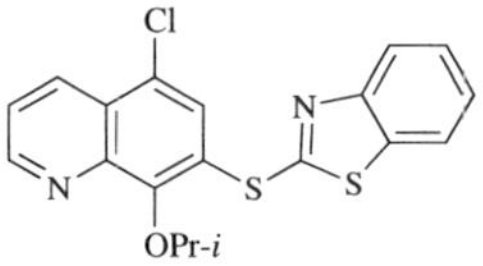 (100)	244

[a] The compound decomposes.

[b] The compound is at least 95% enantiomerically pure.

[c] The compound is completely racemized.

[d] A catalytic amount of benzenediazonium tetrafluoroborate was added.

[e] A catalytic amount of *p*-bromobenzenediazonium tetrafluoroborate was added.

[f] TBAH is tetrabutylammonium hydroxide.

TABLE VIII. OTHER NUCLEOPHILES (*Continued*)

F. Nucleophiles Derived from Selenium and Tellurium

Substrate	Nucleophile	Conditions	Product(s) and Yield(s) (%)	Refs.
PhI	Se^{2-}	1. *hν*, NH_3, 4 h 2. [O]	Ph_2Se (12) + Ph_2Se_2 (78)	268
		1. *hν*, NH_3, 4 h 2. Na 3. [O]	Ph_2Se_2 (92)	268
		1. *hν*, NH_3, 4 h 2. Na 3. MeI	PhSeMe (67)	268
		1. *hν*, NH_3, 4 h 2. Na 3. *hν*, 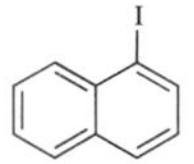, 4 h	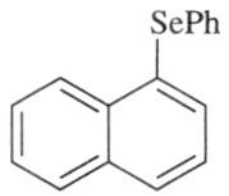 (98)	268
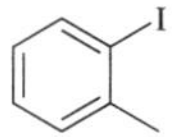		1. *hν*, NH_3, 4 h 2. Na 3. MeI	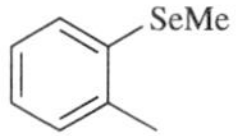 (87)	268
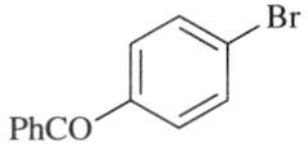		e⁻, MeCN	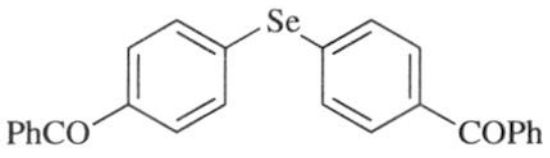 (20) + 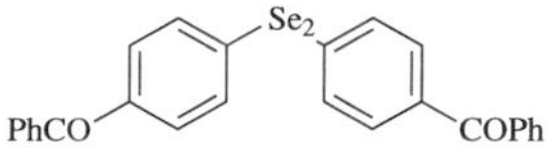 (40) + Ph_2CO (6)	277
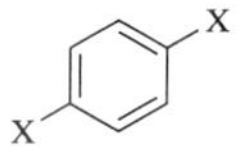		DMF, 120-140°, 20 h	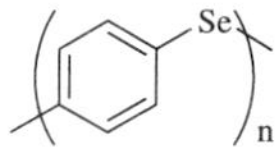 X = Cl (5) X = Br (65-80)	272 271, 272

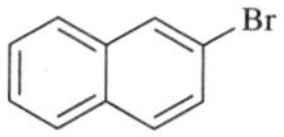		HMPA, 130-170°, 16-24 h	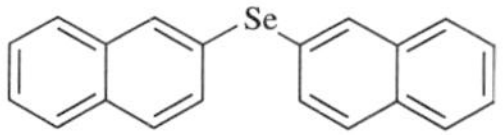 (30)	269
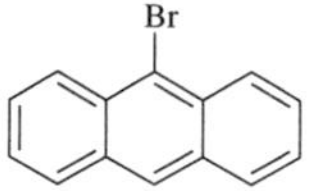		1. DMF, 115-120°, 72 h 2. DMF, 150°, 24 h	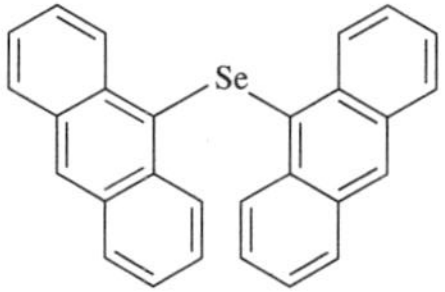 (4)	364
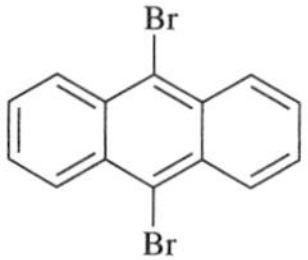		DMF, 120-130°, 20 h	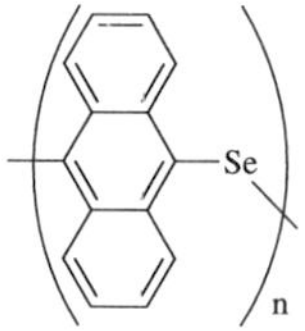 (38)	364
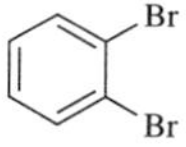	Se_2^{2-}	DMF, 140°, 40 h	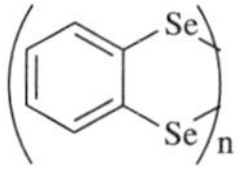 (26)	364
		1. DMF, 140°, 40 h 2. THF, 0° 3. $NaBH_4$, MeOH, Cl_2CS	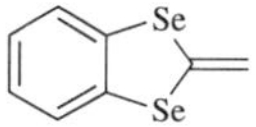 (23)	365
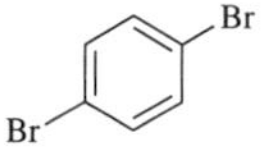		DMF, 130°, 72 h	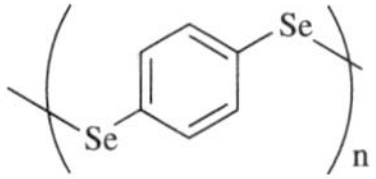 (85)	364
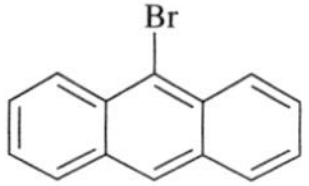		e^-, MeCN	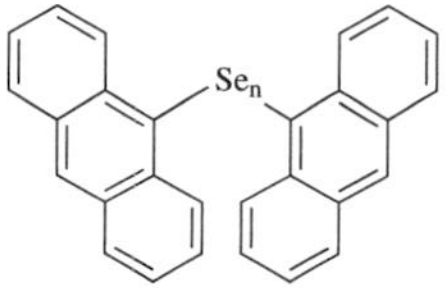 **I** n: 2 (54); 3 (10)	276

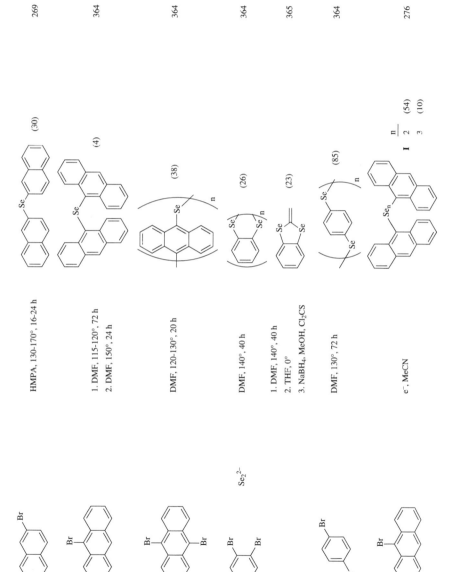

TABLE VIII. OTHER NUCLEOPHILES (*Continued*)

F. Nucleophiles Derived from Selenium and Tellurium (*Continued*)

Substrate	Nucleophile	Conditions	Product(s) and Yield(s) (%)	Refs.
	Se_2^{2-}	1. DMF, 110-115°, 16 h 2. DMF, 140-145°, 8 h	n = 1, **I** (22); n = 2, **I** (28)	364
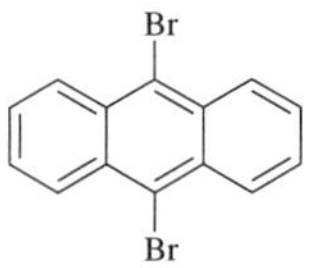		DMF, 120°, 3.5 h	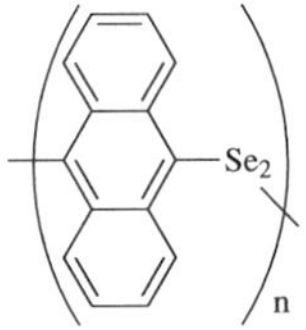 (69)	364
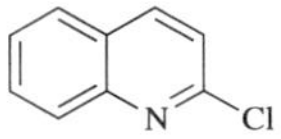		e^-, MeCN, mediator	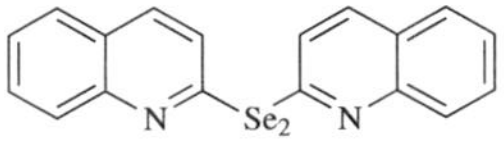 **I** (70)	276
		e^-, DMSO, mediator	**I** (46)	276
		e^-, MeCN	**I** (79)	278
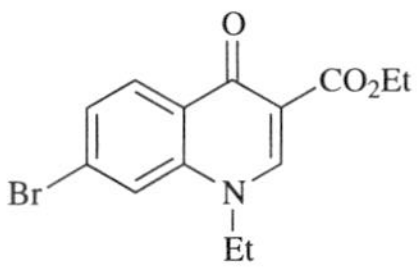		e^-, MeCN, *t*-BuOH, mediator	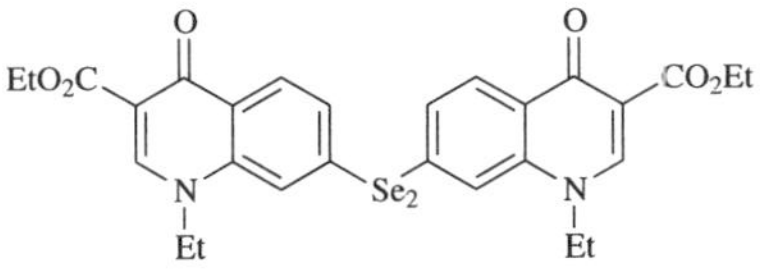 (70)	278
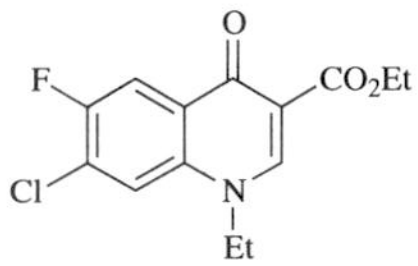		e^-, MeCN, mediator, diethyl malonate	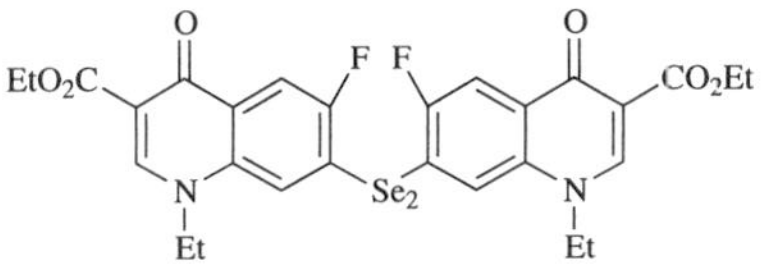 (82)	278, 279[a]
PhI	$PhSe^-$	1. *hν*, NH_3, 3.7 h 2. MeI	Ph_2Se (73) + PhSeMe (26)	122

	$h\nu$, NH_3, 2.3 h $PhSe^-$:PhI = 1	Ph_2Se (55) + C_6H_6 (4)	366
	$h\nu$, NH_3, 2.3 h $PhSe^-$:PhI = 8	Ph_2Se (13) + C_6H_6 (91)	366
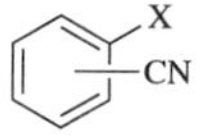X CN		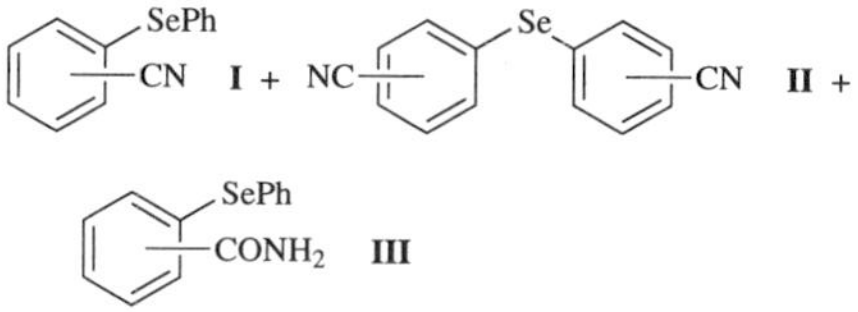SePh CN I + NC Se CN II + SePh CONH₂ III	
X			
Cl 2-CN	e^-, MeCN	**I** (48) + **II** (3) + **III** (10)	284
Cl 2-CN	e^-, MeCN, mediator	**I** (61) + **II** (5) + **III** (4)	284, 281
Cl 3-CN	e^-, MeCN, mediator	**I** (68) + **II** (9) + **III** (4)	284, 281
Cl 4-CN	e^-, MeCN	**I** (70) + **II** (8)	282, 281
Cl 4-CN	e^-, MeCN, *t*-BuOH	**I** (73) + **II** (16)	257
Br 2-CN	e^-, MeCN	**I** (36)	282
Br 3-CN	e^-, MeCN	**I** (42)	282
Br 4-CN	e^-, MeCN	**I** (58) + **II** (7)	282, 283
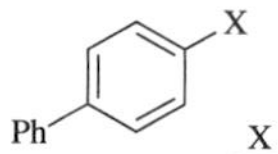Ph X		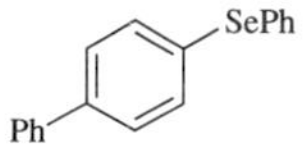Ph SePh	
X			
Cl	$h\nu$, NH_3, 4 h	(52)	367, 122
Br	e^-, MeCN	(65) + PhPh (18)	285
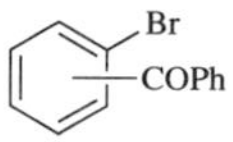Br COPh	e^-, MeCN, mediator	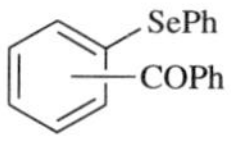SePh COPh	
		2-COPh, (72)	368
		3-COPh, (62)	280, 281
		4-COPh, (86)	280, 281

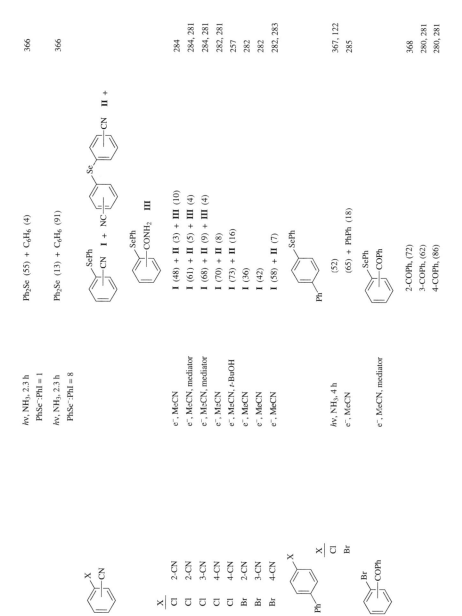

TABLE VIII. OTHER NUCLEOPHILES (*Continued*)

F. Nucleophiles Derived from Selenium and Tellurium (*Continued*)

Substrate	Nucleophile	Conditions	Product(s) and Yield(s) (%)	Refs.
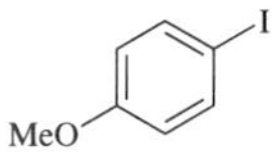	$PhSe^-$		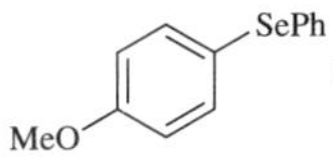 **I** + 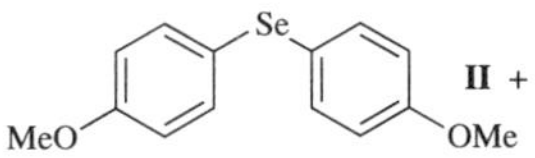 **II** + Ph_2Se **III**	
		$h\nu$, NH_3, 4.3 h	**I** (25) + **II** (19) + **III** (20)	117
		$h\nu$, dioxane, DMSO, 4 h	**I** (18) + **II** (10) + **III** (22)	250
		$h\nu$, MeCN, 4 h[b]	**I** (17) + **II** (7) + **III** (8)	250
		$h\nu$, MeCN, 4 h[c]	**I** (67) + PhOMe (10)	250
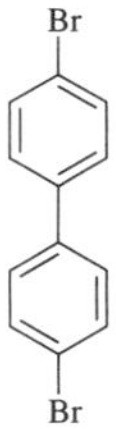		e^-, MeCN	 (36) + 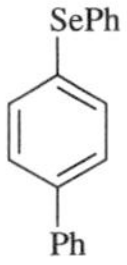 (9) + 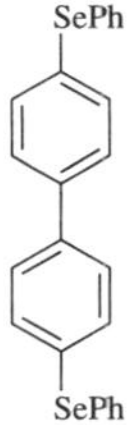 (46)	285
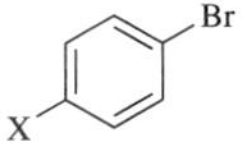		$h\nu$, NH_3, 3.6 h	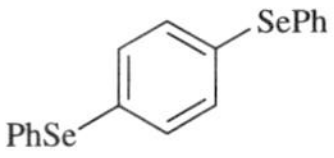 X: Br (13); I (70)	122
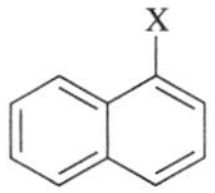			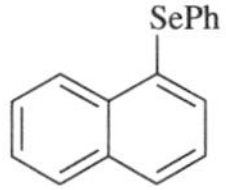	
X: Cl		$h\nu$, NH_3, 2.8 h	(73)	367. 122
X: Br		$h\nu$, MeCN, 4 h	(88)	250

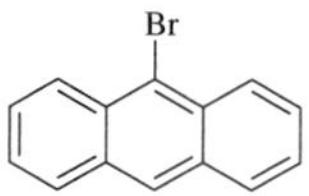			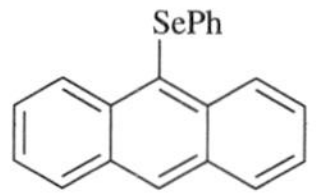 + 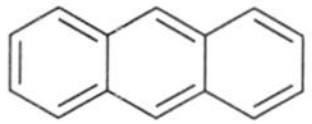	285
		e⁻, MeCN	(74) (18)	
		e⁻, DMSO	(43) (36)	
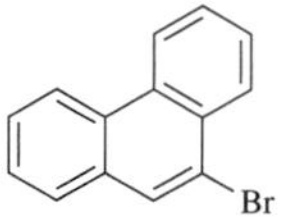		*h*ν, NH_3, 3.6 h	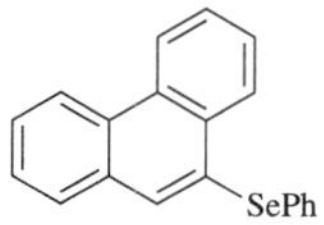 (72)	367, 122
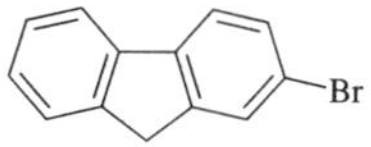		e⁻, MeCN	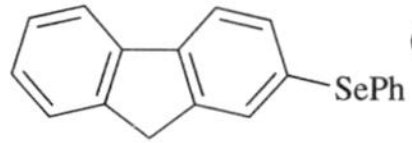 (53) + 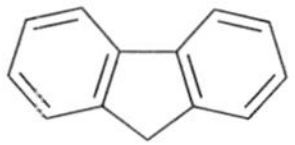 (23)	285
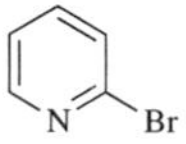		*h*ν, NH_3, 3.6 h	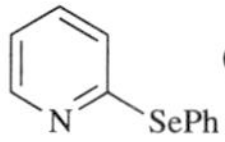 (72) + 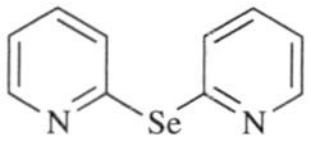 (2) + Ph_2Se (5)	117
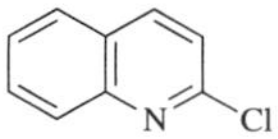		*h*ν, NH_3, 3.6 h	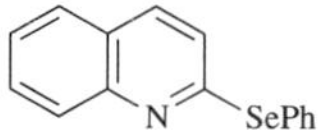 **I** (46)	117, 122
		*h*ν, NH_3, 2.2 h	**I** (98) $PhSe^-$:substrate = 4.8	366
		*h*ν, MeCN, 4 h	**I** (70)	250
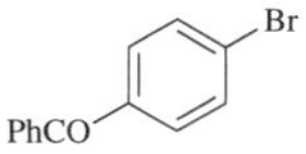	$BnSe^-$	e⁻, MeCN, mediator	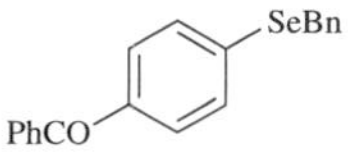 (35)	368
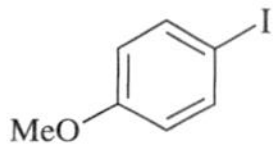	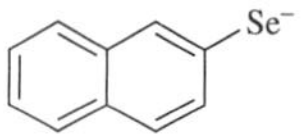	*h*ν, NH_3, 3 h	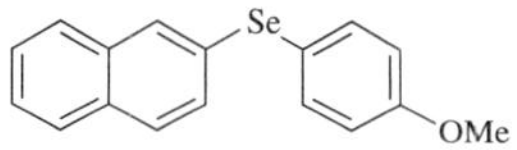 (85)	125

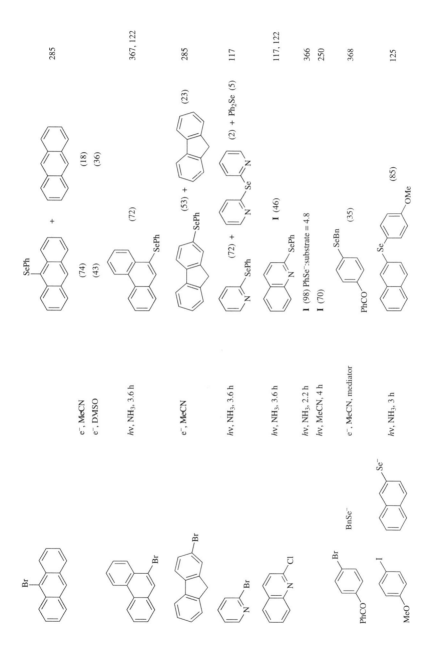

TABLE VIII. OTHER NUCLEOPHILES (*Continued*)

F. Nucleophiles Derived from Selenium and Tellurium (*Continued*)

Substrate	Nucleophile	Conditions	Product(s) and Yield(s) (%)	Refs.
PhI	Te^{2-}	1. *h*ν, NH_3, 4 h 2. [O]	Ph_2Te (17) + Ph_2Te_2 (17)	268
		1. *h*ν, NH_3, 4 h 2. MeI	Ph_2Te (21) + PhTeMe (29)	268
		HMPA, 130-170°, 16-24 h	Ph_2Te (35)	269
		DMF, 60°, 10 h	Ph_2Te (71)	273
		NMP,[d] 60°, 8 h	Ph_2Te (77)	275
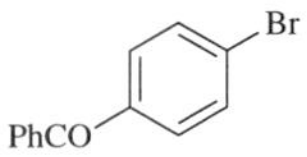		e^-, MeCN	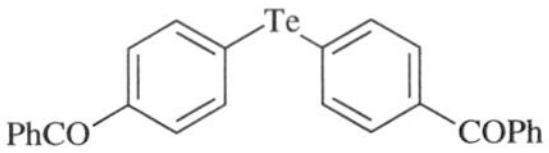 (6) + Ph_2CO (57) + 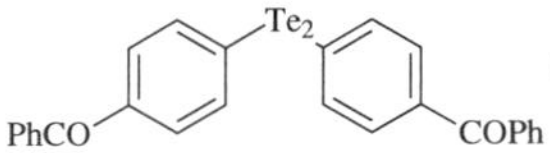 (14)	277
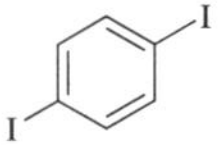		DMF, 110-120°, 20 h	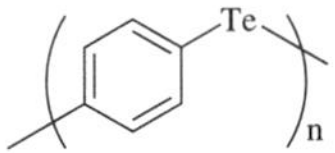 (70-75)	272
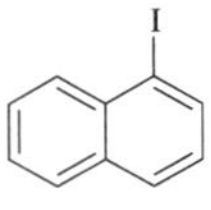		*h*ν, NH_3, 4 h	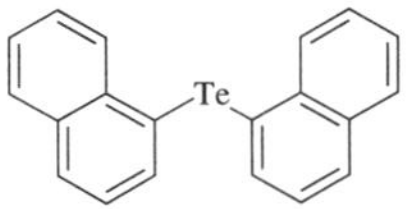 (35)	268
		DMF, 60°, 10 h	(70)	274
		DMF, 100°, 19 h	(32)	275
		DMF, 140°, 1 h	(59)	274
		NMP,[d] 60°, 10 h	(45)	275

Reactant	Conditions	Product(s) and Yield(s) (%)	Refs.
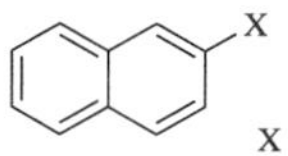		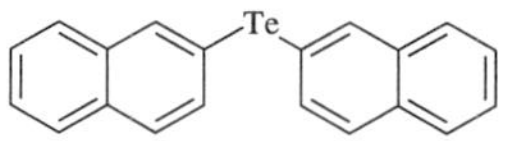	
X			
Cl	HMPA, 160-175°, 22 h	(27)	270
Br	HMPA, 130-170°, 16-24 h	(30)	269
I	DMF, 60°, 10 h	(61)	274
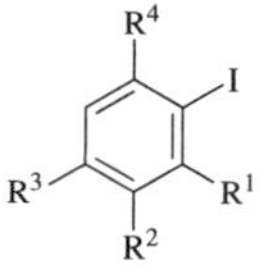	DMF, 60°, 10 h	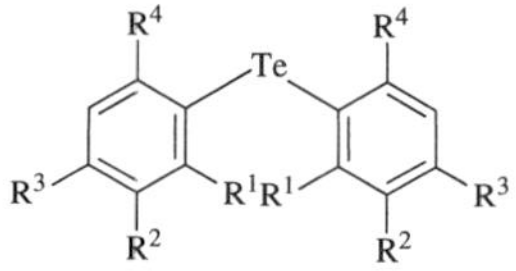	273

R^1	R^2	R^3	R^4	Yield (%)
H	Me	H	H	(84)
H	H	Me	H	(81)
Me	H	Me	H	(94)
Me	H	Me	Me	(77)
H	H	OMe	H	(74)

Reactant	Conditions	Product(s) and Yield(s) (%)	Refs.
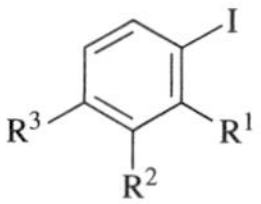	1. NMP[d] 2. R^4X	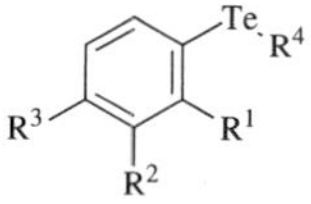	275

R^1	R^2	R^3	R^4	Conditions	Yield (%)
H	H	Me	n-C_7H_{15}	1. 120°, 4 h 50-60°, 2 h	(63)
$(CH{=}CH)_2$		H	n-C_7H_{15}	1. 110°, 2 h 2. 60°, 3 h	(52)
H	$(CH{=}CH)_2$		Me	1. 130°, 4 h 50°, 2 h	(67)

TABLE VIII. OTHER NUCLEOPHILES (*Continued*)

F. Nucleophiles Derived from Selenium and Tellurium (*Continued*)

Substrate					Nucleophile	Conditions	Product(s) and Yield(s) (%)	Refs.
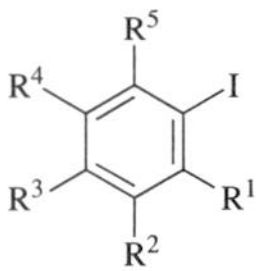					Te^{2-}	NMP[d]	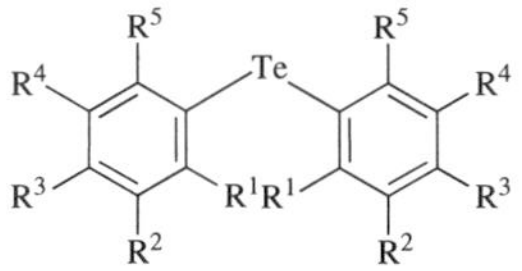	275
R^1	R^2	R^3	R^4	R^5				
Me	H	H	H	H		70°, 14 h	(55)	
H	H	Me	H	H		70°, 12 h	(42)	
H	H	OMe	H	H		80°, 16 h	(58)	
$(CH{=}CH)_2$	$(CH{=}CH)_2$	H	H	H		100°, 21 h	(76)	
H	$(CH{=}CH)_2$	$(CH{=}CH)_2$	H	H		100°, 24 h	(61)	
Me	H	H	$(CH{=}CH)_2$	$(CH{=}CH)_2$		90°, 20 h	(53)	
F	F	F	F	F		50°, 2 h	(56)	
						1. NMP[d] 2. [O]	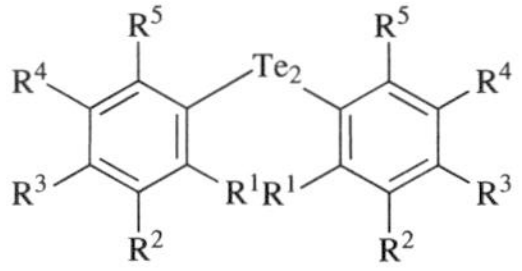	275
R^1	R^2	R^3	R^4	R^5				
H	H	H	H	H		120°, 5 h	(66)	
H	H	Me	H	H		110°, 3 h	(64)	
H	H	OMe	H	H		100°, 7 h	(57)	
Me	H	Me	H	Me		120°, 5 h	(71)	
$(CH{=}CH)_2$	$(CH{=}CH)_2$	H	H	H		90°, 8 h	(59)	
Me	H	H	$(CH{=}CH)_2$	$(CH{=}CH)_2$		110°, 8 h	(51)	

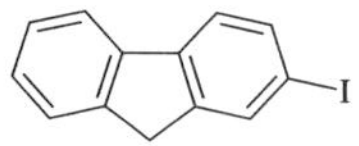		DMF, 140°, 1 h	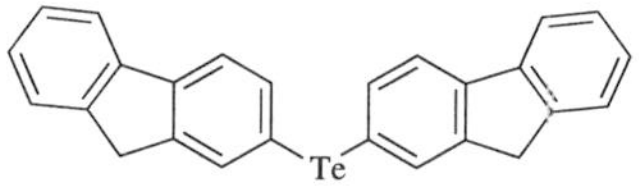 (48)	274
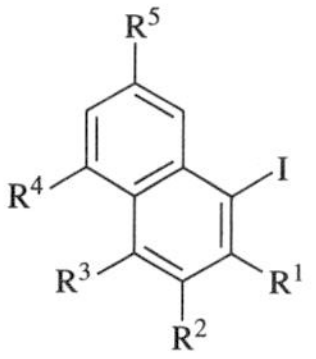		DMF	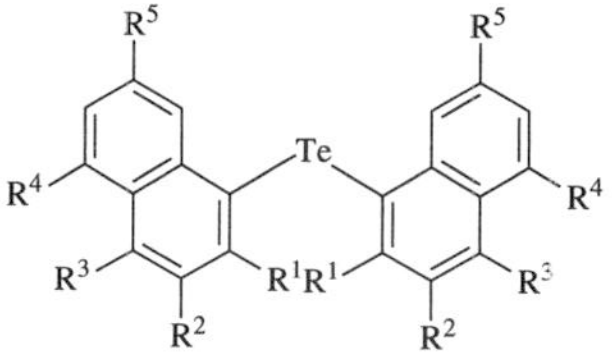	274

R^1	R^2	R^3	R^4	R^5	Conditions	Yield
Me	H	H	H	H	140°, 1 h	(48)
Me	Me	H	H	H	140°, 1 h	(45)
H	H	Me	H	H	140°, 1 h	(47)
H	H	OMe	H	H	140°, 1 h	(43)
H	H	F	H	H	140°, 1 h	(41)
H	*t*-Bu	H	H	*t*-Bu	140°, 1 h	(45)
H	H	$(CH=CH)_2$		H	140°, 1 h	(42)
$(CH=CH)_2$		H	H	H	60°, 10 h	(89)
H	$(CH=CH)_2$		H	H	60°, 10 h	(55)

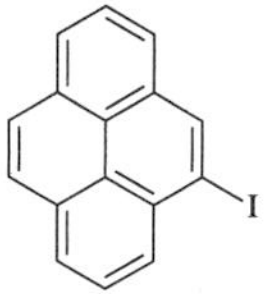		DMF, 60°, 10 h	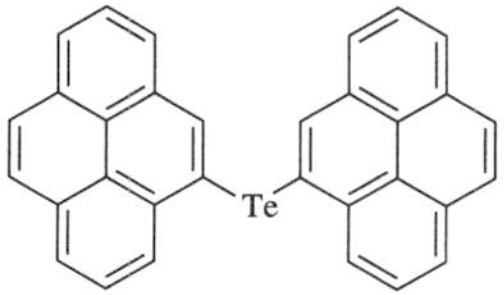 (60)	274
PhBr	Te_2^{2-}	HMPA, 130-170°, 16-24 h	Ph_2Te_2 (1)	269
PhI		HMPA, 130-170°, 16-24 h	Ph_2Te_2 (5) + Ph_2Te (10)	269

TABLE VIII. OTHER NUCLEOPHILES (*Continued*)

F. Nucleophiles Derived from Selenium and Tellurium (*Continued*)

Substrate	Nucleophile	Conditions	Product(s) and Yield(s) (%)	Refs.
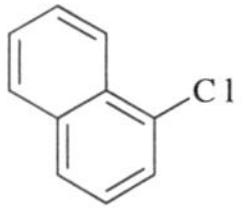	Te_2^{2-}	HMPA, 130-170°, 16-24 h	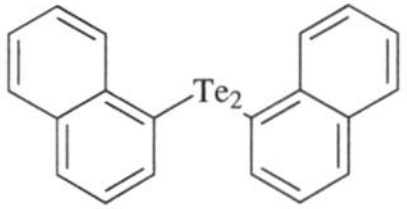 (7)	269
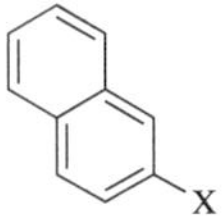 X			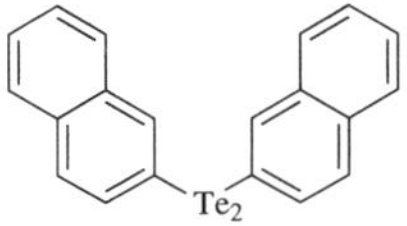	
Cl		HMPA, 130-170°, 16-24 h	(20)	269
Cl or Br		$Naph^{-\bullet}$ Na^+	(22)	364
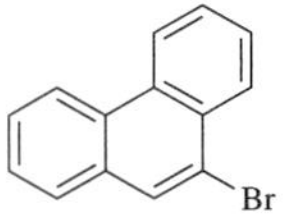		DMF, 110°, 16-24 h	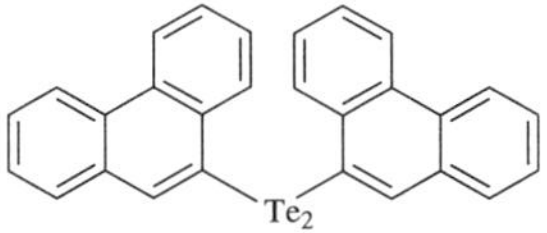 (40)	269
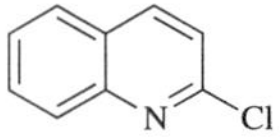		e^-, MeCN, mediator	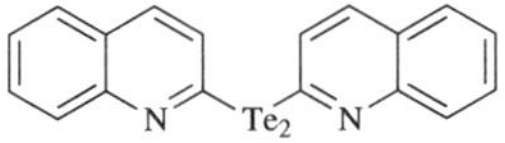 **I** (50)	276
		e^-, MeCN	**I** (48)	257
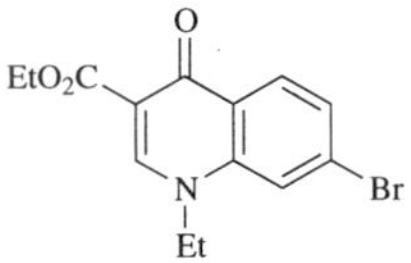		e^-, MeCN, *t*-BuOH, mediator	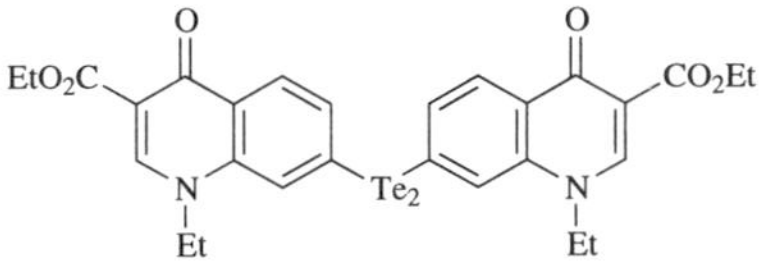 (35)	278
PhBr	$PhTe^-$	*hν*, NH_3, 3.7 h	Ph_2Te (20)	122
PhI		*hν*, NH_3, 3.7 h	Ph_2Te (90)	120, 122

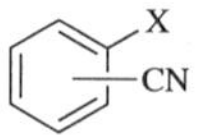		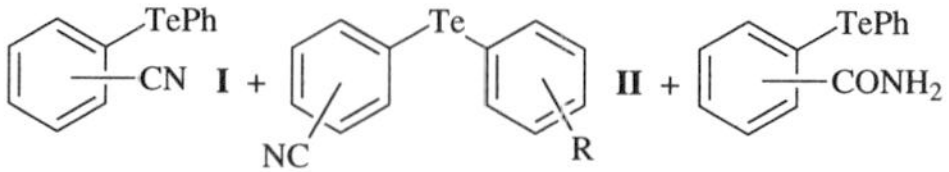	
X			
Cl 2-CN	e^-, MeCN	**I** (20) + **II**, R = $CONH_2$ (23) + **III** (21)	284
Cl 2-CN	e^-, MeCN, mediator	**I** (39) + **II**, R = $CONH_2$ (31) + **III** (8)	284, 281
Cl 3-CN	e^-, MeCN, mediator	**I** (48) + **II**, R = CN (4) + **III** (5)	281
Cl 4-CN	e^-, MeCN	**I** (53) + **II**, R = CN (22)	284, 281
Br 4-CN	e^-, MeCN	**I** (42) + **II**, R = CN (13)	283
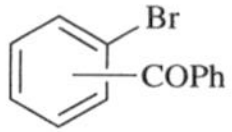	e^-, MeCN, mediator	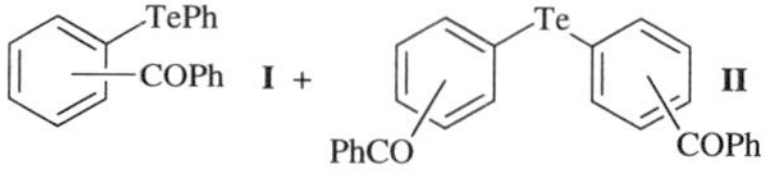 2-COPh, **I** (75) 3-COPh, **I** (48) + **II** (31) 4-COPh, **I** (45) + **II** (36)	281, 280
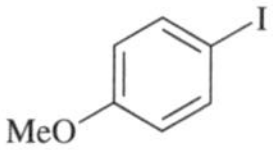	$h\nu$, NH_3, 3.7 h	Ph_2Te **I** (15) + 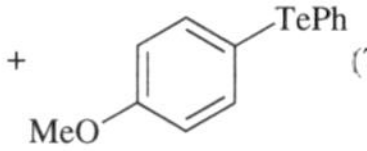 (73) + 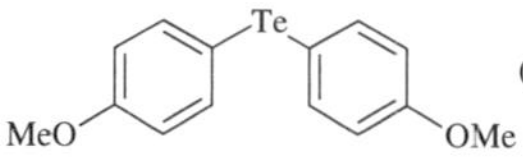 (11)	122
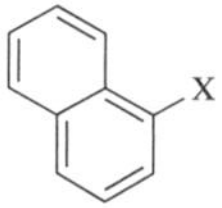 X Cl Br	$h\nu$, NH_3, 3.7 h	**I** + 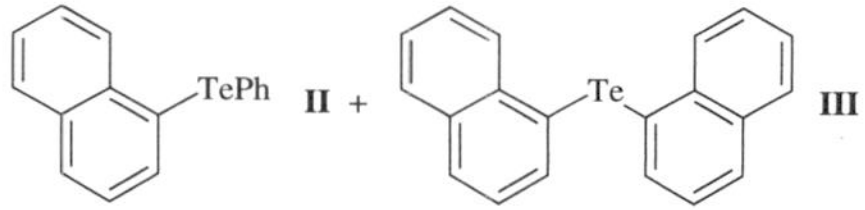 **I** (9) + **II** (41) + **III** (6) **I** (16) + **II** (53) + **III** (10)	122, 120

TABLE VIII. OTHER NUCLEOPHILES (*Continued*)

F. Nucleophiles Derived from Selenium and Tellurium (*Continued*)

Substrate	Nucleophile	Conditions	Product(s) and Yield(s) (%)	Refs.
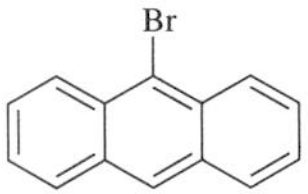	$PhTe^-$	e^-, MeCN	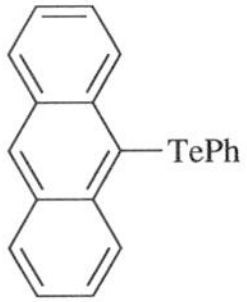 + 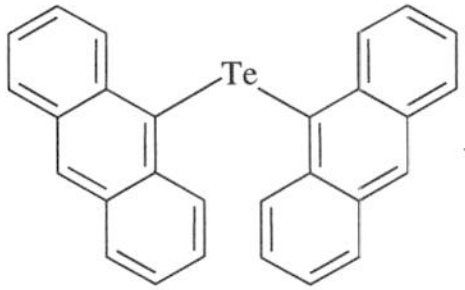 + $C_{14}H_{10}$ **I** (68) **II** (9) **III** (11)	285
		e^-, DMSO	**I** (49) + **II** (6) + **III** (42)	285
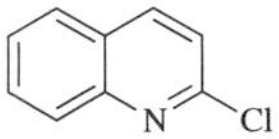		$h\nu$, NH_3, 3.7 h	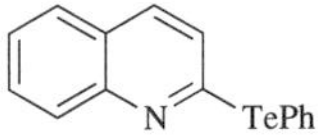 (43) + Ph_2Te (11)	117
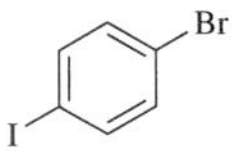		$h\nu$, NH_3, 3.7 h	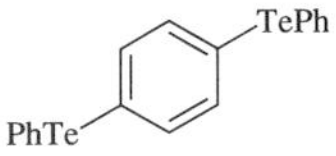 (40) + Ph_2Te (20)	122
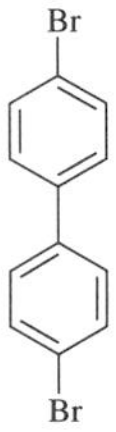		e^-, MeCN	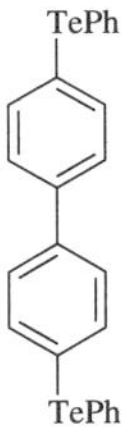 (16) + 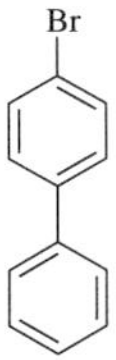 (52)	285

[a] Cathodic reduction of this compound followed by addition of MeI and hydrolysis lead to 1-ethyl-6-fluoro-7-methylseleno-4-(1*H*)-oxoquinoline-3-carboxylic acid.

[b] The substrate concentration was 0.019 M.

[c] The substrate concentration was 2.5 M.

[d] NMP = *N*-Methylpyrrolidinone.

TABLE IX. RING CLOSURE REACTIONS

A. $S_{RN}1$ Substitution Followed by a Ring Closure Reaction with an ortho Substituent

Substrate	Nucleophile	Conditions	Product(s) and Yield(s) (%)	Refs.
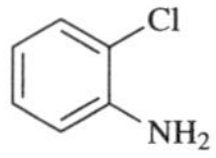	$^{-}CH_2COMe$	$FeSO_4$, NH_3, 4 h	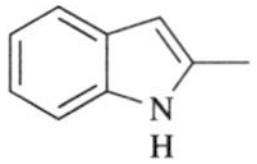 (51)	75
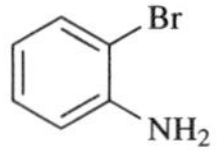	$^{-}CH_2COR$	*hν*, NH_3	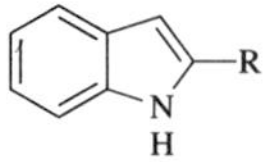	146
	R			
	Me	2 h	(93)	
	Pr-*i*	4 h	(84)	
	Bu-*t*	3 h	(94)	
	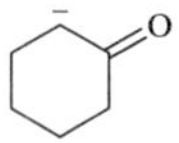	*hν*, NH_3, 4 h	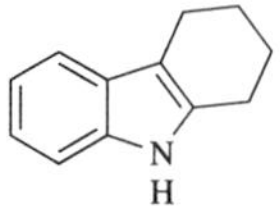 (14)	146
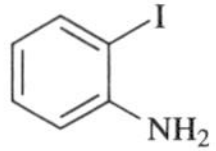	$^{-}CH_2CHO$		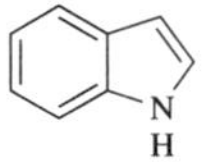	
		hν, NH_3, 13 min	(75) + $PhNH_2$ (25)	286, 106
		e^{-}, NH_3	(93)	287
	$^{-}CH_2COR$		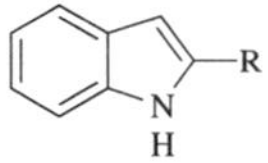	
	R			
	Me	*hν*, NH_3, 15 min	(100)	286, 106
	Me	e^{-}, NH_3	(87) + $PhNH_2$ (13)	287
	Pr-*i*	*hν*, NH_3, 10 min	(100)	286, 106
	Pr-*i*	e^{-}, NH_3	(75) + $PhNH_2$ (22)	287
	Bu-*t*	*hν*, NH_3, 20 min	(53)	106

TABLE IX. RING CLOSURE REACTIONS (*Continued*)

A. $S_{RN}1$ Substitution Followed by a Ring Closure Reaction with an ortho Substituent (*Continued*)

Substrate	Nucleophile	Conditions	Product(s) and Yield(s) (%)	Refs.
	$^-CH_2COEt \rightleftharpoons {}^-CHMeCOMe$	$h\nu$, NH_3, 25 min	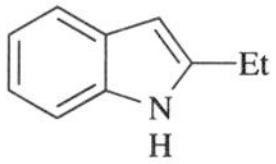 (20) + 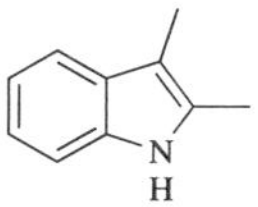 (40)	106
	$^-CH_2COCR(OMe)_2$	1. $h\nu$, NH_3 2. SiO_2 column	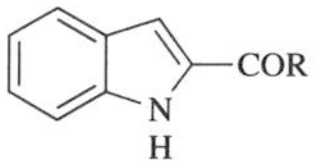	106
	R			
	H	1. 20 min	(45)	
	Me	1. 25 min	(55)	
	^-CHRCHO	$h\nu$, NH_3	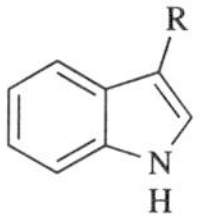	106
	R			
	Me	20 min	(49)	
	Et	23 min	(33)	
	i-Bu	42 min	(26)	
	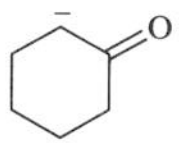	$h\nu$, NH_3	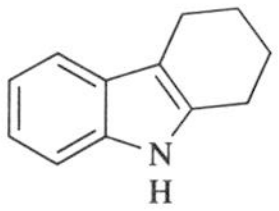	
		30 min	(33)	106
		8 min	(40) + $PhNH_2$ (59)	286
	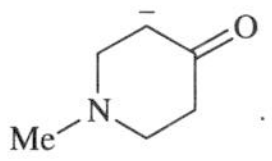	$h\nu$, NH_3, 10 min	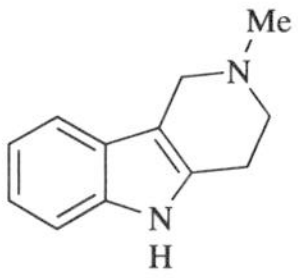 (22)	106

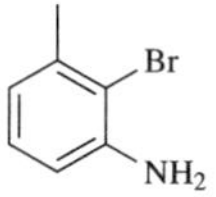	$^{-}CH_2COMe$	$h\nu$, NH_3, 4 h	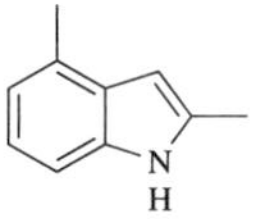 (80)	146
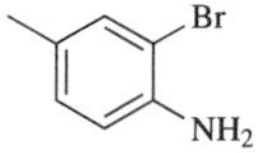		$h\nu$, NH_3, 2 h	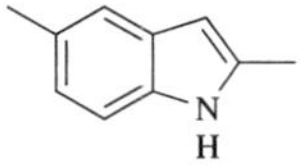 (88)	146
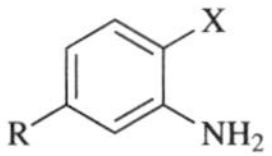		$h\nu$, NH_3	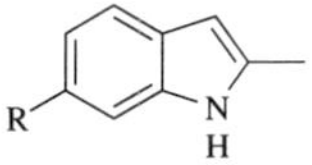	146
X R				
Cl Ph		12 h	(88)	
Cl OMe		10.5 h	(42)	
Cl CO_2H		12 h	(89)	
Br Me		2 h	(82)	
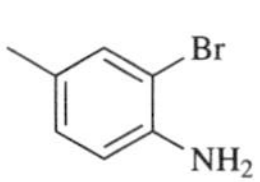	$^{-}CHMeCOEt$	$h\nu$, NH_3, 2 h	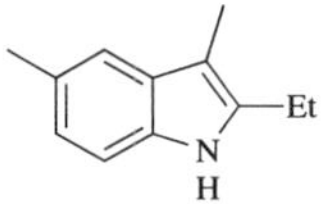 (73)	146
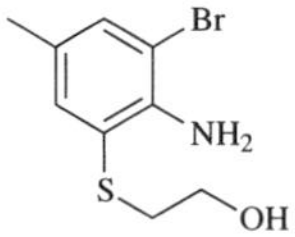	$^{-}CH_2COBu$-*t*	$h\nu$, NH_3, 2 h	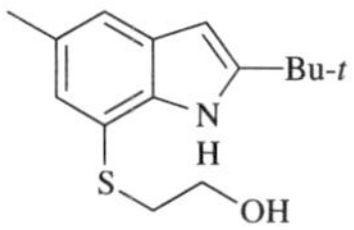 (75)	233
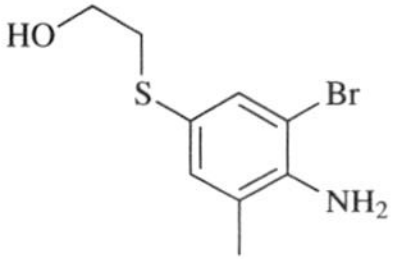		$h\nu$, NH_3, 2 h	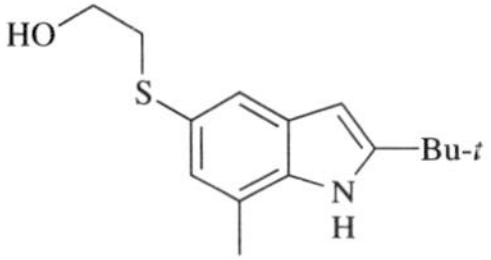 (60)	233

TABLE IX. RING CLOSURE REACTIONS (*Continued*)

A. $S_{RN}1$ Substitution Followed by a Ring Closure Reaction with an ortho Substituent (*Continued*)

Substrate	Nucleophile	Conditions	Product(s) and Yield(s) (%)	Refs.
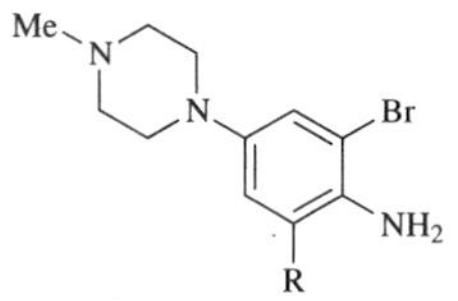 R: H; P(O)(OEt)$_2$	$^-CH_2COBu$-*t*	*h*ν, NH_3, 1 h	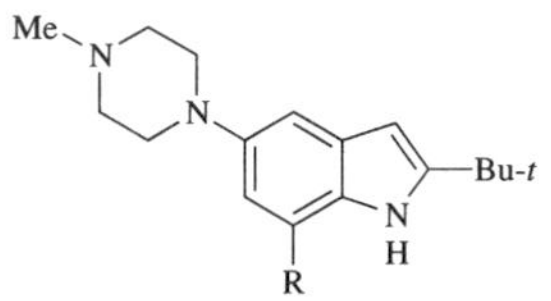 (85) (15)	233
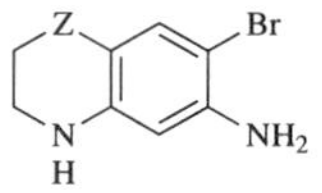		*h*ν, NH_3, 0.5 h	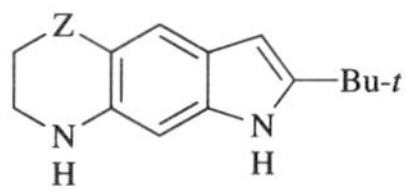 Z: S (52); NH (43); NMe (46)	233
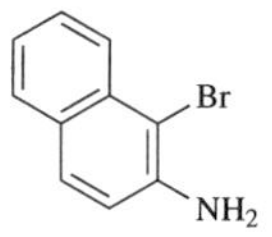	$^-CH_2COR$ R: Bu-*t*; CH(OMe)$_2$	*h*ν, NH_3, 1 h	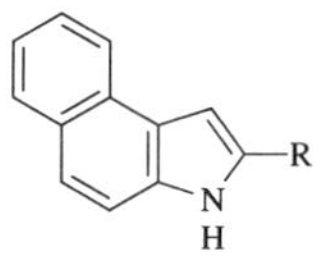 (67) (52)[a]	237
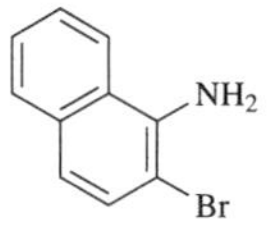	$^-CH_2COR$ R: Bu-*t*; CH(OMe)$_2$	*h*ν, NH_3, 2 h	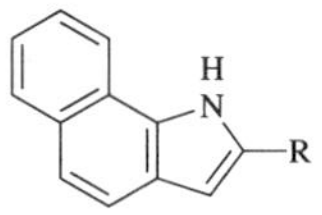 (82) (65)[a]	237

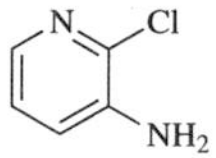	$^{-}CHRCHO$	$h\nu$, NH_3, 1.7 h	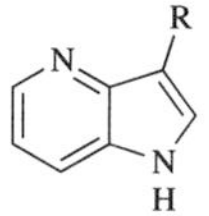	R: H (62); Me (59)	288
	$^{-}CH_2COR$	$h\nu$, NH_3	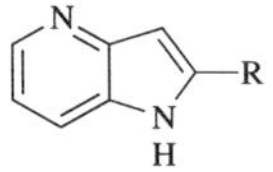		
	R				
	Me	10.5 h	(45)		146,148, 288
	Pr-*i*	1 h	(61)		148
	Bu-*t*	2.5 h	(100)		148
	$^{-}CHMeCOPh$	$h\nu$, NH_3, 1.7 h	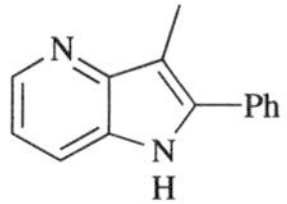	(30)	288
	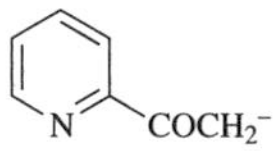	$h\nu$, NH_3, 1.7 h	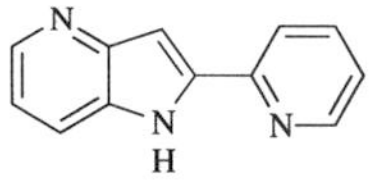	(21)	288
	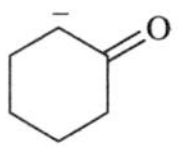	$h\nu$, NH_3	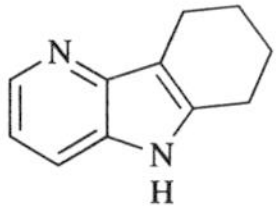		
		1.7 h	(52)		288
		3.5 h	(22)		148
	$^{-}CH_2COEt \rightleftharpoons {}^{-}CHMeCOMe$	$h\nu$, NH_3, 1.5 h	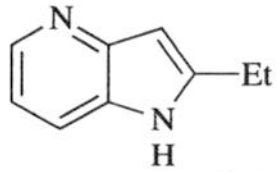 **I** + 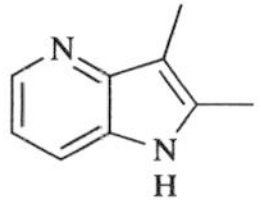 **II**	**I** + **II** (45)	148

TABLE IX. RING CLOSURE REACTIONS (*Continued*)

A. $S_{RN}1$ Substitution Followed by a Ring Closure Reaction with an ortho Substituent (*Continued*)

Substrate	Nucleophile	Conditions	Product(s) and Yield(s) (%)	Refs.
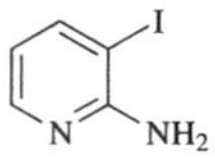	$^{-}CH_2COR$	*hν*, NH_3	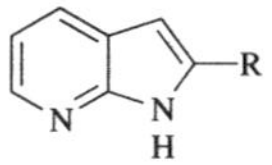 R: Me (75); Bu-*t* (78)	139
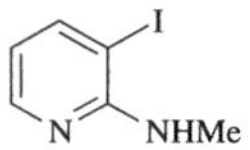		*hν*, NH_3	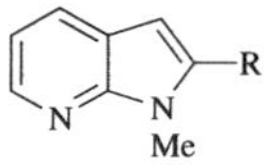 R: Me (95); Bu-*t* (70)	139
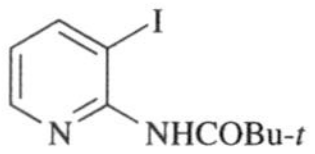		1. *hν*, NH_3 2. H_3O^+	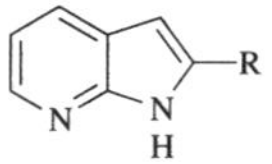 R: Me (80); Bu-*t* (90)	139
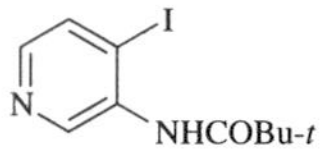		1. *hν*, NH_3 2. H_3O^+	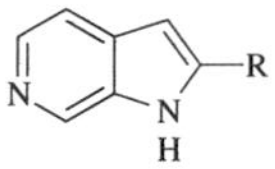 R: Me (98); Bu-*t* (99)	139
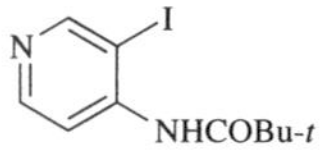		1. *hν*, NH_3 2. H_3O^+	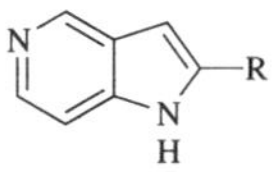 R: Me (98); Bu-*t* (98)	139
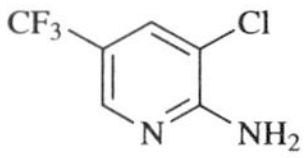	$^{-}CMe(CO_2Et)_2$	*hν*, NH_3, 2 h	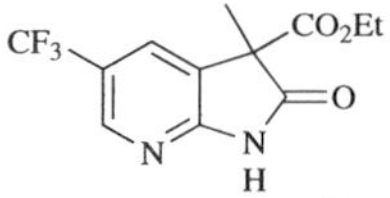 (10)	181
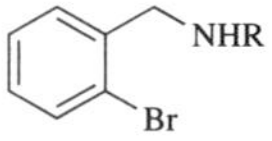	$Co_2(CO)_8$, CO, 1 atm	Sunlamp, NaOH, H_2O, C_6H_6, Bu_4NBr, 65°, 5.5 h	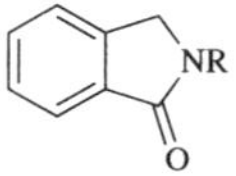 R: H (60); Bn (78)	189
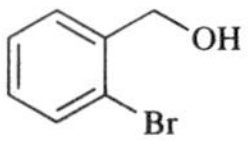	$Co_2(CO)_8$, CO, 1 atm	Sunlamp, NaOH, H_2O, C_6H_6, Bu_4NBr, 65°, 2.5 h	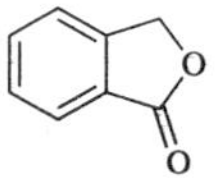 (95)	189

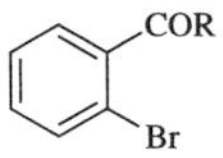COR Br	$^-SCH_2CO_2Et$	1. $h\nu$, *t*-BuOK, NH_3 2. MeI	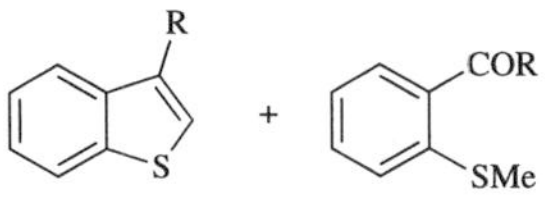R S + COR SMe	49
R				
H		1. 3 h	(55) (10)	
Me		1. 6 h	(40) (48)	
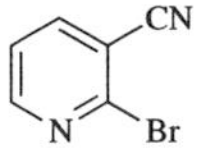CN N Br		$h\nu$, *t*-BuOK, NH_3, 1 h	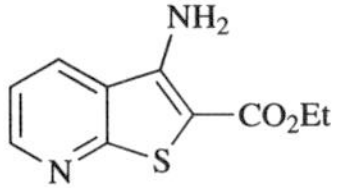NH2 CO2Et N S (90)	49
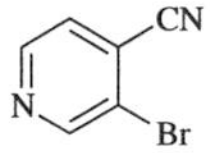CN N Br		$h\nu$, *t*-BuOK, NH_3, 1.25 h	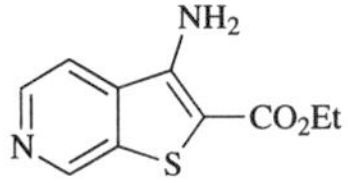NH2 CO2Et N S (98)	49
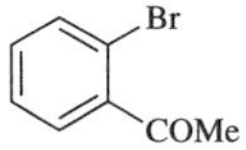Br COMe	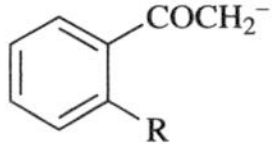COCH2− R	1. $h\nu$, DMSO, 1 h 2. *i*-PrBr	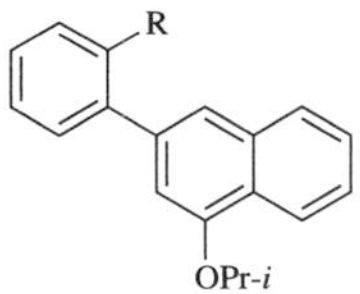R OPr-i R: Me (82); OMe (75)	293
	$^-CHRCOPh$	1. $h\nu$, DMSO, 1 h 2. *i*-PrBr	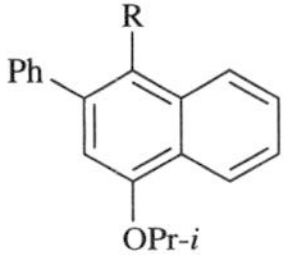R Ph OPr-i R: Me (50); OMe (80)	293
	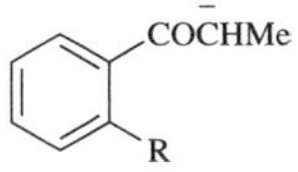COCHMe R	1. $h\nu$, DMSO, 1 h 2. *i*-PrBr	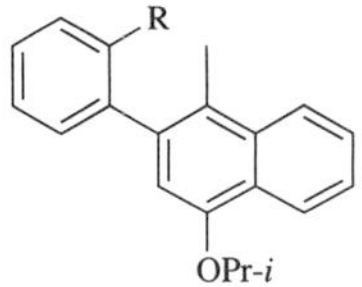R OPr-i R: Me (47); OMe (52)	293

TABLE IX. RING CLOSURE REACTIONS (*Continued*)

A. $S_{RN}1$ Substitution Followed by a Ring Closure Reaction with an ortho Substituent (*Continued*)

Substrate	Nucleophile	Conditions	Product(s) and Yield(s) (%)	Refs.
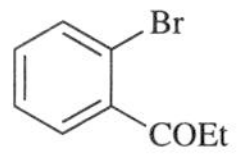	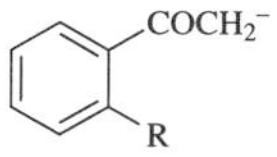	1. *hν*, DMSO, 1 h 2. *i*-PrBr	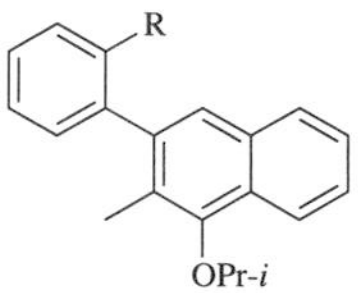 R: Me (58); OMe (57)	293
	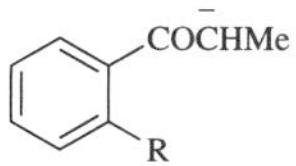	1. *hν*, DMSO, 1 h 2. *i*-PrBr	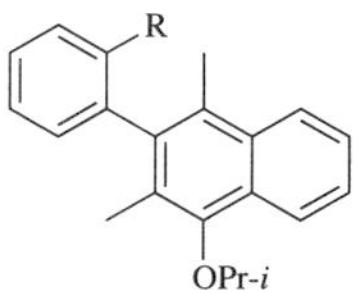 R: Me (36); OMe (41)	293
	$^{-}$CH(OMe)COPh	1. *hν*, DMSO, 1 h 2. *i*-PrBr	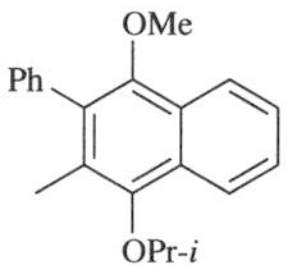 (50)	293
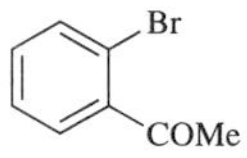	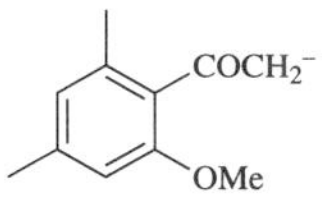	1. *hν*, DMSO, 1 h 2. *i*-PrBr	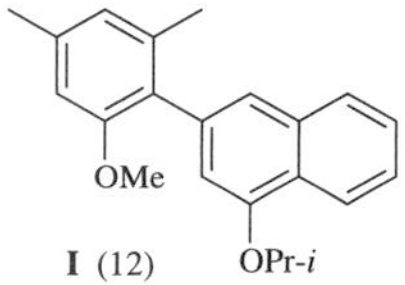 **I** (12) + 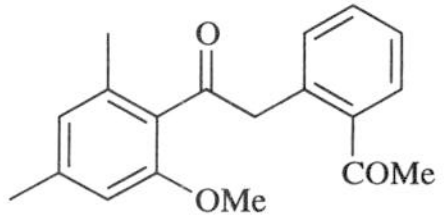 (4)	293
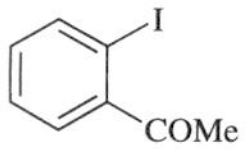		1. *hν*, DMSO, 1.5 h 2. *i*-PrBr	**I** (34)	293

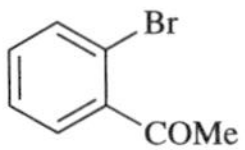	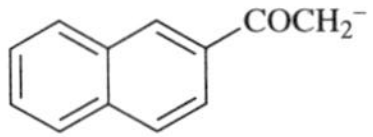		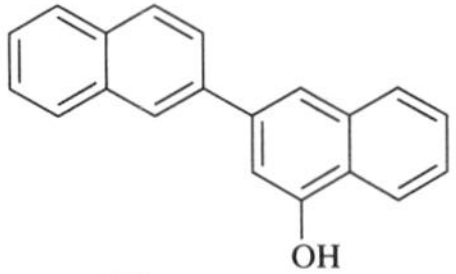	
		$h\nu$, NH_3, 1.5 h	(72)	158
		$h\nu$, DMSO, 1.5 h	(82)	158
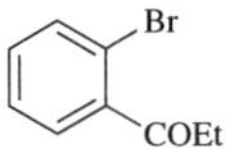		1. $h\nu$, DMSO, 1 h 2. *i*-PrBr	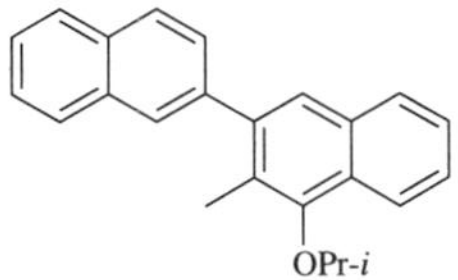 (62)	293
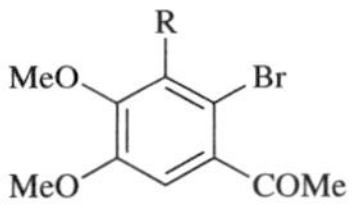		1. $h\nu$, DMSO, 1 h 2. *i*-PrBr	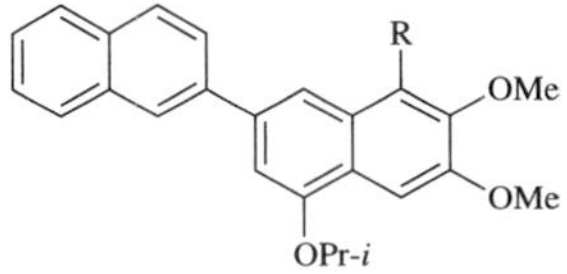 R: H (80); OMe (52)	158
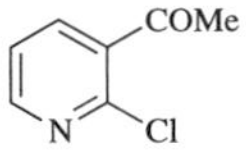	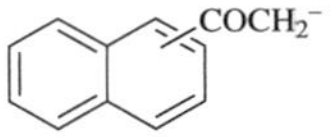	1. $h\nu$, DMSO 2. *i*-PrBr	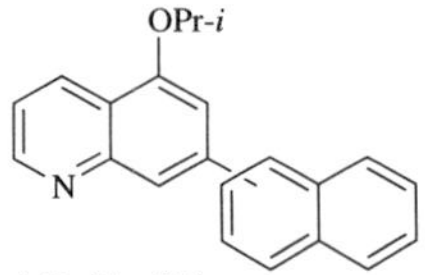	215
		1. 0.5 h	1-$C_{10}H_7$ (87)	
		1. 1 h	2-$C_{10}H_7$ (75)	
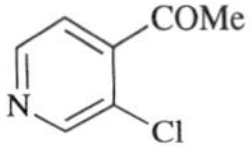		1. $h\nu$, DMSO, 1 h 2. *i*-PrBr	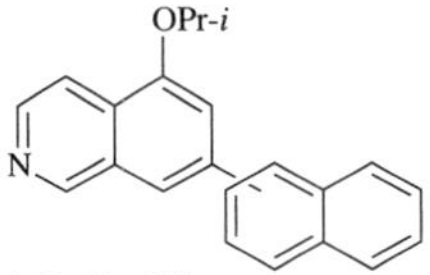 1-$C_{10}H_7$ (70) 2-$C_{10}H_7$ (63)	215

TABLE IX. RING CLOSURE REACTIONS (*Continued*)

A. $S_{RN}1$ Substitution Followed by a Ring Closure Reaction with an ortho Substituent (*Continued*)

Substrate	Nucleophile	Conditions	Product(s) and Yield(s) (%)	Refs.
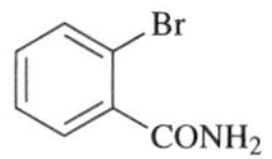	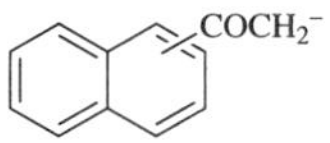	1. *hν*, DMSO 2. *i*-PrBr 1. 2 h 1. 1 h	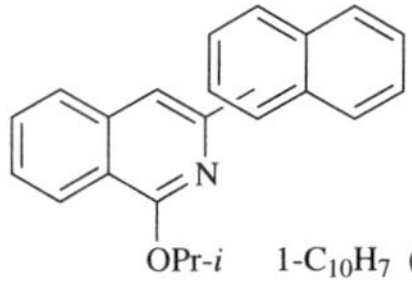 1-$C_{10}H_7$ (70) 2-$C_{10}H_7$ (60)	215
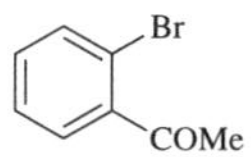	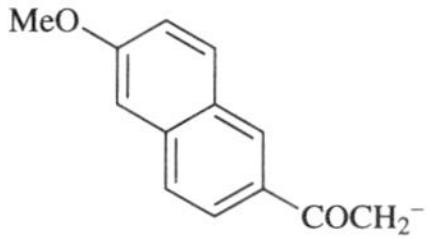	1. *hν*, DMSO, 1 h 2. *i*-PrBr	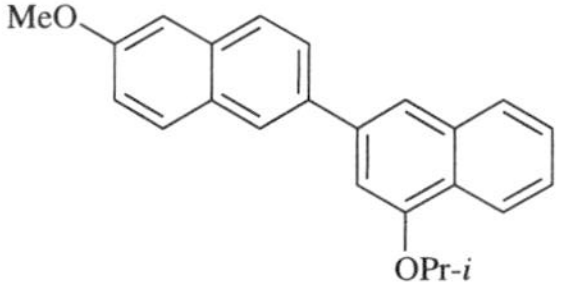 (60)	158
	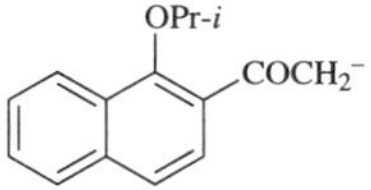	1. *hν*, DMSO, 1 h 2. *i*-PrBr	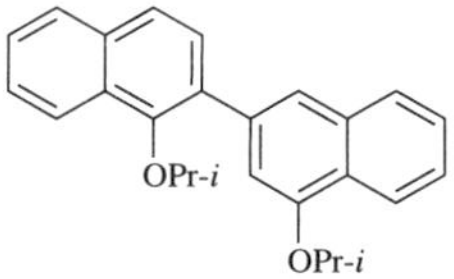 (83)	158
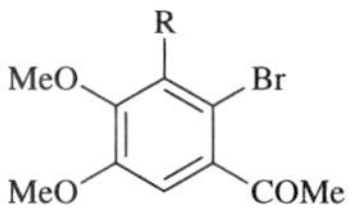	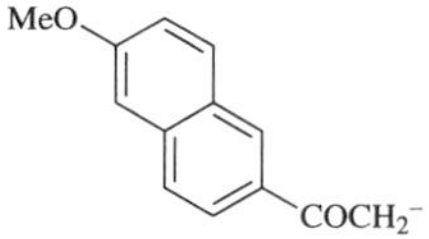	1. *hν*, DMSO, 1 h 2. *i*-PrBr	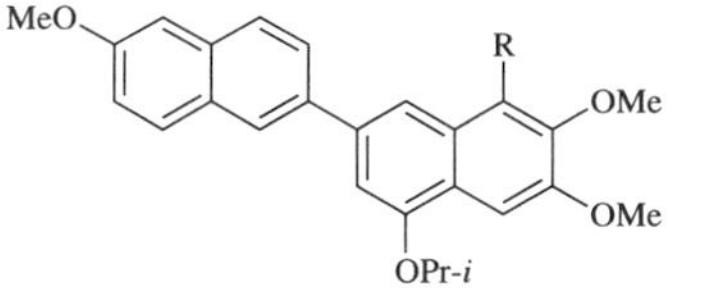 R: H (59); OMe (73)	158
	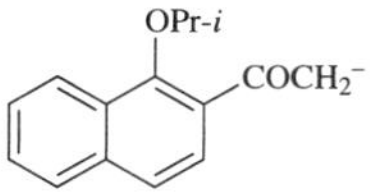	1. *hν*, DMSO, 1 h 2. *i*-PrBr	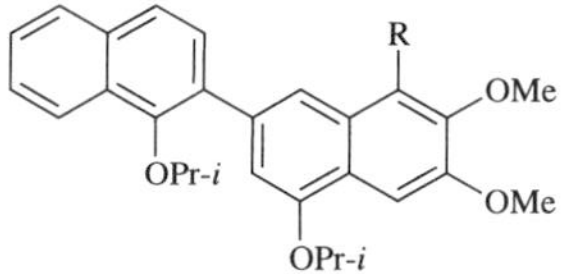 R: H (74); OMe (72)	158

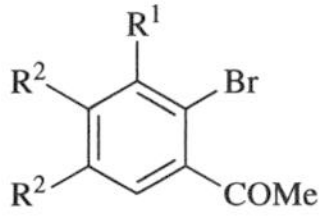R1 R2 Br R2 COMe R¹ R² H H H OMe OMe OMe	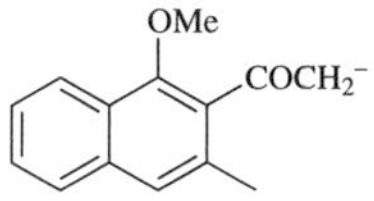OMe COCH2⁻	1. *h*ν, DMSO, 1 h 2. *i*-PrBr	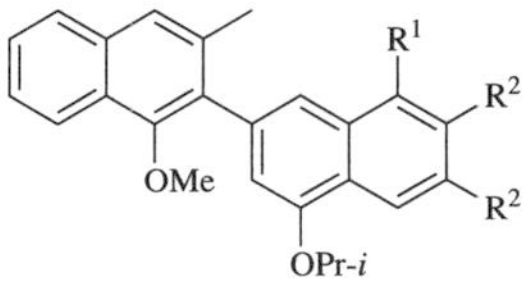R1 R2 R2 OMe OPr-i (30) (25) (26)	293
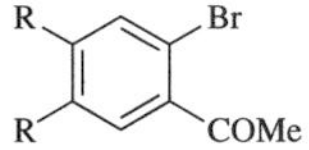R Br R COMe	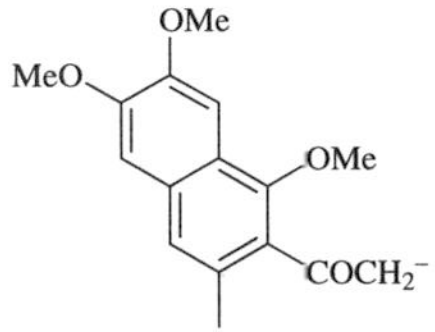OMe MeO OMe COCH2⁻	1. *h*ν, DMSO, 1 h 2. *i*-PrBr	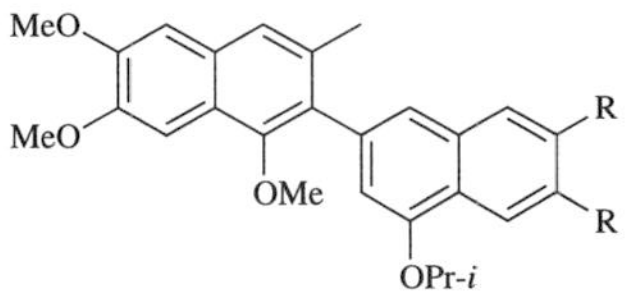MeO MeO OMe R R OPr-i R: H (36) R: OMe (40)	293
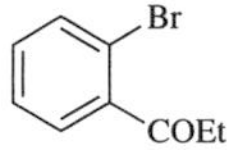Br COEt		1. *h*ν, DMSO, 1 h 2. *i*-PrBr	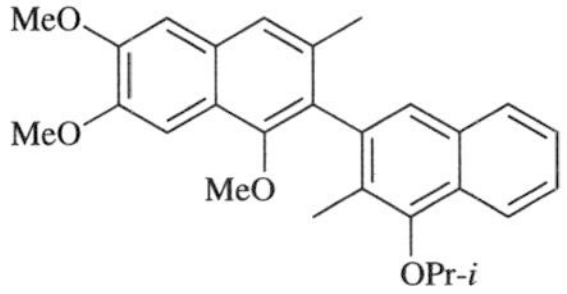MeO MeO MeO OPr-i (42)	293
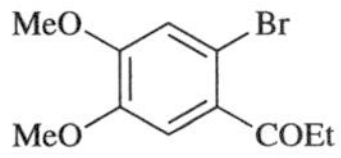MeO Br MeO COEt	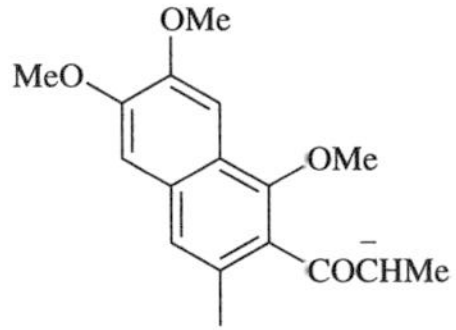OMe MeO OMe COCHMe	1. *h*ν, DMSO, 1 h 2. *i*-PrBr	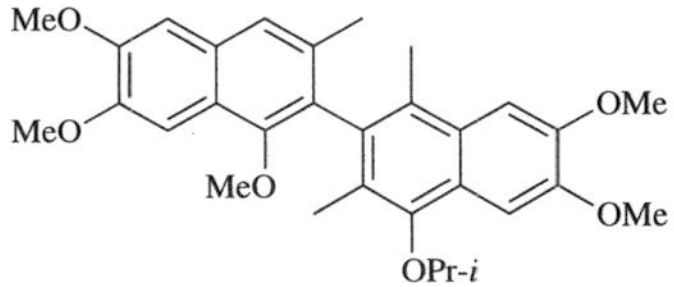MeO MeO MeO OMe OMe OPr-i (17)	293
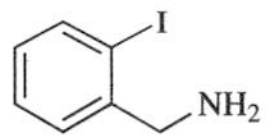I NH2	$^-$CHRCHO	*h*ν, NH_3, 10 min	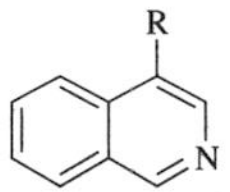R N **I** + $BnNH_2$ **II** R: H, I (50), II (10) R: Me, I (30), II (—)	291

TABLE IX. RING CLOSURE REACTIONS (*Continued*)

A. $S_{RN}1$ Substitution Followed by a Ring Closure Reaction with an ortho Substituent (*Continued*)

Substrate	Nucleophile	Conditions	Product(s) and Yield(s) (%)	Refs.
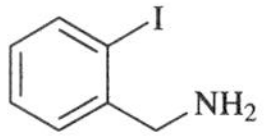	$^{-}CH_2COR$	$h\nu$, NH_3, 5 min	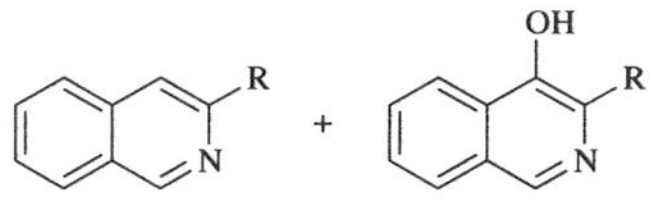	
	R			
	Me		(23) (60)	291, 290
	Pr-*i*		(80)	291
	Bu-*t*		(20) (60)	291
		1. $h\nu$, NH_3 2. $NaBH_4$, MeOH	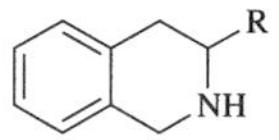	
	R			
	Me	1. 10 min	(90)	291, 290
	Pr-*i*	1. 6 min	(60)	291
		1. $h\nu$, NH_3 2. Pd/C, MeOH, reflux	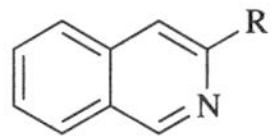	
	R			
	Me		(100)	291, 290
	Pr-*i*		(94)	291
	Bu-*t*		(90-100)	290, 291
	$CH(OMe)_2$		(38)	291
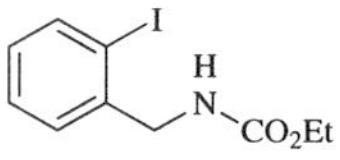	$^{-}CH_2COMe$	$h\nu$, NH_3	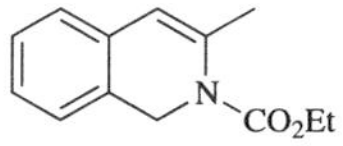 (41)	290
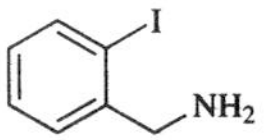	$^{-}CHMeCOEt$	1. $h\nu$, NH_3, 25 min 2. $NaBH_4$, MeOH	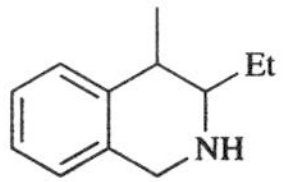 (90-100)	290, 291

		1. $h\nu$, NH_3, 25 min 2. Pd/C, MeOH, reflux	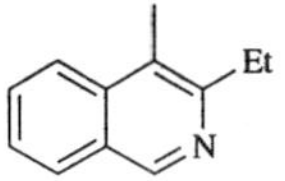 (77)	291, 290
	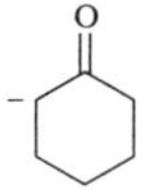	1. $h\nu$, NH_3, 20 min 2. Pd/C, MeOH, reflux	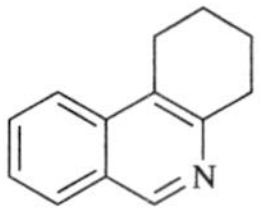 (43)	291
	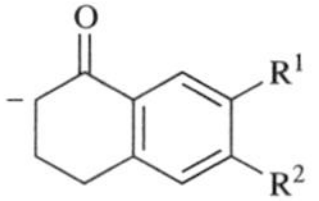	1. $h\nu$, NH_3 2. O_2 (air)	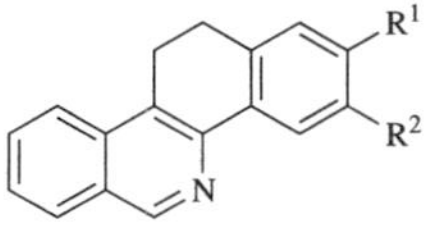	292
	R^1 / R^2: H / H	1.3 h	(38)	
	OMe / H	1 h	(79)	
	OCH_2O	1 h	(65)	
	OMe / OMe	1 h	(35)	
	OPr-*i* / OMe	1 h	(39)	
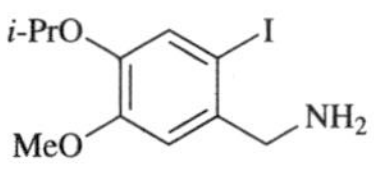	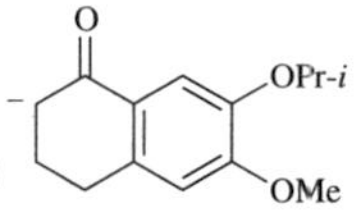	1. $h\nu$, NH_3, 0.7-1.3 h 2. O_2 (air)	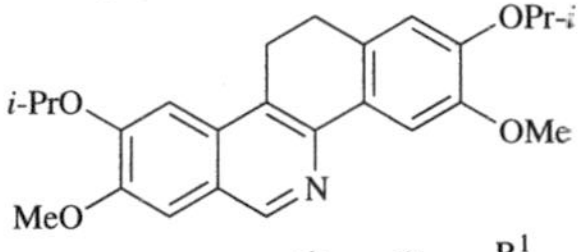 (65)	292
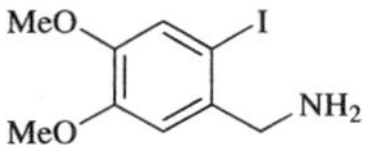	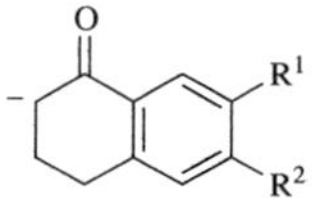	1. $h\nu$, NH_3, 0.7-1.3 h 2. O_2 (air)	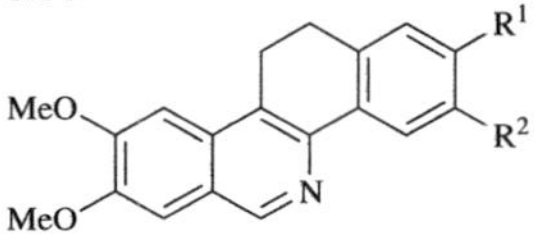	292
	R^1 / R^2: H / H		(45)	
	OMe / H		(56)	
	OCH_2O		(48)	
	OMe / OMe		(68)	
	OPr-*i* / OMe		(42)	

TABLE IX. RING CLOSURE REACTIONS (*Continued*)

A. $S_{RN}1$ Substitution Followed by a Ring Closure Reaction with an ortho Substituent (*Continued*)

Substrate	Nucleophile	Conditions	Product(s) and Yield(s) (%)	Refs.
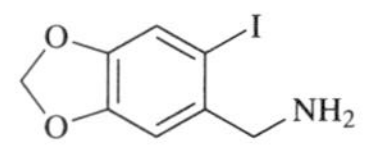	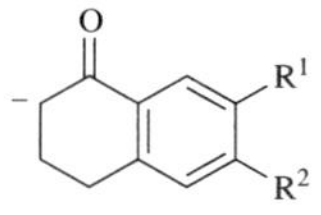	1. $h\nu$, NH_3, 0.7-1.3 h 2. O_2 (air)	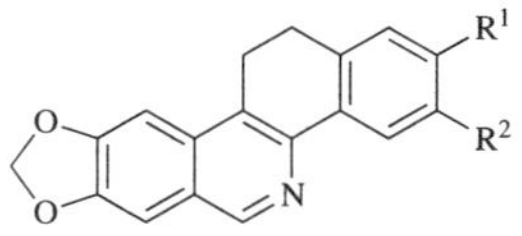	292
	R^1 R^2			
	H H		(76)	
	OMe H		(52)	
	OCH_2O		(38)	
	OMe OMe		(72)	
	OPr-*i* OMe		(48)	
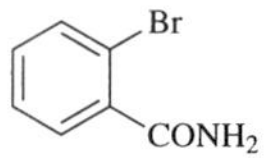	$^-CH_2COR$	$h\nu$, NH_3, 0.5 h	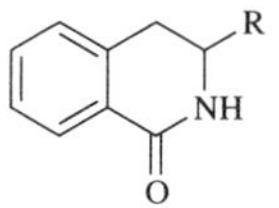 R: Me (90), Pr-*i* (90), Bu-*t* (90)	147
	$^-CH_2COCH(OMe)_2$	$h\nu$, NH_3, 6 h	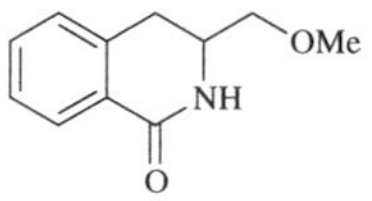 (10)	147
	$^-CH_2COCH_2OMe$	$h\nu$, NH_3, 5 h	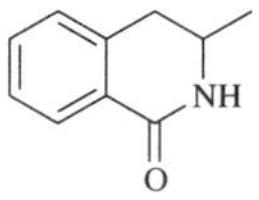 (35)	147
	$^-CHMeCOEt$	$h\nu$, NH_3, 5 h	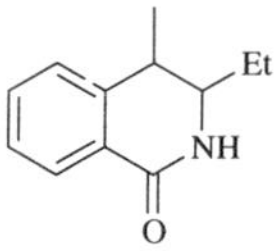 (70)	147

	$^{-}CH_2COCH_2Bn \rightleftharpoons$ $^{-}CH(Bn)COMe$	$h\nu$, NH_3, 5 h	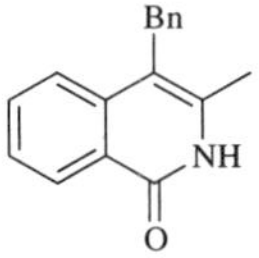 (30) + 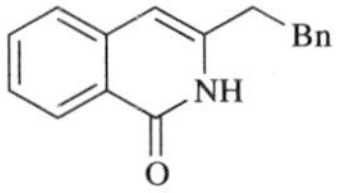 (15)	147
	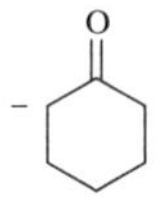	$h\nu$, NH_3, 3 h	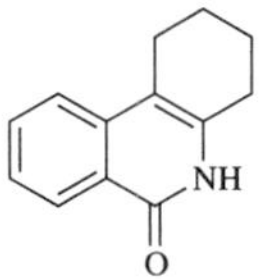 (30)	147
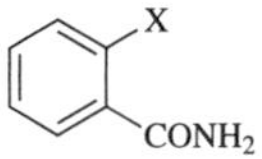	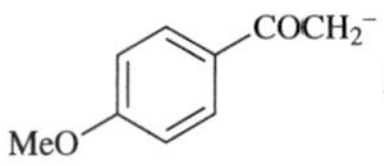	$h\nu$, NH_3, 6 h	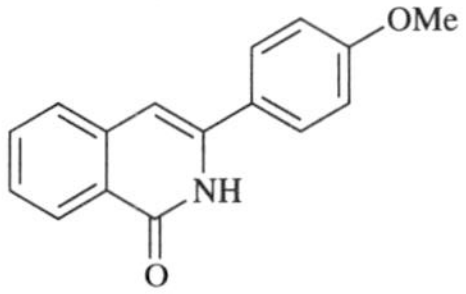 X: Br (4); I (80)	147
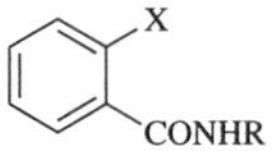	$^{-}CH_2COMe$	$h\nu$, NH_3	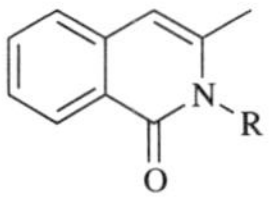	147
X = Br, R = Me		5 h	(40)	
X = Br, R = Et		4 h	(40)	
X = I, R = Me		3 h	(45)	
X = I, R = Et		3 h	(43)	
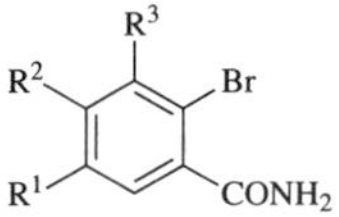		$h\nu$, NH_3, 1 h	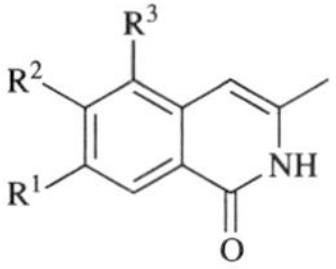	168
R^1 = OMe, R^2 = OMe, R^3 = OMe			(80)	
R^1 = OMe, R^2 = OMe, R^3 = H			(75)	
R^1R^2 = OCH_2O, R^3 = H			(70)	

TABLE IX. RING CLOSURE REACTIONS (*Continued*)

A. $S_{RN}1$ Substitution Followed by a Ring Closure Reaction with an ortho Substituent (*Continued*)

Substrate	Nucleophile	Conditions	Product(s) and Yield(s) (%)	Refs.
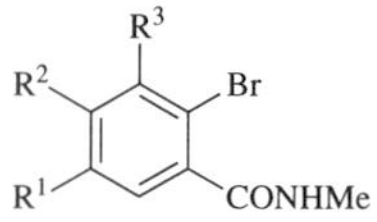	$^{-}CH_2COMe$	$h\nu$, NH_3	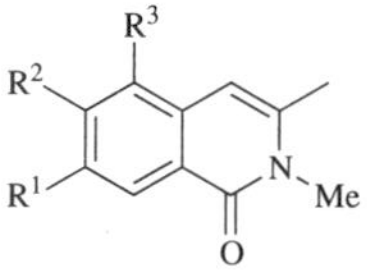	168
R^1 / R^2 / R^3				
OMe / OMe / OMe		6 h	(45)	
OMe / OMe / H		4 h	(40)	
OCH_2O / H		1.3 h	(60)	
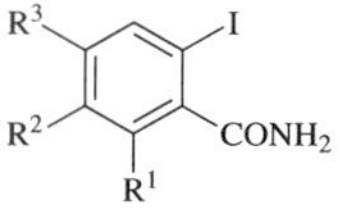	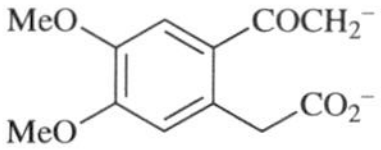	1. $h\nu$, NH_3, 1 h 2. BF_3, MeOH, reflux, 1 h	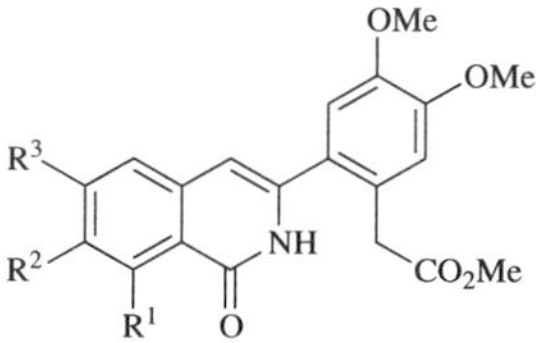	289
R^1 / R^2 / R^3				
H / H / H			(85)	
H / OMe / OMe			(72)	
OMe / OMe / H			(68)	
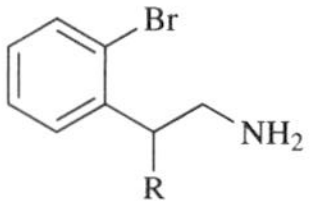	$Co_2(CO)_8$, CO, 1 atm	Sunlamp, NaOH, C_6H_6, H_2O, Bu_4NBr, 65°, 5.5 h	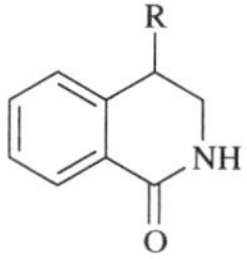 R: H (72); Et (82)	189
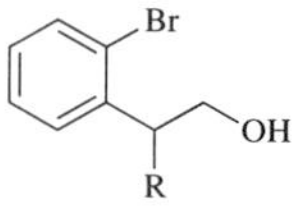	$Co_2(CO)_8$, CO, 1 atm	Sunlamp, NaOH, C_6H_6, H_2O, Bu_4NBr, 65°, 2.5 h	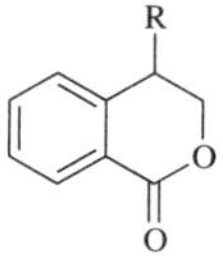 R: H (95); Et (92)	189

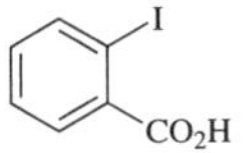	$^{-}CH_2COMe$	1. *hν*, NH_3 2. H_3O^+	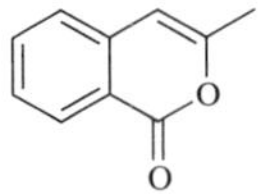 (80)	168
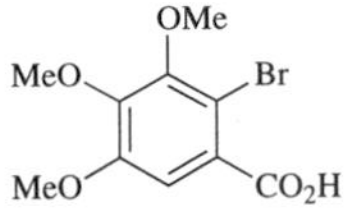		1. *hν*, NH_3, 2 h 2. TsOH, C_6H_6, reflux	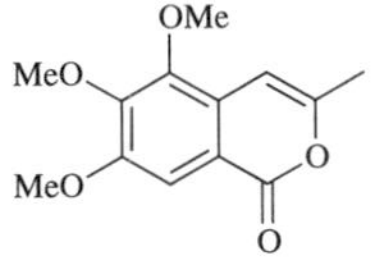 (69)	168
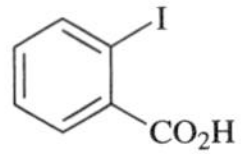	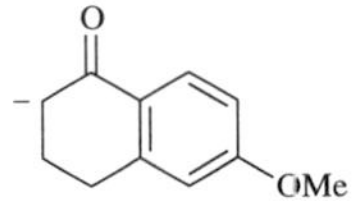	1. *hν*, NH_3, 1.25 h 2. TsOH, C_6H_6, reflux	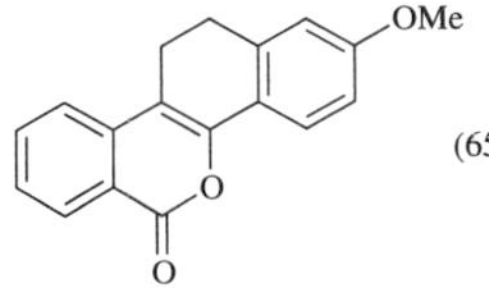 (65)	292
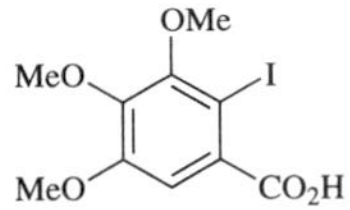	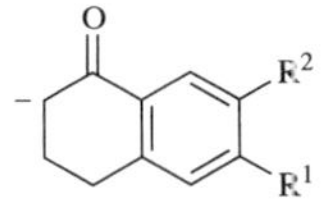	1. *hν*, NH_3, 1.25 h 2. TsOH, C_6H_6, reflux	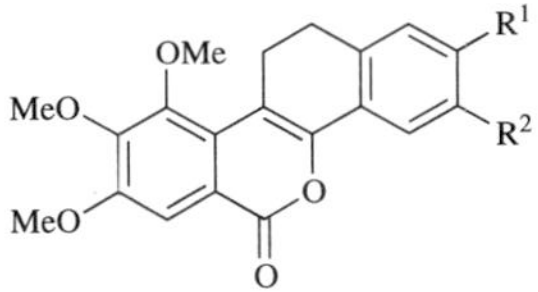 R^1 / R^2 OMe / H (72) OPr-*i* / OMe (70)	292
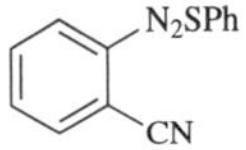	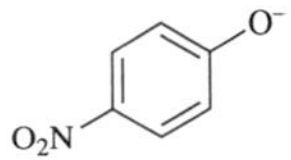	1. *hν*, DMSO, 1.5-2 h 2. SiO_2, CH_2Cl_2	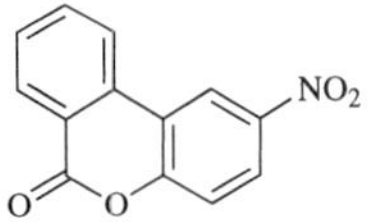 (59)	294
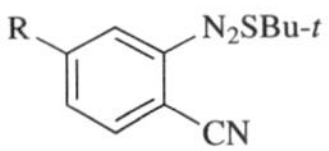	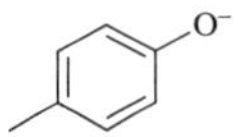	1. *hν*, DMSO, 1.5-2 h 2. SiO_2, CH_2Cl_2	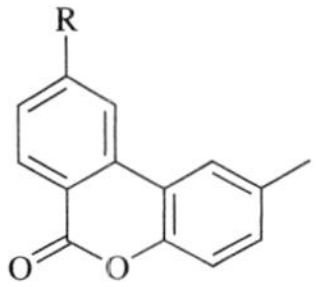 R Me (40) OMe (38)	294

TABLE IX. RING CLOSURE REACTIONS (*Continued*)

A. $S_{RN}1$ Substitution Followed by a Ring Closure Reaction with an ortho Substituent (*Continued*)

Substrate	Nucleophile	Conditions	Product(s) and Yield(s) (%)	Refs.
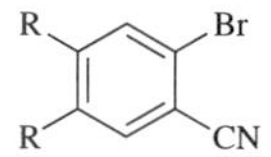	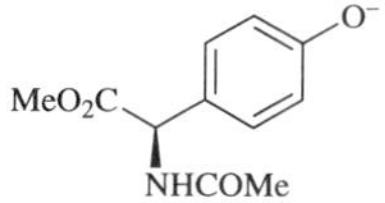	1. $h\nu$, NH_3 2. SiO_2, $CHCl_3$, reflux	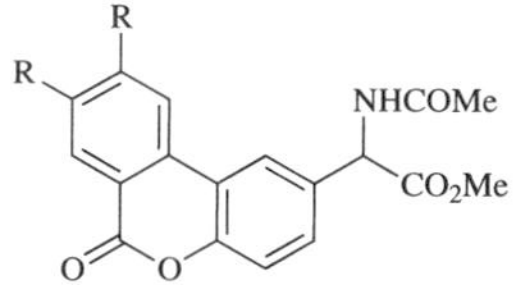 R: H (65)[b]; OMe (79)[b]	214
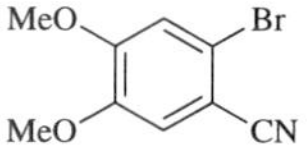	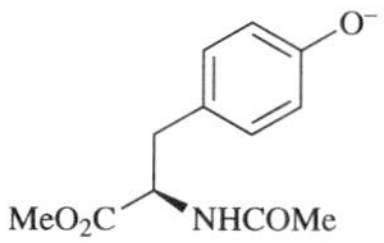	1. $h\nu$, NH_3 2. SiO_2, $CHCl_3$, reflux	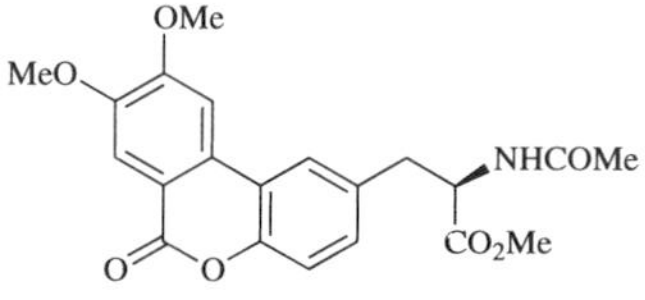 (52)	214
	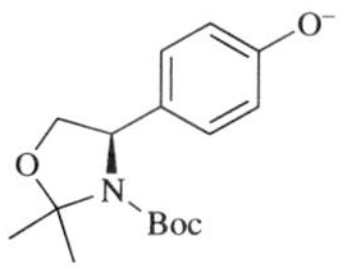	1. $h\nu$, NH_3 2. SiO_2, $CHCl_3$, reflux	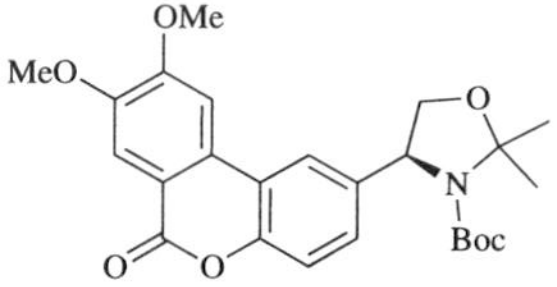 (77)	214
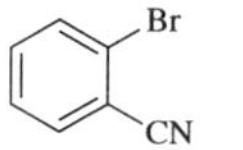	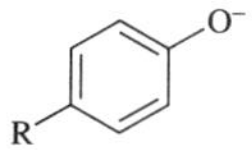	1. $h\nu$, NH_3, 1 h 2. SiO_2, $CHCl_3$, reflux	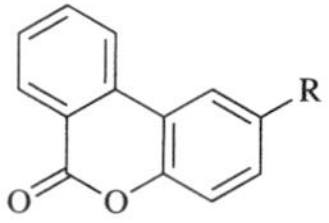	214
	R			
	Bu-*t*		(80)	
	OCF_3		(86)	
	F		(49)	
	CN		(51)	
	NO_2	7 h	(0)	

	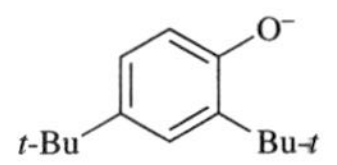	1. $h\nu$, NH_3, 1 h 2. SiO_2, $CHCl_3$, reflux	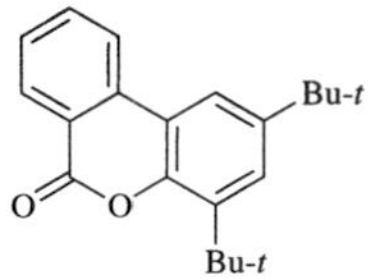 (78)	214
	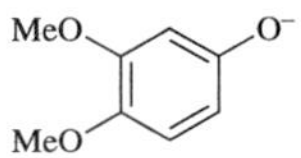	1. $h\nu$, NH_3, 1 h 2. SiO_2, $CHCl_3$, reflux	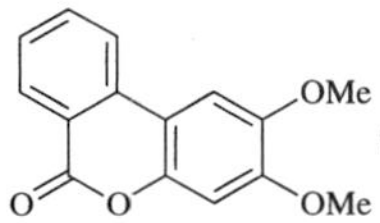 (65) + 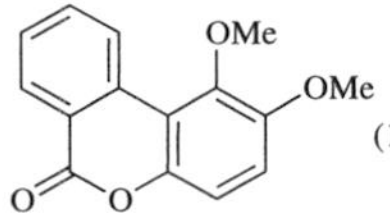 (13)	214
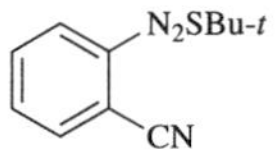	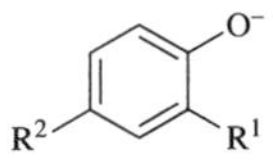	1. $h\nu$, DMSO 2. SiO_2, $CHCl_3$, reflux	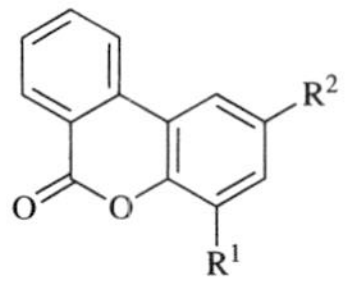	294
	R^1 R^2			
	H H		(31)	
	H Me		(48)	
	Me Me		(20)	
	H OMe		(56)	
	H Br		(57)	
	H CF_3		(55)	
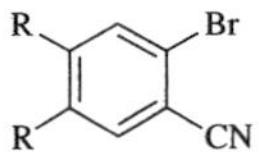	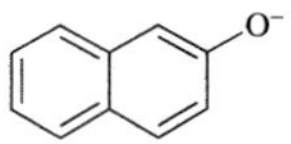	1. $h\nu$, NH_3, 1 h 2. SiO_2, $CHCl_3$, reflux	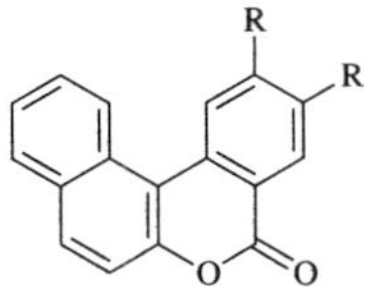 + 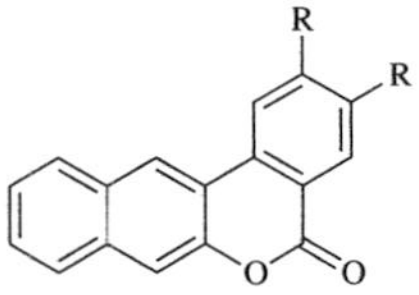	214
R				
H			(76) + (5)	
OMe			(84) + (5)	

TABLE IX. RING CLOSURE REACTIONS (*Continued*)

A. $S_{RN}1$ Substitution Followed by a Ring Closure Reaction with an ortho Substituent (*Continued*)

Substrate	Nucleophile	Conditions	Product(s) and Yield(s) (%)	Refs.
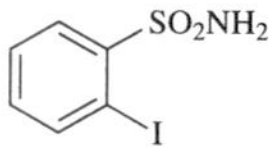	$^{-}CH_2COR$	$h\nu$, NH_3, 15 min	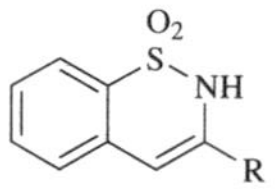 + $PhSO_2NH_2$ **I**	140
	R			
	Me		(90)	
	Et		(27) + **I** (9)	
	Pr-*i*		(67) + **I** (6)	
	Bu-*t*		(80)	
	$^{-}CHMeCOR$	$h\nu$, NH_3, 15 min	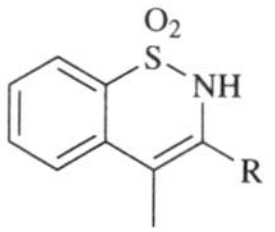 R: Et (20) + **I** (42); Pr-*i* (9) + **I** (58)	140
	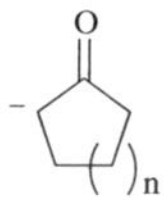	$h\nu$, NH_3, 15 min	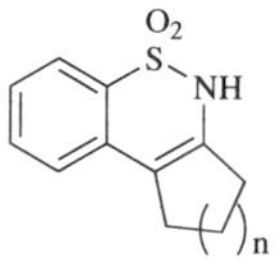 n: 1 (45) + **I** (10); 2 (45) + **I** (42)	140
	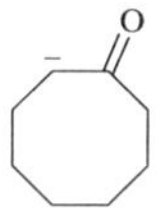	1. $h\nu$, NH_3, 15 min 2. [O] (air)	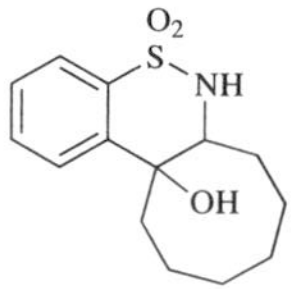 (65)	140

[a] The product was isolated as the ethyl ester ($R = CO_2Et$) after spontaneous hydrolysis and oxidation of the aldehyde function.

[b] This was a racemic mixture.

TABLE IX. RING CLOSURE REACTIONS (*Continued*)

B. Intramolecular $S_{RN}1$ Reactions

Substrate	Conditions	Product(s) and Yield(s) (%)	Refs.
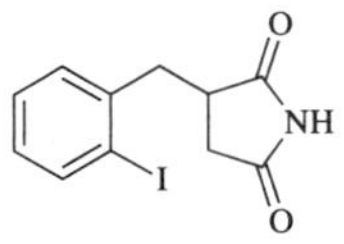	*hν*, KNH_2, NH_3, 1 h	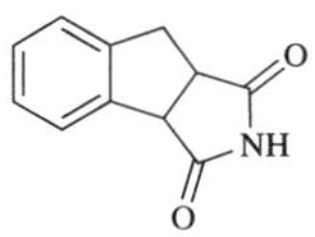 (40) + 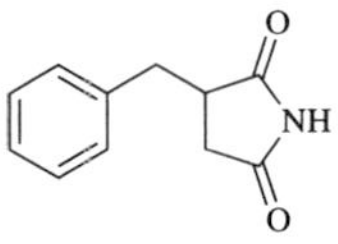 (30)	140
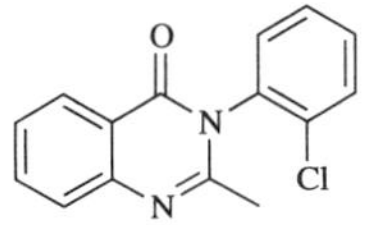	*hν*, $NaNH_2$, NH_3, 1 h	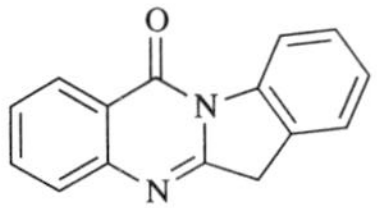 (60)	369
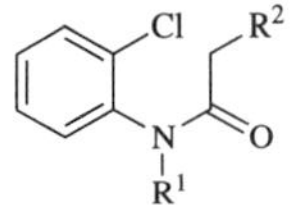	*hν*, LDA, THF	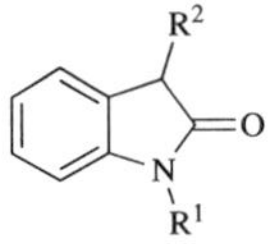	
R^1 / R^2			
Me / H	3 h	(82)	298
Bn / H	0.5 h	(32)	298
Me / *n*-Bu	3 h	(73)	298
Me / Ph	3 h	(64)	298
n-Pr / Et	1 h	(66)	370
n-C_9H_{19} / Me	1 h	(68)	370
n-$C_{18}H_{37}$ / Me	1 h	(8)	370

TABLE IX. RING CLOSURE REACTIONS (*Continued*)

B. Intramolecular $S_{RN}1$ Reactions (*Continued*)

Substrate	Conditions	Product(s) and Yield(s) (%)	Refs.
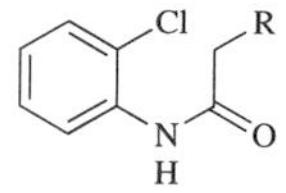	$h\nu$, LDA, THF, C_6H_{14}, 3 h	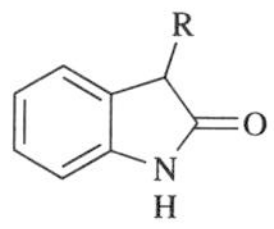 R: H (74); Me (73); Ph (63)	298
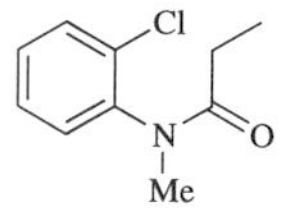	1. $h\nu$, LDA, THF, 3 h 2. MeI	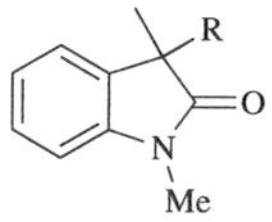 R: Me (66); Bn (60)	299
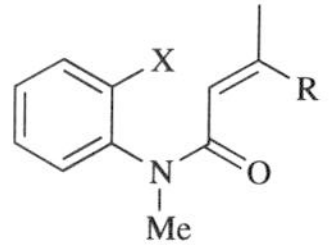		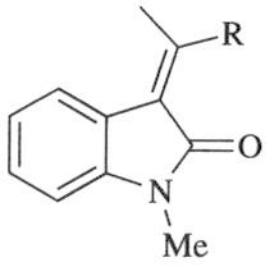	299
X = Cl, R = H	KNH_2, NH_3, 3 h	(65)	
X = Cl, R = H	LDA, THF, 3 h	(43)	
X = Cl, R = Me	KNH_2, NH_3, 15 min	(100)	
X = Cl, R = Me	$h\nu$, LDA, THF, 3 h	(50)	
X = Cl, R = Me	$h\nu$, LDA, THF, 15 h	(70)	
X = I, R = Me	KNH_2, NH_3, 10 min	(92)	
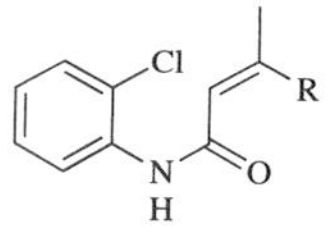	$h\nu$, LDA, THF, 3 h	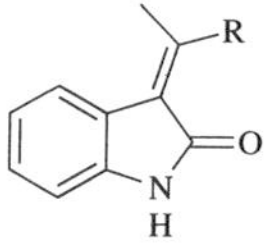 R: H (48); Me (89)	299

Reactant	Conditions	Product(s) and Yield(s) (%)	Refs.
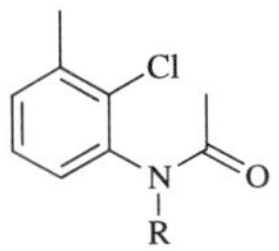		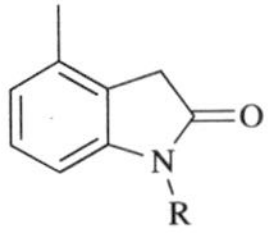	
R = H	$h\nu$, LDA, THF, C_6H_{14}, 3 h	(76)	298
R = Me	$h\nu$, LDA, THF, C_6H_{14}, 3 h	(87)	298
R = Me	KNH_2, NH_3, 0.5 h	(80)	299
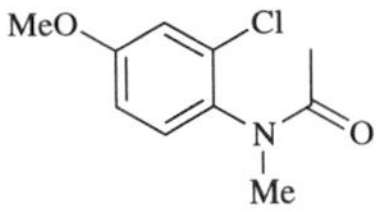	$h\nu$, LDA, THF, 1 h	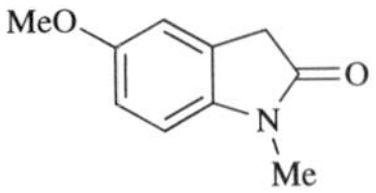 (91)	299
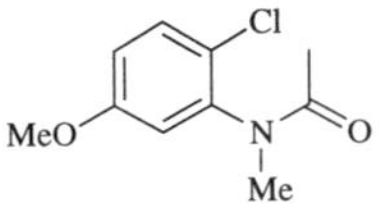	$h\nu$, LDA, THF, 1 h	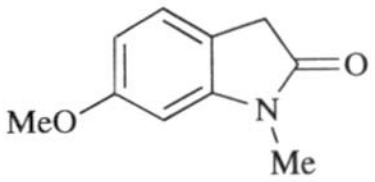 (55)	299
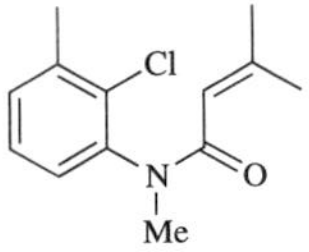	$h\nu$, KNH_2, NH_3, 15 min	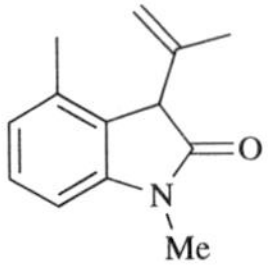 (100)	299
	1. $h\nu$, KNH_2, NH_3, 3 h 2. MeI	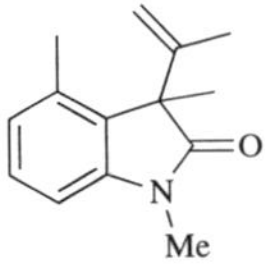 (73)	299
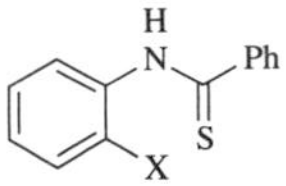	$h\nu$, t-BuOK, Me_2CO, DMSO	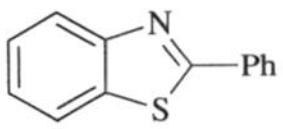	74
X = Cl		5 h (5)	
X = Br		6 h (22)	
X = I		3.5 h (100)	

TABLE IX. RING CLOSURE REACTIONS (*Continued*)

B. Intramolecular $S_{RN}1$ Reactions (*Continued*)

Substrate	Conditions	Product(s) and Yield(s) (%)	Refs.
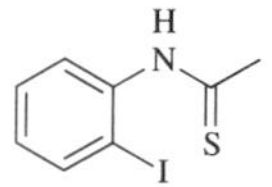	*h*ν, NaH, Me_2CO, DMSO, 6.5 h	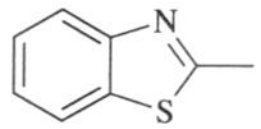 (100)	74
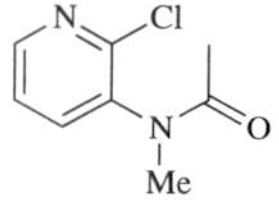		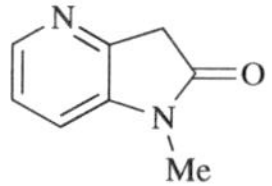	
	*h*ν, LDA, THF, C_6H_{14}, −78°, 3 h	(83)	298
	*h*ν, KNH_2, NH_3, 0.5 h	(62)	299
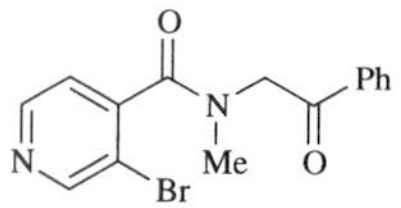	*h*ν, LDA, THF, 0°	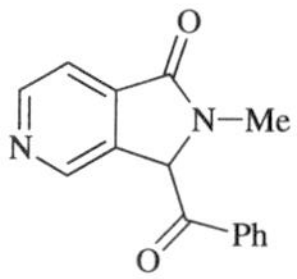 (87)	300
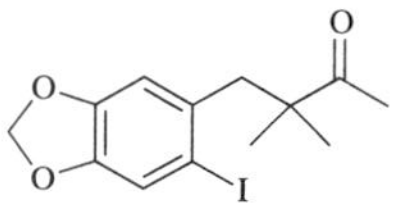	*h*ν, *t*-BuOK, NH_3	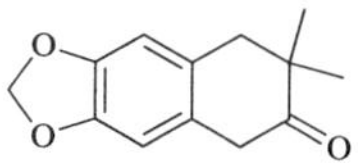 (99)	170
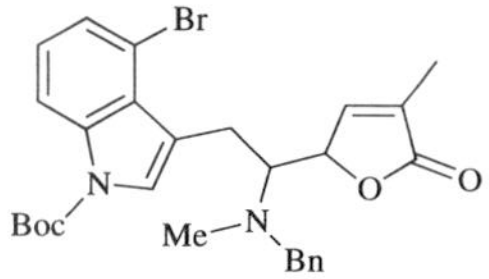	*h*ν, *t*-BuOK, NH_3, 1 h	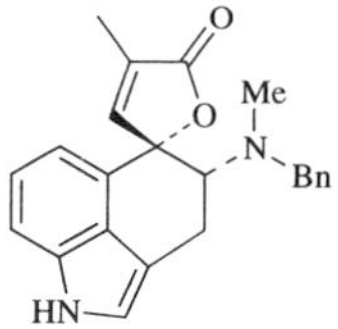 (17) + 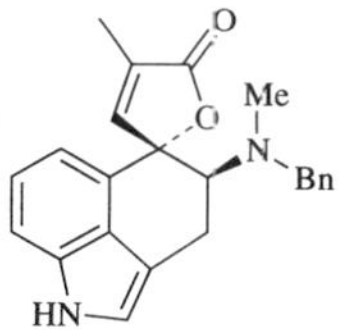 (34)	302[a]

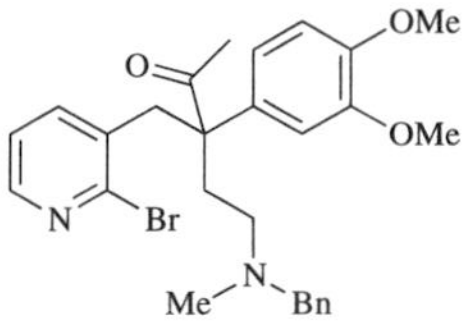	1. $h\nu$, LDA, THF 2. CF_3SO_2NPh	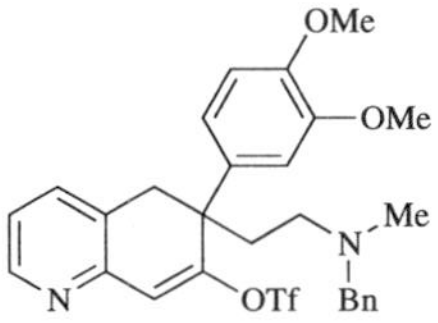 (59)	301
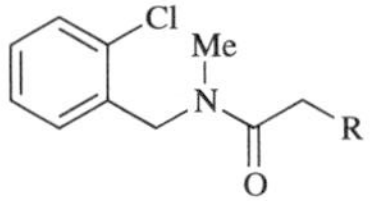	$h\nu$, KNH_2, NH_3, 1 h	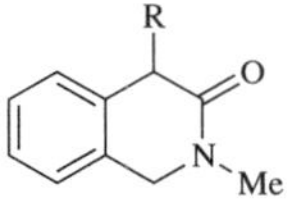 R: H (62); Me (60); Et (54)	299
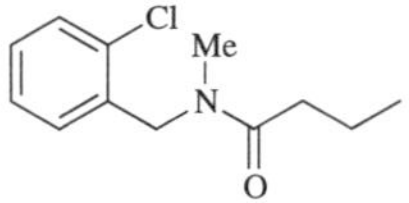	1. $h\nu$, KNH_2, NH_3, 1 h 2. BnCl	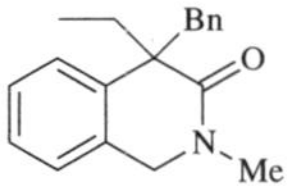 (32)	299
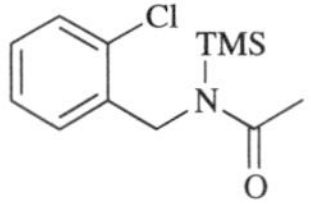	1. $h\nu$, KNH_2, NH_3, 1 h 2. H_2O	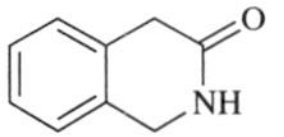 (43)	299
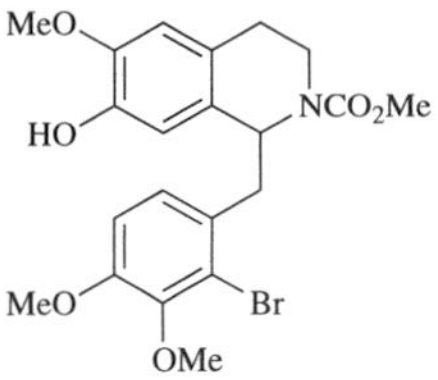	$h\nu$, t-BuOK, NH_3, 1.25 h	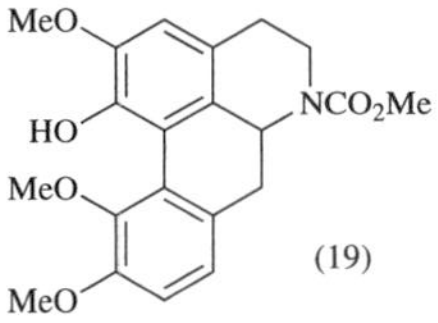 (19) + 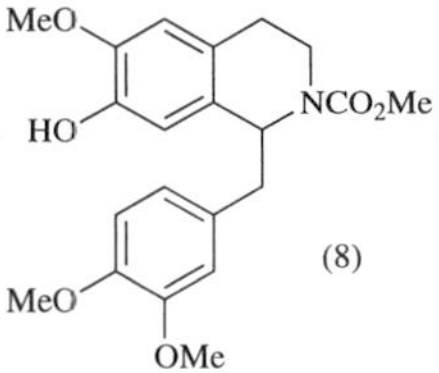 (8) + 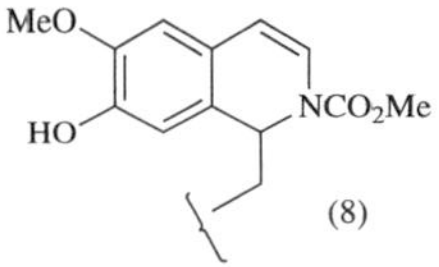 (8)	303

TABLE IX. RING CLOSURE REACTIONS (*Continued*)

B. Intramolecular $S_{RN}1$ Reactions (*Continued*)

Substrate	Conditions	Product(s) and Yield(s) (%)	Refs.
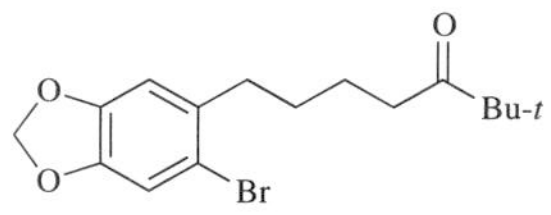	hν, *t*-BuOK, NH_3	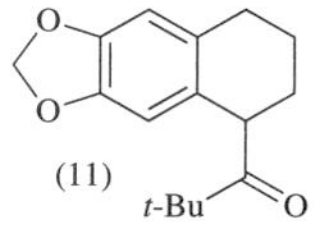 (11) + 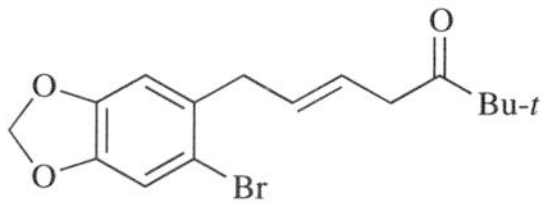 (67)	170
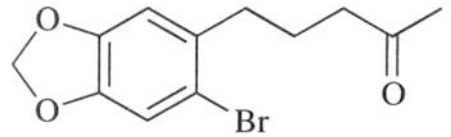	hν, *t*-BuOK, NH_3	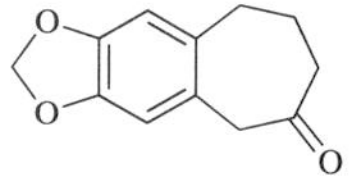 (42) + 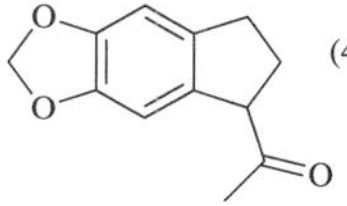 (4)	170[b]
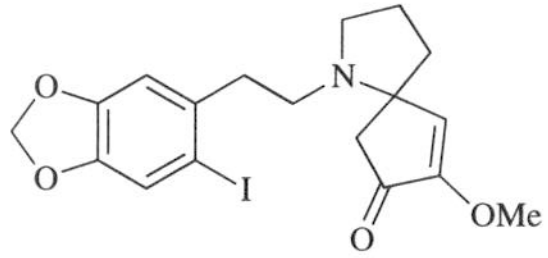	1. KNH_2, NH_3 2. Na/K hν, *t*-BuOK, NH_3	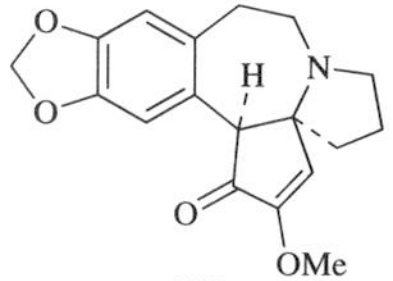 (45) (94)	295-297
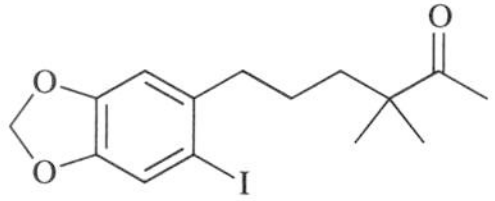	hν, *t*-BuOK, NH_3	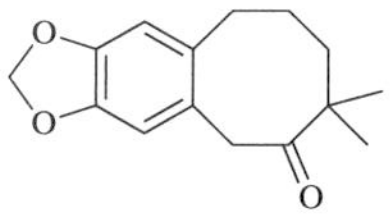 Substrate: 50 x 10^{-3} M (28) Substrate: 1.25 x 10^{-3} M (71)	170
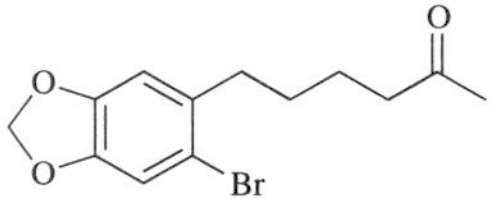	hν, *t*-BuOK, NH_3	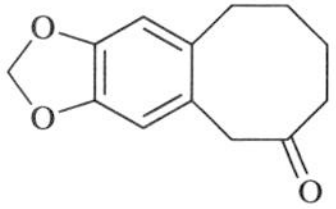 (19) + 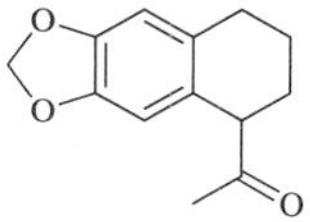 (3)	170[c]

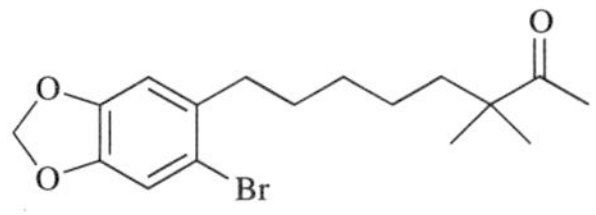	$h\nu$, *t*-BuOK, NH_3	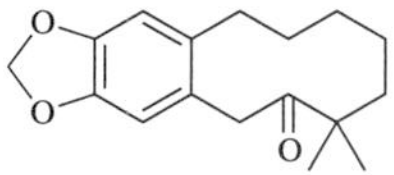 (35)	170
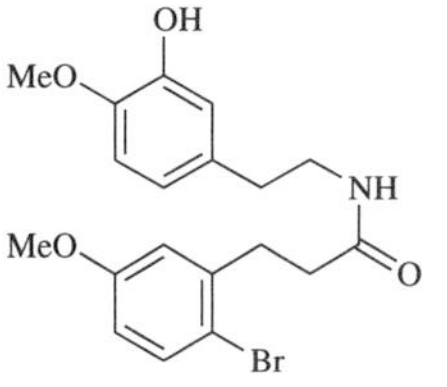	$h\nu$, NaOH, MeOH, 0.75 h	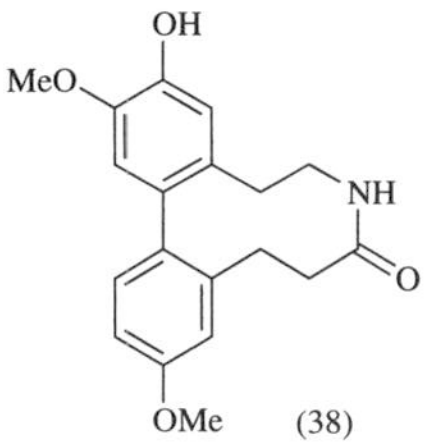 (38) + 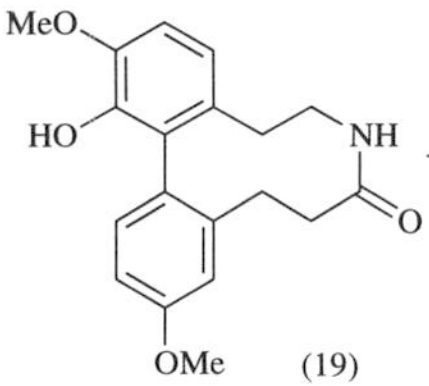 (19) + 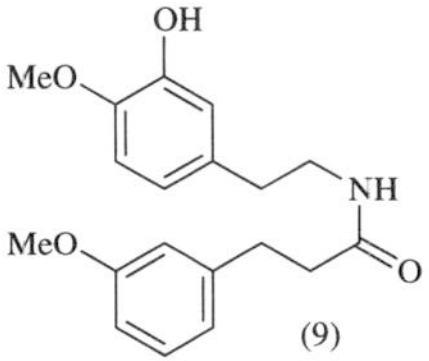 (9)	304

[a] The substrate was a 1:2 mixture of diastereomers.

[b] The dehalogenated, open, reduced α,β-unsaturated ketone and the dehalogenated, open, reduced ketone were formed in a combined yield of 4%.

[c] Also formed were the dehalogenated α,β-unsaturated ketone (7%) and the dehalogenated reduced ketone (3%).

TABLE IX. RING CLOSURE REACTIONS (*Continued*)

C. Miscellaneous

Substrate	Nucleophile	Conditions	Product(s) and Yield(s) (%)	Refs.
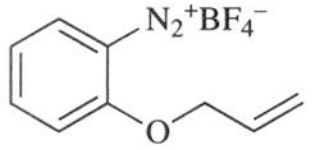	PhS^-	DMSO, 0.5 h	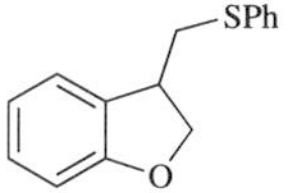 (53)	36
	$MeCOS^-$	DMSO, 1 h	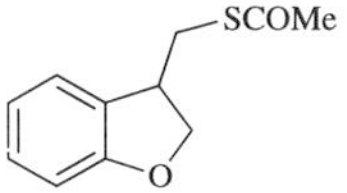 (41)	34
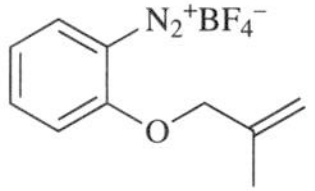	PhS^-	DMSO, 1 h	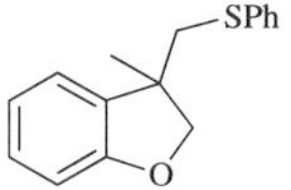 (60)	36
	$EtOCS_2^-$	1. Me_2CO, 10 min 2. reflux, 3 min	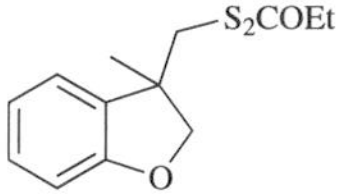 (75)	36
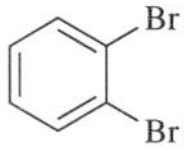	$^-CH_2COMe$	$h\nu$, NH_3, 3 h	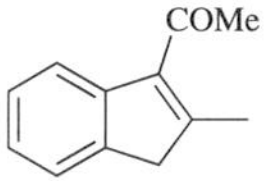 **I** + 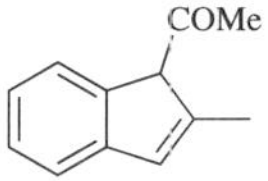 **II** **I** + **II** (64)	165, 30
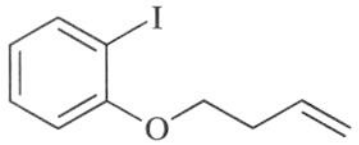	Ph_2P^-	1. $h\nu$, MeCN, 3 h 2. [O]	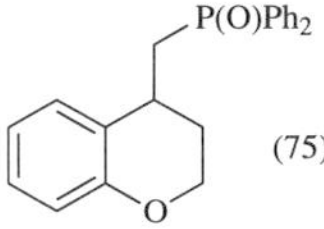 (75) + 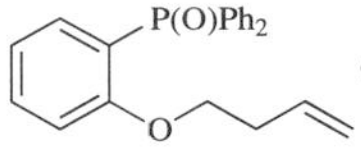 (24)	33
	PhSNa	$h\nu$, NH_3, 4 h	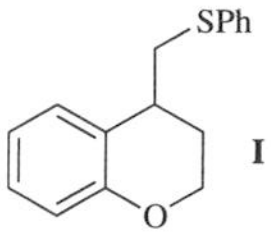 **I** (75) + 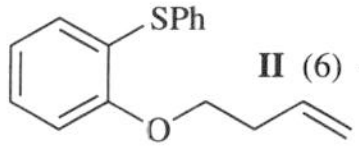 **II** (6) +	33

			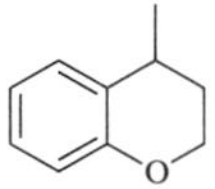 **III** (10) + 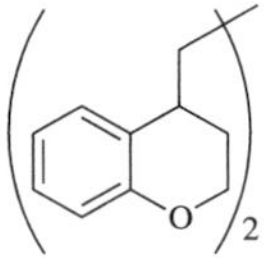 **IV** (9)	
	PhSLi	*h*ν, DMSO, 3 h	**I** (15) + **II** (76) + **III** (9)	33
		*h*ν, DMSO, 5 h	**I** (50) + **II** (35) + **III** (15)	33
	$(EtO)_2PO^-$	*h*ν, NH_3, 6 h	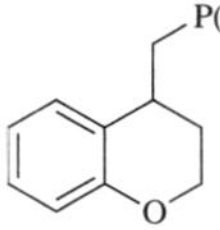 (15) + 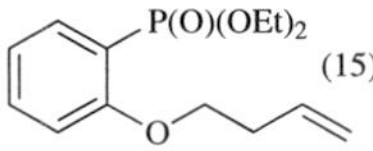 (15) + **III** (9) + **IV** (10)	33
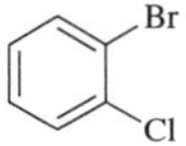	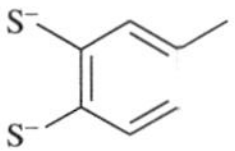	*h*ν, NH_3, 3 h	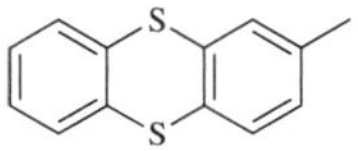 **I** (55)	305
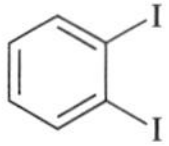		*h*ν, NH_3, 3 h	**I** (64)	305
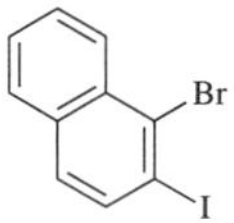		*h*ν, NH_3, 3 h	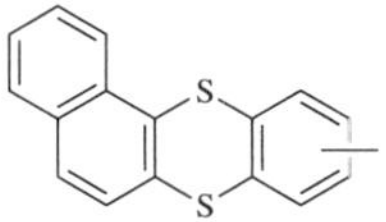 (24)	305
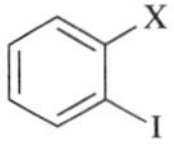	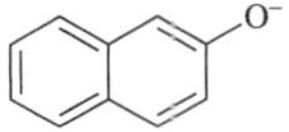	*h*ν, NH_3	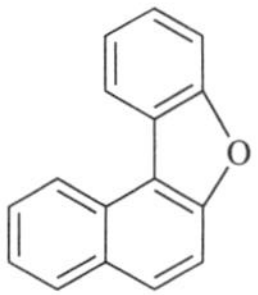 + 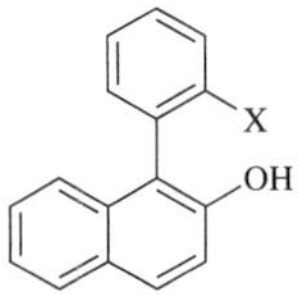	371, 308
X = Br		3 h	(24) + (32)	
X = I		5 h	(44) + (3)	

TABLE IX. RING CLOSURE REACTIONS (*Continued*)

C. Miscellaneous (*Continued*)

Substrate	Nucleophile	Conditions	Product(s) and Yield(s) (%)	Refs.
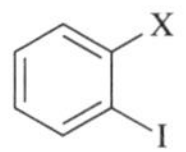 X: Br, I	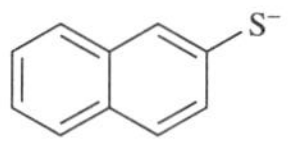	$h\nu$, NH_3, 3 h	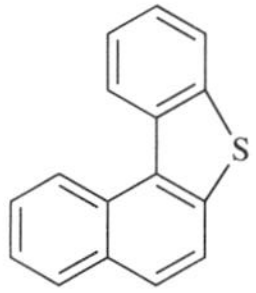 + 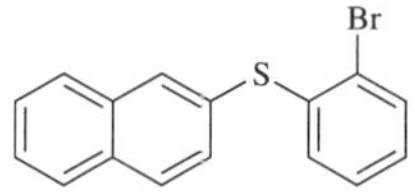 X = Br: (46) + (56); X = I: (62) + (—)	308
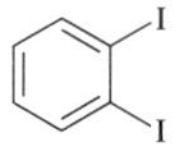	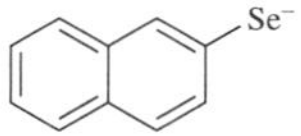	$h\nu$, NH_3, 3 h	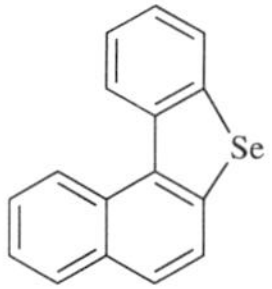 (15) + 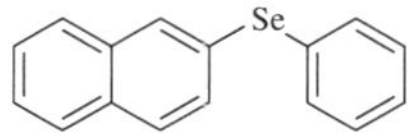 (15)	308
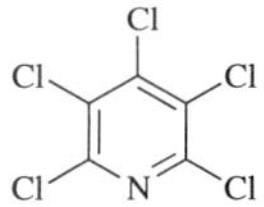	PhS^-	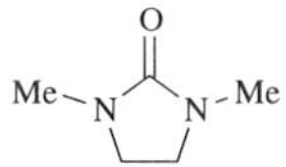	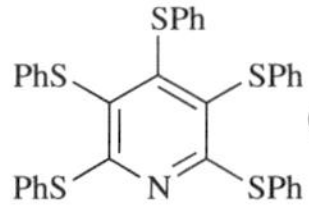 (~70) + 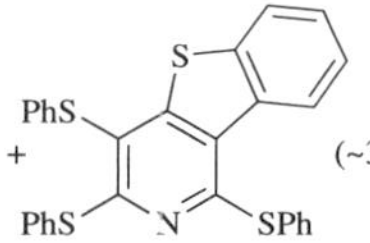 (~30)	372
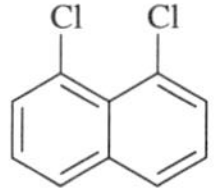	Na_2X_2	HMPA	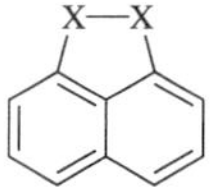 X = S, 150° (46); X = Se, 100° (69)	373
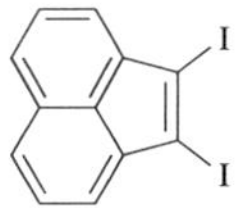	Na_2Te	DMF, 24 h	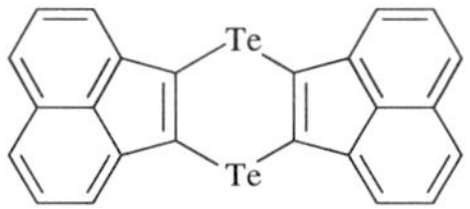 (35)	274

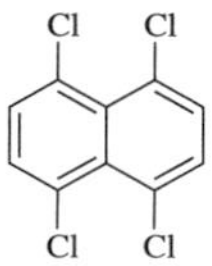	Na_2Se_2	DMF, 100-110°, 17 h	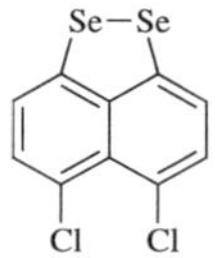 (55) + 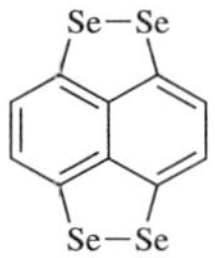 (10)	374, 375
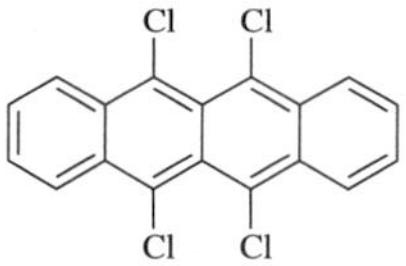	Na_2X_2	DMF	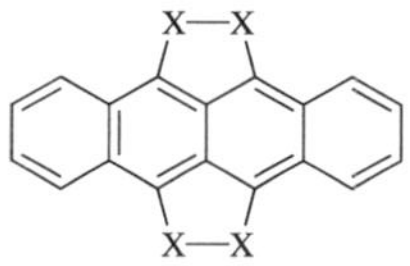	
	X			
	Se	100°, 4-5 h	(50)	376
	Te	45-55°, 20 h	(13)	377
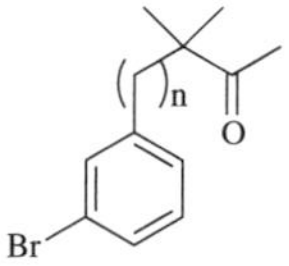	—	$h\nu$, *t*-BuOK, NH_3, 0.25-1 h	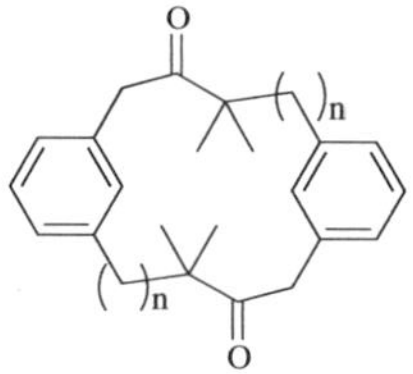 n: 0 (33); 1 (25); 2 (23); 3 (21); 4 (22); 5 (24)	378
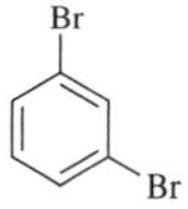	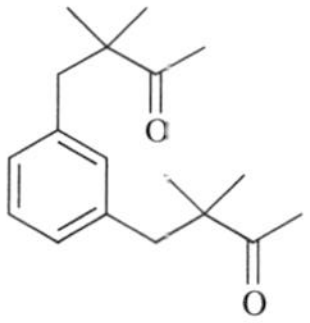	$h\nu$, *t*-BuOK, NH_3, 15 min	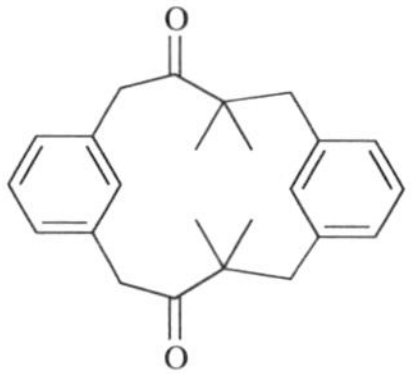 (18)	306

TABLE IX. RING CLOSURE REACTIONS (*Continued*)

C. Miscellaneous (*Continued*)

Substrate	Nucleophile	Conditions	Product(s) and Yield(s) (%)	Refs.
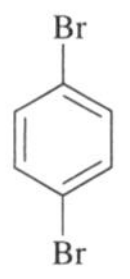	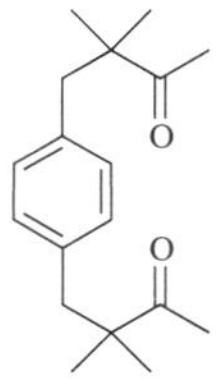	$h\nu$, *t*-BuOK, NH_3, 15 min	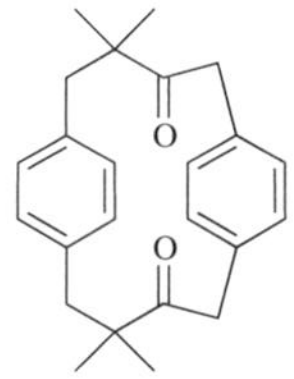 (10)	307
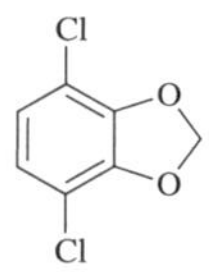		$h\nu$, *t*-BuOK, NH_3, 15 min	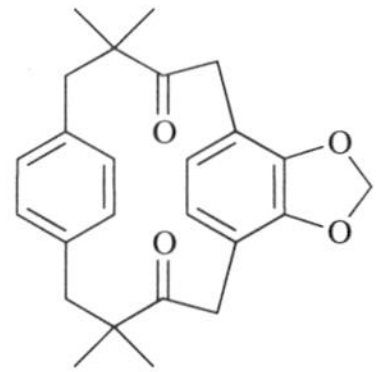 (3)	307
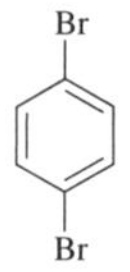	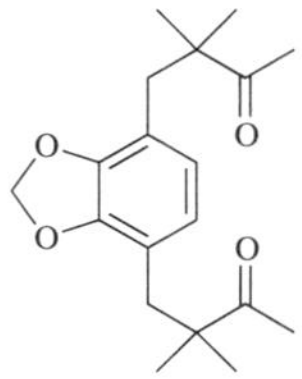	$h\nu$, *t*-BuOK, NH_3, 15 min	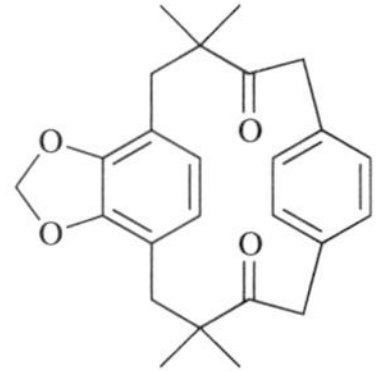 (7)	307
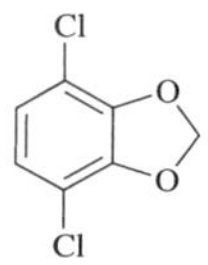		$h\nu$, *t*-BuOK, NH_3, 15 min	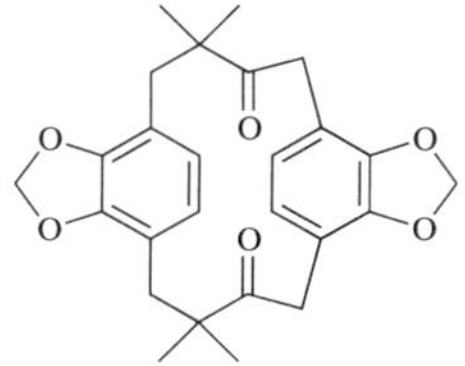 (5)	307

REFERENCES

[1] Kornblum, N.; Michel, R. E.; Kerber, R. C. *J. Am. Chem. Soc.* **1966**, *88*, 5662.
[2] Russell, G. A.; Danen, W. C. *J. Am. Chem. Soc.* **1966**, *88*, 5663.
[3] Kim, J. K.; Bunnett, J. F. *J. Am. Chem. Soc.* **1970**, *92*, 7463.
[4] Kim, J. K.; Bunnett, J. F. *J. Am. Chem. Soc.* **1970**, *92*, 7464.
[5] Rossi, R. A.; Pierini A. B.; Peñéñory, A. B. In *The Chemistry of Functional Groups*, Patai S.; Rappoport, Z.; Eds., Wiley, Chichester, 1995; Suppl. D2, Ch 24, p. 1395.
[6] Bowman, W. R. *Chem. Soc. Rev.* **1988**, *17*, 283.
[7] Rossi, R. A.; Pierini, A. B.; Palacios, S. M. *Adv. Free Rad. Chem.*, Tanner, D. D., Ed., Jai Press, **1990**, *1*, 193.
[8] Albini, A.; Fasani, E.; Mella, M. *Topics Curr. Chem.* **1993**, *168*, 143.
[9] Soumillion, J. P. *Topics Curr. Chem.* **1993**, *168*, 93.
[10] Bowman, W. R. In *Photoinduced Electron Transfer*, Fox, M. A.; Chanon, M.; Eds., Part C, Elsevier, The Hague, 1988, p. 487.
[11] Lablache-Combier, A. In *Photoinduced Electron Transfer*, Fox, M. A.; Chanon, M.; Eds., Part C, Elsevier, The Hague, 1988, p. 134.
[12] Julliard, M.; Chanon, M. *Chem. Rev.* **1983**, *83*, 425.
[13] Norris, R. K. In *Comprehensive Organic Synthesis*, Trost, B. M.; Ed., Pergamon Press, 1991, Vol. 4, p. 451.
[14] Chanon, M.; Rajzmann, M.; Chanon, F. *Tetrahedron* **1990**, *46*, 6193.
[15] Rossi, R. A.; de Rossi, R. H. In *Aromatic Substitution by the $S_{RN}1$ Mechanism*, ACS, Washington, D.C., 1983, p. 178.
[16] Rossi, R. A. *Acc. Chem. Res.* **1982**, *15*, 164.
[17] Bunnett, J. F. *Acc. Chem. Res.* **1978**, *11*, 413.
[18] Todres, Z. V. *Russian Chem. Rev.* **1978**, *47*, 148.
[19] Savéant, J. M. *Tetrahedron* **1994**, *50*, 10117.
[20] Savéant, J. M. *Acc. Chem. Res.* **1993**, *26*, 455.
[21] Savéant, J. M. *New J. Chem.* **1992**, *16*, 131.
[22] Pinson, J.; Savéant, J. M. *Electrorg. Synth.* **1991**, *29*, 29.
[23] Savéant, J. M. *Adv. Phys. Org. Chem.* **1990**, *26*, 1.
[24] Andrieux, C. P.; Hapiot, P.; Savéant, J. M. *Chem. Rev.* **1990**, *90*, 723.
[25] Evans, D. H. *Chem. Rev.* **1990**, *90*, 739.
[26] Savéant, J. M. *Bull. Soc. Chim. Fr.* **1988**, 225.
[27] Savéant, J. M. *Acc. Chem. Res.* **1980**, *13*, 323.
[28] Kornblum, N. *Aldrichim. Acta* **1990**, *23*, 71.
[29] Degrand, C.; Prest, R.; Compagnon, P. L. *Electrorg. Synth.* **1991**, *29*, 45.
[30] Bunnett, J. F.; Mitchel, E.; Galli, C. *Tetrahedron* **1985**, *41*, 4119.
[31] Beugelmans, R. *Bull. Soc. Chim. Belg.* **1984**, *93*, 547.
[32] Wolfe, J. F.; Carver, D. R. *Org. Prep. Proced. Int.* **1978**, *10*, 225.
[33] Beckwith, A. L. J.; Palacios, S. M. *J. Phys. Org. Chem.* **1991**, *4*, 404.
[34] Petrillo, G.; Novi, M.; Garbarino, G.; Filiberti, M. *Tetrahedron Lett.* **1988**, *29*, 4185.
[35] Abeywickrema, A. N.; Beckwith, A. L. J. *J. Am. Chem. Soc.* **1986**, *108*, 8227.
[36] Meijs, G. F.; Beckwith, A. L. J. *J. Am. Chem. Soc.* **1986**, *108*, 5890.
[37] Galli, C.; Bunnett, J. F. *J. Am. Chem. Soc.* **1981**, *103*, 7140.
[38] Galli, C.; Bunnett, J. F. *J. Am. Chem. Soc.* **1979**, *101*, 6137.
[39] Ettayeb, R.; Savéant, J. M.; Thiébault, A. *J. Am. Chem. Soc.* **1992**, *114*, 10990.
[40] Amatore, C.; Gareil, M.; Oturan, M. A.; Pinson, J.; Savéant, J. M.; Thiébault, A. *J. Org. Chem.* **1986**, *51*, 3757.
[41] Borosky, G. L.; Pierini, A. B.; Rossi, R. A. *J. Org. Chem.* **1992**, *57*, 247.
[42] Rossi, R. A.; de Rossi, R. H.; Pierini, A. B. *J. Org. Chem.* **1979**, *44*, 2662.
[43] Bunnett, J. F.; Gloor, B. F. *J. Org. Chem.* **1973**, *38*, 4156.
[44] Rossi, R. A.; de Rossi, R. H.; Lopez, A. F. *J. Org. Chem.* **1976**, *41*, 3367.
[45] Rossi, R. A.; de Rossi, R. H.; Lopez, A. F. *J. Am. Chem. Soc.* **1976**, *98*, 1252.

[46] Rossi, R. A.; de Rossi, R. H.; Lopez, A. F. *J. Org. Chem.* **1976**, *41*, 3371.
[47] Yakubov, A. P.; Belen'kii, L. I.; Goldfarb, Y. L. *Izv. Akad. Nauk SSSR, Ser. Khim.* **1981**, 2812; *Chem. Abstr.* **1982**, *96*, 104044r.
[48] Rossi, R. A.; Palacios, S. M. *J. Org. Chem.* **1981**, *46*, 5300.
[49] Beugelmans, R.; Bois-Choussy, M.; Boudet, B. *Tetrahedron* **1983**, *39*, 4153.
[50] Denney, D. B.; Denney, D. Z. *Tetrahedron* **1991**, *47*, 6577.
[51] Marquet, J.; Jiang, Z.; Gallardo, I.; Batlle, A.; Cayon, E. *Tetrahedron Lett.* **1993**, *34*, 2801.
[52] Rossi, R. A.; Palacios, S. M. *Tetrahedron* **1993**, *49*, 4485.
[53] Bunnett, J. F. *Tetrahedron* **1993**, *49*, 4477.
[54] Oostvee, E. A.; van der Plas, H. C. *Recl. Trav. Chim. Pays-Bas* **1979**, *98*, 441.
[55] Carver, D. R.; Komin, A. P.; Hubbard, J. S.; Wolfe, J. F. *J. Org. Chem.* **1981**, *46*, 294.
[56] Carver, D. R.; Hubbard, J. S.; Wolfe, J. F. *J. Org. Chem.* **1982**, *47*, 1036.
[57] Dell'Erba, C.; Novi, M.; Petrillo, G.; Tavani, C. *Tetrahedron* **1993**, *49*, 235.
[58] Dell'Erba, C.; Novi, M.; Petrillo, G.; Tavani, C. *Tetrahedron* **1992**, *48*, 325.
[59] Dell'Erba, C.; Novi, M.; Petrillo, G.; Tavani, C.; Bellandi, P. *Tetrahedron* **1991**, *47*, 333.
[60] Bordwell, F. G.; Zhang, X-M.; Filler, R. *J. Org. Chem.* **1993**, *58*, 6067.
[61] Bordwell, F. G.; Zhang, X-M. *Acc. Chem. Res.* **1993**, *26*, 510.
[62] Scamehorn, R. G.; Hardacre, J. M.; Lukanich, J. M.; Sharpe, L. R. *J. Org. Chem.* **1984**, *49*, 4881.
[63] Scamehorn, R. G.; Bunnett, J. F. *J. Org. Chem.* **1977**, *42*, 1449.
[64] Beugelmans, R.; Chbani, M. *New J. Chem.* **1994**, *18*, 949.
[65] Boumekouez, A.; About-Jaudet, E.; Collignon, N. *J. Organomet. Chem.* **1992**, *440*, 297.
[66] Wu, B. Q.; Zeng, F. W.; Ge, M.-J.; Cheng, X-Z.; Wu, G.-S. *Sci. China* **1991**, *34*, 777; *Chem. Abstr.* **1992**, *116*, 58463h.
[67] Cheng, C.; Stock, L. M. *J. Org. Chem.* **1991**, *56*, 2436.
[68] Hoz, S.; Bunnett, J. F. *J. Am. Chem. Soc.* **1977**, *99*, 4690.
[69] Fox, M. A.; Yaunathan, J.; Fryxell, G. E. *J. Org. Chem.* **1983**, *48*, 3109.
[70] Tolbert, L. M.; Siddiqui, S. *J. Org. Chem.* **1984**, *49*, 1744.
[71] Swartz, J. E.; Bunnett, J. F. *J. Org. Chem.* **1979**, *44*, 340.
[72] Swartz, J. E.; Bunnett, J. F. *J. Org. Chem.* **1979**, *44*, 4673.
[73] Scamehorn, R. G.; Bunnett, J. F. *J. Org. Chem.* **1979**, *44*, 2604.
[74] Bowman, W. R.; Heaney, H.; Smith, P. H. G. *Tetrahedron Lett.* **1982**, *23*, 5093.
[75] Galli, C.; Bunnett, J. F. *J. Org. Chem.* **1984**, *49*, 3041.
[76] van Leeuwen, M.; McKillop, A. *J. Chem. Soc., Perkin Trans. 1* **1993**, 2433.
[77] Galli, C.; Gentili, P. *J. Chem. Soc., Perkin Trans. 2* **1993**, 1135.
[78] Murguía, M. C.; Rossi, R. A. *Tetrahedron Lett.* **1997**, *38*, 1355.
[79] Curran, D. P.; Fevig, T. L.; Jasperse, C. P.; Totleeben, M. J. *Synlett* **1992**, 943, and references cited therein.
[80] Nazareno, M. A.; Rossi, R. A. *Tetrahedron Lett.* **1994**, *35*, 5185.
[81] Savéant, J. M. *Adv. Electron Transfer Chem.* **1994**, *4*, 53.
[82] Swartz, J. E.; Stenzel, T. T. *J. Am. Chem. Soc.* **1984**, *106*, 2520.
[83] Amatore, C.; Oturan, M. A.; Pinson, J.; Savéant, J. M.; Thiébault, A. *J. Am. Chem. Soc.* **1984**, *106*, 6318.
[84] Alam, N.; Amatore, C.; Combellas, C.; Thiébault, A.; Verpeaux, J. N. *J. Org. Chem.* **1990**, *55*, 6347.
[85] Amatore, C.; Combellas, C.; Lebbar, N. E.; Thiébault, A.; Verpeaux, J. N. *J. Org. Chem.* **1995**, *60*, 18.
[86] Amatore, C.; Pinson, J.; Savéant, J. M.; Thiébault, A. *J. Am. Chem. Soc.* **1981**, *103*, 6930.
[87] Amatore, C.; Savéant, J. M.; Combellas, C.; Robveille, S.; Thiébault, A. *J. Electroanal. Chem.* **1985**, *184*, 25.
[88] Pinson, J.; Savéant, J. M. *J. Am. Chem. Soc.* **1978**, *100*, 1506.
[89] Amatore, C.; Pinson, J.; Savéant, J. M.; Thiébault, A. *J. Am. Chem. Soc.* **1982**, *104*, 817.
[90] Andrieux, C. P.; Savéant, J. M.; Su, K. B. *J. Phys. Chem.* **1986**, *90*, 3815.

[91] Amatore, C.; Oturan, M. A.; Pinson, J.; Savéant, J. M.; Thiébault, A. *J. Am. Chem. Soc.* **1985**, *107*, 3451.
[92] Amatore, C.; Combellas, C.; Pinson, J.; Oturan, M. A.; Robveille, S.; Savéant, J. M.; Thiébault, A. *J. Am. Chem. Soc.* **1985**, *107*, 4846.
[93] M'Halla, F.; Pinson, J.; Savéant, J. M. *J. Am. Chem. Soc.* **1980**, *102*, 4120.
[94] Rossi, R. A.; Bunnett, J. F. *J. Am. Chem. Soc.* **1972**, *94*, 683.
[95] Andrieux, C. P.; Savéant, J. M. *J. Am. Chem. Soc.* **1993**, *115*, 8044.
[96] Bard, R. R.; Bunnett, J. F.; Creary, X.; Tremelling, M. J. *J. Am. Chem. Soc.* **1980**, *102*, 2852.
[97] Tremelling, M. J.; Bunnett, J. F. *J. Am. Chem. Soc.* **1980**, *102*, 7375.
[98] Palacios, S. M.; Asis, S. E.; Rossi, R. A. *Bull. Soc. Chim. Fr.* **1993**, *130*, 111.
[99] Austin, E.; Alonso, R. A.; Rossi, R. A. *J. Chem. Res.* **1990**, 190.
[100] Astruc, D. *Angew. Chem., Int. Ed. Engl.* **1988**, *27*, 643, and references cited therein.
[101] Austin, E.; Ferrayoli, C. G.; Alonso, R. A.; Rossi, R. A. *Tetrahedron* **1993**, *49*, 4495.
[102] Austin, E.; Alonso, R. A.; Rossi, R. A. *J. Org. Chem.* **1991**, *56*, 4486.
[103] Galli, C. *Gazz. Chim. Ital.* **1988**, *118*, 365.
[104] Bartak, D. E.; Danen, W. C.; Hawley, M. D. *J. Org. Chem.* **1970**, *35*, 1206.
[105] Beugelmans, R.; Bois-Choussy, M. *Tetrahedron Lett.* **1988**, *29*, 1289.
[106] Beugelmans, R.; Roussi, G. *Tetrahedron* **1981**, *37*, 393.
[107] Beugelmans, R.; Bois-Choussy, M.; Boudet, B. *Tetrahedron* **1982**, *38*, 3479.
[108] Bunnett, J. F.; Sundberg, J. E. *Chem. Pharm. Bull.* **1975**, *23*, 2620.
[109] Danen, W. C.; Kenslet, T. T.; Lawless, J. C.; Marcus, M. F.; Hawley, M. D. *J. Phys. Chem.* **1969**, *73*, 4389.
[110] Compton, R. G.; Dryfe, R. A. W.; Fisher, A. C. *J. Electroanal. Chem.* **1993**, *361*, 275.
[111] Compton, R. G.; Dryfe, R. A. W.; Fisher, A. C. *J. Chem. Soc., Perkin Trans. 2* **1994**, 1581.
[112] Compton, R. G.; Dryfe, R. A. W.; Eklund, J. C.; Page, S. D.; Hirst, J.; Nei, L. B.; Fleet, G. W. J.; Hsia, K. Y.; Bethell, D.; Martingale, L. J. *J. Chem. Soc., Perkin Trans. 2* **1995**, 1673.
[113] Galli, C. *Tetrahedron* **1988**, *44*, 5205.
[114] Novi, M.; Petrillo, G.; Dell'Erba, C. *Tetrahedron Lett.* **1987**, *28*, 1345.
[115] Petrillo, G.; Novi, M.; Garbarino, G.; Dell'Erba, C. *Tetrahedron* **1987**, *43*, 4625.
[116] Bunnett, J. F.; Shafer, S. J. *J. Org. Chem.* **1978**, *43*, 1873.
[117] Pierini, A. B.; Peñéñory, A. B.; Rossi, R. A. *J. Org. Chem.* **1984**, *49*, 486.
[118] Rossi, R. A.; Alonso, R. A.; Palacios, S. M. *J. Org. Chem.* **1981**, *46*, 2498.
[119] Alonso, R. A.; Rossi, R. A. *J. Org. Chem.* **1982**, *47*, 77.
[120] Pierini, A. B.; Rossi, R. A. *J. Organomet. Chem.* **1979**, *168*, 163.
[121] Rossi, R. A. *J. Chem. Educ.* **1982**, *59*, 310.
[122] Pierini, A. B.; Rossi, R. A. *J. Org. Chem.* **1979**, *44*, 4667.
[123] Galli, C.; Gentili, P.; Guarnieri, A. *Gazz. Chim. Ital.* **1995**, *125*, 409.
[124] Rossi, R. A.; Bunnett, J. F. *J. Org. Chem.* **1973**, *38*, 3020.
[125] Pierini, A. B.; Baumgartner, M. T.; Rossi, R. A. *J. Org. Chem.* **1991**, *56*, 580.
[126] Alonso, R. A.; Bardon, A.; Rossi, R. A. *J. Org. Chem.* **1984**, *49*, 3584.
[127] Novi, M.; Garbarino, G.; Petrillo, G.; Dell'Erba, C. *Tetrahedron* **1990**, *46*, 2205.
[128] Tolbert, L. M.; Siddiqui, S. *Tetrahedron* **1982**, *38*, 1079.
[129] Tolbert, L. M.; Martone, D. P. *J. Org. Chem.* **1983**, *48*, 1185.
[130] Fox, M. A.; Singletary, N. J. *J. Org. Chem.* **1982**, *47*, 3412.
[131] Konigsberg, I.; Jagur-Grodzinski, J. *J. Polym. Sci. Polym. Chem. Ed.* **1984**, *22*, 2713.
[132] Tour, J. M.; Stephens, E. B. *J. Am. Chem. Soc.* **1991**, *113*, 2309.
[133] Tour, J. M.; Stephens, E. B.; Davis, J. F. *Macromolecules* **1992**, *25*, 499.
[134] Moon, M. P.; Komin, A. P.; Wolfe, J. F.; Morris, G. F. *J. Org. Chem.* **1983**, *48*, 2392.
[135] Semmelhack, M. F.; Bargar, T. M. *J. Am. Chem. Soc.* **1980**, *102*, 7765.
[136] Rossi, R. A.; Bunnett, J. F. *J. Org. Chem.* **1973**, *38*, 1407.
[137] Komin, A. P.; Wolfe, J. F. *J. Org. Chem.* **1977**, *42*, 2481.
[138] Scamehorn, R. G.; Bunnett, J. F. *J. Org. Chem.* **1977**, *42*, 1457.

[139] Estel, L.; Marsais, F.; Queguiner, G. *J. Org. Chem.* **1988**, *53*, 2740.
[140] Wolfe, J. F., Virginia Polytechnic Institute and State University, unpublished results.
[141] Dillender, S. C.; Greenwood, T. D.; Hendi, M. S.; Wolfe, J. F. *J. Org. Chem.* **1986**, *51*, 1184.
[142] Hay, J. V.; Wolfe, J. F. *J. Am. Chem. Soc.* **1975**, *97*, 3702.
[143] Bunnett, J. F.; Sundberg, J. E. *J. Org. Chem.* **1976**, *41*, 1702.
[144] Moon, M. P.; Wolfe, J. F. *J. Org. Chem.* **1979**, *44*, 4081.
[145] Beugelmans, R.; Ginsburg, H. *J. Chem. Soc., Chem. Commun.* **1980**, 508.
[146] Bard, R. R.; Bunnett, J. F. *J. Org. Chem.* **1980**, *45*, 1546.
[147] Beugelmans, R.; Bois-Choussy, M. *Synthesis* **1981**, 730.
[148] Beugelmans, R.; Boudet, B.; Quintero, L. *Tetrahedron Lett.* **1980**, *21*, 1943.
[149] Beugelmans, R.; Ginsburg, H. *Heterocycles* **1985**, *23*, 1197.
[150] Beltran, L.; Galvez, C.; Prats, M.; Salgado, J. *J. Heterocyclic Chem.* **1992**, *29*, 905.
[151] Prats, M.; Galvez, C.; Beltran, L. *Heterocycles* **1992**, *34*, 1039.
[152] Bunnett, J. F.; Gloor, B. F. *Heterocycles* **1976**, *5*, 377.
[153] Nair, V.; Chamberlain, S. D. *J. Am. Chem. Soc.* **1985**, *107*, 2183.
[154] Nair, V.; Chamberlain, S. D. *J. Org. Chem.* **1985**, *50*, 5069.
[155] Beugelmans, R.; Bois-Choussy, M. *Heterocycles* **1987**, *26*, 1863.
[156] Wu, B. Q.; Zeng, F. W.; Zhao, Y.; Wu, G. S. *Chin. J. Chem.* **1992**, *10*, 253; *Chem. Abstr.* **1993**, *118*, 124792j.
[157] Hay, J. V.; Hudlicky, T.; Wolfe, J. F. *J. Am. Chem. Soc.* **1975**, *97*, 374.
[158] Beugelmans, R.; Bois-Choussy, M.; Tang, Q. *J. Org. Chem.* **1987**, *52*, 3880.
[159] Baumgartner, M. T.; Gallego, M. L.; Pierini, A. B. *J. Org. Chem.* **1998**, *63*, 6394.
[160] Bunnett, J. F.; Creary, X. *J. Org. Chem.* **1975**, *40*, 3740.
[161] Dell'Erba, C.; Novi, M.; Petrillo, G.; Tavani, C. *Phosphorus, Sulfur, and Silicon* **1993**, *74*, 409.
[162] Dell'Erba, C.; Novi, M.; Petrillo, G.; Tavani, C. *Tetrahedron* **1994**, *50*, 11239.
[163] Dell'Erba, C.; Novi, M.; Petrillo, G.; Tavani, C. *Tetrahedron* **1994**, *50*, 3529.
[164] Ferrayoli, C. G.; Palacios, S. M.; Alonso, R. A. *J. Chem. Soc., Perkin Trans. 1* **1995**, 1635.
[165] Bunnett, J. F.; Singh, P. *J. Org. Chem.* **1981**, *46*, 5022.
[166] Alonso, R. A.; Rossi, R. A. *J. Org. Chem.* **1980**, *45*, 4760.
[167] Carver, D. R.; Greenwood, T. D.; Hubbard, J. S.; Komin, A. P.; Sachdeva, Y. P.; Wolfe, J. F. *J. Org. Chem.* **1983**, *48*, 1180.
[168] Beugelmans, R.; Ginsburg, H.; Bois-Choussy, M. *J. Chem. Soc., Perkin Trans. 1* **1982**, 1149.
[169] Wong, J.-W.; Natalie, K. J., Jr.; Nwokogu, G. C.; Pisipati, J. S.; Flaherty, P. T.; Greenwood, T. D.; Wolfe, J. F. *J. Org. Chem.* **1997**, *62*, 6152.
[170] Semmelhack, M. F.; Bargar, T. M. *J. Org. Chem.* **1977**, *42*, 1481.
[171] Rossi, R. A.; Alonso, R. A. *J. Org. Chem.* **1980**, *45*, 1239.
[172] Alonso, R. A.; Rodriguez, C. H.; Rossi, R. A. *J. Org. Chem.* **1989**, *54*, 5983.
[173] Lotz, G. A.; Palacios, S. M.; Rossi, R. A. *Tetrahedron Lett.* **1994**, *35*, 7711.
[174] Goldfarb, Y. L.; Ikubov, A. P.; Belen'kii, L. I. *Khim. Geterotsikl. Soedin.* **1979**, 1044; *Chem. Abstr.* **1979**, *91*, 193081n.
[175] Wu, B. Q.; Fanwen, Z.; Li, W.; Zhao, Y.; Wu, G. S. *Chin. J. Chem.* **1992**, *10*, 73; *Chem. Abstr.* **1992**, *117*, 89658a.
[176] Du, R.; Huang, W. *Jinan Daxue Xuebao, Ziran Kexue Yu Yixueban* **1991**, *12*, 46; *Chem. Abstr.* **1992**, *117*, 48037d.
[177] Hermann, C. K. F.; Sachdeva, Y. P.; Wolfe, J. F. *J. Heterocyclic Chem.* **1987**, *24*, 1061.
[178] Wolfe, J. F.; Greene, J. C.; Hudlicky, T. *J. Org. Chem.* **1972**, *37*, 3199.
[179] Combellas, C.; Lequan, M.; Lequan, R. M.; Simon, J.; Thiébault, A. *J. Chem. Soc., Chem. Commun.* **1990**, 542.
[180] Oturan, M. A.; Pinson, J.; Savéant, J. M.; Thiébault, A. *Tetrahedron Lett.* **1989**, *30*, 1373.
[181] Beugelmans, R.; Chastanet, J. *Tetrahedron* **1993**, *49*, 7883.
[182] Beugelmans, R.; Bois-Choussy, M.; Gayral, P.; Rigothier, M. C. *Eur. J. Med. Chem.* **1988**, *23*, 539.
[183] Tona, M.; Sanchez-Baeza, F.; Messeguer, A. *Tetrahedron* **1994**, *50*, 8117.
[184] Brunet, J-J.; Sidot, C.; Caubere, P. *J. Organomet. Chem.* **1980**, *204*, 229.

[185] Dneprovskii, A. S.; Tuchkin, A. I. *Russian J. Org. Chem.* **1994**, *30*, 435.
[186] Brunet, J.-J.; Sidot, C.; Loubinoux, B.; Caubere, P. *J. Org. Chem.* **1979**, *44*, 2199.
[187] Loubinoux, B.; Fixari, B.; Brunet, J.-J.; Caubere, P. *J. Organometal. Chem.* **1976**, *105*, C22.
[188] Brunet, J.-J.; Sidot, C.; Caubere, P. *Tetrahedron Lett.* **1981**, *22*, 1013.
[189] Brunet, J.-J.; Sidot, C.; Caubere, P. *J. Org. Chem.* **1983**, *48*, 1166.
[190] Brunet, J.-J.; Sidot, C.; Caubere, P. *J. Org. Chem.* **1983**, *48*, 1919.
[191] Kashimura, T.; Kudo, K.; Mori, S.; Sugita, N. *Chem. Lett.* **1986**, 299.
[192] Kashimura, T.; Kudo, K.; Mori, S.; Sugita, N. *Chem. Lett.* **1986**, 483.
[193] Kashimura, T.; Kudo, K.; Mori, S.; Sugita, N. *Chem. Lett.* **1986**, 851.
[194] Kudo, K.; Shibata, T.; Kashimura, T.; Mori, S.; Sugita, N. *Chem. Lett.* **1987**, 577.
[195] Brunet, J. J.; Taillefer, M. *J. Organomet. Chem.* **1989**, *361*, C1.
[196] Brunet, J. J.; de Montauzon, D.; Taillefer, M. *Organometallics* **1991**, *10*, 341.
[197] Brunet, J. J.; El Zaizi, A. *J. Organomet. Chem.* **1995**, *486*, 275.
[198] Brunet, J. J.; Chauvin, R. *Chem. Soc. Rev.* **1995**, 89.
[199] Brunet, J. J.; Taillefer, M. *J. Organomet. Chem.* **1990**, *384*, 193.
[200] Brunet, J. J.; El Zaizi, A. *Bull. Soc. Chim. Fr.* **1996**, *133*, 75.
[201] Tashiro, M. *Synthesis* **1979**, 921.
[202] Boy, P.; Combellas, C.; Thiébault, A. *Synlett* **1991**, 923.
[203] Combellas, C.; Gautier, H.; Simon, J.; Thiébault, A.; Tournilhac, F.; Barzoukas, M.; Josse, D.; Ledoux, I.; Amatore, C.; Verpeaux, J.-N. *J. Chem. Soc., Chem. Commun.* **1988**, 203.
[204] Alam, N.; Amatore, C.; Combellas, C.; Thiébault, A.; Verpeaux, J. N. *Tetrahedron Lett.* **1987**, *28*, 6171.
[205] Alam, N.; Amatore, C.; Combellas, C.; Pinson, J.; Savéant, J. M.; Thiébault, A.; Verpeaux, J. N. *J. Org. Chem.* **1988**, *53*, 1496.
[206] Boy, P.; Combellas, C.; Suba, C.; Thiébault, A. *J. Org. Chem.* **1994**, *59*, 4482.
[207] Boy, P.; Combellas, C.; Thiébault, A.; Amatore, C.; Jutand, A. *Tetrahedron Lett.* **1992**, *33*, 491.
[208] Boy, P.; Combellas, C.; Fielding, S.; Thiébault, A. *Tetrahedron Lett.* **1991**, *32*, 6705.
[209] Beugelmans, R.; Chastanet, J. *Tetrahedron Lett.* **1991**, *32*, 3487.
[210] Combellas, C.; Marzouk, H.; Suba, C.; Thiébault, A. *Synthesis* **1993**, 788.
[211] Beugelmans, R.; Bois-Choussy, M.; Tang, Q. *Tetrahedron Lett.* **1988**, *29*, 1705.
[212] Pierini, A. B.; Baumgartner, M. T.; Rossi, R. A. *Tetrahedron Lett.* **1988**, *29*, 3451.
[213] Combellas, C.; Suba, C.; Thiébault, A. *Tetrahedron Lett.* **1994**, *35*, 5217.
[214] Beugelmans, R.; Bois-Choussy, M.; Chastanet, J.; Legleuher, M.; Zhu, J. *Heterocycles* **1993**, *36*, 2723.
[215] Beugelmans, R.; Bois-Choussy, M. *J. Org. Chem.* **1991**, *56*, 2518.
[216] Petrillo, G.; Novi, M.; Dell'Erba, C. *Tetrahedron Lett.* **1989**, *30*, 6911.
[217] Petrillo, G.; Novi, M.; Dell'Erba, C.; Tavani, C.; Berta, G. *Tetrahedron* **1990**, *46*, 7977.
[218] Amatore, C.; Combellas, C.; Robveille, S.; Savéant, J. M.; Thiébault, A. *J. Am. Chem. Soc.* **1990**, *108*, 4754.
[219] Dell'Erba, C.; Houmam, A.; Novi, M.; Petrillo, G.; Pinson, J. *J. Org. Chem.* **1993**, *58*, 2670.
[220] Bunnett, J. F.; Gloor, B. F. *J. Org. Chem.* **1974**, *39*, 382.
[221] Wursthorn, K. R.; Kuivila, H. G.; Smith, G. F. *J. Am. Chem. Soc.* **1978**, *100*, 2789.
[222] Wursthorn, K. R.; Kuivila, H. G. *J. Organomet. Chem.* **1977**, *140*, 29.
[223] Yammal, C. C.; Podesta, J. C.; Rossi, R. A. *J. Org. Chem.* **1992**, *57*, 5720.
[224] Yammal, C. C.; Podesta, J. C.; Rossi, R. A. *J. Organomet. Chem.* **1996**, *509*, 1.
[225] Rossi, R. A.; Bunnett, J. F. *J. Org. Chem.* **1972**, *37*, 3570.
[226] Pierini, A. B.; Baumgartner, M. T.; Rossi, R. A. *Tetrahedron Lett.* **1987**, *28*, 4653.
[227] Chahma, M.; Combellas, C.; Thiébault, A. *Synthesis* **1994**, 366.
[228] Chahma, M.; Combellas, C.; Marzouk, H.; Thiébault, A. *Tetrahedron Lett.* **1991**, *32*, 6121.
[229] Chahma, M.; Combellas, C.; Thiébault, A. *J. Org. Chem.* **1995**, *60*, 8015.
[230] Medebielle, M.; Oturan, M. A.; Pinson, J.; Savéant, J. M. *Tetrahedron Lett.* **1993**, *34*, 3409.
[231] Bunnett, J. F.; Traber, R. P. *J. Org. Chem.* **1978**, *43*, 1867.
[232] Bunnett, J. F.; Creary, X. *J. Org. Chem.* **1974**, *39*, 3612.
[233] Beugelmans, R.; Chbani, M. *Bull. Soc. Chim. Fr.* **1995**, *132*, 306.

[234] Beugelmans, R.; Chbani, M. *Bull. Soc. Chim. Fr.* **1995**, *132*, 290.
[235] Bulot, J. J.; Aboujaoude, E. E.; Collignon, N.; Savignac, P. *Phosphorus Sulfur* **1984**, *21*, 197.
[236] Bowman, W. R.; Taylor, P. F. *J. Chem. Soc., Perkin Trans. 1* **1990**, 919.
[237] Beugelmans, R.; Chbani, M. *Bull. Soc. Chim. Fr.* **1995**, *132*, 729.
[238] Bard, R. R.; Bunnett, J. F.; Traber, R. P. *J. Org. Chem.* **1979**, *44*, 4918.
[239] Bunnett, J. F.; Shafer, S. J. *J. Org. Chem.* **1978**, *43*, 1877.
[240] Bornancini, E. R. N.; Rossi, R. A. *J. Org. Chem.* **1990**, *55*, 2332.
[241] Manzo, P. G.; Palacios, S. M.; Alonso, R. A. *Tetrahedron Lett.* **1994**, *35*, 677.
[242] Bornancini, E. R. N.; Alonso, R. A.; Rossi, R. A. *J. Organomet. Chem.* **1984**, *270*, 177.
[243] Zoltewicz, J. A.; Oestreich, T. M. *J. Org. Chem.* **1991**, *56*, 2805.
[244] Beugelmans, R.; Bois-Choussy, M. *Tetrahedron* **1986**, *42*, 1381.
[245] Chbani, M.; Bouillon, J-P.; Chastanet, J.; Souflaoui, M.; Beugelmans, R. *Bull. Soc. Chim. Fr.* **1995**, *132*, 1053.
[246] Beugelmans, R.; Ginsburg, H. *Tetrahedron Lett.* **1987**, *28*, 413.
[247] Bunnett, J. F.; Creary, X. *J. Org. Chem.* **1974**, *39*, 3173.
[248] Julliard, M.; Chanon, M. *J. Photochem.* **1986**, *34*, 231.
[249] Kondo, S.; Nakanishi, M.; Tsuda, K. *J. Heterocyclic Chem.* **1984**, *21*, 1243.
[250] Peñéñory, A. B.; Rossi, R. A. *J. Phys. Org. Chem.* **1990**, *3*, 266.
[251] Zoltewicz, J. A.; Oestreich, T. M. *J. Am. Chem. Soc.* **1973**, *95*, 6863.
[252] Zoltewicz, J. A.; Locko, G. A. *J. Org. Chem.* **1983**, *48*, 4214.
[253] Novi, M.; Garbarino, G.; Petrillo, G.; Dell'Erba, C. *J. Org. Chem.* **1987**, *52*, 5382.
[254] van Tilborg, W. J. M.; Smit, C. J. *Tetrahedron Lett.* **1977**, *41*, 3651.
[255] van Tilborg, W. J. M.; Smit, C. J.; Scheele, J. J. *Tetrahedron Lett.* **1977**, 2113.
[256] Pinson, J.; Savéant, J. M. *J. Chem. Soc., Chem. Commun.* **1974**, 923.
[257] Genesty, M.; Thobie, C.; Gautier, A.; Degrand, C. *J. Appl. Electrochem.* **1993**, *23*, 1125.
[258] Gautier, C. T.; Genesty, M.; Degrand, C. *J. Org. Chem.* **1991**, *56*, 3452.
[259] Amatore, C.; Pinson, J.; Savéant, J. M.; Thiébault, A. *J. Electroanal. Chem.* **1981**, *123*, 231.
[260] Amatore, C.; Chaussard, J.; Pinson, J.; Savéant, J. M.; Thiébault, A. *J. Am. Chem. Soc.* **1979**, *101*, 6012.
[261] Petrillo, G.; Novi, M.; Garbarino, G.; Dell'Erba, C. *Tetrahedron Lett.* **1985**, *26*, 6365.
[262] Petrillo, G.; Novi, M.; Garbarino, G.; Dell'Erba, C. *Tetrahedron* **1986**, *42*, 4007.
[263] Amatore, C.; Beugelmans, R.; Bois-Choussy, M.; Combellas, C.; Thiébault, A. *J. Org. Chem.* **1989**, *54*, 5688.
[264] Bunnett, J. F.; Creary, X. *J. Org. Chem.* **1974**, *39*, 3611.
[265] Hobbs, D. W.; Still, W. C. *Tetrahedron Lett.* **1987**, *28*, 2805.
[266] Petrillo, G.; Novi, M.; Garbarino, G.; Filiberti, M. *Tetrahedron* **1989**, *45*, 7411.
[267] Cevasco, G.; Novi, M.; Petrillo, G.; Thea, S. *Gazz. Chim. Ital.* **1990**, *120*, 131.
[268] Rossi, R. A.; Peñéñory, A. B. *J. Org. Chem.* **1981**, *46*, 4580.
[269] Sandman, D. J.; Stark, J. C.; Acampora, L. A.; Gagne, P. *Organometallics* **1983**, *2*, 549.
[270] Sandman, D. J.; Li, L.; Tripathy, S.; Stark, J. C.; Acampora, L. A.; Foxman, B. M. *Organometallics* **1994**, *13*, 348.
[271] Sandman, D. J.; Rubner, M.; Samuelson, L. *J. Chem. Soc., Chem. Commun.* **1982**, 1133.
[272] Acampora, L. A.; Dugger, D. L.; Emma, T.; Mohammed, J.; Rubner, M. F.; Samuelson, L.; Sandman, D. J.; Tripathy, S. K. *Polymers in Electronics, ACS Symposium Series No. 242*, Davidson, T., Ed., 1995, p. 461.
[273] Suzuki, H.; Inouye, M. *Chem. Lett.* **1985**, 389.
[274] Suzuki, H.; Padmanabhan, S.; Inouye, M.; Ogawa, T. *Synthesis* **1989**, 468.
[275] Suzuki, H.; Nakamura, T. *Synthesis* **1992**, 549.
[276] Thobie-Gautier, C.; Degrand, C. *J. Org. Chem.* **1991**, *56*, 5703.
[277] Degrand, C.; Prest, R. *J. Electroanal. Chem.* **1990**, *282*, 281.
[278] Thobie-Gautier, C.; Degrand, C. *J. Electroanal. Chem.* **1993**, *344*, 383.
[279] Genesty, M.; Merle, O.; Degrand, C.; Nour, M.; Compagnon, P. L.; Lemaitre, J. P. *Denki Kagaku* **1994**, *62*, 1158; *Chem. Abstr.* **1995**, *122*, 225147r.
[280] Degrand, C.; Prest, R.; Compagnon, P. L. *J. Org. Chem.* **1987**, *52*, 5229.
[281] Degrand, C.; Prest, R.; Nour, M. *Phosphorus Sulfur* **1988**, *38*, 201.

[282] Degrand, C. *J. Org. Chem.* **1987**, *52*, 1421.
[283] Degrand, C. *J. Chem. Soc., Chem. Commun.* **1986**, 1113.
[284] Degrand, C. *J. Electroanal. Chem.* **1987**, *238*, 239.
[285] Degrand, C. *Tetrahedron* **1990**, *46*, 5237.
[286] Beugelmans, R.; Roussi, G. *J. Chem. Soc., Chem. Commun.* **1979**, 950.
[287] Boujlet, K.; Simonet, J.; Roussi, G.; Beugelmans, R. *Tetrahedron Lett.* **1982**, *23*, 173.
[288] Fontan, R.; Galvez, C.; Viladoms, P. *Heterocycles* **1981**, *16*, 1473.
[289] Beugelmans, R.; Bois-Choussy, M. *Tetrahedron* **1992**, *48*, 8285.
[290] Beugelmans, R.; Chastanet, J.; Roussi, G. *Tetrahedron Lett.* **1982**, *23*, 2313.
[291] Beugelmans, R.; Chastanet, J.; Roussi, G. *Tetrahedron* **1984**, *40*, 311.
[292] Beugelmans, R.; Chastanet, J.; Ginsburg, H.; Quinteros-Cortes, L.; Roussi, G. *J. Org. Chem.* **1985**, *50*, 4933.
[293] Beugelmans, R.; Bois-Choussy, M.; Tang, Q. *Tetrahedron* **1989**, *45*, 4203.
[294] Petrillo, G.; Novi, M.; Dell'Erba, C.; Tavani, C. *Tetrahedron* **1991**, *47*, 9297.
[295] Semmelhack, M. F.; Stauffer, R. D.; Rogerson, T. D. *Tetrahedron Lett.* **1973**, 4519.
[296] Semmelhack, M. F.; Chong, B. P.; Stauffer, R. D.; Rogerson, T. D.; Chong, A.; Jones, L. D. *J. Am. Chem. Soc.* **1975**, *97*, 2507.
[297] Weinreb, S. M.; Semmelhack, M. F. *Acc. Chem. Res.* **1975**, *8*, 158.
[298] Wolfe, J. F.; Sleevi, M. C.; Goehring, R. R. *J. Am. Chem. Soc.* **1980**, *102*, 3646.
[299] Goehring, R. R.; Sachdeva, Y. P.; Pisipati, J. S.; Sleevi, M. C.; Wolfe, J. F. *J. Am. Chem. Soc.* **1985**, *107*, 435.
[300] Goehring, R. R. *Tetrahedron Lett.* **1992**, *33*, 6045.
[301] Goehring, R. R. *Tetrahedron Lett.* **1994**, *35*, 8145.
[302] Martin, S. F.; Liras, S. *J. Am. Chem. Soc.* **1993**, *113*, 10450.
[303] Wiegand, S.; Schafer, H. J. *Tetrahedron* **1995**, *51*, 5341.
[304] Theuns, H. G.; Lenting, H. B. M.; Salemink, C. A.; Tanaka, H.; Shibata, M.; Ito, K.; Lousberg, J. J. C. *Heterocycles* **1984**, *22*, 2007.
[305] Pierini, A. B.; Baumgartner, M. T.; Rossi, R. A. *J. Org. Chem.* **1987**, *52*, 1089.
[306] Fukazawa, Y.; Usui, S.; Tanimoto, K.; Hirai, Y. *J. Am. Chem. Soc.* **1994**, *116*, 8169.
[307] Fukazawa, Y.; Kitayama, H.; Yasuhara, K.; Yoshimura, K.; Usui, S. *J. Org. Chem.* **1995**, *60*, 1696.
[308] Baumgartner, M. T.; Pierini, A. B.; Rossi, R. A. *J. Org. Chem.* **1993**, *58*, 2593.
[309] Raston, C. L.; Salem, G. In *The Chemistry of the Metal-Carbon Bond*, Hartley, F. R.; Patai, S., Eds.; Wiley, New York, 1987; Vol. 4, p. 159.
[309a] Collman, J. P.; Hegedus, L. S.; Norton, J. R.; Finke, R. G. In *Principles and Applications of Organotransition Metal Chemistry*, University Science Books: Mill Valley, 1987.
[310] Stille, J. K. *Pure Appl. Chem.* **1985**, *57*, 1771.
[310a] Stille, J. K. *Angew. Chem., Int. Ed. Engl.* **1986**, *25*, 508.
[311] Kalinin, V. N. *Synthesis* **1992**, 413, and references cited therein.
[312] Grushin, V. V.; Alper, H. *Chem. Rev.* **1994**, *94*, 1047, and references cited therein.
[313] For recent reviews see: Kalinin, V. N. *Russ. Chem. Rev.* **1987**, *56*, 682. Goldberg, Y. in *Phase Transfer Catalysis: Selected Problems and Applications*, Gordon and Breach Science Publishers, Singapore, 1992.
[314] Fanta, P. E. *Synthesis* **1974**, 9.
[315] Lindley, J. *Tetrahedron* **1984**, *40*, 1433, and references cited therein.
[316] Couture, C.; Paine, A. J. *Can. J. Chem.* **1985**, *63*, 111.
[317] Semmelhack, M. F.; Helquist, P. M.; Jones, L. D. *J. Am. Chem. Soc.* **1971**, *94*, 9234.
[318] Grushin, V. V. *Acc. Chem. Res.* **1992**, *25*, 529.
[318a] Grushin, V. V.; Demkina, I. I.; Tolstaya, T. P. *Inorg. Chem.* **1991**, *30*, 1760.
[319] Chen, Z.-C.; Jin, Y.-Y.; Stang, P. J. *J. Org. Chem.* **1987**, *52*, 4115.
[319a] Hampton, K. G.; Harris, T. M.; Hauser, C. R. *J. Org. Chem.* **1964**, *19*, 3511.
[320] Bruggink, A.; McKillop, A. *Tetrahedron* **1975**, *31*, 2607.
[321] Setsune, J.; Matsukawa, K.; Wakemoto, H.; Kitao, T. *Chem. Lett.* **1981**, 367.
[321a] Setsune, J.; Matsukawa, K.; Kitao, T. *Tetrahedron Lett.* **1982**, 663.
[321b] Osuka, A.; Kobayashi, T.; Suzuki, H. *Synthesis* **1983**, 67.

[322] Uno, M.; Seto, K.; Ueda, W.; Masuda, M.; Takahashi, S. *Synthesis* **1985**, 506.
[323] Abramovitch, R. A.; Barton, D. H. R.; Finet, J.-P. *Tetrahedron* **1988**, *44*, 3039 and references cited therein.
[324] Ellis, G. P.; Romney-Alexander, T. M. *Chem. Rev.* **1987**, *87*, 779.
[325] Cassar, L. *J. Organomet. Chem.* **1973**, *54*, C57.
[325a] Cassar, L.; Ferrara, S.; Foa, M. *Adv. Chem. Ser.* **1974**, *132*, 252.
[325b] Cassar, L.; Foa, M.; Montanari, F.; Marinelli, G. P. *J. Organomet. Chem.* **1979**, *173*, 335.
[326] Sakakibara, Y.; Okuda, F.; Shimobayashi, A.; Kirino, K; Sakai, M.; Uchino, N.; Takagi, K. *Bull. Chem. Soc. Jpn.* **1988**, *61*, 1985.
[327] Takagi, K.; Sasaki, K.; Sakakibara, Y. *Bull. Chem. Soc. Jpn.* **1991**, *64*, 1118, and references cited therein.
[327a] Akita, Y.; Shimazaki, M. *Synthesis* **1981**, 974.
[328] Funabiki, T.; Hosomi, H.; Yoshida, S.; Tanara, K. *J. Am. Chem. Soc.* **1982**, *104*, 1560.
[328a] Funabiki, T.; Nakamura, H.; Yoshida, S. *J. Organomet. Chem.* **1983**, *243*, 95.
[329] Ames, D. E.; Brohi, M. I. *J. Chem. Soc., Perkin Trans 1* **1980**, 1384.
[329a] Ames, D. E.; Bull, D. *Tetrahedron* **1982**, *38*, 383.
[330] Heintz, M.; Sock, O.; Saboureau, C.; Périchon, J.; Troupel, M. *Tetrahedron* **1988**, *44*, 1631.
[331] Testaferri, L.; Tiecco, M.; Tingoli, M.; Chianelli, D.; Montanucci, M. *Synthesis* **1983**, 751.
[331a] Pastor, S. D.; Hessell, E. T. *J. Org. Chem.* **1985**, *50*, 4812.
[331b] Pastor, S. D. *Helv. Chim. Acta* **1988**, *71*, 859.
[331c] Caruso, A. J.; Colley; A. M.; Bryant, G. L. *J. Org. Chem.* **1991**, *56*, 862.
[331d] Shaw, J. E. *J. Org. Chem.* **1991**, *56*, 3728.
[332] Migita, T.; Shimizu, T.; Asami, Y.; Shiobara, J.-I.; Kato, Y.; Kosugi, M. *Bull. Chem. Soc. Jpn.* **1980**, *53*, 1385.
[333] Foa, M.; Santi, R.; Garavaglia, F. *J. Organomet. Chem.* **1981**, *206*, C29.
[334] March, J. In *Advanced Organic Chemistry*, Wiley, New York, Fourth Edition, 1992.
[335] Reich, H. J.; Cohen, M. L.; Clark, P. S. *Org. Synth.* **1980**, *59*, 141.
[336] Tiecco, M.; Testaferri, L.; Tingoli, M.; Chianelli, D.; Montanucci, M. *J. Org. Chem.* **1983**, *48*, 4289.
[337] Cristau, H-J.; Chene, A.; Christol, H. *J. Organomet. Chem.* **1980**, *185*, 283.
[338] Cassar, L.; Foa, M. *J. Organomet. Chem.* **1974**, *74*, 75.
[339] Balthazor, T. M.; Grabiak, R. C. *J. Org. Chem.* **1980**, *45*, 5425.
[340] Bunnett, J. F.; Scamehorn, R. G.; Traber, R. P. *J. Org. Chem.* **1976**, *41*, 3677.
[341] Combellas, C.; Lu, Y.; Thiébault, A. *J. Appl. Electrochem.* **1993**, *23*, 841.
[342] Rabideau, P. W.; Marcinow, Z. *Org. React.* **1992**, *42*, 1.
[343] Sakla, A. B.; Masoud, N. K.; Sawiris, Z.; Ebaid, W. S. *Helv. Chim. Acta* **1974**, *57*, 481.
[344] Bunnett, J. F.; Weiss, R. H. *Org. Synth.* **1978**, *58*, 134.
[345] Galli, C. *Chem. Rev.* **1988**, *88*, 765.
[346] Simonet, J.; Dupuy, H. *J. Electroanal. Chem.* **1992**, *327*, 201.
[347] Zhao, W. Y.; Liu, Y.; Huang, Z. T. *Synth. Commun.* **1993**, *23*, 591.
[348] Hamana, M.; Iwasaki, G.; Saeki, S. *Heterocycles* **1982**, *17*, 177.
[349] Alonso, R. A.; Austin, E.; Rossi, R. A. *J. Org. Chem.* **1988**, *53*, 6065.
[350] Nair, V.; Hettrick, B. J. *Tetrahedron* **1988**, *44*, 7001.
[351] Marchal, J.; Bodiguel, J.; Fort, Y.; Caubere, P. *J. Org. Chem.* **1995**, *60*, 8336.
[352] Amatore, C.; Combellas, C.; Pinson, J.; Savéant, J. M.; Thiébault, A. *J. Chem. Soc., Chem. Commun.* **1988**, 7.
[353] Combellas, C.; Suba, C.; Thiébault, A. *Tetrahedron Lett.* **1992**, *33*, 4923.
[354] Boy, P.; Combellas, C.; Mathey, G.; Palacin, S.; Persoons, A.; Thiébault, A.; Verbiest, T. *Advan. Mater.* **1994**, *6*, 580.
[355] Combellas, C.; Petit, M. A.; Thiébault, A.; Froyer, G.; Bosc, D. *Makromol. Chem.* **1992**, *193*, 2445.
[356] Médebielle, M.; Oturan, M. A.; Pinson, J.; Savéant, J. M. *J. Org. Chem.* **1996**, *61*, 1331.
[357] Manzo, P. G.; Palacios, S. M.; Alonso, R. A.; Rossi, R. A. *Org. Prep. Proced. Int.* **1995**, *27*, 668.

[358] Amatore, C.; Pinson, J.; Savéant, J. M.; Thiébault, A. *J. Electroanal. Chem.* **1980**, *107*, 59.
[359] Kampmeier, J. A.; Nalli, T. W. *J. Org. Chem.* **1993**, *58*, 943.
[360] Beugelmans, R.; Chbani, M.; Soufiaoui, M. *Tetrahedron Lett.* **1996**, *37*, 1603.
[361] Nasielski, J.; Moucheron, C.; Nasielski-Hinkens, R. *Bull. Soc. Chim. Belg.* **1992**, *101*, 491.
[362] Novi, M.; Petrillo, G.; Sartirana, L. *Tetrahedron Lett.* **1986**, *27*, 6129.
[363] Rybakova, I. A.; Shekhtman, R. I.; Prilezhaeva, E. N. *Izv. Akad. Nauk SSSR, Ser. Khim.*, **1987**, 833; *Chem. Abstr.* **1988**, *108*, 221560h.
[364] Sandman, D. J.; Stark, J. C.; Rubner, M.; Hamill, G. P.; Acampora, L. A.; Samuelson, L. A.; McGrath, M. A.; Allen, G. W. *Proceedings* of *the Fourth International Conference on the Organic Chemistry of Selenium and Tellurium*, Berry, F. J.; McWhinnie, W. R., Eds., The University of Aston in Birmingham, 1995, p. 637.
[365] Sandman, D. J.; Allen, G. W.; Acampora, L. A.; Stark, J. C.; Jansen, S.; Jones, M. T.; Ashwell, G. J.; Foxman, B. M. *Inorg. Chem.* **1987**, *26*, 1664.
[366] Peñéñory, A. B.; Pierini, A. B.; Rossi, R. A. *J. Org. Chem.* **1984**, *49*, 3834.
[367] Pierini, A. B.; Rossi, R. A. *J. Organomet. Chem.* **1978**, *144*, C12.
[368] Prest, R.; Degrand, C. *J. Chem. Soc., Perkin Trans. 2* **1989**, 607.
[369] Staskun, B.; Wolfe, J. F. *S. Afr. J. Chem.* **1992**, *45*, 5.
[370] Wu, G. S.; Tao, T.; Cao, J. J.; Wei, X. L. *Acta Chim. Sinica* **1992**, *50*, 614; *Chem. Abstr.* **1992**, *117*, 159860m.
[371] Baumgartner, M. T.; Pierini, A. B.; Rossi, R. A. *Tetrahedron Lett.* **1992**, *33*, 2323.
[372] Gilmore, C. J.; MacNicol, D. D.; Murphy, A.; Russell, M. A. *Tetrahedron Lett.* **1984**, *25*, 4303.
[373] Yui, K.; Aso, Y.; Otsubo, T.; Ogura, F. *Chem. Lett.* **1986**, 551.
[374] Stark, J. C.; Reed, R.; Acampora, L. A.; Sandman, D. J.; Jansen, S.; Jones, M. T.; Foxman, B. M. *Organometallics* **1984**, *3*, 732.
[375] Yamahira, A.; Nogami, T.; Mikawa, H. *J. Chem. Soc., Chem. Commun.* **1983**, 904.
[376] Balodis, K. A.; Livdane, A. D.; Medne, R. S.; Neiland, O. Ya. *Zh. Org. Khim.* **1979**, *15*, 391; *Chem. Abstr.* **1979**, *90*, 203982c.
[377] Sandman, D. J.; Stark, J. C.; Foxman, B. M. *Organometallics* **1982**, *1*, 739.
[378] Usui, S.; Fukazawa, Y. *Tetrahedron Lett.* **1987**, *28*, 91.

CHAPTER 2

OXIDATION OF CARBONYL COMPOUNDS WITH ORGANOHYPERVALENT IODINE REAGENTS

Robert M. Moriarty and Om Prakash[†]

Department of Chemistry
The University of Illinois at Chicago, Chicago, USA

CONTENTS

[†](On sabbatical leave from Kurukshetra University, Kurukshetra, Haryana, 136119 INDIA)

Organic Reactions, Vol. 54, Edited by Leo A. Paquette et al.
ISBN 0-471-34888-0

ACKNOWLEDGMENT

We thank Dr. Jerry Kosmeder and Dr. Neena Rani for their help in preparing this Chapter.

INTRODUCTION

The search for novel oxidizing agents, especially those that not only transform a broad range of diverse functional groups, but also do so with a high degree of selectivity, remains the focus of intense exploration. In recent years, the use of hypervalent iodine compounds as oxidizing agents has gained attention in synthetic organic chemistry.[1–21] The α-oxidation of carbonyl compounds by hypervalent iodine oxidants has been most widely investigated and has broad synthetic utility. Although this area has recently been reviewed in various forms and scope, a comprehensive coverage that focuses on synthetic applications is absent and the present chapter addresses this need.[7,11,13–15,19,20,22]

In this chapter we discuss the oxidation of carbonyl compounds with hypervalent iodine reagents. These reagents effect the electrophilic α-oxidation of large numbers of ketones, β-diketones, α,β-unsaturated ketones, and their derivatives such as silyl enol ethers. We restrict ourselves mainly to the reactions of organoiodine(III) reagents; inclusion of other hypervalent iodine reagents such as organoiodine(V) reagents, periodic acid, or purely inorganic iodate reagents is beyond the scope of this chapter. Furthermore, reactions brought about with other oxidizing agents in combination with a catalytic amount of a hypervalent iodine reagent are not considered here. The chapter also excludes discussion of a considerable amount of work based upon the reaction of iodonium ylides, a distinct feature of organoiodine(III) reagents.[3,11,18a] The common abbreviations and names of the reagents used in this chapter are listed in Table A.

A general feature of the reactions of carbonyl compounds described in this chapter is electrophilic attack of the organoiodine(III) reagent at the α-carbon atom of a carbonyl group to yield a presumed tricoordinate iodine(III) intermediate 1 (Eq. 1).

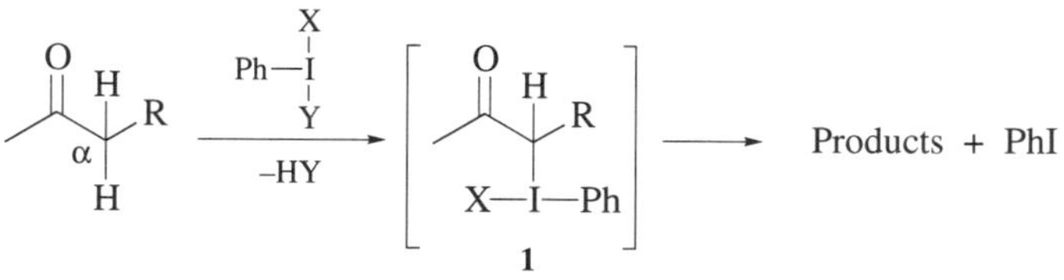
O
H
R
α
H
X
Ph—I
Y
–HY
O
H
R
X—I—Ph
1
Products + PhI

(Eq. 1)

Intermediate **1** reacts further to yield products via various pathways depending upon the reaction conditions. The major processes that occur are summarized (Eqs. 2–7). These include oxidation of ketones under basic conditions leading to α-hydroxydimethylacetals (Eq. 2), oxidation of α,β-unsaturated ketones leading

TABLE A. Hypervalent Iodine Reagents

Formula	Name(s)	Abbreviation
$PhI(OAc)_2$	Iodobenzene diacetate or (Diacetoxyiodo)benzene	IBD
$PhI(O_2CCF_3)_2$	Iodobenzene bis(trifluoroacetate) or Bis(trifluoroacetoxy)iodobenzene	IBTA
PhI(OH)OTs	Hydroxy(tosyloxy)iodobenzene or Koser's reagent	HTIB
PhI(OH)OMs	Hydroxy(mesyloxy)iodobenzene	HMIB
$PhI(OH)OSO_2CH_2$ 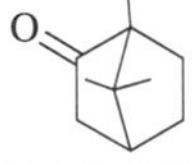	Hydroxy[(+)-(10-camphorsulfonyl)oxy]iodobenzene	HCIB
$PhI(OH)OPO(OPh)_2$	Hydroxy[bis(phenoxy)phosphoryloxy]iodobenzene	
PhI(OH)OPO(OPh)Me	Hydroxy[(phenoxy)(methyl)phosphonyloxy]iodobenzene	
$(PhIO)_n$	Iodosylbenzene or Iodosobenzene	
$PhIF_2$	(Difluoroiodo)benzene	
$PhICl_2$	(Dichloroiodo)benzene	
$R_fI(Ph)OTf$	(Perfluoroalkyl)phenyliodonium triflates (R_f = perfluoroalkyl)	
$R_fCH_2I(Ph)OTf$	(1*H*,1*H*-Perfluoroalkyl)phenyliodonium triflates	
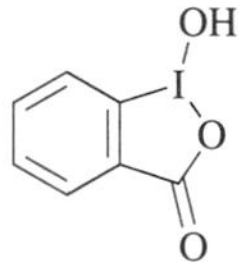	1-Hydroxy-1,2-benziodoxolin-3-(1*H*)one or *o*-Iodosylbenzoic acid	
$Ph_2I^+X^-$	Diphenyliodonium halide	
$(Mes)_2I^+Cl^-$	Dimesityliodonium chloride	
$(PhI^+)_2O(BF_4^-)_2$	Iodosobenzene tetrafluoroborate	
$(PhI^+)_2O(PF_6^-)_2$	Iodosobenzene hexafluorophosphate	
$BnMe_3N^+Cl_2I^-$	Benzyltrimethylammonium dichloroiodate	

to α-hydroxy-β-methoxydimethylacetals (Eq. 3), oxidation of ketones under neutral conditions (Eq. 4), and related oxidation of silyl enol ethers under various

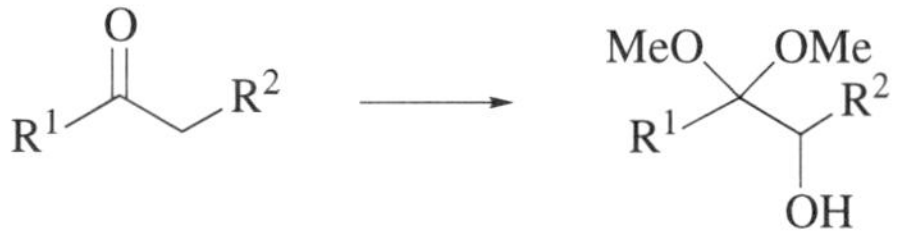

(Eq. 2)

(Eq. 3)

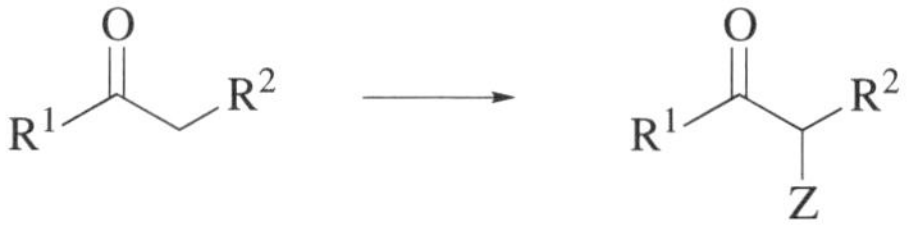

Z = OH, OR, O_2CR, OSO_2R, $P(O)(OR)_2$, CH_2CF_3, Ar (Eq. 4)

conditions leading to α-functionalization (Eq. 5), intramolecular participation reactions leading to oxygen-containing heterocyclic compounds (Eq. 6) including

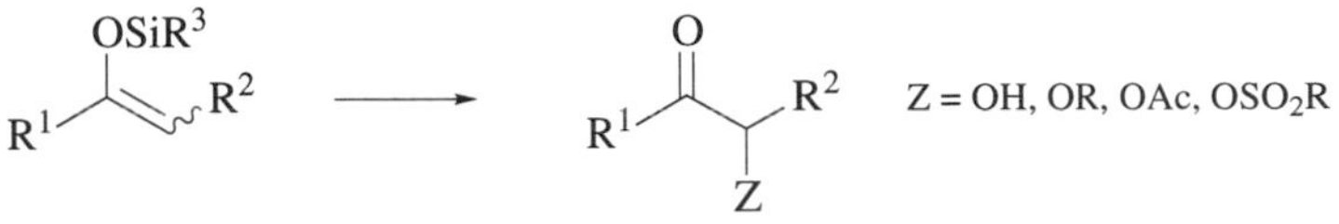
OSiR3
R1
R2
O
R1
R2
Z
Z = OH, OR, OAc, OSO2R

(Eq. 5)

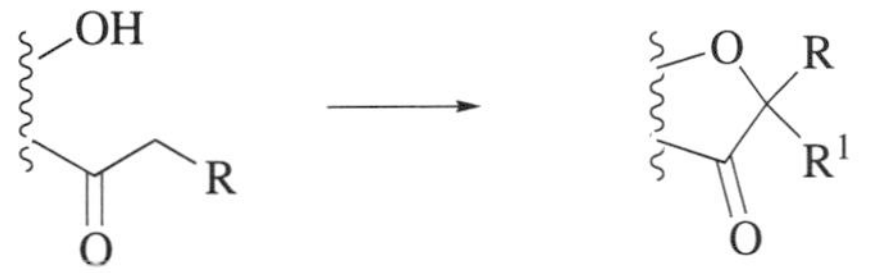
OH
R
O
O
R
R1
O

(Eq. 6)

cyclic α-keto ethers and lactones, and rearrangement processes such as 1,2-aryl shifts in alkyl aryl ketones leading to 2-arylalkanoates (Eq. 7). All of these reactions are driven by the reduction of iodine(III) to iodine(I) (iodobenzene).

O
Ar
R
R
Ar
CO2Me

(Eq. 7)

MECHANISM

General Pathways

The common thread connecting reactions of carbonyl compounds with hypervalent iodine reagents is the generation of an electrophilic center α to a carbonyl group by addition of a hypervalent reagent [Ph–I(X)Y] to give phenyliodinated intermediate **3** (Eq. 8) analogous to **1** mentioned earlier. Subsequent steps lead to various products depending upon the reaction conditions (reagents, solvent, temperature, etc.) and the structure of the substrate. This sequence may be viewed as an umpolung in reactivity of an enol, enolate, or silyl enol ether **2a-c** (Eq. 8).

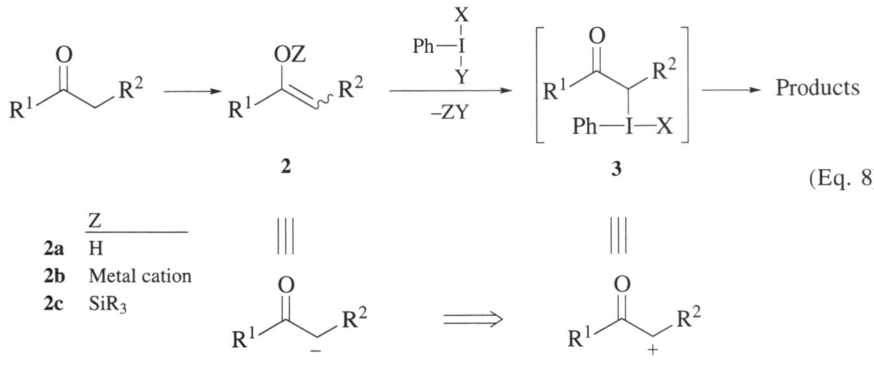
O
R1
R2
OZ
R1
R2
X
Ph—I
Y
–ZY
O
R1
R2
Ph—I—X
Products
2
3
Z
2a H
2b Metal cation
2c SiR3
O
R1
R2
O
R1
R2

(Eq. 8)

Since the fate of intermediate **3** is controlled both by the reaction conditions as well as by its structure, mechanistic aspects are discussed on the basis of these considerations. The stereochemical aspects of the reaction are discussed where appropriate under Scope and Limitations.

Oxidation Under Basic Conditions

The oxidation of an enolizable ketone leading to α-hydroxydimethylacetal formation is a major synthetic application of organohypervalent iodine reagents. The reaction of acetophenone is typical (Eq. 9).[23]

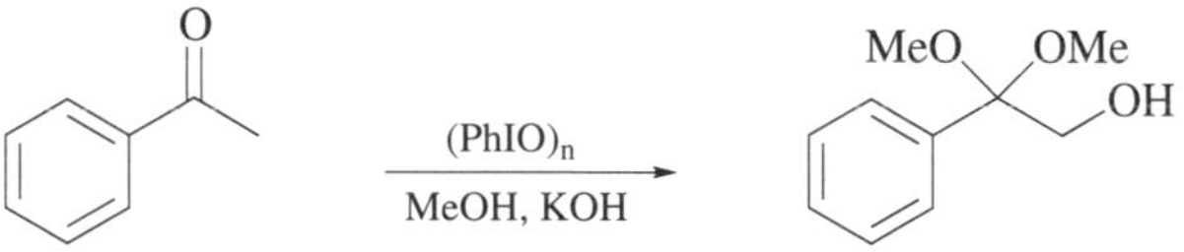
O
(PhIO)n
MeOH, KOH
MeO
OMe
OH

(Eq. 9)

The various steps that are considered to occur in this reaction are: (Eq. 10) step (a) dissociation of the polymeric iodosobenzene or hydrolysis of $PhI(OAc)_2$

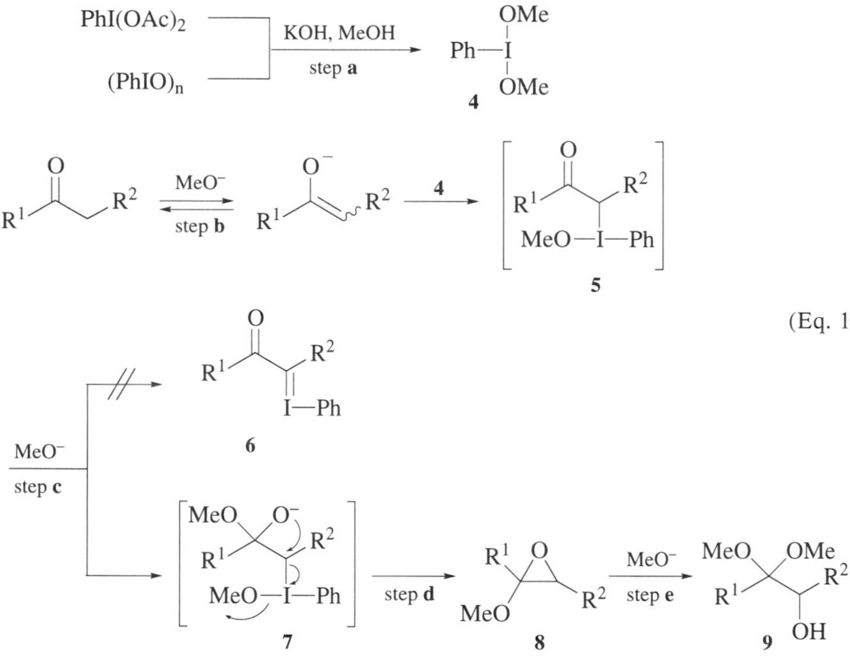
PhI(OAc)2
(PhIO)n
KOH, MeOH
step a
OMe
Ph—I
OMe
4
MeO⁻
step b
4
MeO—I—Ph
5
6
I—Ph
MeO⁻
step c
7
step d
8
MeO⁻
step e
9
OH

(Eq. 10)

to generate the active reagent (dimethoxyiodo)benzene (**4**);[24] step (b) the concomitant formation of the enolate anion and subsequent displacement of methoxide ion from reagent **4** by the enolate anion to yield intermediate **5**; step (c) attack of methoxide anion upon the carbonyl group of intermediate **5** to yield a tetrahedral intermediate **7** (step c), and finally intramolecular nucleophilic displacement by alkoxide anion to give the oxirane **8** accompanied by reductive elimination of PhI (step d). The reaction is completed by attack of a second methoxide ion on the oxirane **8** to yield product **9** (step e).

The presumed intermediate **5**, which has never been isolated, could conceivably react further with loss of methanol to generate an iodonium ylide **6**. However, this result is not observed although one example of a monocarbonyl ylide is known.[18b] Iodonium ylides **10** formed from β-dicarbonyl compounds are widely reported (Eq. 11).[3,11,18a]

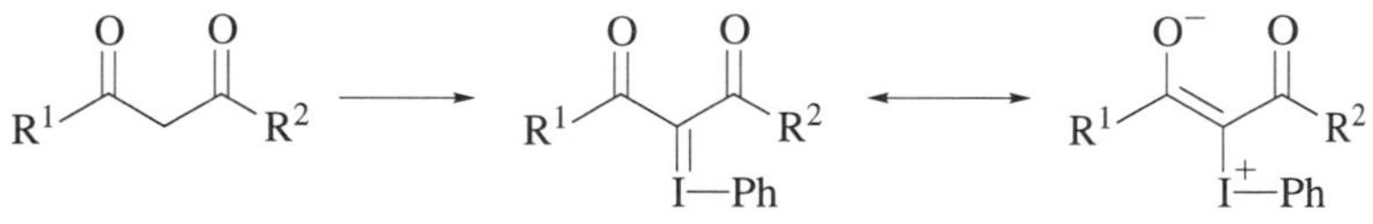
O
O
R1
R2
O
O
R1
R2
I—Ph
O⁻
O
R1
R2
I—Ph

10 **10a**

(Eq. 11)

Formation of the stable iodonium ylide **10** may result from the additional resonance stabilization that is provided by the second carbonyl group of the β-dicarbonyl compound (**10** ↔ **10a**). This dichotomy between the behavior of monocarbonyl and β-dicarbonyl substrates is taken as indirect proof of the intermediacy of **7**.

When the pathway shown in Eq. 10 step e is not favored for steric reasons, the other carbon atom of the oxirane ring is attacked by methoxide ion and an α-methoxyketone is formed; this route is observed in sterically hindered ketones, such as 2,2,6,6-tetramethyl-4-piperidone (Eq. 12).[25]

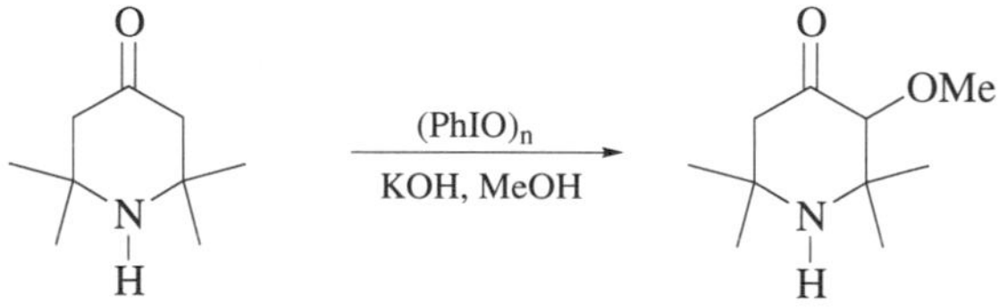
O
N
H
(PhIO)n
KOH, MeOH
O
OMe
N
H

(Eq. 12)

Intermediate **5** (Eq. 10) may be viewed as a highly reactive (hypervalent) α-haloketone and as such can undergo Favorskii rearrangement. Also, neighboring group interactions may occur in the decomposition of **5** in appropriately substituted examples.[26]

A requirement of the pathway outlined in Eq. 10 is that the original carbonyl oxygen atom of the substrate becomes the oxygen atom of the hydroxy group of the α-hydroxydimethylacetal product. This has been established experimentally by isotopic labeling using acetophenone, $C_6H_5C^{18}OCH_3$ (Eq. 13).[27]

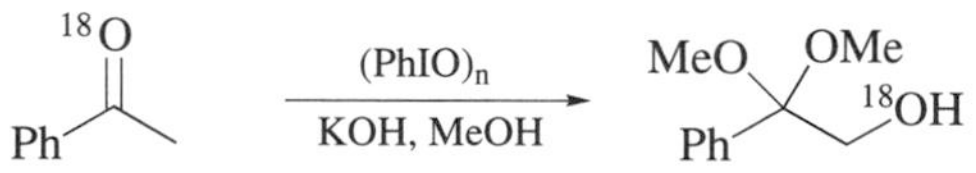
18O
Ph
(PhIO)n
KOH, MeOH
MeO
OMe
18OH
Ph

(Eq. 13)

Furthermore, the pathway shown in Eq. 10 has stereochemical consequences. In step d, an inversion of configuration must occur at the carbon at which the C-iodine(III) cleavage occurs. In step e, opening of the epoxide ring results in an inversion of configuration at the carbon atom attacked by MeO^-. Both of these stereochemical features are manifested in various cyclic systems that are discussed in Scope and Limitations.

The oxidation of α,β-unsaturated ketones such as **11** which do not contain an enolizable carbonyl group occurs by generation of an anionic intermediate **12** (Eq.14) via a Michael addition of methoxide ion. The formation of the α-hydroxy-β-methoxydimethylacetal **13** then occurs via steps analogous to those of Eq. 10.

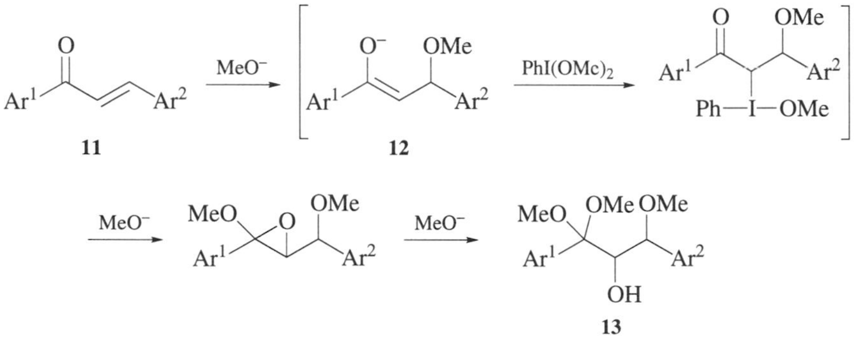

O
Ar1
Ar2
MeO−
O−
OMe
PhI(OMe)2
Ph—I—OMe
MeO
11
12
13
OH

(Eq. 14)

Oxidation under Neutral Conditions in Aprotic Solvents

The oxidation of ketones under neutral conditions, for example, α-sulfonyloxylation in dichloromethane or acetonitrile, results in direct α-functionalization. A possible pathway for α-sulfonyloxylation of ketones outlined in Eq. 15 involves dissociation of the hypervalent iodine reagent[28] to give the reactive species $(PhIOH)^+(OSO_2R^3)^-$ followed by electrophilic attack of the I(III) species on the enol tautomer of the ketone and subsequent loss of water to give ion pair **14**. Attack of the sulfonate anion at the α position with displacement of iodobenzene gives the α-sulfonyloxy ketone **15**.

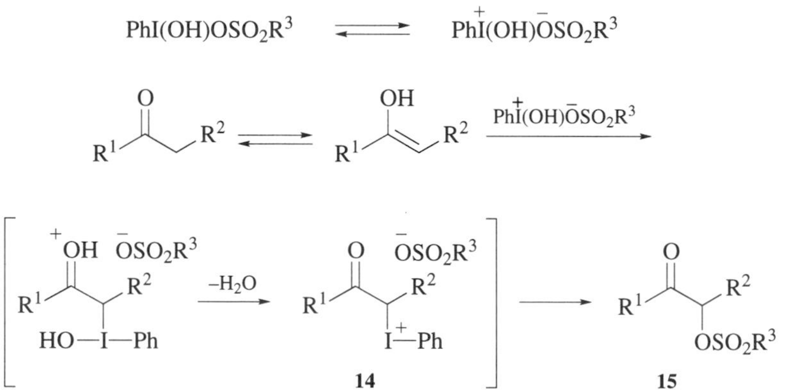

PhI(OH)OSO2R3
O
OH
R1
R2
OSO2R3
–H2O
HO—I—Ph
I—Ph
14
15

(Eq. 15)

The course of these oxidation reactions is greatly influenced by both the structure of the carbonyl compound and the reaction conditions. For example, the oxidation of flavanone with HTIB leads to rearrangement giving isoflavone rather than the expected 3-tosyloxyflavanone (see Eq. 20).

Oxidation of Silyl Enol Ethers

α-Functionalization using silyl enol ethers **16** involves electrophilic addition of $PhI^+\ OBF_3^-$ (generated from iodosobenzene and boron trifluoride etherate) to give an intermediate **17** (Eq. 16) which is analogous to iodonium ion **14** (Eq. 15). This intermediate is the synthetic equivalent of an α-keto carbocation, and it reacts with external nucleophiles to give the products **18**. In the absence of an external nucleophile, a second molecule of the silyl enol ether **16** attacks intermediate **17** resulting in carbon-carbon bond formation to afford 1,4-diketone **19**.

$$(PhIO)_n + BF_3{\bullet}OEt_2 \longrightarrow PhI^+OBF_3^-$$

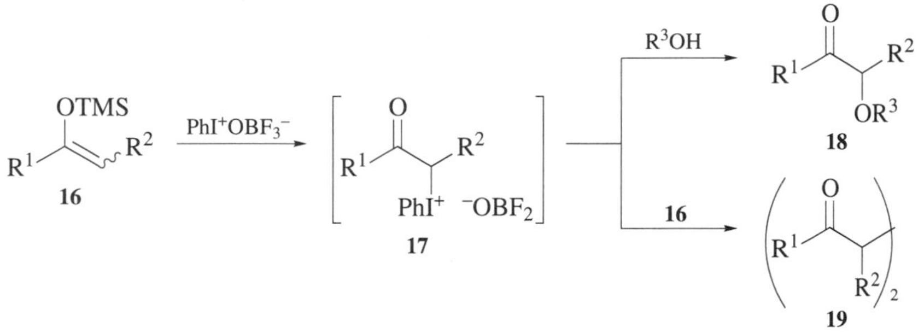

OTMS
R1
R2
16
PhI+OBF3−
O
R1
R2
PhI+ −OBF2
17
R3OH
O
R1
R2
OR3
18
16
O
R1
R2
2
19

(Eq. 16)

Intramolecular Participation

Intramolecular participation occurs when certain ketones bearing an appropriately positioned hydroxy or carboxy group are subjected to hypervalent iodine oxidation leading to cyclic α-keto ether and lactone formation, respectively. For example, conversion of *o*-hydroxyacetophenone (**20**) to 2,2-dimethoxycoumaran-3-one (**23**) results from neighboring group participation of the *ortho* phenoxy group of intermediate **21** to yield coumaran-3-one (**22**), followed by further oxidation of **22** to give final product **23** (Eq. 17).[26]

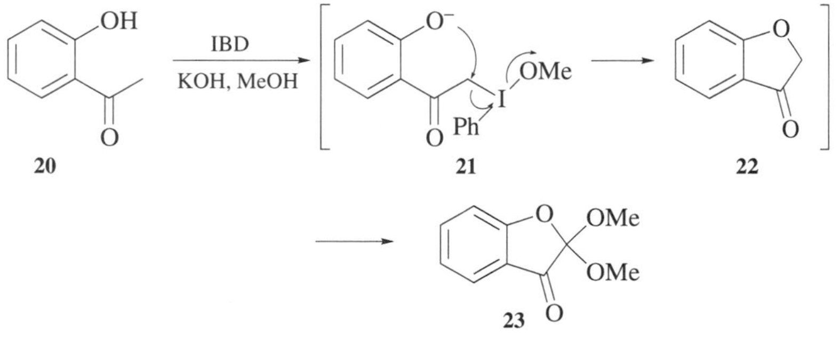

OH
O
20
IBD
KOH, MeOH
O−
OMe
I
Ph
O
21
O
O
22
O
OMe
OMe
O
23

(Eq. 17)

Rearrangement Processes

Upon hypervalent iodine oxidation, a number of alkyl aryl ketones, α,β-unsaturated ketones and cyclic ketones undergo rearrangement processes such as 1,2-aryl shift[20] and Favorskii type ring contraction.[25] The conversion of alkyl aryl ketones **24** in methanol or trimethyl orthoformate (in the presence or absence of acid) to 2-arylalkanoates **28** is shown in Eq. 18. This may be explained in terms of initial enol ether formation (**24** → **25**, Eq. 18), followed by addition of $PhI(OMe)_2$ to give **26**. Aryl group migration and hydrolysis via **27** yields the rearranged ester **28**.[20]

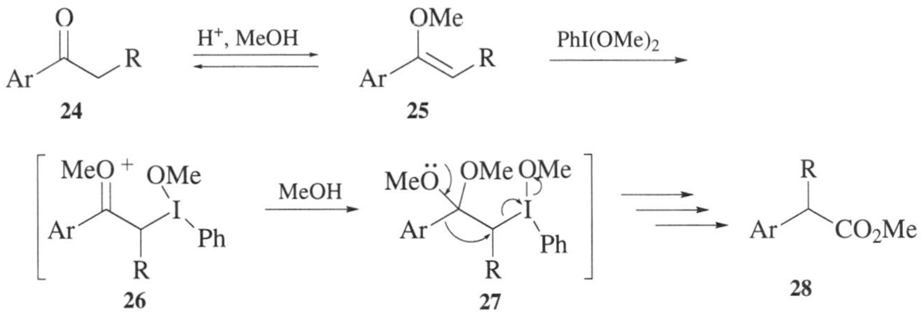

(Eq. 18)

The oxidation of α,β-unsaturated ketones leading to a 1,2-aryl shift may also be explained by a similar reaction pathway (Eq. 19). Initial attack of methanol at the β position of the α,β-unsaturated ketone **29** coupled with electrophilic addition of the iodine(III) reagent results in the formation of intermediate **30**. A 1,2-aryl shift accompanied by reductive elimination of iodobenzene leads to intermediate **31**, which reacts further with methanol to yield the dimethylacetal **32**.

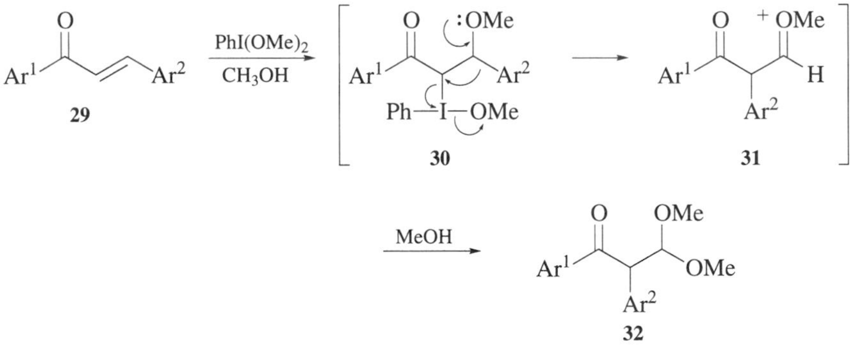

(Eq. 19)

Conversions of flavanone **33** to isoflavone **34** and to methyl 2-aryl-2,2-dihydrobenzofuran-3-carboxylate **35** are further examples of 1,2-aryl migrations. In the first case when oxidation is effected by using HTIB in acetonitrile, phenyl migration ("*anti*") to the electrophilic center occurs because of the stability of the resulting oxonium ion intermediate **36** and the aromaticity of the product **34** formed by deprotonation (Eq. 20). Under acetalizing conditions using trimethyl orthoformate (TMOF) and IBD/H_2SO_4 or HTIB/TMOF, an alternative pathway involving ring contraction via enol ether **37** occurs to give **35** (Eq. 21).

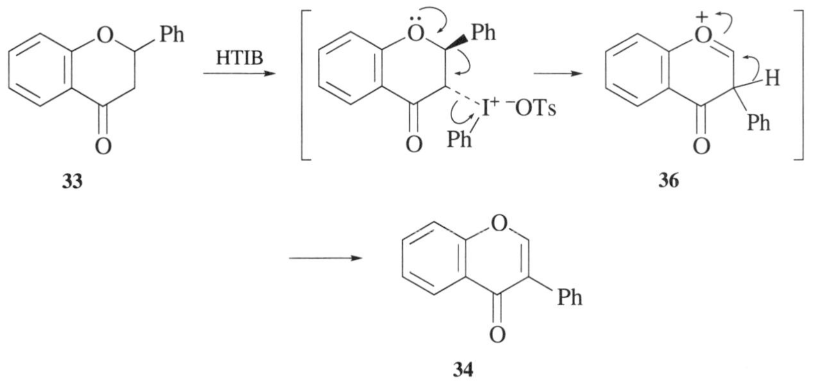
HTIB
Ph
I⁺ ⁻OTs
H
33
36
34

(Eq. 20)

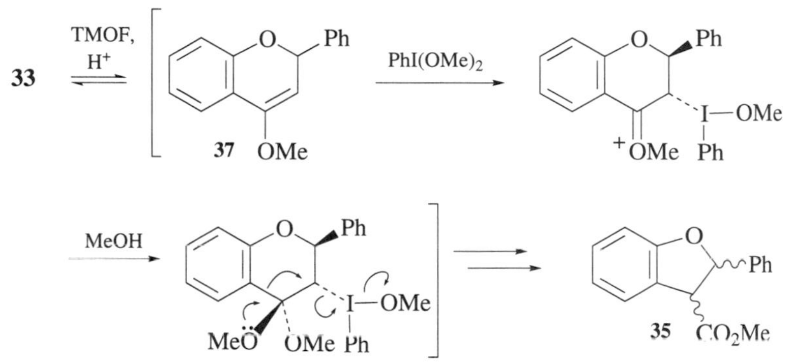
TMOF,
H⁺
33
PhI(OMe)₂
37
OMe
MeOH
35
CO₂Me

(Eq. 21)

SCOPE AND LIMITATIONS

α-Functionalization of Carbonyl Compounds Under Basic Conditions

α-Hydroxylation of Ketones via α-Hydroxydimethylacetal Formation. The oxidation of acetophenones with iodosobenzene or IBD in methanolic sodium or potassium hydroxide provides an efficient route to α-hydroxydimethylacetals. This method offers an indirect approach for the α-hydroxylation of ketones, since the dimethylacetal can be hydrolyzed under acidic conditions in overall yields of 40–70% (Eq. 22).[7,23]

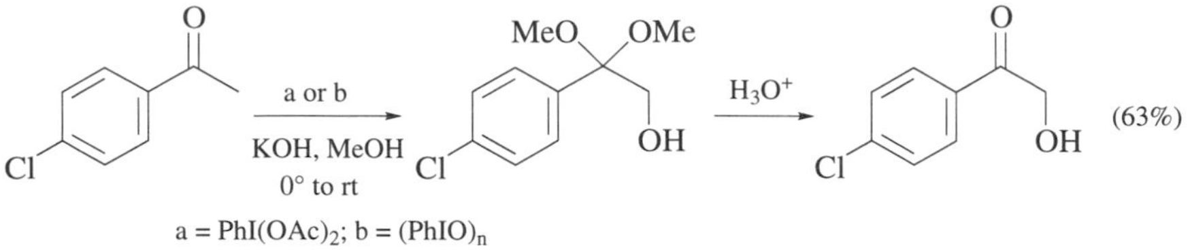

(Eq. 22)

Under similar conditions, *o*-iodosylbenzoic acid also effects conversion of ketones to α-hydroxydimethylacetals in high yields with the advantage that the reduction product *o*-iodobenzoic acid is removed in the workup by a base extraction.[27] Application of this procedure to a series of cyclic ketones affords products in yields ranging from 61–83%. Eq. 23 illustrates this reaction with cyclohexanone. In the case of norbornanone, the reaction is stereoselective and gives 3,3-dimethoxy-2-*endo*-norbornanol (Eq. 24).

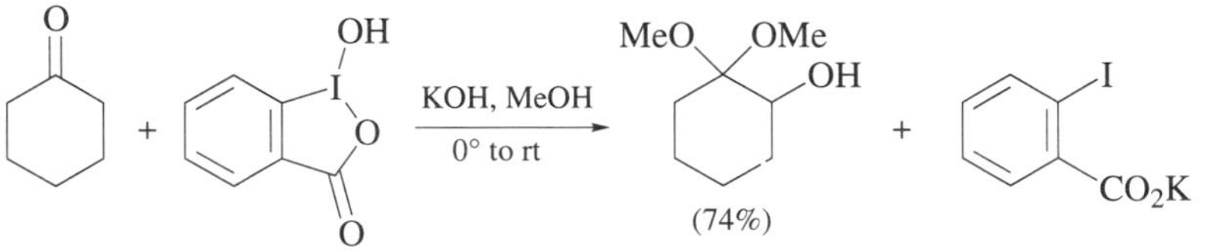

(Eq. 23)

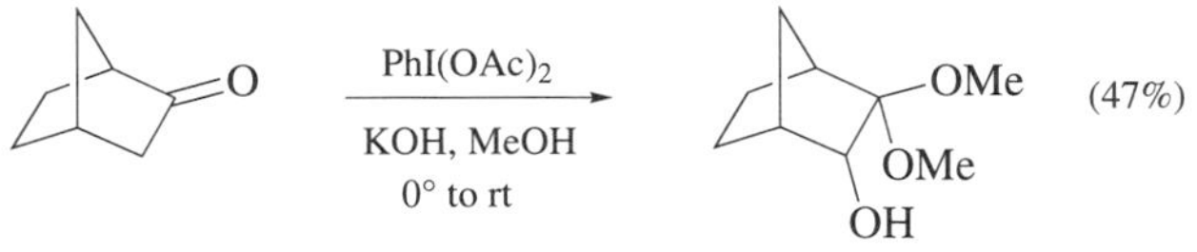

(Eq. 24)

This procedure can be used with ketones containing other potentially oxidizable functional groups such as secondary and tertiary amines, thioethers, and alkenes (Eq. 25).[29,30] In addition, various heterocycles containing both nitrogen

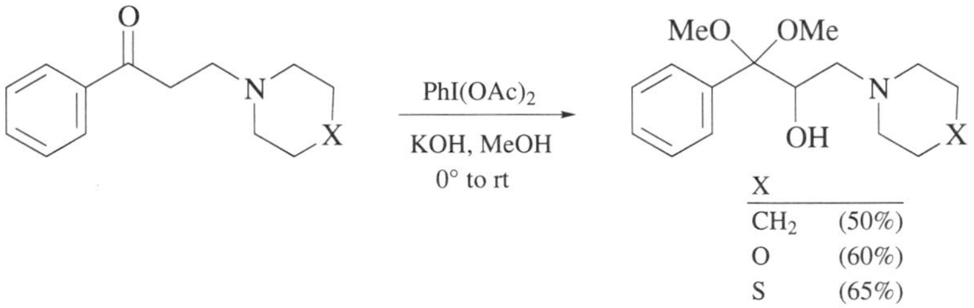

(Eq. 25)

and sulfur have also been transformed into the corresponding α-hydroxydimethylacetals (Eqs. 26 and 27).[31,32] The regio- and stereoselective conversion of **38** to **39** finds an interesting application in the stereoselective synthesis of analog **40** of the antitumor agent (±)-cephalotaxine (Eq. 27).[32] However, 2-acetylthiophene gives a complex mixture of products under these conditions.[31a]

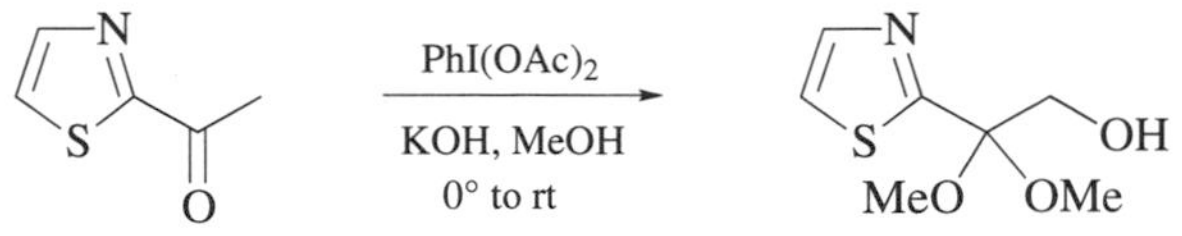

N
S
O
PhI(OAc)2
KOH, MeOH
0° to rt
N
S
OH
MeO
OMe

(Eq. 26)

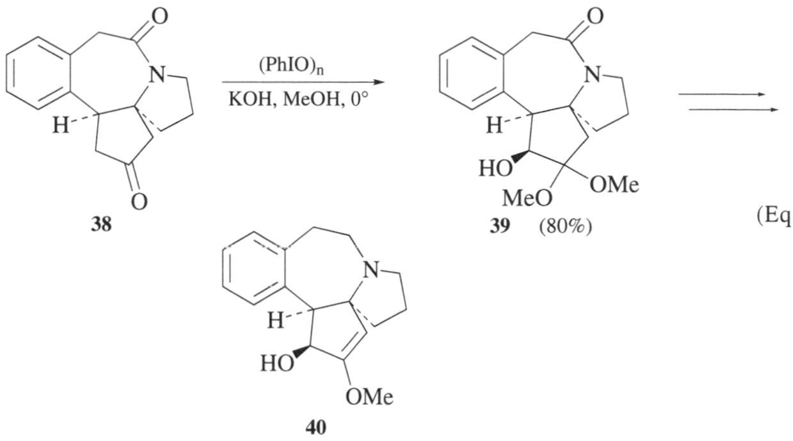

O
N
H
O
38
(PhIO)n
KOH, MeOH, 0°
O
N
H
HO
OMe
MeO
39 (80%)
N
H
HO
OMe
40

(Eq. 27)

Steroidal ketones such as **41** that contain a keto group in the side chain are also converted to the α-hydroxydimethylacetal **42** without oxidation of the secondary alcohol group present at C3. This approach is useful for the construction of the dihydroxyacetone side chain at the 17β-position in the pregnane series (Eq. 28).[33] Other steroids containing a ketone group at C3 in the A-ring do not give the dimethylacetal product, presumably due to steric constraints in the steroidal ring system.[25,34]

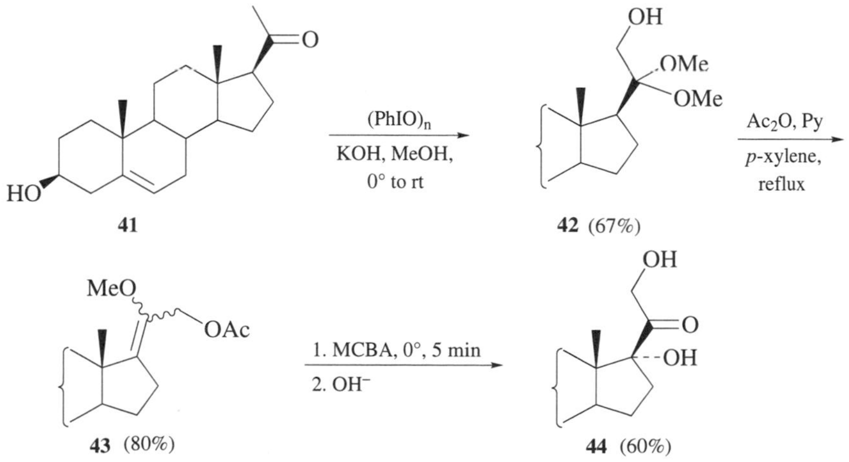

O
HO
41
(PhIO)n
KOH, MeOH,
0° to rt
OH
OMe
OMe
42 (67%)
Ac2O, Py
p-xylene,
reflux
MeO
OAc
43 (80%)
1. MCBA, 0°, 5 min
2. OH−
OH
O
OH
44 (60%)

(Eq. 28)

The control of stereochemistry of α-hydroxydimethylacetal formation has been investigated with the prochiral ketone 1-tetralone, which forms the chiral substrate **45** upon complexation with $Cr(CO)_3$ (Eq. 29). Application of the hyper-

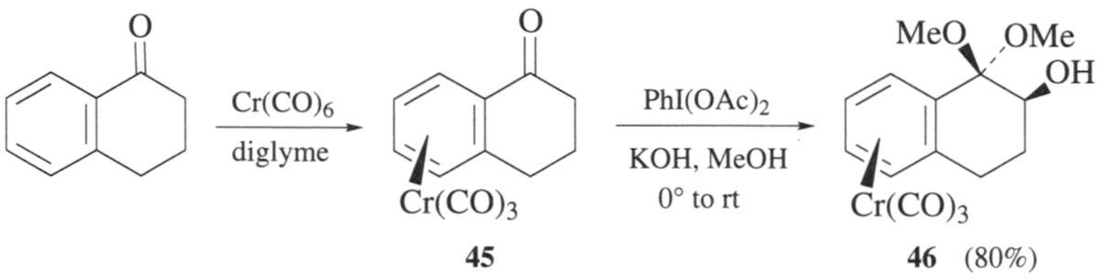
O
Cr(CO)6
diglyme
Cr(CO)3
45
PhI(OAc)2
KOH, MeOH
0° to rt
MeO
OMe
OH
Cr(CO)3
46 (80%)

(Eq. 29)

iodination reaction yields only product **46** in which the hydroxy group is syn with respect to the $Cr(CO)_3$ tripod (Eq. 29).[35,36] These stereochemical results can be understood in terms of addition of $PhI(OMe)_2$ *anti* to the $Cr(CO)_3$ group due to a steric effect. The relative configuration of the C–I bond in the molecule determines the final stereochemical result, and this is shown in Eq. 30.

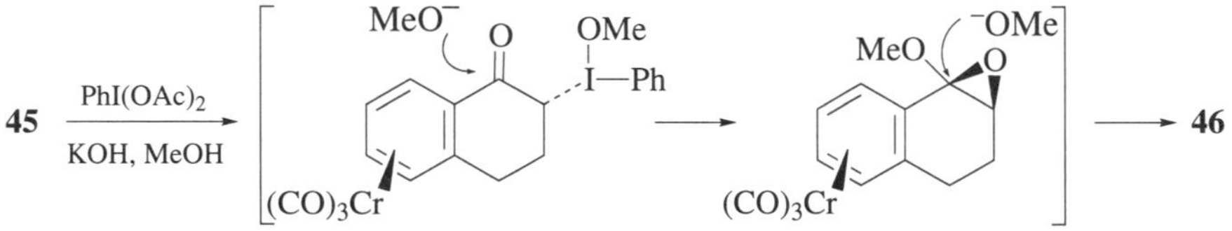
45
PhI(OAc)2
KOH, MeOH
MeO−
O
OMe
I—Ph
(CO)3Cr
MeO
−OMe
O
(CO)3Cr
46

(Eq. 30)

Since **45** can be resolved, application of the hyperiodination reaction offers a way of making a pure enantiomer of **47** of known configuration (*R*) after disengagement of the metal ligand from complex (−)-**46** (Eq. 31).

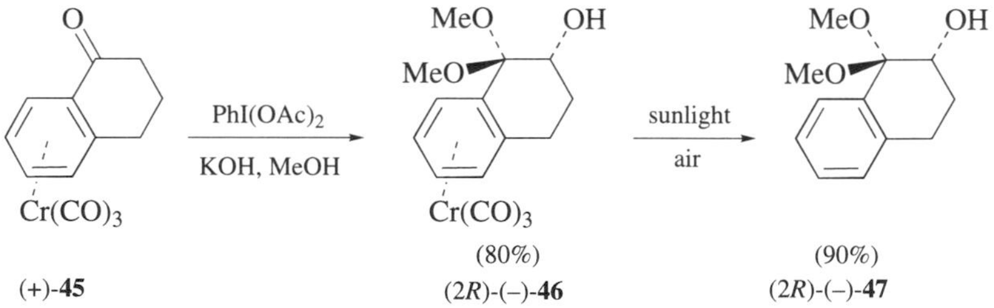
O
Cr(CO)3
(+)-45
PhI(OAc)2
KOH, MeOH
MeO
OH
MeO
Cr(CO)3
(80%)
(2R)-(−)-46
sunlight
air
MeO
OH
MeO
(90%)
(2R)-(−)-47

(Eq. 31)

Although oxidation with IBD-KOH/MeOH constitutes a quite general approach for α-hydroxylation of ketones via α-hydroxydimethylacetal formation, it still suffers from certain limitations. Hydrolysis of some acetals (e.g., that of 2-acetylpyridine) is difficult, and functional groups incompatible with aqueous base present a limitation.

α-Hydroxylation of α,β-Unsaturated Carbonyl Compounds. α,β-Unsaturated ketones such as chalcones (Eq. 32), chromones, and flavones (Eq. 33), which do not contain enolizable ketonic groups, also undergo oxidation with IBD-KOH/MeOH under standard conditions.[37] Interestingly, the reaction occurs regiospecifically and stereospecifically to give the corresponding α-hydroxy-β-methoxydimethylacetals **49a** and **49b**, which can be hydrolyzed with hydrochloric acid to give C-3 hydroxylated products **50a** and **50b** (Eq. 33).[38,39]

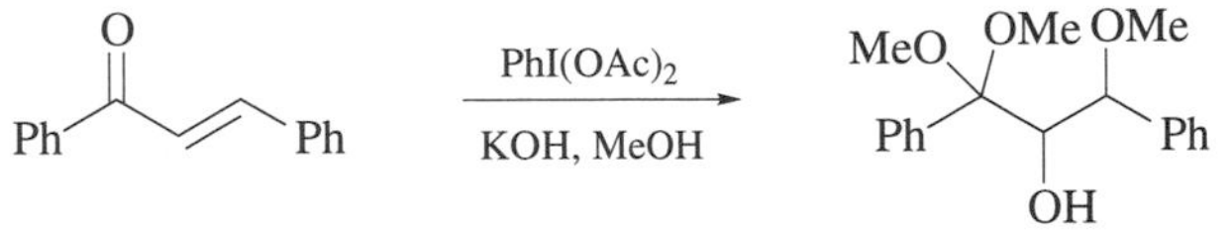
O
Ph
Ph
PhI(OAc)2
KOH, MeOH
MeO
OMe
OMe
Ph
Ph
OH

(60%) (Eq. 32)

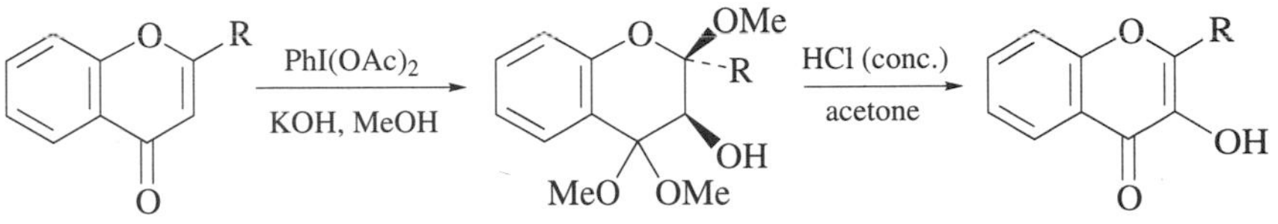
O
R
O
PhI(OAc)2
KOH, MeOH
OMe
R
OH
MeO
OMe
HCl (conc.)
acetone
OH

48a R = H
48b R = Ph

49a (53%)
49b (67%)

50a (90%)
50b (85%)

(Eq. 33)

The stereochemistry of intermediates **49a** and **49b** in each case results from initial conjugate addition of methoxide ion on the chromone **48**. Ensuing attack of the so-formed enolate **51** upon $PhI(OMe)_2$ occurs in an *anti* manner to methoxy group because of steric interaction. Sequential addition of MeO^- to the carbonyl group of **52** gives **53**, and intramolecular reductive elimination of C_6H_5I then occurs with inversion of configuration, **53** → **54**. The reaction is completed by a second addition of methoxide ion to the oxirane ring (Eq. 34).

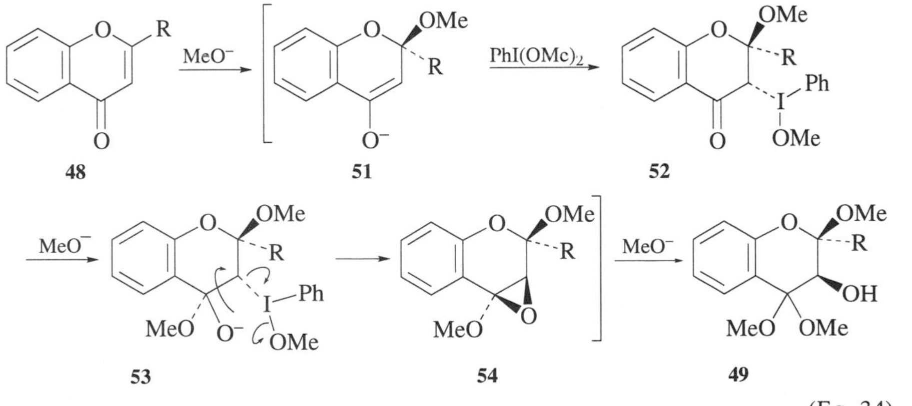
O
R
O
MeO⁻
OMe
R
O⁻
PhI(OMe)2
OMe
R
I
Ph
OMe
48
51
52
MeO⁻
OMe
R
I
Ph
MeO
O⁻
OMe
OMe
R
MeO
O
MeO⁻
OMe
R
OH
MeO
OMe
53
54
49

(Eq. 34)

The oxidation of enolizable β,β-disubstituted-α,β-unsaturated carbonyl compounds, however, yields α-hydroxylated ketones (Eq. 35).[40]

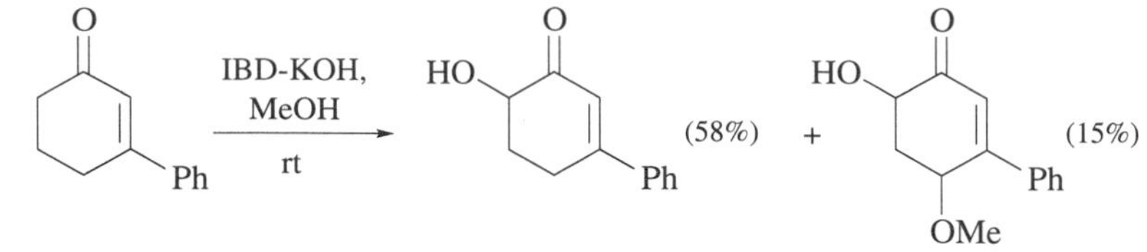

(Eq. 35)

In contrast, reaction of steroidal β,β-disubstituted-α,β-unsaturated ketones, for example, androst-4-ene-3,17-dione, with *o*-iodosylbenzoic acid in methanolic KOH gives the C4 and C6 methoxylated products along with the dehydrogenated product androst-4,6-dien-3,17-dione (Eq. 36). 17-β-Hydroxy-4-androsten-3-one also gives similar results.[41,42]

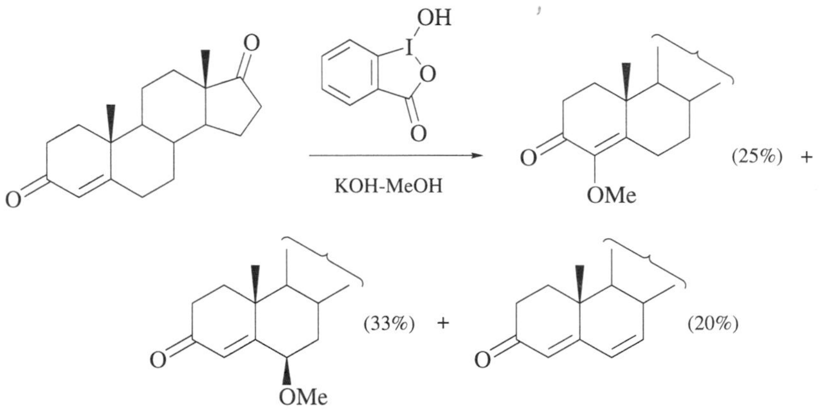

(Eq. 36)

The oxidation of α,β-unsaturated ketones becomes of special interest when an *o*-hydroxy group is present in the aroyl ring of chalcones. For example, oxidation of *o*-hydroxychalcone (**55**) affords the dimethylacetal **56**, which on hydrolysis under mild acidic conditions gives *cis*-3-hydroxyflavanone (**57**) (Eq. 37).[43,44] Under more vigorous hydrolytic conditions, isomerization accompanies hydrolysis of the acetal with the formation of the *trans* isomer **58**. The stereochemistry of acetal **56** is in agreement with the reaction pathway presented in Eq. 34 with the phenolic hydroxy group acting as an intramolecular Michael donor.

The reaction is quite general in that not only a variety of substituted *o*-hydroxychalcones are transformed to *cis*-3-hydroxyflavanone dimethylacetals but also 2-furyl analogs are smoothly converted to the desired products.[45] However, when the *o*-hydroxychalcone (**59**) contains a methoxy group at the C4 position of the phenol ring, the yield of the expected dimethyl acetal **61** is low and the major product is the flavone derivative **60** (Eq. 38).[44]

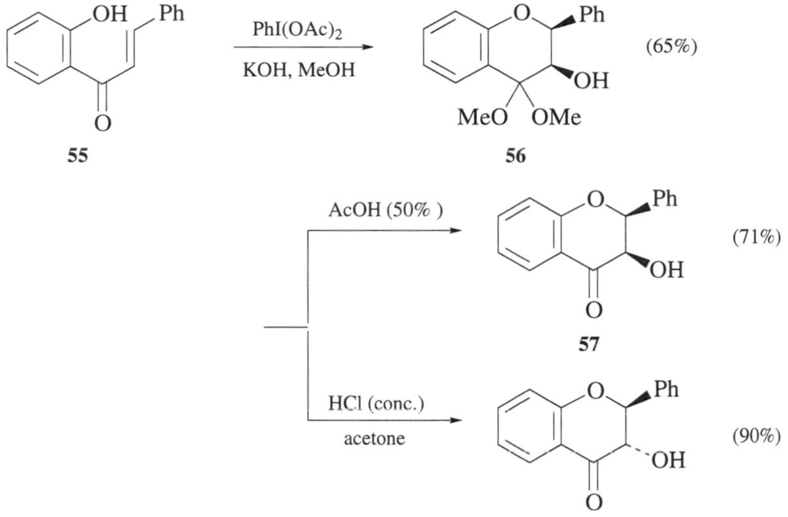

OH
Ph
O
$PhI(OAc)_2$
KOH, MeOH
O
Ph
OH
MeO
OMe
(65%)
AcOH (50%)
O
Ph
OH
O
(71%)
HCl (conc.)
acetone
O
Ph
OH
O
(90%)

55 56 57 58

(Eq. 37)

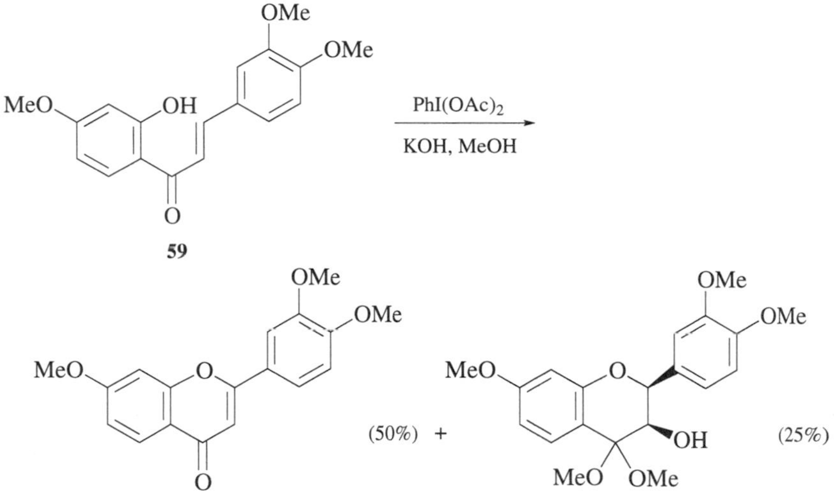

OMe
OMe
MeO
OH
O
$PhI(OAc)_2$
KOH, MeOH
OMe
OMe
MeO
O
O
(50%) +
OMe
OMe
MeO
O
OH
MeO
OMe
(25%)

59 60 61

(Eq. 38)

Arylation. Carbanions generated from various carbonyl compounds are arylated at their α or γ positions with diaryl iodonium salts under suitable conditions. Carbonyl compounds that have been α-arylated include monoketones (Eq. 39),[46] esters (Eq. 40),[47] β-diketones (Eq. 41),[48] malonates, cyclic malonates (Eq. 42),[47,49] and β-keto esters (Eq. 43).[47,50]

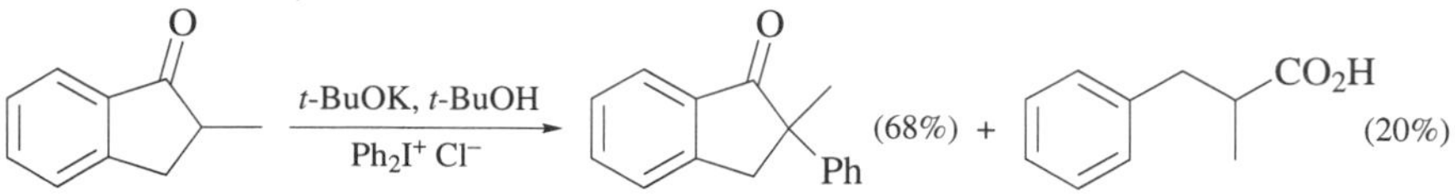

(Eq. 39)

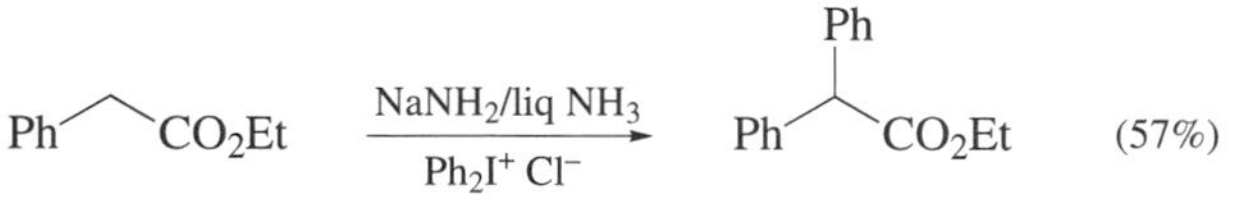

(Eq. 40)

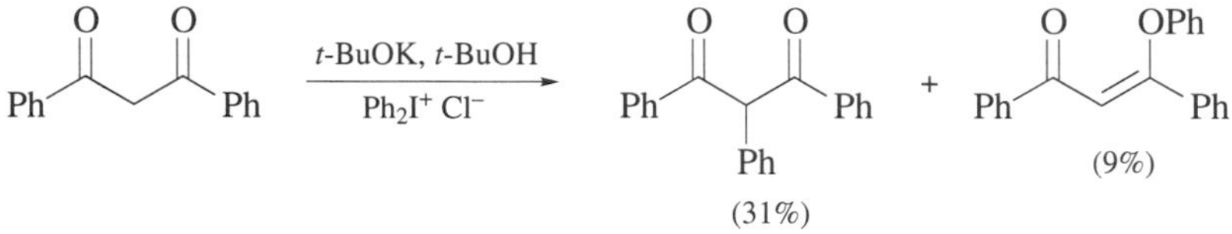

(Eq. 41)

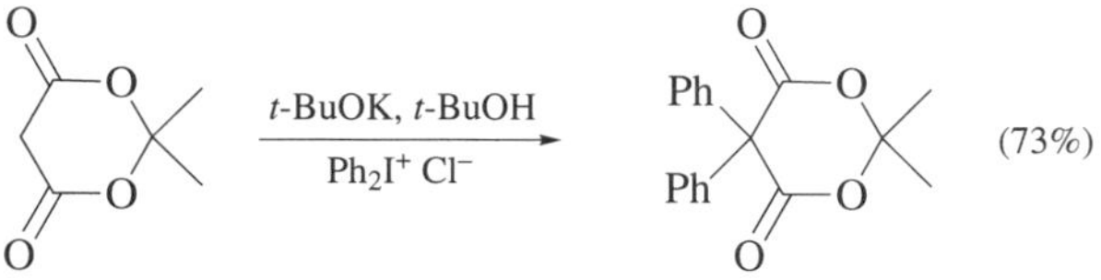

(Eq. 42)

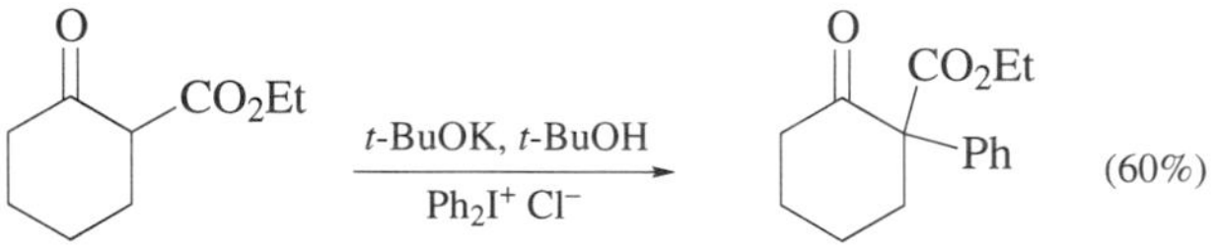

(Eq. 43)

Although arylation occurs fairly cleanly, the structure of the diaryliodonium salt and the reaction conditions employed can play important roles in these reactions. For example, while the anion generated from 2-phenyl-1,3-indandione (**62**) and sodium *tert*-butoxide in *tert*-butyl alcohol on treatment with diphenyliodonium chloride gives 52% of 2,2-diphenyl-1,3-indandione (**63**), it gives with dimesityliodonium chloride [$(Mes)_2I^+Cl^-$] only 23% of the analogous 2-mesityl-2-phenyl-1,3-indandione (**64**) and a small amount of 2-(*o*-mesitylphenyl)-1,3-indandione (**65**). The major product in this reaction is the dehydrodimer **66** (Eq. 44).[51] On the other hand, when reactions of β-diketones with diaryliodonium chlorides are carried out in the presence of 2 equivalents of sodium amide in liquid ammonia, γ-arylated products are formed regioselectively (Eq. 45). The γ position of the dianion generated under these conditions is more reactive than the α position.[52]

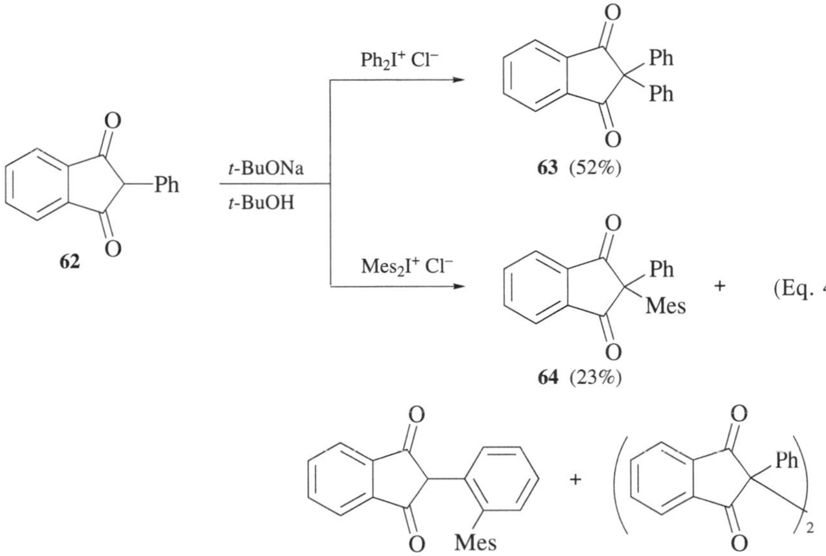

O
O
Ph
$Ph_2I^+ Cl^-$
t-BuONa
t-BuOH
$Mes_2I^+ Cl^-$
Ph
Ph
Ph
Mes
+
Mes
+
Ph
2

62 63 (52%) 64 (23%) 65 66 (40%)

(Eq. 44)

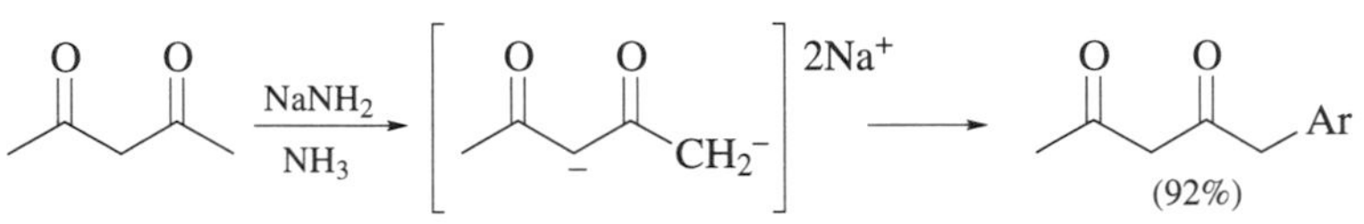

O O
NaNH2
NH3
2Na+
CH2−
Ar

(92%)

(Eq. 45)

α-Hydroxylation and α-Alkoxylation of Esters. Iodobenzene diacetate under basic conditions in an appropriate solvent can be employed for effective α-hydroxylation and α-alkoxylation of several esters (Eq. 46).[53] This oxidative

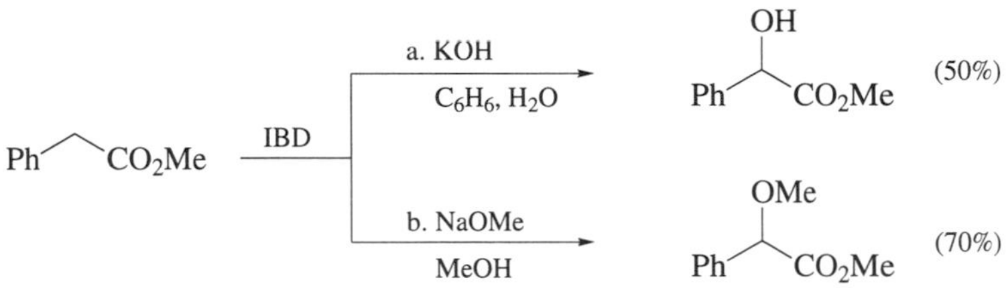

Ph
CO2Me
IBD
a. KOH
C6H6, H2O
b. NaOMe
MeOH
OH
OMe
Ph
CO2Me
(50%)
(70%)

(Eq. 46)

approach can be extended to the α-methoxylation of dimethyl homophthalate.[54] The oxidation product, methyl 2-(2-methoxycarbonylphenyl)acetate, on hydrolysis followed by dehydrative cyclization gives 4-methoxyhomophthalic anhydride (Eq. 47), which is a key intermediate in the synthesis of anthracyclinones. α-Arylation of esters with diaryliodonium salts is covered in the previous section (Eq. 40).

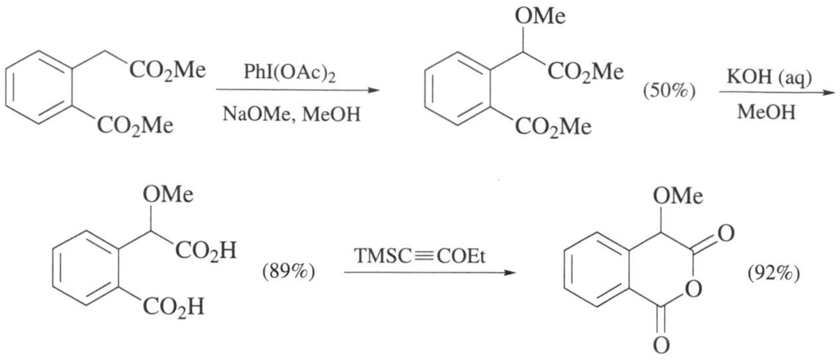

CO2Me
CO2Me
PhI(OAc)2
NaOMe, MeOH
OMe
CO2Me
CO2Me
(50%)
KOH (aq)
MeOH
OMe
CO2H
CO2H
(89%)
TMSC≡COEt
OMe
O
O
O
(92%)

(Eq. 47)

Under Neutral and Acidic Conditions

α-Acetoxylation and α-Hydroxylation. These reactions are complementary to the IBD-KOH/MeOH reactions in that base is avoided. Several *p*-substituted acetophenones and *β*-diketones undergo *α*-acetoxylation when treated with one equivalent of IBD in a mixed solvent Ac_2O/AcOH using sulfuric acid as a catalyst (Eq. 48).[55] The acetoxylation also occurs when acetic acid and water are used

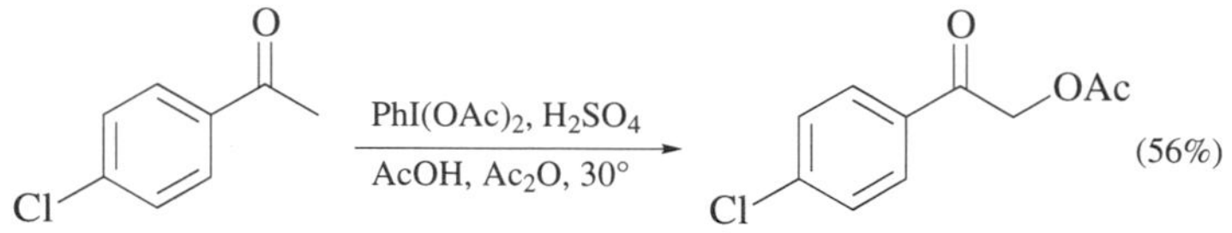

O
Cl
PhI(OAc)2, H2SO4
AcOH, Ac2O, 30°
O
OAc
Cl
(56%)

(Eq. 48)

in the absence of sulfuric acid. However, during the oxidation of various *β*-diketones with 4 equivalents of IBD in AcOH/H_2O at 80–100°, carbon–carbon bond cleavage occurs (Eq. 49).[56]

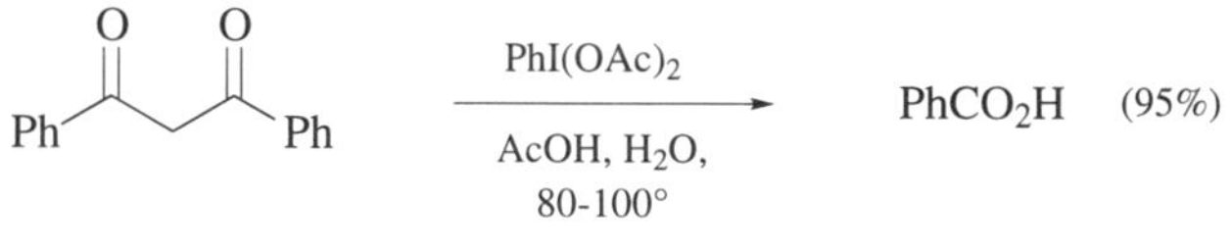

O O
Ph Ph
PhI(OAc)2
AcOH, H2O,
80-100°
PhCO2H (95%)

(Eq. 49)

A simple and direct route for the *α*-hydroxylation of a large variety of ketones including aliphatic, aromatic, and heteroaromatic ketones involves the use of bis(trifluoroacetoxy)iodobenzene and trifluoroacetic acid in acetonitrile/H_2O (Eq. 50).[57] The procedure gives good yields with methyl ketones, but *α*-methylene ketones are hydroxylated in lower yields.

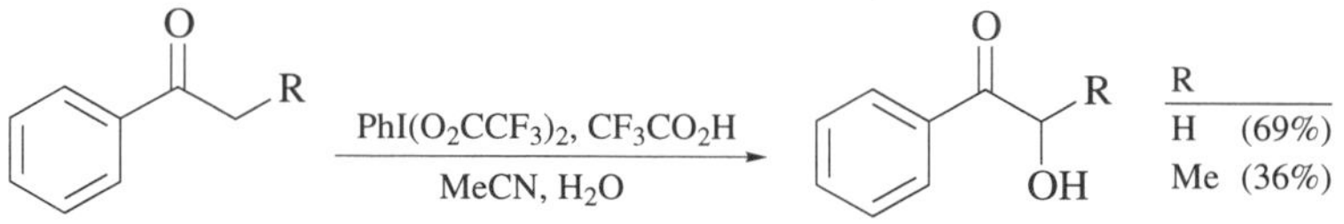

O
R
PhI(O2CCF3)2, CF3CO2H
MeCN, H2O
O
R
OH
R
H (69%)
Me (36%)

(Eq. 50)

The oxidation of 2-methylcyclohexanone (**67**) provides 2-methylcyclohex-2-enone (**68**) as the major product; the expected α-hydroxyketone **69** is obtained in a minor amount (Eq. 51). The formation of the other regioisomer of the α-hydroxyketone is not observed.

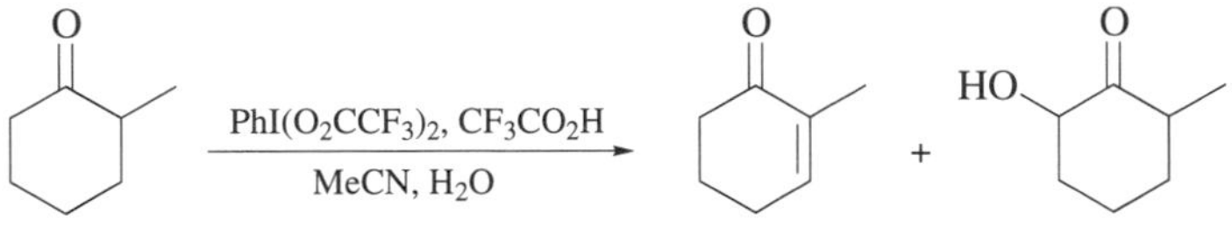
O
PhI(O2CCF3)2, CF3CO2H
MeCN, H2O
O
+
HO
O

(Eq. 51)

67 **68** (55%) **69** (23%)

α-Sulfonyloxylation. The direct introduction of the α-tosyloxy (**70** → **71**) or mesyloxy groups (**70** → **72**) into ketones and β-dicarbonyl compounds has been effected with HTIB[58] or HMIB,[59,60] respectively, in MeCN or CH_2Cl_2. The reagents HTIB and HMIB can be generated in situ from IBD-*p*-TsOH and iodosobenzene-methanesulfonic acid, respectively.[61] A similar approach starting from hydroxy[(+)-(10-camphorsulfonyl)oxy]iodobenzene (HCIB) has been used to prepare α-(10-camphorsulfonyl)oxyketones **73** (Eq. 52).[62]

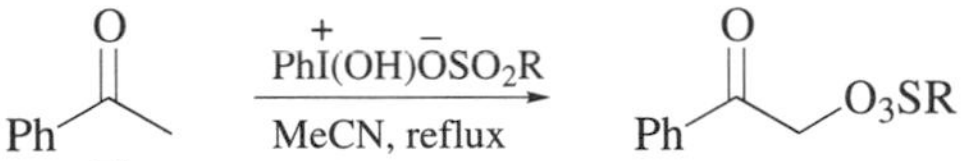
O
Ph
PhI(OH)OSO2R
MeCN, reflux
O
Ph
O3SR

70

71 R = *p*-MeC_6H_4 (73%)
72 R = Me (62%)
73 R = (+)-10-camphoryl (95%)

(Eq. 52)

In the case of benzoylacetone, the crude product is a 3:1 mixture of diastereomers **74** (Eq. 53). However, an attempt to separate the diastereomers by column

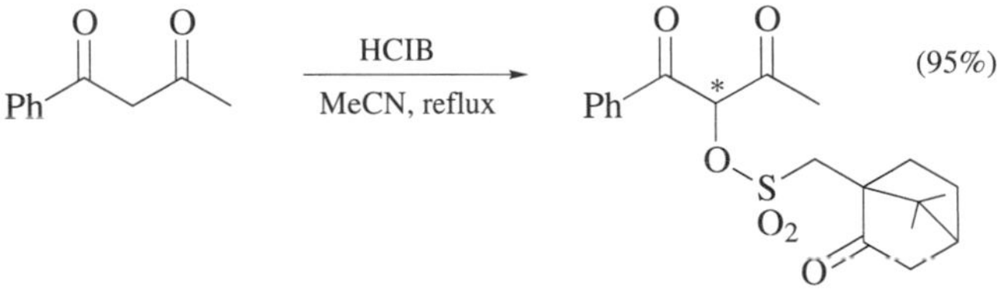
O O
Ph
HCIB
MeCN, reflux
O O
Ph
*
O
S
O2
O
(95%)

74

(Eq. 53)

chromatography (SiO_2) resulted in isomerization, thereby yielding a nearly 1:1 mixture of the two diastereomers. An interesting feature of this reagent is the steric bulk of the camphorsulfonate ligand, which allows the regioselective formation of the less hindered product. Thus the sterically differentiated substrate 4-methyl-2-pentanone apparently gives only the C-1 camphorsulfonate **75** (Eq. 54).[62]

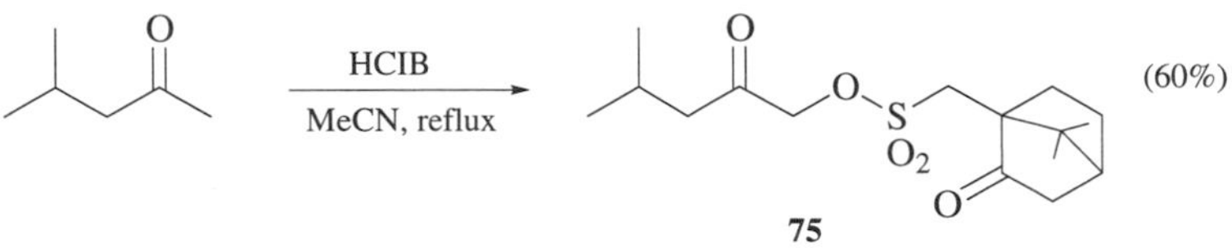
O
HCIB
MeCN, reflux
O
O
S
O2
O
(60%)

75

(Eq. 54)

α-Functionalized Carbonyl Compounds via α-Tosyloxylation. Since α-tosyloxyketones undergo nucleophilic substitution reactions analogous to α-haloketones, various α-substituted ketones can be synthesized via α-tosyloxyketones derived from the hypervalent iodine oxidation. It is generally not necessary to isolate the α-tosyloxyketones, and this approach provides a one-pot synthesis of α-functionalized ketones (Eq. 55).

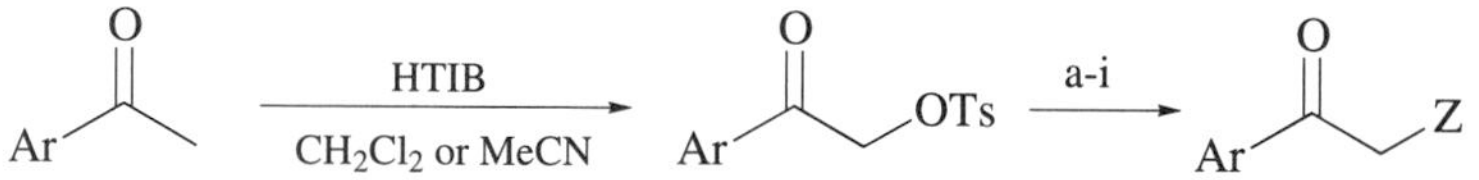
HTIB
CH_2Cl_2 or MeCN
a-i
Ar
OTs
Z

Conditions	Z	Ref.
a. PPh_3	P^+Ph_3 ^-OTs	63
b. Me_2S	S^+Me_2 ^-OTs	63
c. $ArNH_2$	NHAr	64
d. 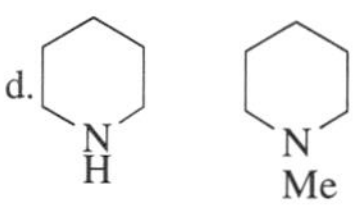		63

Conditions	Z	Ref.
e. KSCN	SCN	65
f. $NaNO_2$, H_2O	OH	66
g. ArOH, K_2CO_3, EtOH	OAr	61
h. $ArCO_2H$, Et_3N	O_2CAr	61
i. KOH, MeOH, H_3O^+	OH	67

(Eq. 55)

α-Phosphoryloxylation and α-Phosphonyloxylation. The α-phosphoryloxylation of ketones can be achieved by using hydroxy[bis(phenoxy)phosphoryloxy]-iodobenzene (**76**) in acetonitrile or dichloromethane (Eq. 57). The reaction proceeds through a pathway similar to α-sulfonyloxylation, but in this instance the nucleophile is $^-OP(O)(OPh)_2$.[68] Reagent **76** is obtained by the treatment of IBD with diphenyl phosphate and water in acetonitrile (Eq. 56).

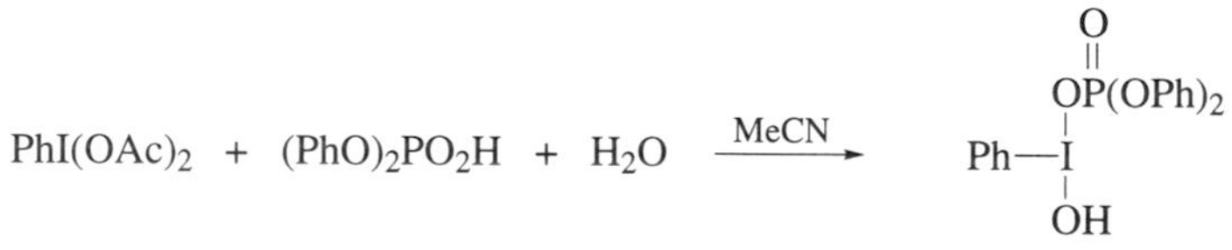
PhI(OAc)2 + (PhO)2PO2H + H2O
MeCN
OP(OPh)2
Ph—I
OH

76 (90%)

(Eq. 56)

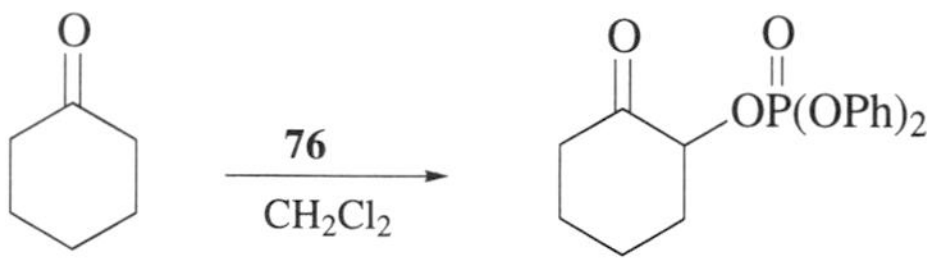
76
CH2Cl2
OP(OPh)2

(62%) (Eq. 57)

The reaction of 4-pentenoic acids with **76**, however, gives 5-bis(phenoxy)-phosphoryloxy-4-pentanolactone (Eq. 58).

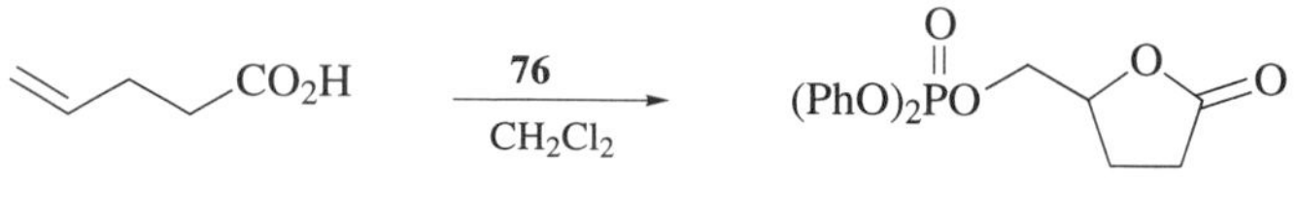
CO2H
76
CH2Cl2
(PhO)2PO

(55%) (Eq. 58)

In a related study, α-phosphonyloxylation of several ketones is effected by using hydroxy[(phenoxy)(methyl)phosphonyloxy]iodobenzene. The latter is prepared by the reaction of methyl phenyloxyphosphonic acid with iodosobenzene in acetonitrile (Eq. 59).[69]

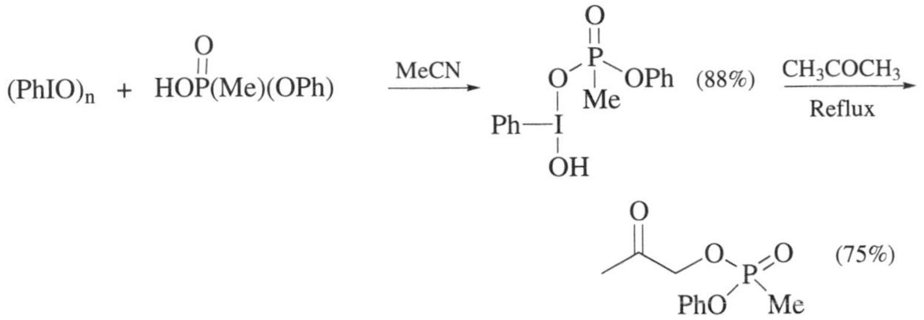

(Eq. 59)

Under Various Conditions via Silyl Enol Ethers and Ketene Silyl Acetals
Hypervalent iodine oxidation of silyl enol ethers under neutral or Lewis acid conditions provides an alternative approach for α-hydroxylation of ketones. Limitations encountered under the previous methods are overcome by using this alternative. The most significant aspect of this approach is its versatility in functionalizing various carbonyl compounds that include ketones, esters, and lactones at the α position with various substituents.

α-Hydroxylation. Treatment of silyl enol ethers derived from various ketones with iodosobenzene and water under different conditions affords the corresponding α-hydroxyketones (Eq. 60).[70] Boron trifluoride etherate has been

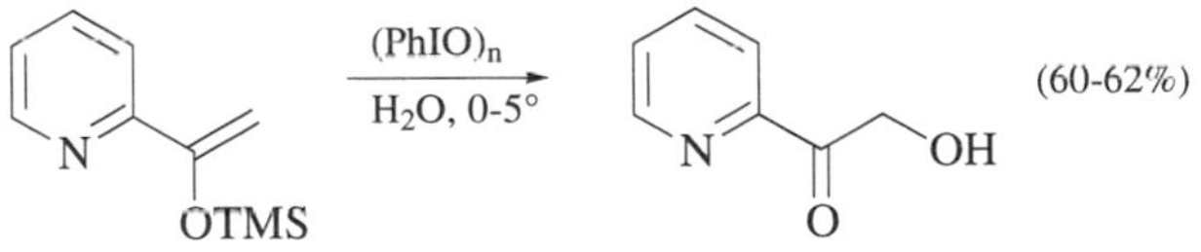

(Eq. 60)

employed to effect these α-hydroxylations,[71] but it is not necessary. The method of α-hydroxylation with iodosobenzene in water gives excellent results in the reaction of the silyl enol ether of *tert*-butyl methyl ketone (Eq. 61).

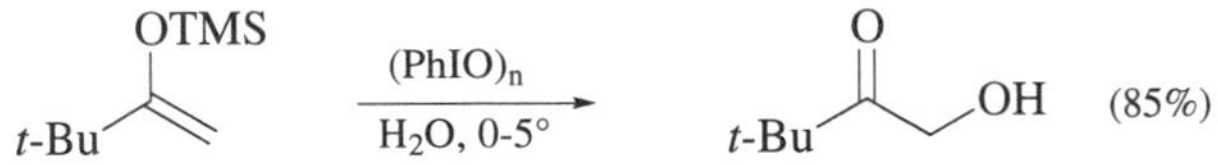

(Eq. 61)

However, ethyl trimethylsilyl phenylketene acetal derived from ethyl phenylacetate gives the α-hydroxylated product in only 20% yield. The alternative procedure using iodosobenzene, boron trifluoride etherate, and water in dichloromethane at $-40°$ successfully converts silyl enol ethers of esters (Eq. 62) to α-hydroxylated derivatives, but lactones do not give acceptable yields of α-hydroxylactones (Eq. 63).[72]

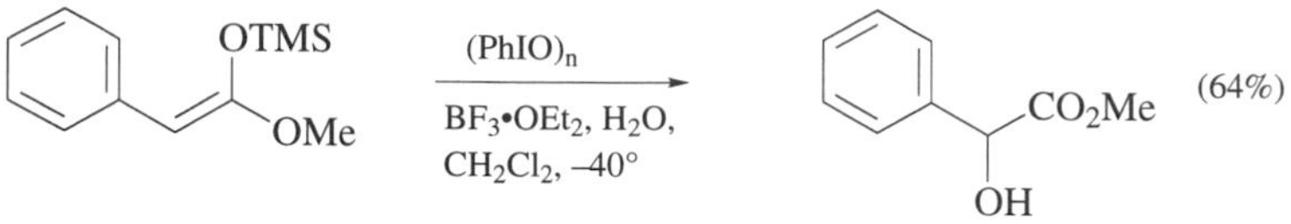

(Eq. 62)

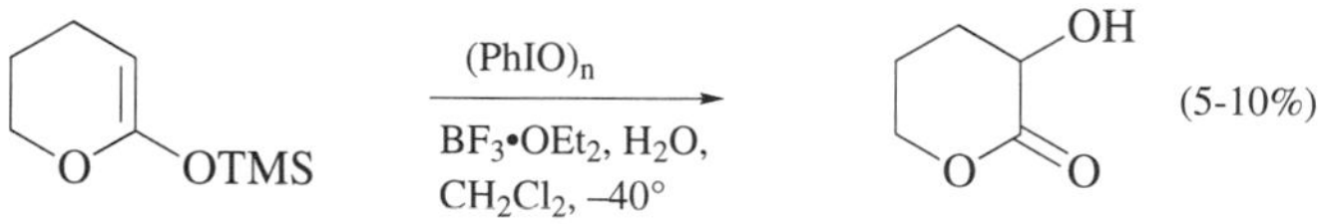

(Eq. 63)

α-Alkoxylation. Silyl enol ethers derived from ketones (Eqs. 64 and 65), and trimethylsilyl ketene acetals derived from esters (Eq. 66) and lactones (Eq. 67)

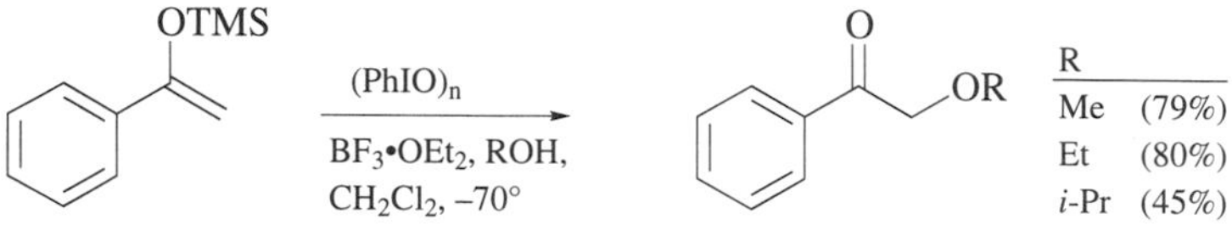

(Eq. 64)

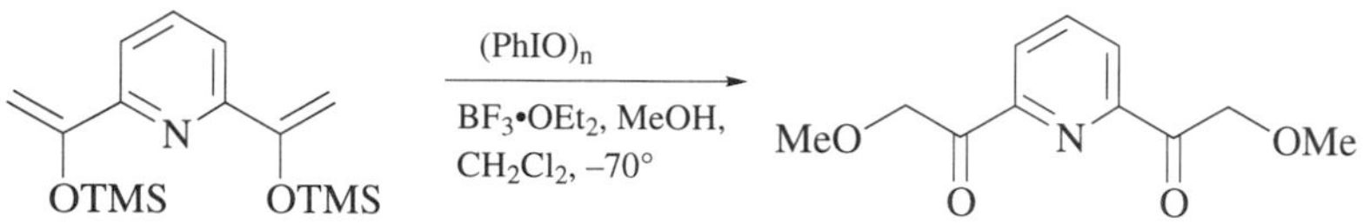

77 **78** (71%)

(Eq. 65)

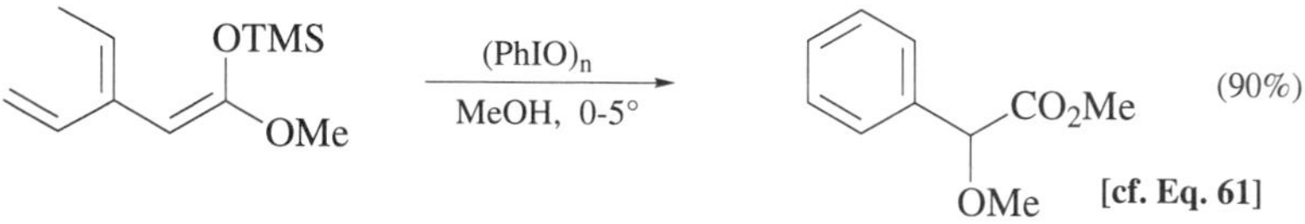

(Eq. 66)

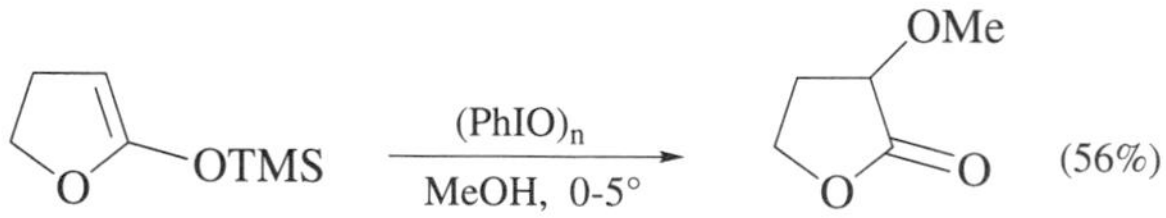

(Eq. 67)

are smoothly converted to the corresponding α-alkoxylated carbonyl compounds by iodosobenzene and an alcohol under suitable conditions.[72,73] The oxidation of bis-silyl enol ether **77** derived from 2,6-diacetylpyridine yields 2,6-di(methoxy-acetyl)pyridine (**78**) (Eq. 65). In this reaction, the ratio of **77**, iodosobenzene, boron trifluoride etherate, and methanol is 1:2:4:10.

In the alkoxylations of ketones, neither α-hydroxydimethylacetals nor rearranged products such as 2-arylalkanoates (as observed in the hypervalent iodine oxidation of alkyl aryl ketones in MeOH or trimethyl orthoformate) are formed under these reaction conditions.

α-Sulfonyloxylation. Silyl enol ethers of ketones **79** can be converted into tosyloxyketones **80** and mesyloxyketones **81** by using HTIB and HMIB, respectively (Eq. 68). Reactions of carbonyl compounds with HTIB, HMIB, or HCIB

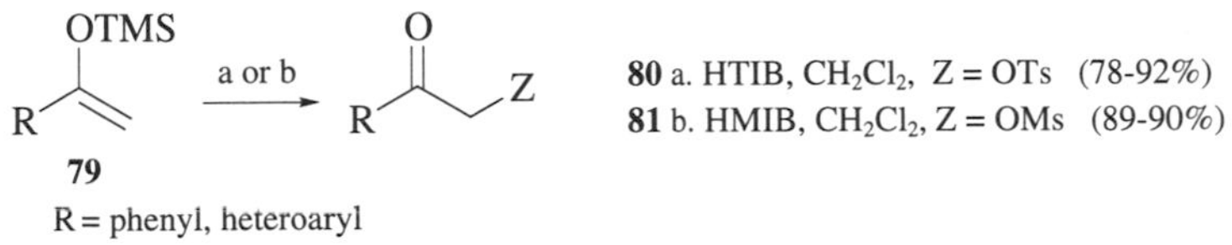
OTMS
a or b
O
Z
R
R
79
R = phenyl, heteroaryl
80 a. HTIB, CH2Cl2, Z = OTs (78-92%)
81 b. HMIB, CH2Cl2, Z = OMs (89-90%)

(Eq. 68)

provide a direct approach for the α-sulfonyloxylation of ketones in dichloromethane or acetonitrile. However, the major drawback of these reactions is that they proceed with relatively low regioselectivity. Alternatively, sulfonyloxylation involving silyl enol ethers is regioselective. For example, 2-methyl-6-tosyloxycyclohexanone can be prepared regioselectively from 1-trimethylsilyloxy-6-methylcyclohex-1-ene with HTIB in dichloromethane (Eq. 69). This approach is also used for the α-sulfonyloxylation of esters (Eq. 70) and lactones (Eq. 71).[74]

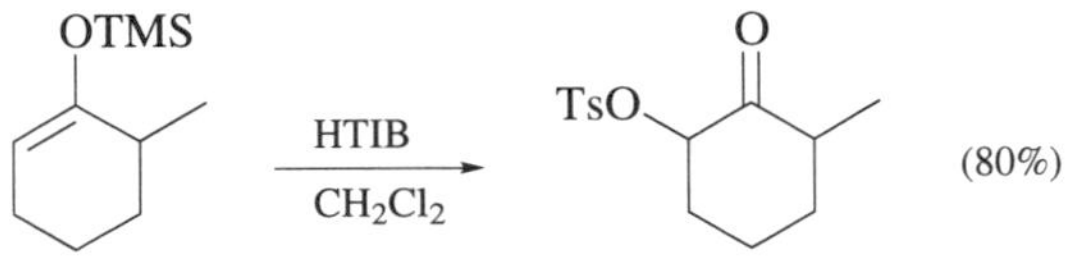
OTMS
HTIB
CH2Cl2
TsO
O
(80%)

(Eq. 69)

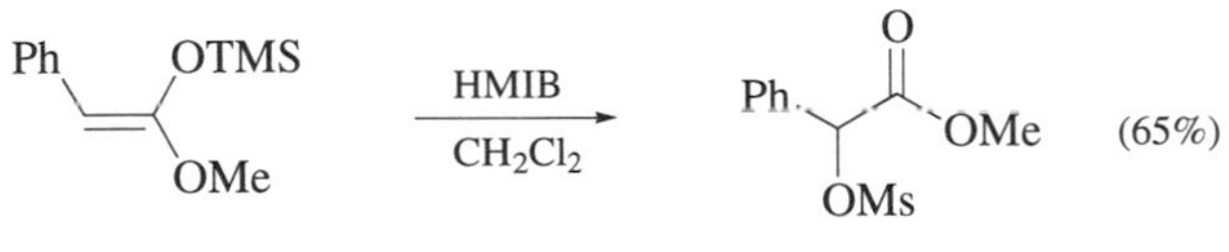
Ph
OTMS
OMe
HMIB
CH2Cl2
Ph
O
OMe
OMs
(65%)

(Eq. 70)

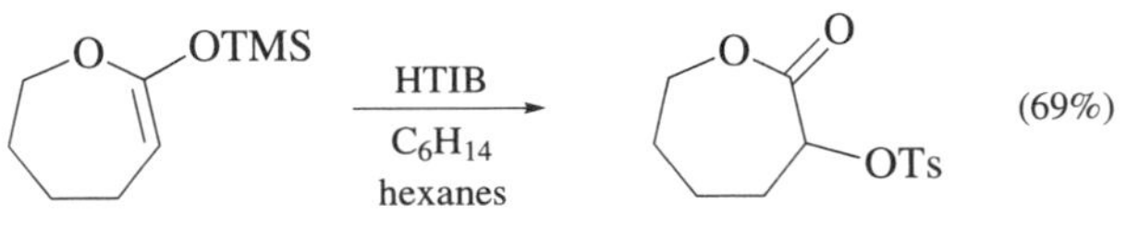
O
OTMS
HTIB
C6H14
hexanes
O
O
OTs
(69%)

(Eq. 71)

This reaction also yields α-(trifluoromethanesulfonyloxy)ketones when silyl enol ethers are treated with iodosobenzene in the presence of trimethylsilyl trifluoromethanesulfonate (**82**) in dichloromethane (Eq. 72).[75] It seems likely that the active reagent in this transformation is trimethylsilyloxy(trifluoromethanesulfonyloxy)iodobenzene (**83**) generated from the reaction between iodosobenzene and **82** (Eq. 73).

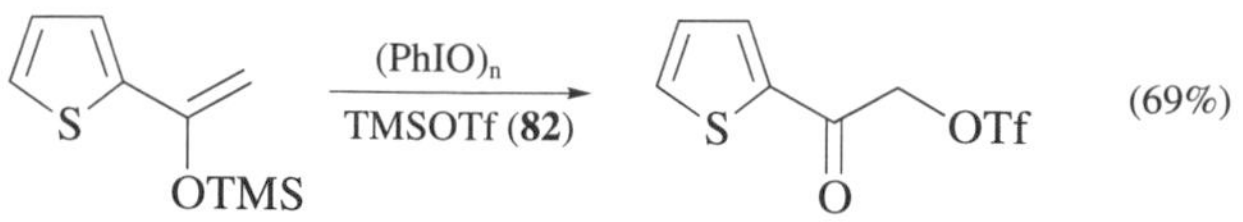
(PhIO)n
TMSOTf (82)
S
OTMS
S
O
OTf
(69%)

(Eq. 72)

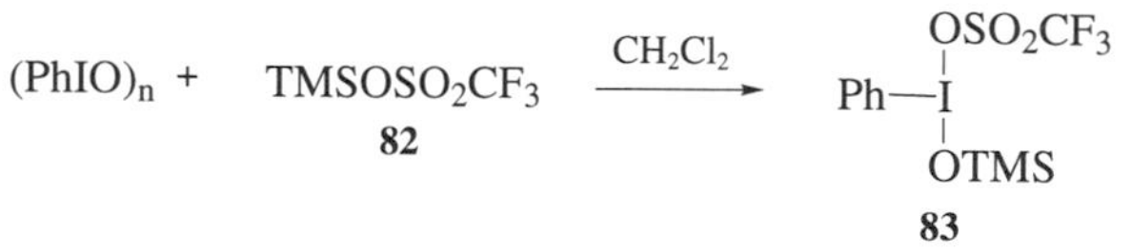
(PhIO)n +
TMSOSO2CF3
82
CH2Cl2
OSO2CF3
Ph—I
OTMS
83

(Eq. 73)

α-Acetoxylation and α-Trifluoroacetoxylation. Oxidation of silyl enol ethers derived from saturated and α,β-unsaturated ketones with IBD in the presence of $BF_3 \cdot OEt_2$ gives α-acetoxylated products directly. However, the reaction of a silyl enol ether derived from a ketone in the absence of $BF_3 \cdot OEt_2$ takes place with retention of the trimethylsilyl group and effects either substitution of the vinylic hydrogen or diacetoxylation (Eq. 74). This reaction can be used for the regiose-

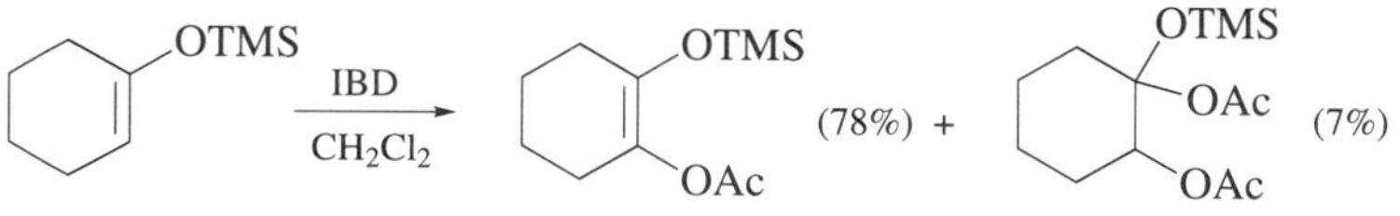
OTMS
IBD
CH2Cl2
OTMS
OAc
(78%) +
OTMS
OAc
OAc
(7%)

(Eq. 74)

lective synthesis of α-acetoxyketones since the trimethylsilyl group is readily eliminated from both products by the action of $BF_3 \cdot OEt_2$ (Eq. 75).[76]

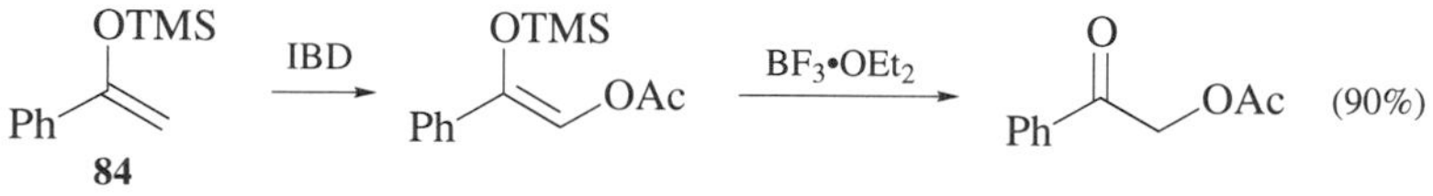
OTMS
Ph
84
IBD
OTMS
Ph
OAc
BF3•OEt2
O
Ph
OAc
(90%)

(Eq. 75)

Treatment of the silyl enol ether of cyclohexanone with an equimolar ratio of bis(trifluoroacetoxyiodo)benzene and pyridine in chloroform gives α-trifluoroacetoxycyclohexanone as the sole product (Eq. 76). Unlike other silyl enol ethers, the silyl enol ether of camphor is inert to the action of IBD and IBTA.

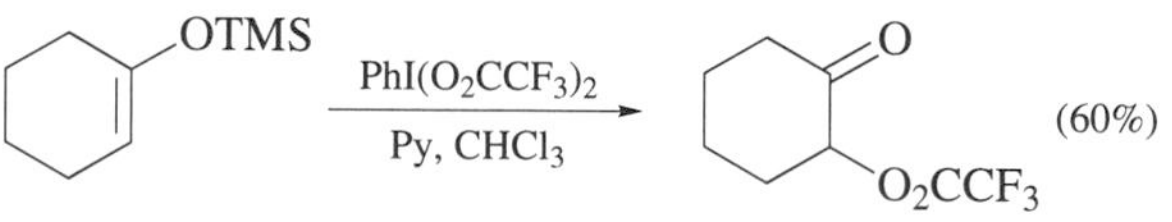
OTMS
PhI(O2CCF3)2
Py, CHCl3
O
O2CCF3
(60%)

(Eq. 76)

α-Phosphorylation and α-Phosphoryloxylation. The reaction of triethyl phosphite with silyl enol ethers of aromatic ketones using iodosobenzene under Lewis acid conditions provides a direct route for the α-phosphorylation of ketones. For example, oxidation of acetophenone silyl enol ether (**84**) leads to the formation of diethyl (2-oxo-2-phenylethyl)phosphonate (Eq. 77).[77]

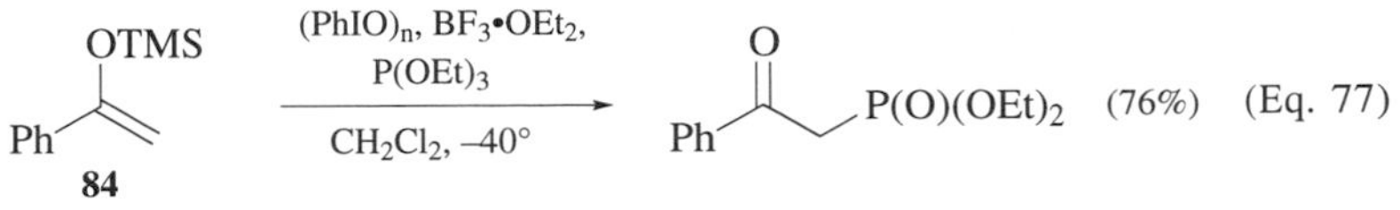

Hypervalent iodine methodology is also applicable to the α-phosphoryloxylation of silyl enol ethers. Upon oxidation with *p*-(difluoroiodo)toluene in the presence of phosphoric acid and *tert*-butyl alcohol, ketone trimethylsilyl enol ethers afford *tris*-ketol phosphates in good yields (Eq. 78).[78]

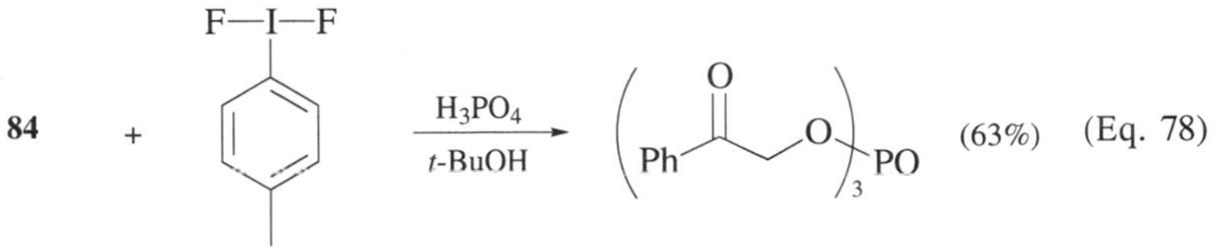

Carbon-Carbon Bond Formation. In the absence of any external nucleophile, hypervalent iodine oxidations of silyl enol ethers of ketones with iodosobenzene/boron trifluoride etherate in dichloromethane result in carbon-carbon coupling reactions leading to the formation of the corresponding 1,4-disubstituted butane-1,4-diones (Eq. 79).[79,80] The method has been applied to the synthesis of 3,2′,5′,3″-terthiophene (**85**) and 2,5-di(3′-thienyl)furan (**86**) (Eq. 80).[81]

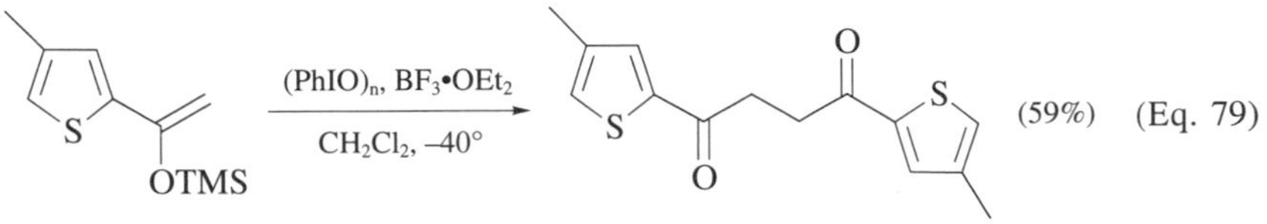

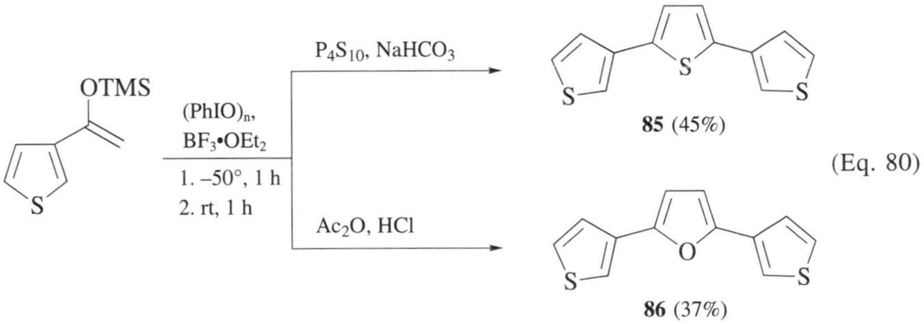

The complex $PhIF_2 \cdot BF_3$, prepared by reaction between equimolar amounts of (difluoroiodo)benzene and boron trifluoride etherate at −78° to −10°, is also effective for the preparation of 1,4-diketones from aliphatic and aromatic silyl enol ethers (Eq. 81).[82] (Difluoroiodo)benzene, which is known to give α-fluorination of silyl enol ethers (see Eq. 115), does not yield fluorination products in these reactions.

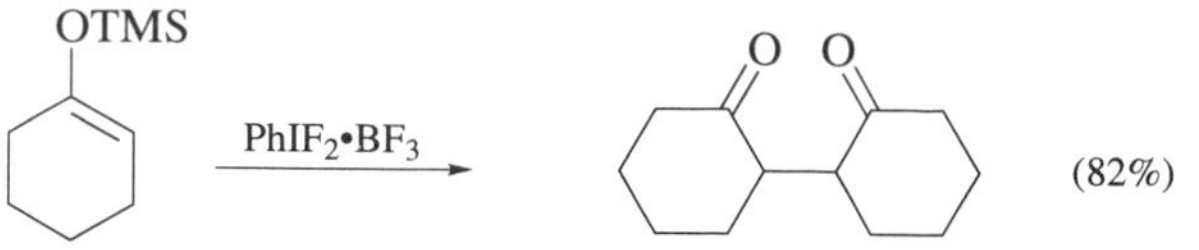
OTMS
PhIF2•BF3
O O
(82%)

(Eq. 81)

A limitation of this oxidative coupling is that only symmetrical 1,4-diketones can be synthesized with $(PhIO)_n$. Unsymmetrical carbon-carbon coupling can be accomplished by using the highly reactive iodonium salt phenacyl(phenyl)-iodonium tetrafluoroborate (**87**) which is generated in situ by the reaction of the silyl enol ether **84** with the electrophilic complex $PhIO{\cdot}HBF_4$ at −78° (Eq. 82). Reaction of 1-trimethylsilyloxycyclohexene with iodonium salt **87** yields the unsymmetrical 1,4-diketone **88**. Similarly, reactions of alkenes and cycloalkenes also result in carbon-carbon bond formation (for example **89**).[83,84]

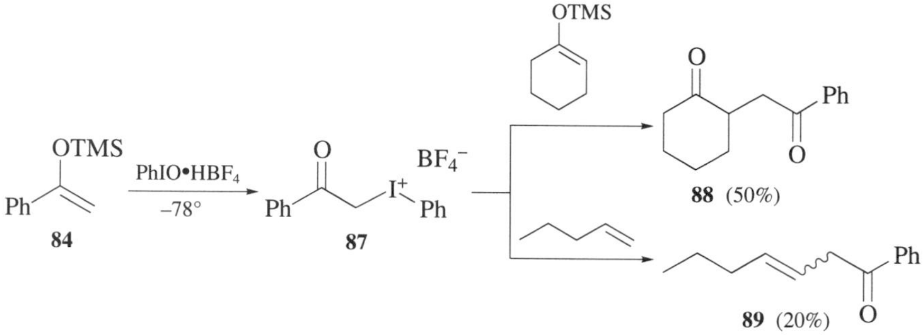
OTMS
Ph
84
PhIO•HBF4
-78°
O
BF4−
Ph
I+
Ph
87
OTMS
O
Ph
O
88 (50%)
Ph
89 (20%)
O

(Eq. 82)

α-Perfluoroalkylation and 1H,1H-Perfluoroalkylation. (Perfluoroalkyl)-phenyliodonium trifluoromethanesulfonates and sulfates smoothly react with silyl enol ethers in the presence of pyridine (1.1 and 2.2 equivalents, respectively) to give α-perfluoroalkyl derivatives of ketones (Eq. 83) and α,β-unsaturated carbonyl compounds (Eq. 84).[85]

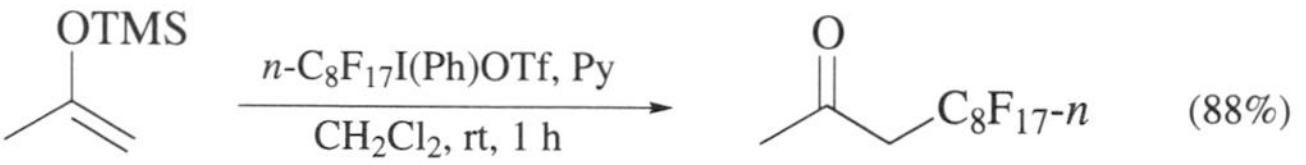
OTMS
n-C8F17I(Ph)OTf, Py
CH2Cl2, rt, 1 h
O
C8F17-n
(88%)

(Eq. 83)

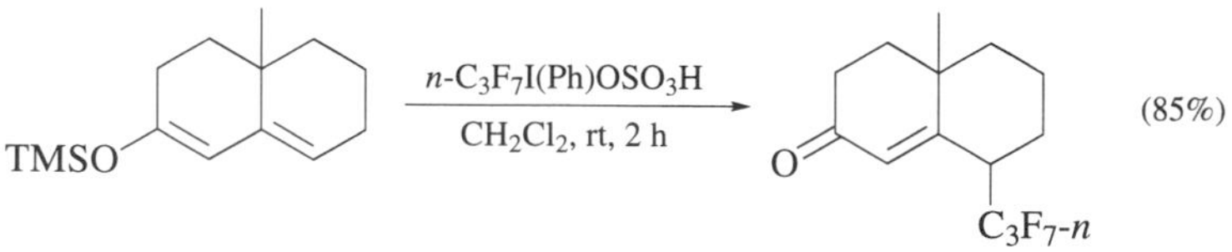
TMSO
n-C3F7I(Ph)OSO3H
CH2Cl2, rt, 2 h
O
C3F7-n
(85%)

(Eq. 84)

In contrast, the methyl enol ether of cyclohexanone (**90**) reacts with *n*-$C_6F_{13}I(Ph)OSO_3H$ in the presence of base to produce a 1:1:1 mixture of products **91**, **92**, and **93**. Acidic hydrolysis converts this mixture to α-(perfluoroalkyl)-cyclohexanone (**93**) in 78% overall yield (Eq. 85).

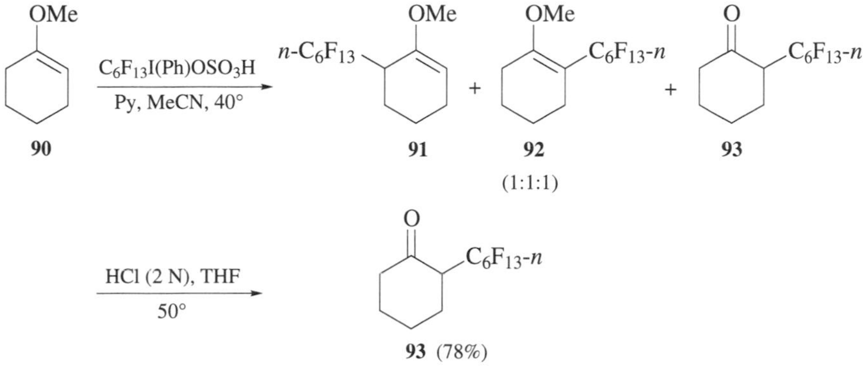

(Eq. 85)

The reaction of silyl enol ethers of carbonyl compounds with (1*H*,1*H*-perfluoroalkyl)phenyliodonium triflates in the presence of potassium fluoride in dichloromethane at room temperature affords β-perfluoroalkyl carbonyl compounds in good yield (Eq. 86). Silyl enol ethers of α,β-unsaturated carbonyl compounds give δ-perfluoroalkyl α,β-unsaturated carbonyl compounds selectively (Eq. 87).[86]

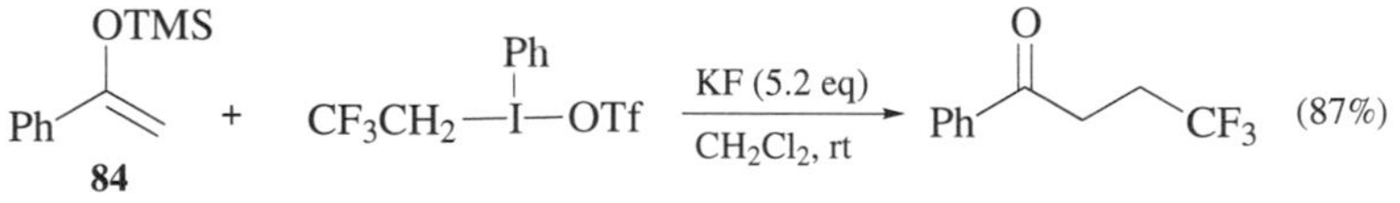

(Eq. 86)

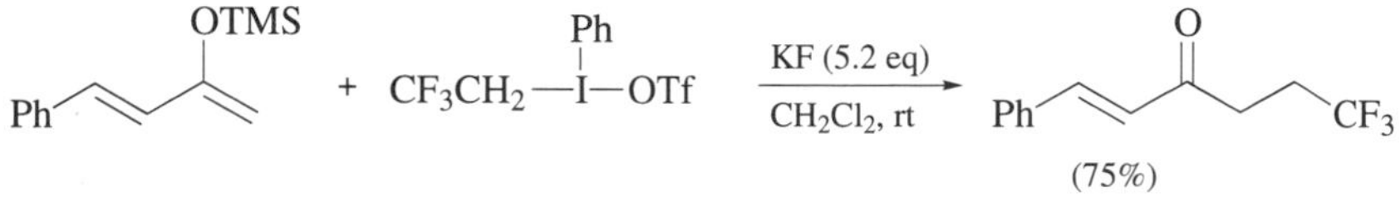

(Eq. 87)

α-Phenylation. Various cyclic and acyclic silyl enol ethers react with diphenyliodonium fluoride in tetrahydrofuran to give α-phenyl or α,α-diphenyl ketones, depending on the structure of the substrate (Eq. 88).[87] The regiochemistry of phenylation at the α-carbon atom can be controlled by an appropriate choice of silyl enol ether.

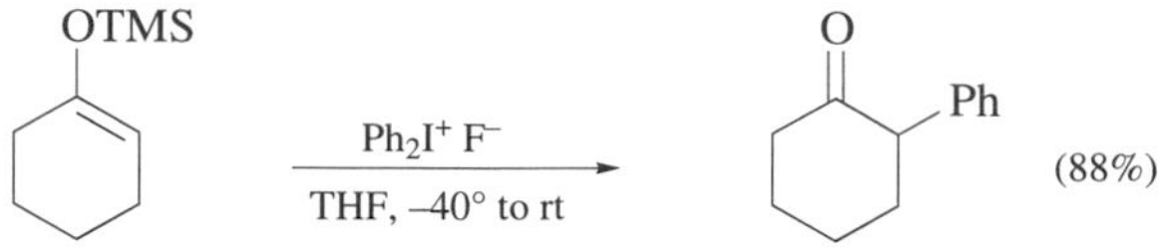
OTMS
$Ph_2I^+ F^-$
THF, –40° to rt
O
Ph
(88%)

(Eq. 88)

A similar approach is also used to effect α-phenylation of hydrocodone (**94**), but the yield of the desired product **95** starting from the silyl enol ether is only 11%. However, the reaction of the lithium enolate of hydrocodone with diphenyliodonium iodide affords **95** in good yield (71%) together with minor amounts of diphenylated product **96** (4%) and the starting ketone **94** (1%) (Eq. 89).[88]

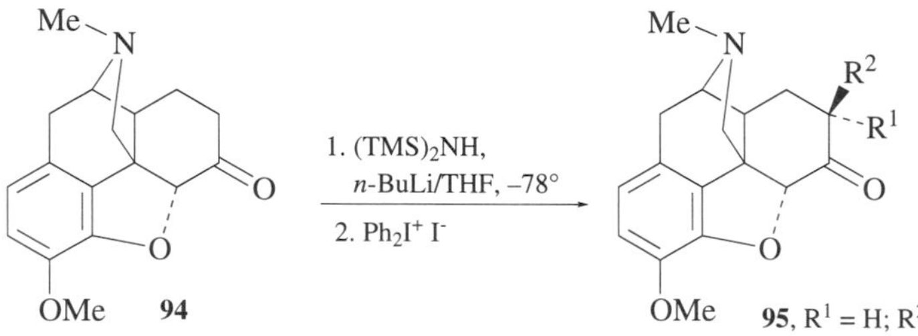
Me
N
O
O
OMe
94
1. $(TMS)_2NH$,
n-BuLi/THF, –78°
2. $Ph_2I^+ I^-$
Me
N
R2
R1
O
O
OMe

95, R^1 = H; R^2 = Ph (71%)
96, R^1, R^2 = Ph (4%)

(Eq. 89)

α-Amination. [*N*-(*p*-Toluenesulfonyl)imino]phenyliodinane (**97**) reacts with silyl enol ethers of ketones in the presence of catalytic $CuClO_4$ to give α-(*p*-toluenesulfonyl)aminoketones. The reaction probably proceeds through the intermediacy of aziridines **98** (Eq. 90). The influence of the substituents on silicon on

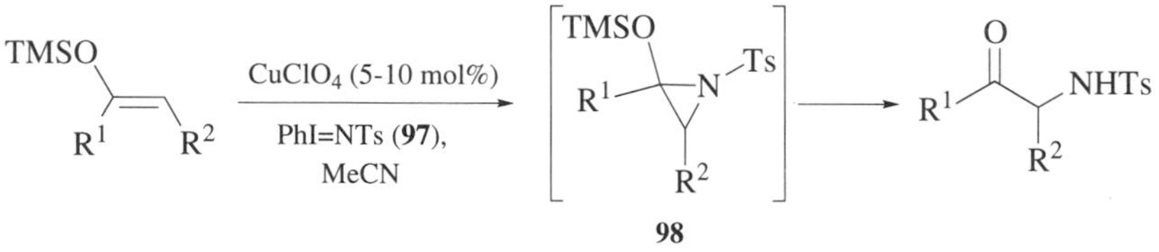
TMSO
R1
R2
$CuClO_4$ (5-10 mol%)
PhI=NTs (97),
MeCN
TMSO
R1
N
Ts
R2
98
O
R1
NHTs
R2

(Eq. 90)

reaction yields is negligible. For example, the Cu(I)-catalyzed amination of 1-[(*tert*-butyldimethylsilyl)oxy]cyclohexene (**99**) with reagent **97** in acetonitrile gives a 65% yield of the α-(*p*-toluenesulfonyl)aminoketone **101**, which is comparable to the 64% yield obtained when the trimethylsilyl enol ether (**100**) is employed (Eq. 91).[89]

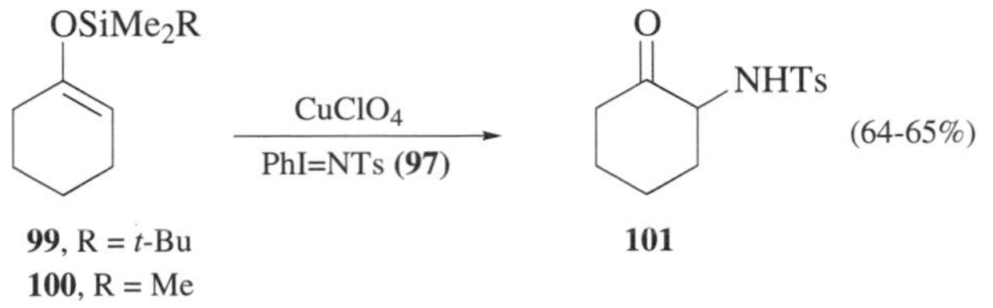

(Eq. 91)

Silyl ketene acetals derived from esters likewise undergo α-amination under these conditions. The yields are generally not good and vary depending on both the conditions and the ester employed. For example, trimethylsilyl ketene acetal **102**, derived from phenyl hexanoate, gives product **104** in 50% yield, while the analogous derivative of methyl hexanoate (**103**) affords product **105** only in 10% yield (Eq. 92).

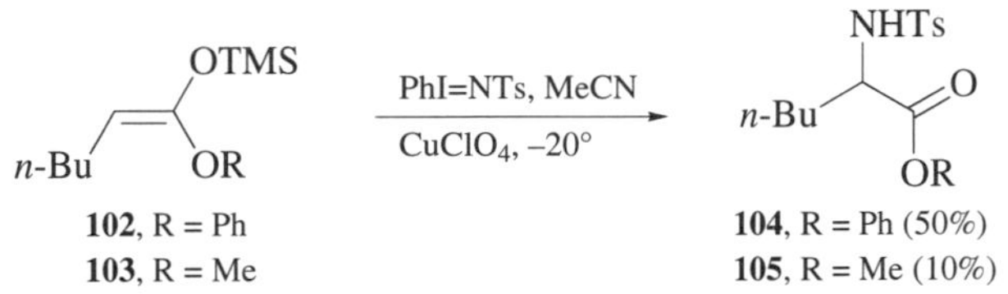

(Eq. 92)

Functionalization of β-Dicarbonyl Compounds

In addition to the standard procedures for the α-functionalization of β-dicarbonyl compounds, there are a variety of other conditions that have been used. Iodosobenzene can be employed to effect one-pot α-functionalizations of various β-dicarbonyl compounds.[90] For example, treatment of β-dicarbonyl compounds with iodosobenzene and azidotrimethylsilane in chloroform under reflux gives α-azido-β-dicarbonyl compounds (Eq. 93). Similarly, use of methanesulfonic acid or an alcohol provides α-mesyloxy (Eq. 94) or α-alkoxy-β-dicarbonyl compounds (Eq. 95), respectively. When two equivalents of the β-dicarbonyl com-

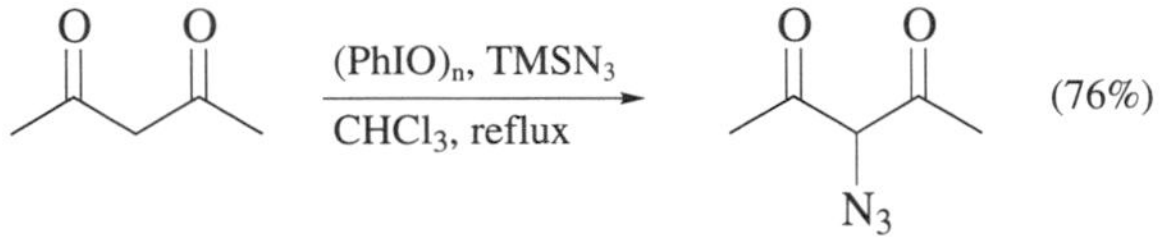

(Eq. 93)

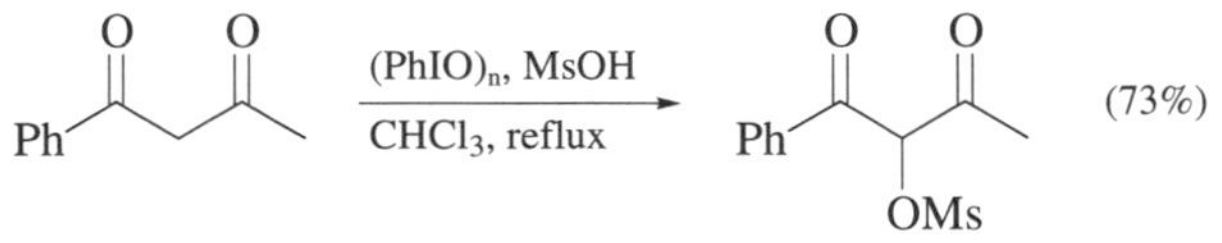

(Eq. 94)

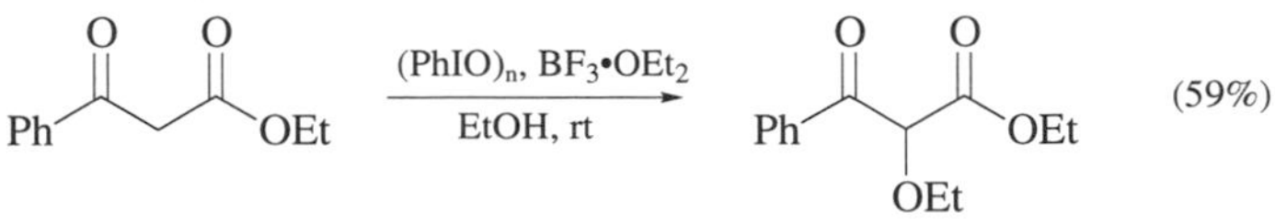

(Eq. 95)

pound are treated with 1.2 equivalents of $(PhIO)_n$-$BF_3 \cdot OEt_2$ in chloroform, self-coupling at the α position occurs, leading to dimers (Eq. 96). The reaction of 4-

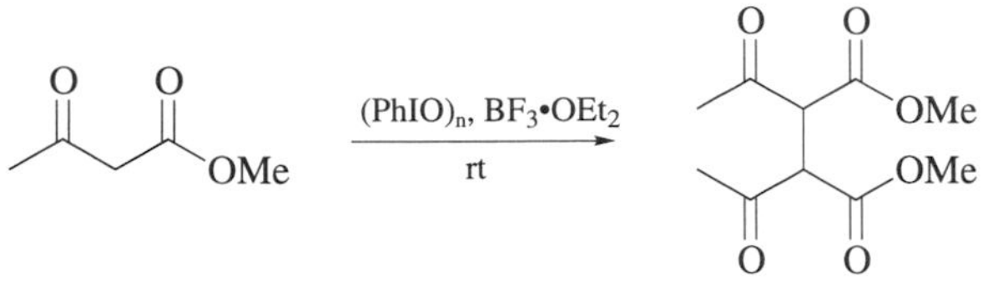

(46%) (Eq. 96)

aryl-2,4-dioxobutanoic acids **106** with IBD in acetic acid also leads to self-coupling with the formation of intermediate **107**, which finally undergoes cyclization to yield 3-aroyl-5-aryl-2-hydroxyfurans **108** (Eq. 97).[91] However,

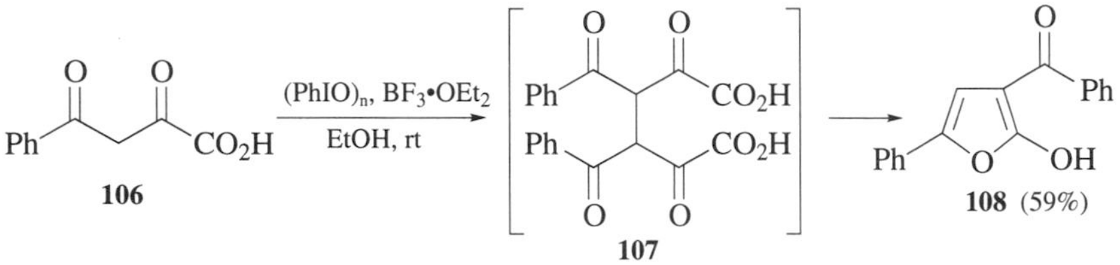

106 **107** **108** (59%)

(Eq. 97)

the reaction of benzoylacetone with iodosobenzene with or without $BF_3 \cdot OEt_2$ in methanol leads to carbon-carbon bond cleavage rather than normal α-functionalization (Eq. 98).[90]

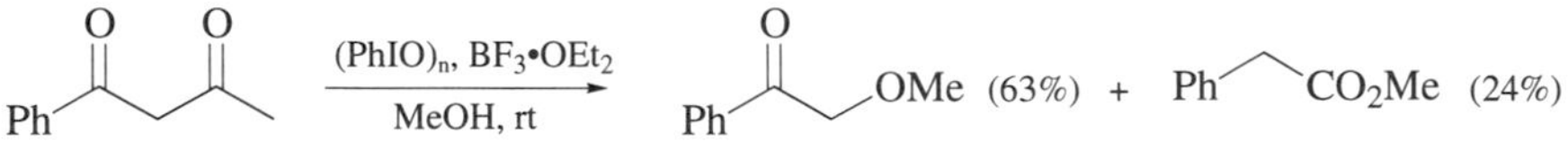

(Eq. 98)

1,3-Indanedione is transformed to 2,2-dialkoxy derivatives **109** by using IBD–H_2SO_4/ROH (Eq. 99). In addition, the reaction also yields a minor product **110** due to self-condensation.[92]

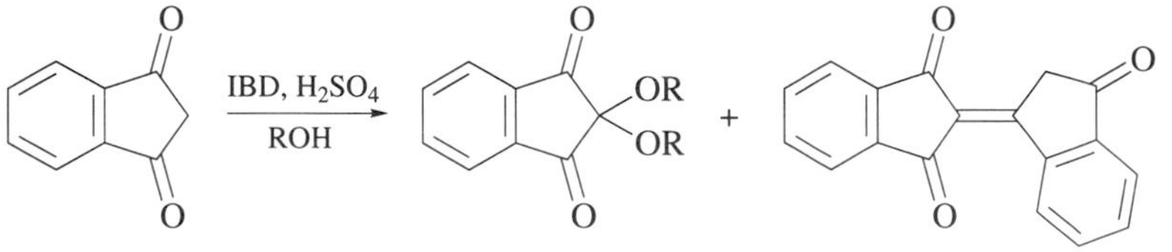

109 R	
Me	(44%)
Et	(49%)

110

(Eq. 99)

Meldrum's acid (isopropylidene malonate) and its alkyl and benzyl derivatives undergo oxidative dimerization by treatment with IBD-K_2CO_3 when benzyltrimethyl ammonium chloride (TEBA) is used as a phase-transfer catalyst (Eq. 100).[93]

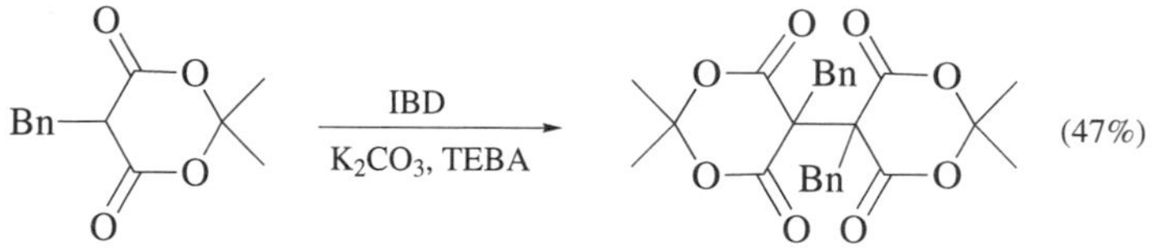
Bn
IBD
K_2CO_3, TEBA
Bn
Bn
(47%)

(Eq. 100)

Perfluoroalkylphenyl iodonium triflates react with the enolate anions of *β*-diketones to afford *O*- and *C*-perfluoroalkylated products in moderate yields. The ratio of *O*- to *C*-perfluoroalkylation products depends on the reaction temperature. Lower temperatures favor *O*-perfluoroalkylation products in the case of *β*-diketones (Eq. 101). The sodium salt of diethyl 2-methylmalonate produces the *C*-perfluoroalkylation product only (Eq. 102).[94]

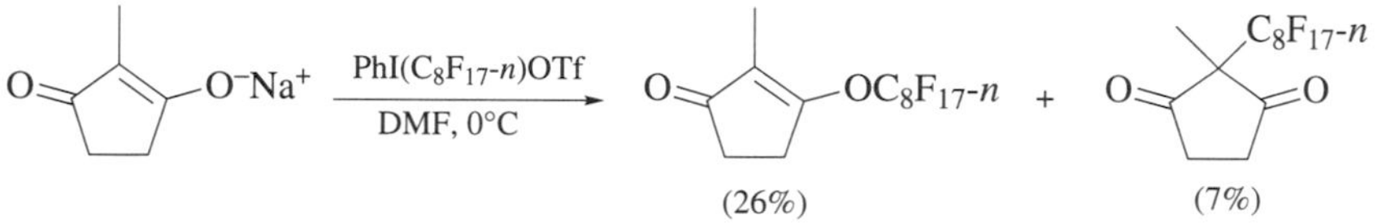
O^-Na^+
PhI(C_8F_{17}-*n*)OTf
DMF, 0°C
OC_8F_{17}-*n*
+
C_8F_{17}-*n*
(26%)
(7%)

(Eq. 101)

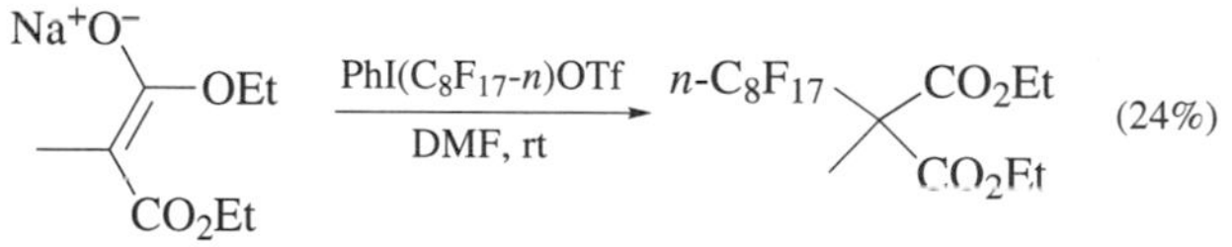
Na^+O^-
OEt
CO_2Et
PhI(C_8F_{17}-*n*)OTf
DMF, rt
n-C_8F_{17}
CO_2Et
CO_2Et
(24%)

(Eq. 102)

Intramolecular Participation Reactions Leading to Oxygen Heterocyclic Compounds

Hypervalent iodine oxidation of a carbonyl compound containing an oxygen functionality at a suitable position leads to intramolecular participation with subsequent formation of heterocyclic oxygen compounds. These types of reactions are discussed in the following subsections.

Spirooxetan-3-ones. 17*β*-Acetyl-17*α*-hydroxysteroids on oxidation with IBD-KOH/MeOH at 20° give steroidal spiro-oxetan-3-ones (Eq. 103).[95] This is an interesting example of intramolecular participation where the C17 *α*-hydroxy group acts an internal nucleophile (intermediate **111**), and replaces the external nucleophilic methoxide ion (Eq. 103).

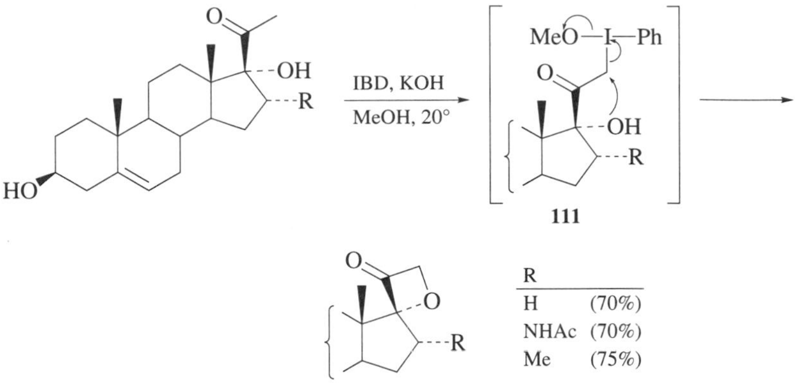
IBD, KOH
MeOH, 20°
MeO—I—Ph
O
OH
R
HO
111
O
R

R	
H	(70%)
NHAc	(70%)
Me	(75%)

(Eq. 103)

Coumaran-3-ones. On reaction with IBD-KOH/MeOH, *o*-hydroxyacetophenones and related compounds give the corresponding 2-methoxycoumaran-3-ones **114** (Eq. 104).[26] The reaction is believed to occur via intramolecular

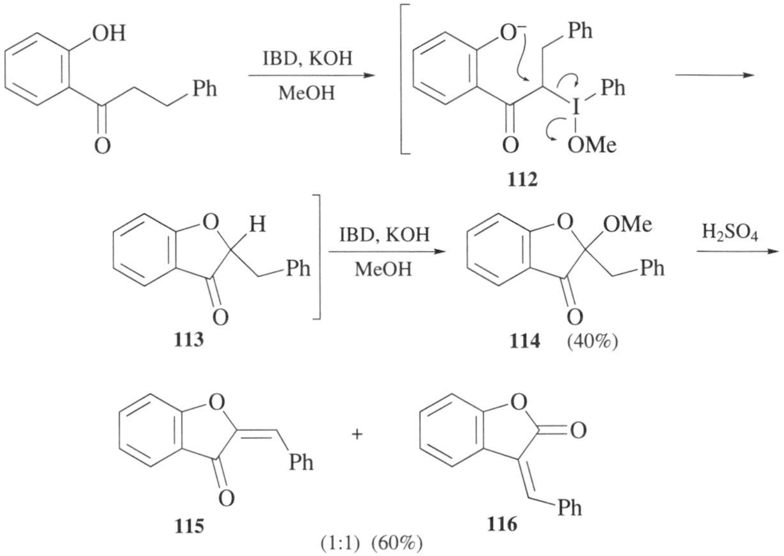
OH
Ph
O
IBD, KOH
MeOH
O⁻
Ph
I
OMe
112
H
113
H_2SO_4
OMe
114 (40%)
115
+
116
(1:1) (60%)

(Eq. 104)

participation of the *ortho* hydroxy group which undergoes nucleophilic attack at the α carbon of the C-iodine(III) intermediate **112**. The resulting coumaranone (**113**) undergoes further oxidation to give product **114**. Heating of 2-benzyl-2-methoxycoumaranone (**114**) with dilute sulfuric acid results in the elimination of a molecule of methanol to yield aurone (**115**) and isoaurone (**116**) in a 1:1 ratio.

In contrast to this observation, 2,4-dihydroxyacetophenones, when oxidized with IBD-KOH/MeOH, lead to the formation of *o*-iodo ethers leaving the keto group intact (Eq. 105).[96]

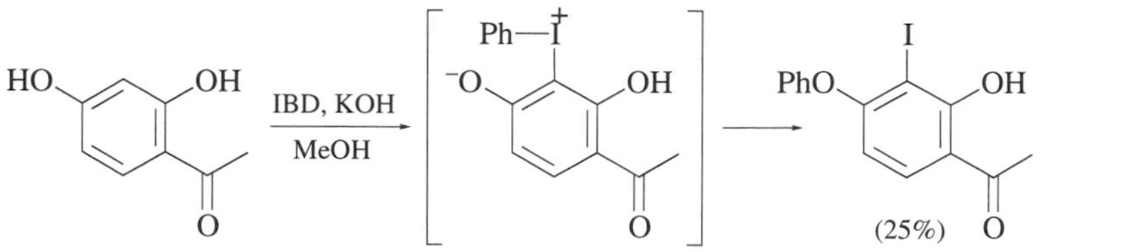

(Eq. 105)

The oxidation of α-aroyl-*o*-hydroxyacetophenones (**117**) under these conditions provides a useful route to 2-aroylcoumaran-3-ones (**118**) (Eq. 106).[97] A notable feature of this reaction is that iodonium ylides **119** are not isolable.

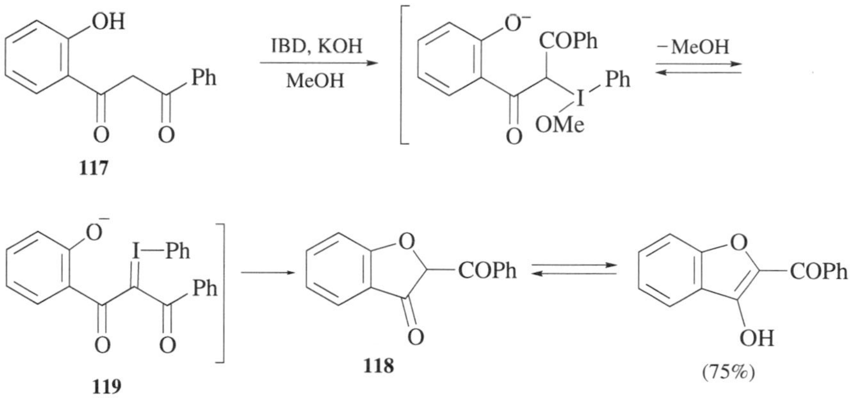

(Eq. 106)

The formation of 2-aroylcoumaran-3-ones (**121**) involving intramolecular participation also occurs when *o*-aroyloxyacetophenones (**120**) are oxidized with HTIB and the resulting *o*-aroyloxy-α-tosyloxyacetophenones (**122**) are subjected to the Baker-Venkataraman rearrangement. This transformation can also be accomplished in one pot without isolating intermediate **122** (Eq. 107).[98]

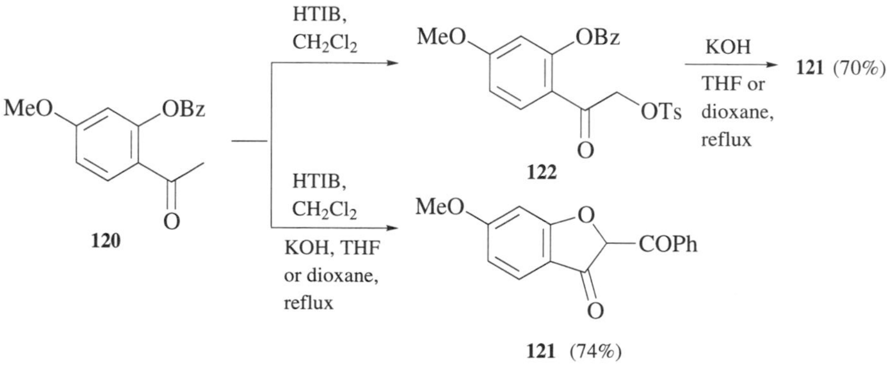

(Eq. 107)

Oxidation of 1-trimethylsilyloxy-1-[(2-trimethylsilyloxy)phenyl]ethene with $(PhIO)_nBF_3{\cdot}OEt_2$, H_2O results in the formation of coumaran-3-one (**123**) by a route involving intramolecular participation of the *o*-hydroxy group. In addition, α-hydroxyketone **124** is also formed in 25% yield (Eq. 108).[99]

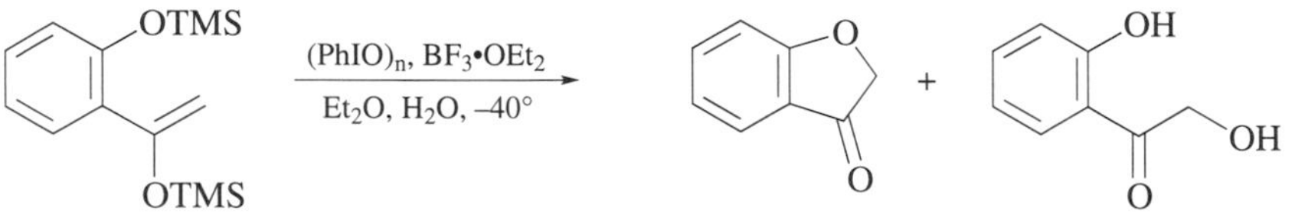
OTMS
OTMS
(PhIO)n, BF3•OEt2
Et2O, H2O, –40°
O
O
+
OH
OH
O

123 (31%) **124** (25%)

(Eq. 108)

Lactones. When 5-oxocarboxylic acids (Eq. 109) or 4,6-dioxocarboxylic acids (Eq. 110) are treated with HTIB in dichloromethane, intramolecular participation

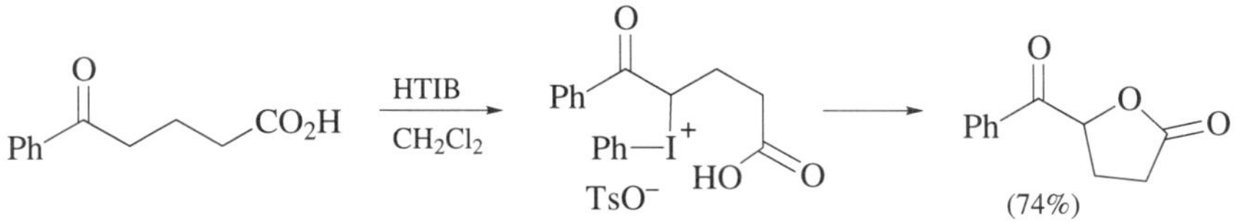
O
Ph
CO2H
HTIB
CH2Cl2
O
Ph
Ph—I+
TsO−
HO
O
O
Ph
O
O
(74%)

(Eq. 109)

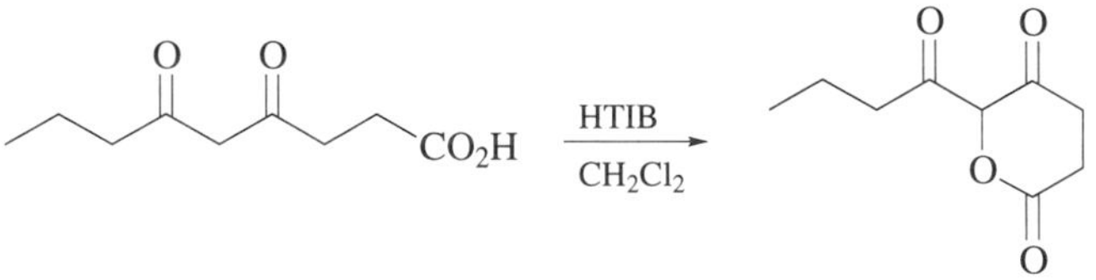
O O
CO2H
HTIB
CH2Cl2
O O
O
O

(68%) (Eq. 110)

by the carboxy groups takes place leading to the formation of the corresponding lactones and oxolactones. Cyclic 4,6-dioxocarboxylic acids such as **125** give spirolactones **126** (Eq. 111).[100]

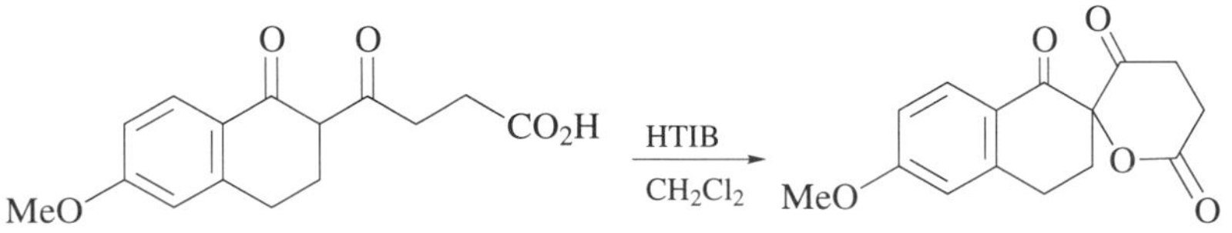
O O
CO2H
MeO
HTIB
CH2Cl2
MeO
O
O
O
O

125 **126** (36%)

(Eq. 111)

Halogenation

The halogenation of carbonyl compounds with hypervalent iodine has limited scope. The reaction of ketones and β-diketones with (dichloroiodo)benzene in acetic acid gives the corresponding α-chlorocarbonyl compounds (Eq. 112). α-

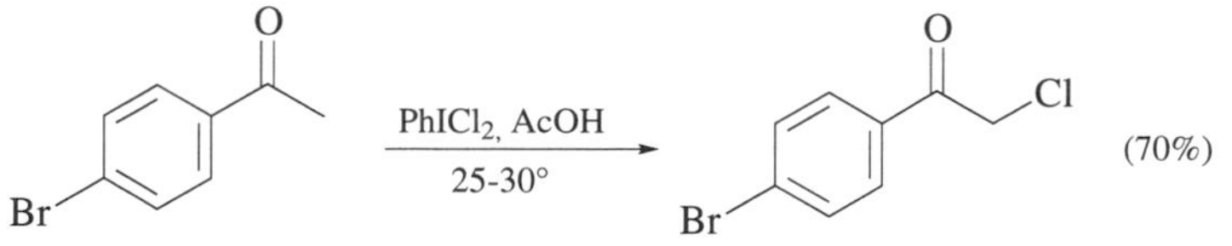

PhICl2, AcOH
25-30°
Br
O
Cl
(70%)

(Eq. 112)

Chlorinated products are also obtained under free radical conditions. For example, benzyl *p*-tolyl ketone on treatment with $PhICl_2$ in benzene in the presence of azobisisobutyronitrile (AIBN) and light gives exclusively the α-chloroketone (Eq. 113).[101]

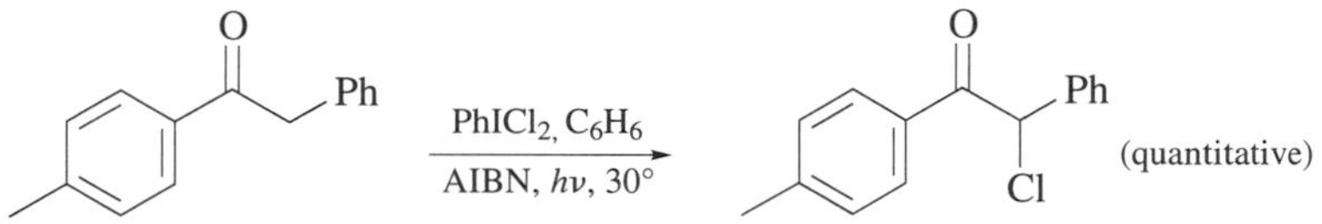

O
Ph
PhICl2, C6H6
AIBN, hν, 30°
Cl
(quantitative)

(Eq. 113)

Reaction of various aryl methyl ketones with benzyltrimethylammonium dichloroiodate in refluxing dichloroethane/methanol gives the corresponding α-chloroketones in good yields (Eq. 114).[102] A similar approach is successful for the α-chlorination of 2-acetylthiophene and acetylpyrroles. In the latter case, reaction is carried out at room temperature using THF as the solvent.[103] A limitation of the procedure is that chlorination of nitroacetophenones is not successful.

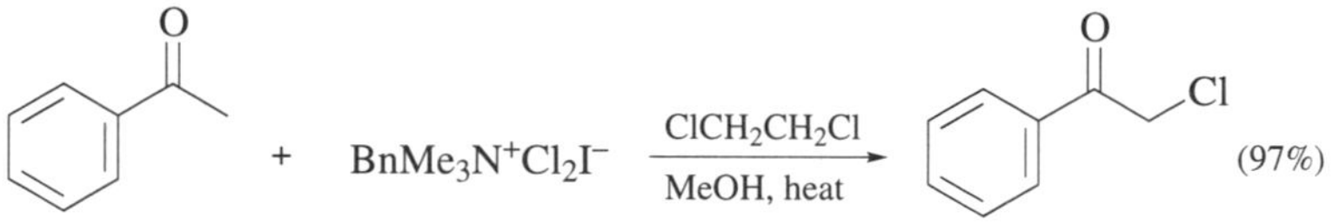

O
+ BnMe3N+Cl2I−
ClCH2CH2Cl
MeOH, heat
Cl
(97%)

(Eq. 114)

Oxidation of silyl enol ethers derived from steroids such as 17β-hydroxy-5-α-androstan-3-one acetate with *p*-(difluoroiodo)toluene produces a mixture of diastereomeric α-fluorinated ketones together with the dehydrogenated product (Eq. 115).[104] However, the silyl enol ether derived from 5-pregnen-3,11,20-trione-3,20-bis(ethyleneketal) is recovered unchanged.

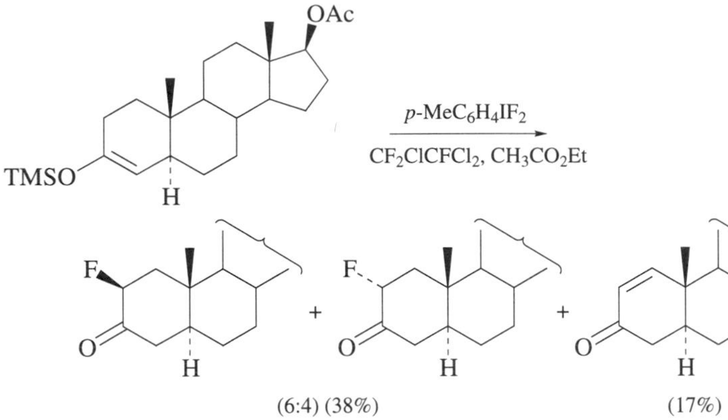
OAc
TMSO
H
p-MeC6H4IF2
CF2ClCFCl2, CH3CO2Et
F
O
H
+
F
O
H
+
O
H
(6:4) (38%)
(17%)

(Eq. 115)

α-Imidylation of Ketones

Organoiodine(III) reagents with two I-N bonds such as **127**[105] and **128**[106] are relatively less common and can be prepared by the reaction of sodium or potassium salts of imides and sulfimides with IBTA (Eq. 116). These reagents are use-

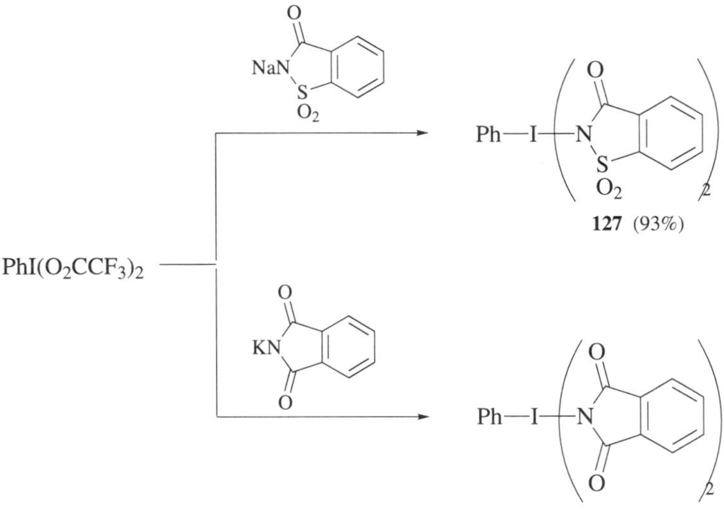
O
NaN
S
O2
Ph—I
O
N
S
O2
2
PhI(O2CCF3)2
O
KN
O
Ph—I
O
N
O
2

127 (93%)

128 (74%)

(Eq. 116)

ful for α-N-imidylation of ketones (Eq. 117).[106,107] While several aliphatic and alicyclic ketones and acetophenones give the corresponding α-N-saccharinyl derivatives in fairly good yield on reaction with **127**, acetophenones bearing electron-withdrawing substituents and benzyl phenyl ketone do not give good results.[107] Reagent **128** also reacts with acetophenone to give N-phenacylphthalimide but the scope of this reaction has not been studied.[106]

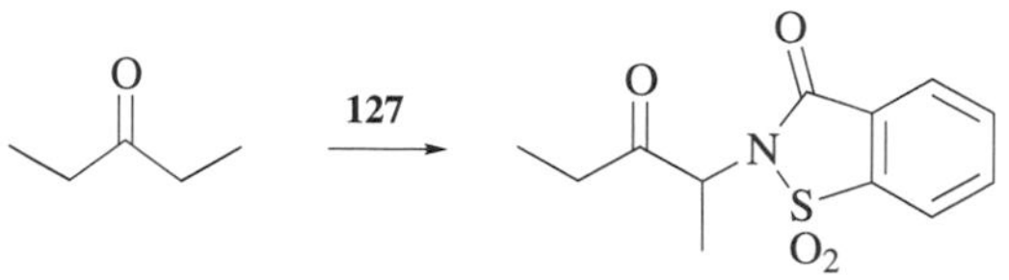
(18%) (Eq. 117)

β- and γ-Functionalization

Although α-functionalizations are more common, both β- and γ-functionalizations can be accomplished with certain substrates. Functionalization of ketones and lactones has been achieved through the agency of substituted cyclopropanols derived from ketones and lactones with the ultimate product being the larger homologous α,β-unsaturated compound (Eq. 118).[108]

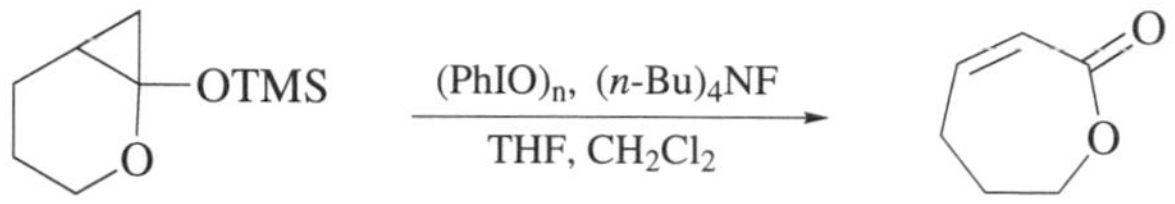
(75%) (Eq. 118)

Further examples of β-functionalization are the formation of β-azidoketones from the reaction of isopropylsilyl enol ethers with $(PhIO)_n$-trimethylsilyl azide (Eq. 119).[109] These azido products allow ready access to various β-functionalized

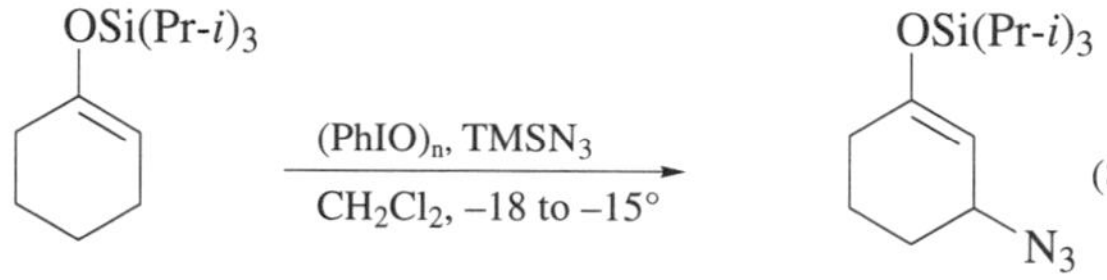
(84%) (Eq. 119)

ketones. The course of this reaction is greatly influenced by reaction temperature[110a] and the presence of an amine.[110b] While a change from β-azidonation to α-bis(azidonation) is observed with decreasing temperature, β-azidonation is completely suppressed when the reaction is carried in the presence of trimethylamine.

γ-Functionalization occurs when 2-(trimethylsilyl)furan reacts with iodosobenzene and boron trifluoride in the presence of a nucleophile (Eq. 120).[111] The products of this reaction are 5-substituted 2-(5*H*)furanones.

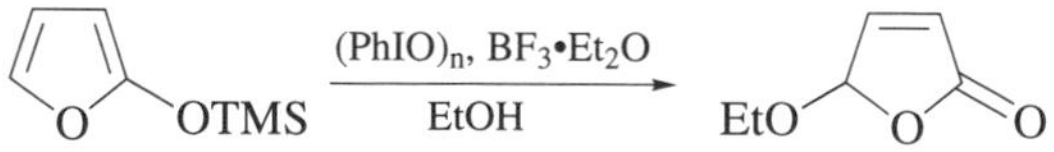
(55%) (Eq. 120)

Rearrangement

Enolizable Aromatic Ketones to 2-Arylalkanoates

A useful feature of hypervalent iodine oxidations is the occurrence of rearrangement processes, as exemplified by 1,2-aryl migration of enolizable aromatic ketones to 2-arylalkanoates.[112–117] Organoiodine(III) reagents such as HTIB/IBD or iodosobenzene and concentrated sulfuric acid or $BF_3{\cdot}OEt_2$ in methanol react with acetophenones to afford methyl arylacetates in good yields with α-methoxyacetophenones as byproducts (Eq. 121).[114,115] The nature of the aryl

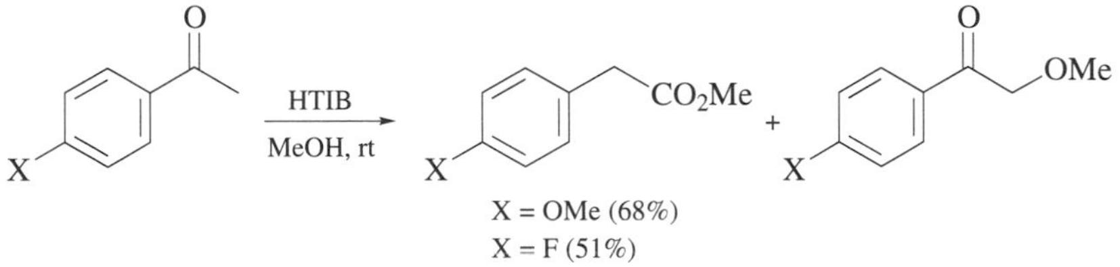
HTIB
MeOH, rt
CO_2Me
OMe
X = OMe (68%)
X = F (51%)

(Eq. 121)

group plays an important role in this migration; substrates with a good migratory aptitude (like $MeOC_6H_4$) give better results. However, oxidation of p-nitroacetophenone under similar conditions mainly gives the α-methoxy derivative (Eq. 122).

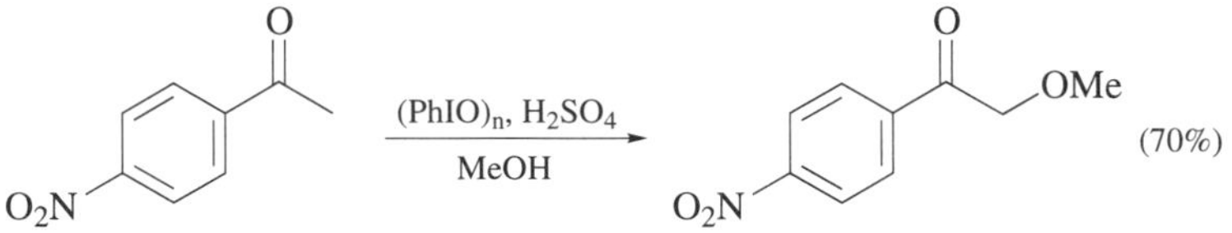
$(PhIO)_n$, H_2SO_4
MeOH
O_2N
OMe
(70%)

(Eq. 122)

The reaction has been applied successfully to the synthesis of methyl 2-substituted-3-methyl-6-chromonyl acetates (Eq. 123).[116] Preparation of thiazolylacetic acids by this approach requires reflux conditions (Eq. 124).[117]

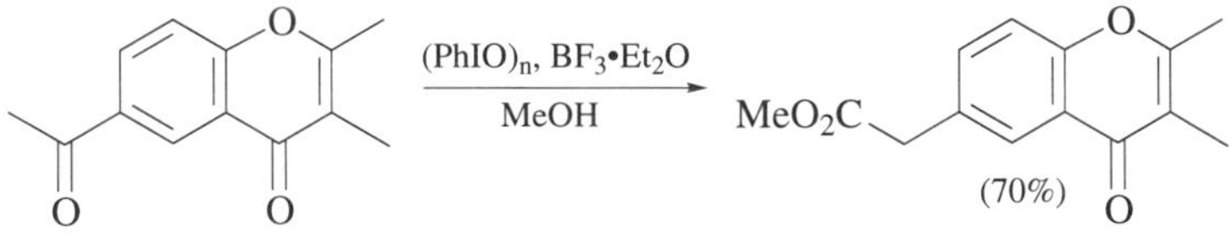
$(PhIO)_n$, $BF_3{\bullet}Et_2O$
MeOH
MeO_2C
(70%)

(Eq. 123)

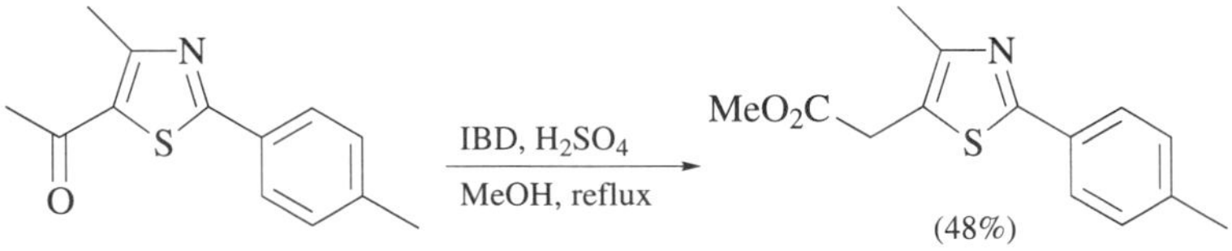
IBD, H_2SO_4
MeOH, reflux
MeO_2C
(48%)

(Eq. 124)

Reaction conditions that employ methanol as solvent do not give satisfactory yields of propionates from the corresponding propiophenones. However, the use of trimethyl orthoformate (TMOF) allows conversion of these ketones to 2-arylpropanoates (Eq. 125). This method is also effective for the synthesis of the

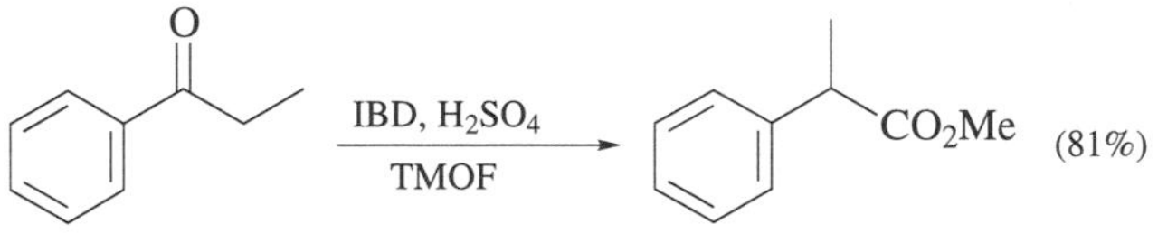
O
IBD, H_2SO_4
TMOF
CO_2Me
(81%)

(Eq. 125)

anti-inflammatory drugs naproxen [2-(6-methoxy-2-naphthyl)propanoic acid] (Eq. 126),[115] and ibuprofen [2-(4-isobutylphenyl)propanoic acid] as well as methyl

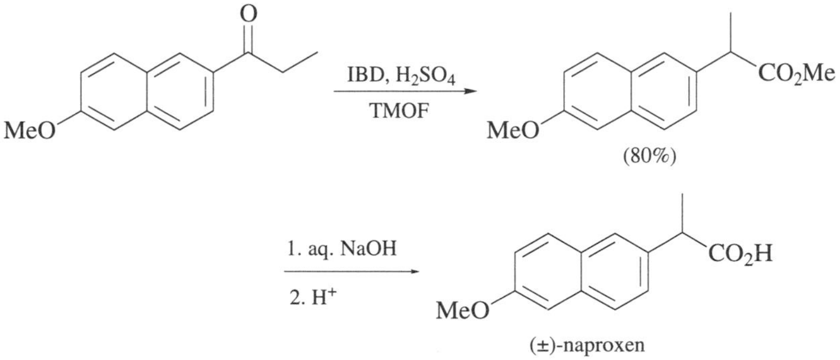
O
MeO
IBD, H_2SO_4
TMOF
CO_2Me
MeO
(80%)
1. aq. NaOH
2. H^+
CO_2H
MeO
(±)-naproxen

(Eq. 126)

2-(2-substituted-3-methylchromonyl)propanoates (Eq. 127).[116] γ-Ketoacids on oxidation under similar conditions afford the rearranged dimethyl arylsuccinates (Eq. 128).[113]

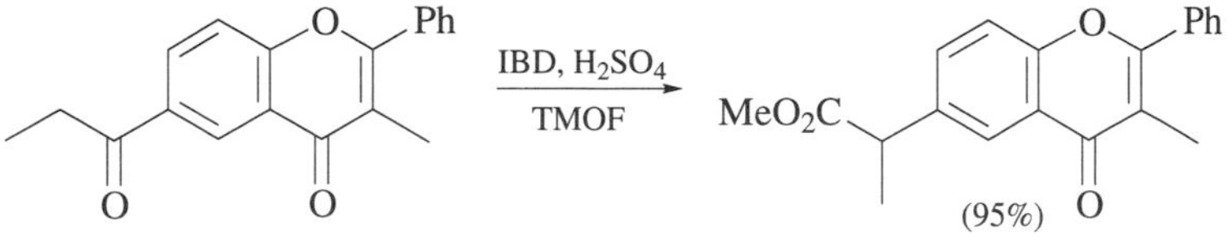
O
Ph
O
O
IBD, H_2SO_4
TMOF
MeO_2C
O
Ph
(95%)
O

(Eq. 127)

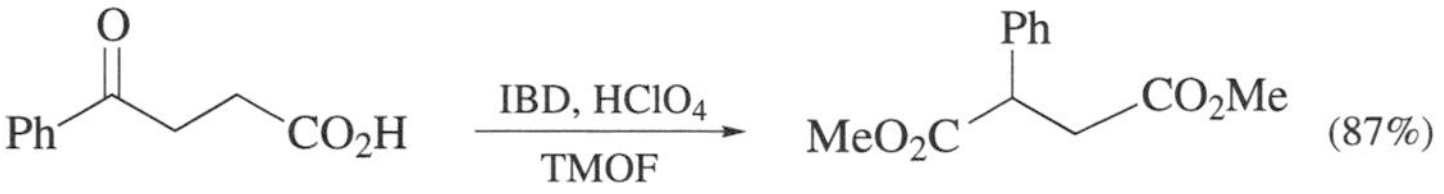
O
Ph
CO_2H
IBD, $HClO_4$
TMOF
Ph
MeO_2C
CO_2Me
(87%)

(Eq. 128)

Oxidation of acetophenones with one equivalent of IBD-H_2SO_4 in TMOF affords a mixture of methyl arylacetates, α-methoxyarylacetates **129**, and α-methoxyacetophenones (Eq. 129). When 2 equivalents of IBD are used, the

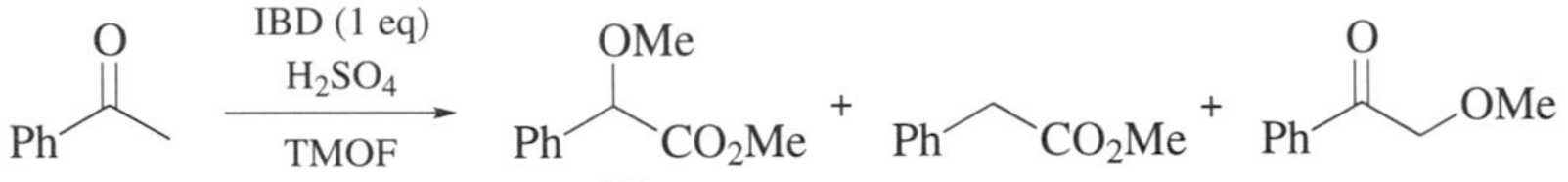
O
IBD (1 eq)
H_2SO_4
TMOF
Ph
OMe
CO_2Me
+
Ph
CO_2Me
+
O
Ph
OMe
129

(Eq. 129)

reaction gives only product **129** in good yield probably via the intermediates **130**, **131a**, and **131b** (Eq. 130).[118]

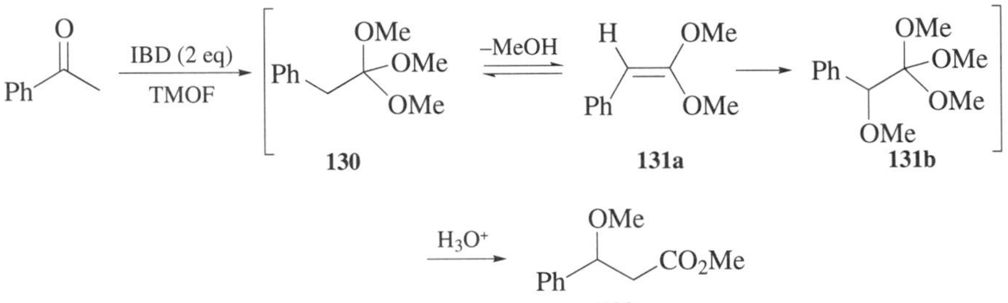
O
Ph
IBD (2 eq)
TMOF
Ph
OMe
OMe
OMe
–MeOH
H
OMe
Ph
OMe
Ph
OMe
OMe
OMe
OMe
130
131a
131b
H_3O^+
OMe
Ph
CO_2Me
129

(Eq. 130)

Chalcones to 1,2-Diaryl-3,3-dimethoxypropan-1-ones and Methyl 2,3-Diaryl-3-methoxypropanoates

Chalcones, a typical class of α,β-unsaturated ketones, are known to undergo 1,2-aryl migrations with organoiodine(III) reagents. These oxidative rearrangements leading to formation of 1,2-diaryl-3,3-dimethoxypropan-1-ones are effected by using conditions similar to those used for acetophenones (Eq. 131, route a).[114] The solvent plays an important role in the course of this reaction. For instance, if TMOF is used as a solvent, methyl 2,3-diaryl-3-methoxypropanoates are obtained in good yields (Eq. 131, route b).[119] On the other hand, the oxidation of chalcones with HTIB in dichloromethane occurs without rearrangement and gives 1,3-diaryl-2,3-ditosyloxypropan-1-one (Eq. 131, route c).[120]

2-Substituted chromones and flavones fail to react with HTIB. In contrast, 2-methyl-4-quinolones on oxidation with HTIB afford 3-iodo-4-phenoxyquinolines via the isolable intermediate α-phenyliodonium tosylates and monocarbonyl ylides (Eq. 132).[121]

Rearrangement of Flavanones and Chromanones

The oxidation of flavanones (**33**) with HTIB in acetonitrile or propionitrile at reflux does not give the expected α-functionalized 3-tosyloxyflavanones (**132**). Instead, 1,2-shift of the C(2)-aryl group to C(3) occurs, thus constituting a novel and useful route to isoflavones (**34**) (Eq. 133).[122,123]

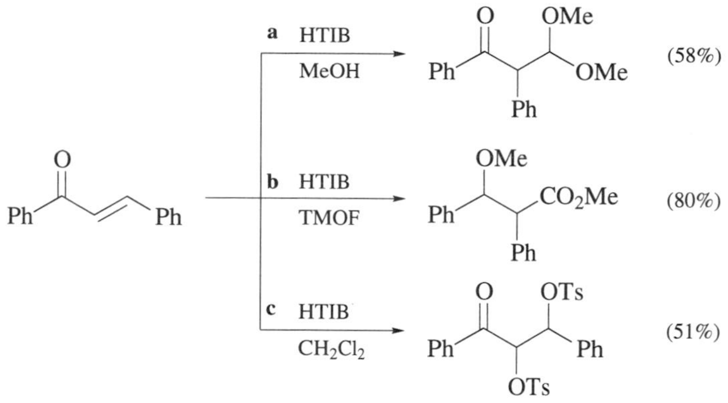
a HTIB
MeOH
(58%)
b HTIB
TMOF
(80%)
c HTIB
CH_2Cl_2
(51%)

(Eq. 131)

(51%)

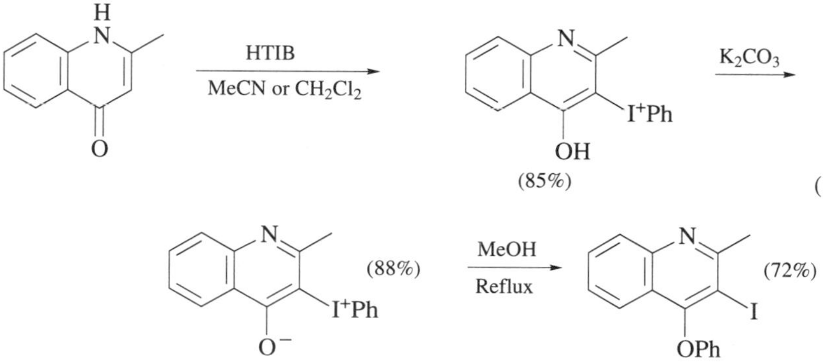
HTIB
MeCN or CH_2Cl_2
(85%)
K_2CO_3
(88%)
MeOH
Reflux
(72%)

(Eq. 132)

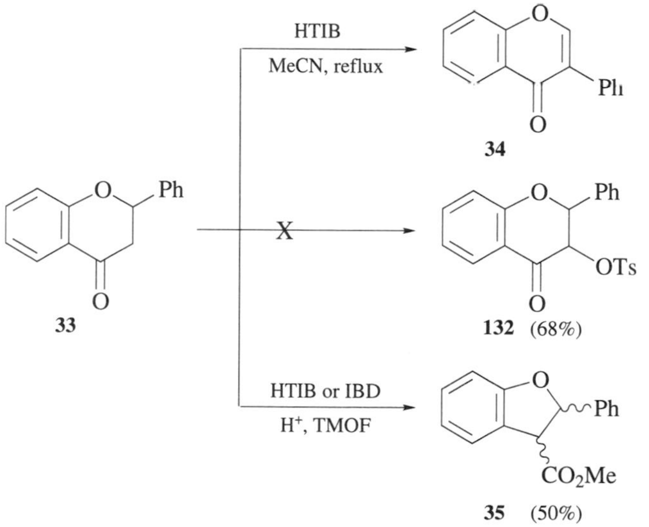
HTIB
MeCN, reflux
34
X
33
132 (68%)
HTIB or IBD
H^+, TMOF
35 (50%)

(Eq. 133)

The course of the oxidative rearrangement is greatly influenced by the reaction conditions, and other products in addition to isoflavone can be isolated. For example, when TMOF is employed as a solvent in this oxidation, ring contraction of the pyran occurs with the formation of methyl 2-aryl-2,3-dihydrobenzofuran-3-carboxylates **35**.[124]

Oxidation of several 2,2-dialkylsubstituted chromanones with HTIB involves 1,2-alkyl shifts. The reaction, which provides a convenient route to chromones (Eq. 134) and tetrahydroxanthones and their higher homologs (Eq. 135), can occur under the influence of both heat as well as ultrasound waves (Eqs. 134 and 135).[125]

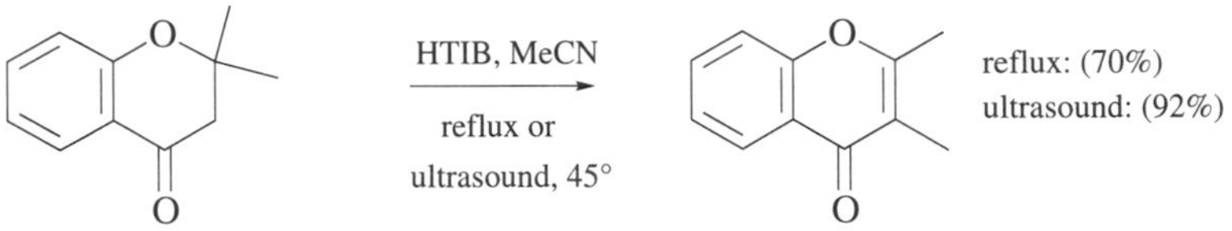
O
HTIB, MeCN
reflux or
ultrasound, 45°
reflux: (70%)
ultrasound: (92%)

(Eq. 134)

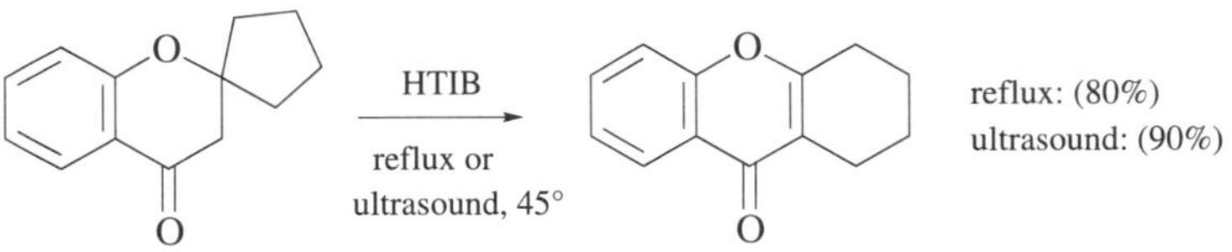
O
HTIB
reflux or
ultrasound, 45°
reflux: (80%)
ultrasound: (90%)

(Eq. 135)

Favorskii-Type Ring Contraction

The mechanism for the formation of the α-hydroxydimethylacetals from ketones imposes some rather stringent steric demands that are not problematic in simple ketones such as acetophenone or cycloalkanones. However, with more highly substituted systems, alternative reaction pathways may be followed. For example, cholestanone (**133**) yields 2-α-carbomethoxy-A-norcholestane (**134**; 60%) and 3-α-carbomethoxy-A-norcholestane (**135**; 10%) (Eq. 136).[25,34] None of

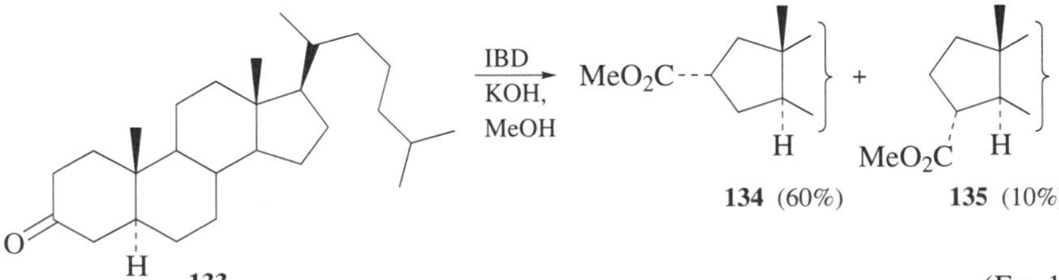
IBD
KOH,
MeOH
MeO_2C
H
+
MeO_2C
H
O
H

133 **134** (60%) **135** (10%)

(Eq. 136)

the expected 2-α-hydroxy-3,3-dimethoxycholestane was observed. Similarly, 4-phenylcyclohexanone undergoes ring contraction to provide a convenient procedure for the preparation of methyl 3-phenylcyclopentylcarboxylate (Eq. 137),[126] which is an important precursor for the synthesis of phenylcyclidine analogs.[127]

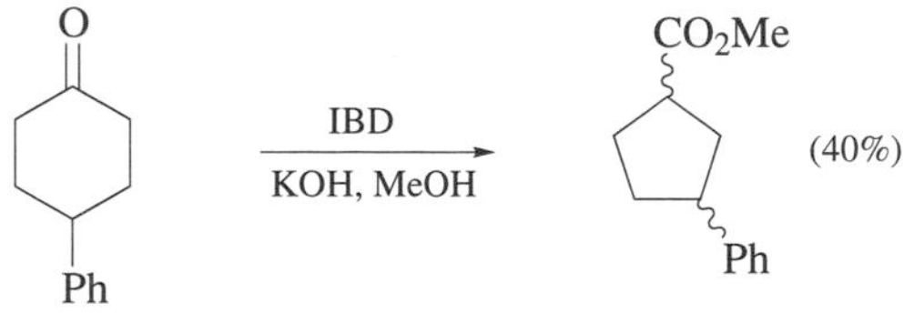
O
IBD
KOH, MeOH
CO_2Me
Ph
Ph
(40%)

(Eq. 137)

The formation of these products may be expected upon conformational and steric grounds. Initially, C(2) axial hyperiodination is considered to occur and the resulting intermediate **136** may convert torsionally to the twist-boat form **137**. The C-I(III) bond is now stereoelectronically incorrect for intramolecular epoxide formation, but does have the correct relationship with the C(3) and C(4) bond in the tetrahedral intermediate for migration of the C(3)–C(4) bond. This occurs with inversion of configuration at C(2) to yield 2-α-carbomethoxy-A-norcholestane (**134**) (Eq. 138).

133 → 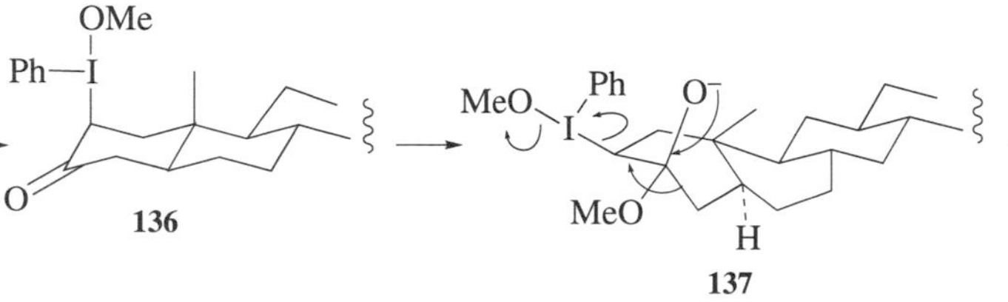 → **134**
OMe
Ph—I
O
136
MeO
Ph
I
O⁻
MeO
H
137

(Eq. 138)

Dehydrogenation

Flavanones to Flavones

In some instances, the oxidation of ketones with hypervalent iodine reagents can lead to dehydrogenation products. This occurs only when structural requirements are fulfilled, such as the oxidation of flavanones with iodine(III) reagents under a variety of conditions to give flavones (Eq. 139).[123,128] The driving force in these reactions is aromaticity of the pyran ring present in the flavones.

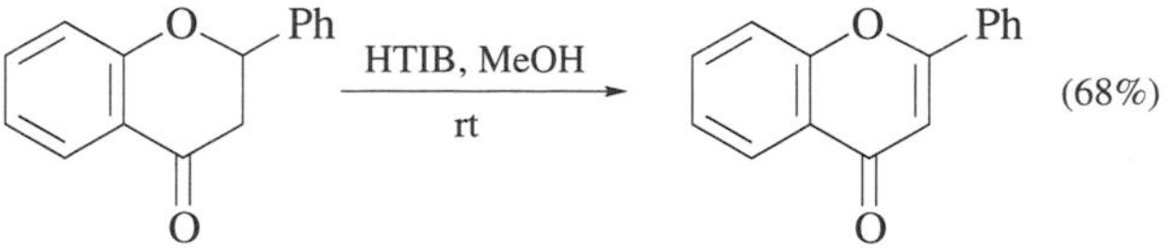
O
Ph
O
HTIB, MeOH
rt
O
Ph
O
(68%)

(Eq. 139)

1,2,3,4-Tetrahydro-4-quinolones to 4-Quinolones

The oxidation of a variety of 2-aryl 1,2,3,4-tetrahydro-4-quinolones with IBD–KOH/MeOH gives the corresponding dehydrogenated products **138** (Eq. 140).[129]

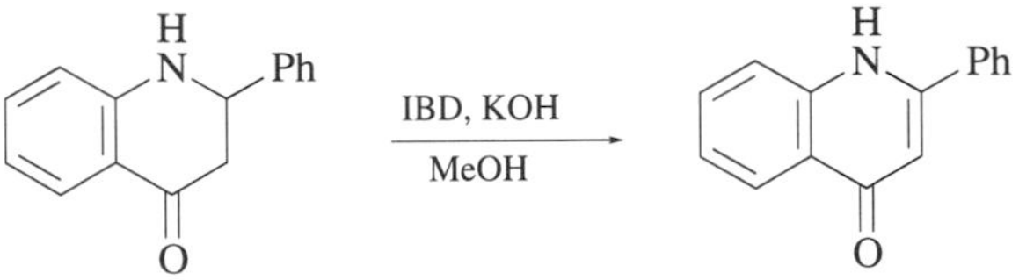
H
N
Ph
O
IBD, KOH
MeOH
H
N
Ph
O

138 (85%)

(Eq. 140)

Steroidal Ketones

The hypervalent-iodine oxidative approach, when applied to certain steroidal ketones and their silyl enol ether derivatives, yields dehydrogenated products. For example, the silyl enol ether of 3-β-hydroxy-5-α-spirostan-12-one acetate on reaction with *p*-(difluoroiodo)toluene gives the spiro-9-sten-12-one derivative (Eq. 141). Other examples are given in Eqs. 36 and 115.

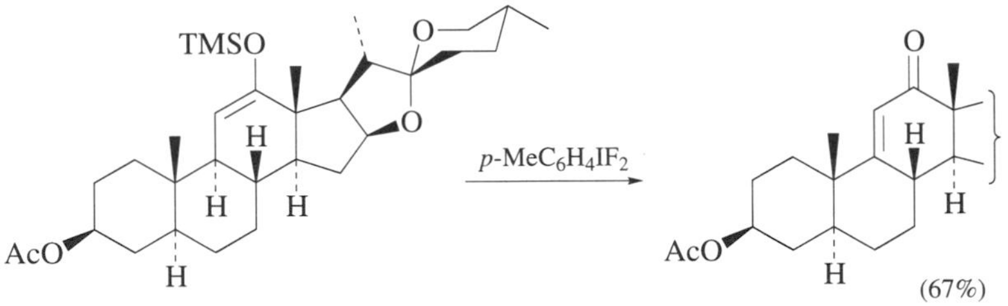
TMSO
O
O
H
H
H
AcO
H
p-MeC6H4IF2
O
H
H
AcO
H
(67%)

(Eq. 141)

COMPARISON WITH OTHER METHODS AND CONCLUSIONS

The two main oxidative processes of organohypervalent iodine reagents reviewed in this chapter are α-functionalization and rearrangement of carbonyl compounds. With respect to α-oxidative functionalization, these reagents are useful not only for α-hydroxylation of carbonyl compounds but also for a large range of α-substitution reactions such as direct α-sulfonyloxylation, α-phosphorylation, α-phosphoryloxylation, and α-amination. Specific β-functionalization can be carried out and in some cases γ-functionalization is possible. This versatility is not matched by other reagents commonly used for direct α-hydroxylation of carbonyl compounds such as oxo(diperoxymolybdenum)pyridine-hexamethylphosphoric triamide (MoOPH)[130–132] and the Davis reagent,[133,134] (camphorylsulfonyl)oxaziridine. Other methods include addition of molecular oxygen (O_2) to a lithium enolate with subsequent reduction of the thus formed α-hydroperoxy group either by triethyl phosphite[135,136] or zinc in acetic acid,[137] oxidation of silyl enol ethers with dioxygen,[138] MCBA,[139–145] osmium tetroxide-*N*-methylmorpholine *N*-oxide,[146] chromyl chloride[147] lead tetraacetate,[148] direct oxidations with *N*-sulfonyloxaziridines,[134,149] benzeneseleninic anhydride,[150] dimethyldioxirane,[151,152] or thallium(III) nitrate.[153] Use of organohypervalent iodine reagents avoids toxic chromium, lead(IV), and thallium(III) reagents; moreover, the iodine(III) reagents are operationally more convenient to work with as compared to these methods.

Similarly, literature procedures for rearrangement processes of carbonyl compounds of the type discussed in this chapter also involve either multistep procedures or make use of toxic reagents. One of the most interesting cases is the conversion of alkyl aryl ketones to 2-arylalkanoates or 2-arylalkanoic acids. The classical Willgerodt-Kindler[154–161] reaction is limited because of the rather drastic reaction conditions which often afford very poor yields. Direct methods include the oxidative rearrangements of ketones or acetals[162] using Tl(III),[163–169] Pb(IV),[170,171] or silver salts.[172,173] Organohypervalent iodine reagents, which are quite stable at room temperature and less toxic, can provide a superior and safer alternative to these toxic reagents. The iodine(III)-based methodology is more effective than other methods because of its ease, simplicity, generality, and efficiency. Another important aspcct of this approach is that it is often possible to proceed directly to the next step without isolation of the α-oxidized ketone. One-pot syntheses of a wide variety of heterocyclic compounds are available as a result of this approach.[13–15] α-Tosyloxyketones, used in such one-pot heterocyclic syntheses, are readily accessible through HTIB-mediated oxidations. They are synthetically equivalent to α-haloketones,[174] but are more stable and are not lachrymatory. Finally, iodobenzene which is produced as a byproduct of the reaction, can be recycled to regenerate the iodine(III) reagent.

EXPERIMENTAL PROCEDURES

Availability of Hypervalent Iodine Reagents. IBD,[175–178] IBTA,[179] HTIB,[28,180] *o*-iodosobenzoic acid,[181,182] and diphenyliodonium chloride[183] are stable commercially available compounds, or can be prepared by standard procedures. Iodosobenzene[177] and dichloroiodobenzene[177] are prone to decomposition and should be stored under refrigeration in dark containers. Difluoroiodobenzene can be prepared by different methods starting from iodobenzene,[184] dichloroiodobenzene,[185] IBD, or iodosobenzene.[186,187] The literature procedures for some rare iodine(III) reagents discussed in this chapter are briefly described.

Hydroxy[(+)-(10-camphorsulfonyl)oxy]iodobenzene (HCIB).[62] To a stirred suspension of iodobenzene diacetate (6.44 g, 20 mmol) in acetonitrile (40 mL) was added a solution of (+)-10-camphorsulfonic acid (6.96 g, 30 mmol) and water (0.6 mL) in acetonitrile (70 mL). The mixture was stirred for 30 minutes until a clear solution resulted. The solution was concentrated to about half its volume in vacuo and left to stand, giving 7.23 g (80%) of HCIB as colorless crystalline solid, mp 118–120° (from MeOH/Et_2O); $[\alpha]_D^{20}$ + 24.5° (MeOH); IR (Nujol) 2950, 1735, 1200, 1250, 1150 cm^{-1}; ^1H NMR [$CDCl_3$/$(CD_3)_2SO$] δ 0.78 (s, 3 H), 0.98 (s, 3 H), 1.28–2.45 (m, 7 H), 2.62–3.45 (m, 2 H), 5.0 (br s, 1 H), 7.42–8.33 (m, 5 H); MS, m/z 384 (3), 216 (31), 204 (90), 151 (7), 77 (100).

Hydroxy[bis(phenoxy)phosphoryloxy]iodobenzene (Eq. 56).[68] A mixture of iodobenzene diacetate (19.32 g, 60 mmol), diphenyl phosphate (15.25 g, 61 mmol), and water (2.16 g, 120 mmol) in acetonitrile (150 mL) was stirred at room temperature for 4 hours. Refrigeration followed by filtration gave 25.38 g (90%) of the reagent as white needles (from MeCN) , mp 102–105°; IR (CH_2Cl_2) 3399 (br, OH) cm^{-1}; ^1H NMR (300 MHz, $CDCl_3$) δ 6.95–7.14 (m, 6 H), 7.14–7.34 (m, 6 H), 7.35–7.43 (apparent t with fine structure, 1 H), 7.75–7.95 (d, 2 H), OH not observed; ^{13}C NMR ($CDCl_3$) δ 120.2 (d, J_{CP} = 5.1 Hz), 124.0 (s), 124.1 (s), 129.4 (s), 130.8 (s), 131.3 (d, J_{CP} = 6.5 Hz), 132.3 (s), 151.8 (d, J_{CP} = 7.3 Hz); ^{31}P NMR ($CDCl_3$) δ -12.4 (s).

Hydroxy[(phenoxy)(methyl)phosphonyloxy]iodobenzene (Eq. 59).[69] To a suspension of iodosobenzene (22.0 g, 100 mmol) in acetonitrile (200 mL) was added a solution of phenyl methylphosphonic acid (17.37 g, 101 mmol) in acetonitrile (50–60 mL) over 2 minutes at room temperature. The resulting mixture was stirred for about 10–15 minutes, and the initial yellow solution slowly changed to a suspension containing a white solid. Refrigeration and vacuum filtration gave 34.49 g (88%) of the title reagent as a white solid, mp 95°–105°; IR (Nujol) 3323, 1592, 1488, 1306, 1221, 1081, 922 cm^{-1}; ^1H NMR ($CDCl_3$) δ 1.34 (d, J_{HP} = 17.03 Hz, 3 H), 7.05–7.44 (m, 10 H); ^{31}P NMR ($CDCl_3$) δ 28.13 (s).

(2,2,2-Trifluoroethyl)phenyliodonium Triflate [General Procedure for the Preparation of (1H, 1H-Perfluoroalkyl)aryliodonium Triflates].[188–189] To a mixture of trifluoroacetic anhydride (82 mL) and trifluoroacetic acid (0.7 mL) was added dropwise hydrogen peroxide (6.45 mL, 60%) with stirring under cooling with an ice bath. After the reaction mixture was stirred for an additional 10 minutes, 1,1,1-trifluoro-2-iodoethane (29.89 g, 143 mmol) was added. The reaction mixture was stirred for 1 day at room temperature, and evaporated to dryness to give 1-[bis(trifluoroacetoxy)iodo]-2,2,2-trifluoroethane as a white solid in almost quantitative yield. ^1H NMR($CDCl_3$) δ 4.92 (q, J = 9 Hz, 2 H); ^{19}F NMR ($CDCl_3$) δ 62.9 (t, J = 9 Hz, 3 F), 73.4 (s, 6 F).

To 1-[bis(trifluoroacetoxy)iodo]-2,2,2-trifluoroethane (54.41 g,122 mmol) in 1,1,2-trichloro-1,2,2-trifluoroethane (150 mL) were added benzene (14.74 g, 189 mmol) and trifluoromethanesulfonic acid (18.30 g, 122 mmol) at 0° and the mixture was stirred for 1 day at 0°. The resulting mixture was evaporated to dryness to give a solid which was washed with chloroform and crystallized from acetonitrile or acetonitrile-ether at room temperature to yield 46.81 g (88%) of pure crystalline reagent, mp 88–89° (dec.); IR (KBr) 3060, 3040, 2970, 1565, 1470, 1440, 1400, 1280, 1240, 1195, 1170, 1120, 1040, 985, 840, 760, 730, 680, 650, 620 cm^{-1}; ^1H NMR (CD_3CN) δ 4.80 (q, J = 10 Hz, 2 H), 7.40–7.90 (m,

3 H), 8.00–8.30 (m, 2 H); ^{19}F NMR (CD_3CN) δ 61.5 (t, J = 10 Hz, 3 F), 77.9 (s, 3 F); MS, m/z 204 (PhI^+, 100%).

A similar procedure was adopted for the preparation of perfluoroalkylaryliodonium sulfonates.[190]

Diphenyliodonium Fluoride.[87,191] To a heterogeneous mixture of diphenyliodonium iodide (50 g, 123 mmol) and water (200 mL) was added silver oxide (24.6 g, 106 mmol) portionwise at 0°. The reaction mixture was stirred for 4 hours at 0° and 9 hours at room temperature. The insoluble material (AgI and excess Ag_2O) was then removed by filtration and washed with water. The filtrate and washings were combined, cooled below 5°, treated with phenolphthalein (5 drops, 0.05%), and acidified with 10% aqueous hydrofluoric acid. Concentration on a rotary evaporator at 50° gave crude diphenyliodonium fluoride as a pale yellow solid. The crude iodonium salt was initially recrystallized from 200 mL of hot acetone. The large crystals and mother liquor together were concentrated to ca. half-volume on a rotary evaporator, the mechanical action converting the crystalline phase to a powder. Refrigeration and vacuum filtration gave 23.53 g (63%) of diphenyliodonium fluoride as a white solid, mp 79–102°.

Benzyltrimethylammonium Dichloroiodate.[107,192] To a stirred black solution of iodine monochloride (16.25 g, 100 mmol) in dichloromethane (200 mL) was added dropwise a solution of benzyltrimethylammonium chloride (18.60 g, 100 mmol) in water. Stirring was continued at room temperature for 30 minutes. The organic layer was separated, dried ($MgSO_4$), and evaporated at reduced pressure. The residue was recrystallized from dichloromethane/ether (3:1) to afford 30.0 g (86%) of the product as brilliant yellow needles, mp 125–126°.

(Disaccharinyliodo)benzene (Eq. 116).[105] The dry sodium salt of saccharin (410 mg, 2 mmol) was suspended in acetonitrile (100 mL) or chloroform and [bis(trifluoroacetoxy)iodo]benzene (430 mg, 1 mmol) was added. After the reaction mixture was stirred at room temperature for 12 hours, the precipitate was filtered and washed abundantly with several solvents to give 528 mg (93%) of the reagent, mp 240° (dec.), IR (Nujol) 1705 cm^{-1}.

2,2-Dimethoxy-2-phenylethanol [α-Hydroxydimethylacetal Formation]

Method A. Using Iodobenzene Diacetate (Eq. 22).[23] Methanol (80 mL) was cooled to 0–5° and potassium hydroxide (8.4 g, 150 mmol) was added with stirring. Acetophenone (6.0 g, 50 mmol) dissolved in 20 mL of methanol was added dropwise over a period of 15 minutes. After the solution was stirred for 15 minutes, iodobenzene diacetate (17.71 g, 55 mmol) was added in several portions over 15 minutes. The ice bath was removed and the resultant yellow solution was stirred overnight at room temperature. The mixture was concentrated under reduced pressure in a rotary evaporator until about half of the methanol was removed, then 30 mL of water was added, and the mixture extracted with four 50-mL portions of dichloromethane. The combined dichloromethane extracts were washed with two 10-mL portions of water and dried ($MgSO_4$). The solvent was evaporated in vacuo and the crude residue was purified by distillation to give 7.37 g (81%) of the title compound, bp 73–76° (0.4 mm); IR (neat) 3470 (OH) cm^{-1}; ^1H NMR ($CDCl_3$) δ 1.83 (s, 1 H), 3.23 (s, 6 H), 3.73 (s, 2 H), 7.27–7.67 (m, 5 H); ^{13}C NMR ($CDCl_3$) δ 139.3 (s), 128.4 (d), 127.4 (d), 102.4 (s), 65.3 (t), 49.1 (s); MS, m/z 151 (M^+-OCH_3, 100) 105 (29.7), 91 (31.7), 77 (7).

Method B. Using o-Iodosylbenzoic Acid (Eq. 23).[27,193] To a solution of acetophenone (6.0 g, 50 mmol) in methanolic potassium hydroxide (8.4 g, 150 mmol) at 0–5° was added *o*-iodosylbenzoic acid (14.52 g, 55 mmol) over a period of 30 minutes. Stirring the mixture overnight, followed by workup according to Method A, resulted in almost pure product. Distillation gave 5.9 g (65%) of pure acetal, bp 73–76° (0.4 mm).

3-Hydroxychromone [α-Hydroxylation of an α,β-Unsaturated Ketone] (Eq. 33).[38]

Step I. cis-3-Hydroxy-2-methoxychromanone Dimethylacetal (**49a**). Chromone (2.92 g, 20 mmol) dissolved in 100 mL of absolute methanol was added dropwise to a stirred solution of potassium hydroxide (3.36 g, 60 mmol) in 50 mL of methanol over a period of 15 minutes at 5–10°. After the solution was stirred for 10 minutes, iodobenzene diacetate (7.09 g, 22 mmol) was added in 4–5 portions during 10 minutes and the resulting mixture was stirred overnight. Most of the methanol was evaportated in vacuo and 100 mL of water was added to the residue. The mixture was extracted with ether (5 × 40 mL) and the combined ether extracts were dried ($MgSO_4$), filtered, and evaporated in vacuo to yield crude product. Column chromatography on silica gel using hexane-ether (1:1) as eluant gave 2.54 g (53%) of pure 3-hydroxy-2-methoxychromanone acetal (**49a**) as an oil, IR (Nujol) 3510 (OH) cm^{-1}; ^1H NMR ($CDCl_3$) δ 3.10 (s, 3 H), 3.40 (s, 3 H), 3.65 (s, 3 H), 4.06 (dd, 1 H), 5.18 (d, $J_{2\text{-}3}$ = 3 Hz, 1 H), 2.5 (d, 1 H), 6.75–7.65 (m, 4 H); MS, m/z 240 (M^+, 6), 209 (17), 177 (11), 167 (76), 166 (100), 134 (53), 121 (59), 105 (97), 77 (7), 75 (33).

Step II. 3-Hydroxychromone (**50a**). To a solution of **49a** (1.20 g, 5 mmol) in acetone (10–15 mL) was added 1 mL of concentrated hydrochloric acid, and the mixture was left at room temperature for 4 hours. Colorless crystals separated from the solution. Filtration followed by washing with cold acetone (5 mL) and drying yielded 0.73 g (90%) of pure **50a**, mp 181–182°.

α,α-Diphenyl-1,3-indanedione (63) **[α-Arylation of a β-Diketone]** (Eq. 44).[194] To a solution of sodium (2.53 g, 110 mmol) in *tert*-butyl alcohol (1.2 L; freshly distilled from calcium hydride) was added with stirring 2-phenyl-1,3-indanedione (20 g; 90 mmol). After the mixture had been heated, the blood-red solution was cooled, and diphenyliodonium chloride (28.44 g, 90 mmol) was added. The reaction mixture was heated at reflux under nitrogen for 6 hours, at which time a sample gave no precipitate with a potassium iodide solution, indicating the absence of unreacted diphenyliodonium ion. The cooled (room temperature) mixture was filtered, the solid (10 g of a mixture of $CaCl_2$ and some organic material) was saved, and the filtrate was concentrated to a small volume in vacuo. The residue was combined with the 10 g of solid and was suspended in 400 mL of 1 N sodium hydroxide to remove any acidic components. This mixture was extracted with dichloromethane (1.5 L), and the organic phase was washed with 1 N sodium hydroxide until the aqueous phase was colorless. The combined aqueous phase was backwashed throughly with dichloromethane, acidified with concentrated hydrochloric acid, and extracted with dichloromethane. This extract, constituting the acidic fraction, was concentrated to dryness in vacuo to give 2.58 g (11.6%) of unchanged starting material. The combined organic phase containing the neutral fraction was dried ($MgSO_4$) and concentrated to dryness in vacuo. Column chromatography on 500 g of Florisil (60/100 mesh) and successive elution with petroleum ether and benzene yielded 21.5 g of crude **63** from the benzene fraction. Two crystallizations from ethanol yielded 14 g (52%) of pure **63**, mp 122–123°; IR 1720, 1700 cm^{-1}; ^{1}H NMR ($CDCl_3$) δ 7.28 (s, 10 H), 7.85–7.88 (m, 2 H), 8.06–8.08 (m, 2 H); MS, m/z 298(M^+).

1-Phenyl-2,4-pentanedione [γ-Arylation of a β-Diketone] (Eq. 45).[52] A 1-L three-necked flask containing a stirred suspension of sodium amide (7.8 g, 200 mmol, prepared from 4.6 g of sodium and 600 mL of liquid ammonia) was cooled in a dry ice-acetone bath under nitrogen. A solution of acetylacetone (10 g, 100 mmol) in ether (20 mL) was added in small portions from a pressure-equalizing addition funnel. The reaction mixture was warmed to room temperature and nitrogen flow was stopped. After 30 minutes, diphenyliodonium chloride (15.8 g, 50 mmol) was added over 5–10 minutes. Ether (500 mL) was added after 1 hour and the suspension was refluxed for 1 hour, cooled in ice, and 100 g of ice and 20 mL of concentrated hydrochloric acid were added. The organic layer was separated and the aqueous layer was extracted 4 times with ether. The organic extracts were dried ($MgSO_4$) and concentrated by distillation. The residual liquid was purified by vacuum distillation to give 8.1 g (92%) of the title compound, bp 138–141°(13 mm) (mp 50–53°); IR ($CHCl_3$) 1705, 1605 cm^{-1}; ^{1}H NMR (CCl_4) δ 2.15 (s, 3 H), 3.45 (s, 2 H), 5.20 (s, 1 H), 7.10–7.40 (m, 5 H), 15.6 (b s, 1 H).

Methyl α-Methoxyphenylacetate [α-Methoxylation of an Ester] (Eq. 46b).[53] A mixture of methyl phenylacetate (1.5 g, 10 mmol), sodium methoxide (30 mmol), and iodobenzene diacetate (3.22 g, 10 mmol) in methanol was stirred for 3 days. The resulting reaction mixture was acidified with dilute hydrochloric acid at 0°, the solvent (MeOH) was evaporated in vacuo, and water (20 mL) was added. Extraction with dichloromethane followed by removal of the solvent and reduced pressure distillation gave 1.26 g (70%) of α-methoxyester, bp 118–120°(10 mm); ^{1}H NMR (CCl_4) δ 3.35 (s, 3 H), 3.65 (s, 3 H), 4.67 (s, 1 H), 7.35 (s, 5 H).

α-Hydroxyphenylacetic acid [α-Hydroxylation of an Ester] (Eq. 46a).[53] A mixture of methyl phenylacetate (1.5 g, 10 mmol), potassium hydroxide

(30 mmol), and iodobenzene diacetate (3.22 g, 10 mmol) in a two-phase system consisting of benzene/water (10 mL each) was stirred until the starting material disappeared. Extraction with diisopropyl ether and crystallization from chloroform gave 0.76 g (50%) of α-hydroxy acid, mp 118–120° (from $CHCl_3$); ^1H NMR (acetone-d_6) δ 5.20 (s, 1 H), 7.00–7.60 (m, 7 H).

α-Acetoxy-*p*-chloroacetophenone [α-Acetoxylation of a Ketone] (Eq. 48).[55] *p*-Chloroacetophenone (9.27 g, 60 mmol) was dissolved in 55 mL of a mixture of acetic acid (50 mL) and acetic anhydride (5 mL) with stirring at 30°. To the vigorously stirred solution was added 5 mL of sulfuric acid, and then eight 2-g portions of iodobenzene diacetate (16.10 g, 50 mmol) were added at 30-minute intervals. The temperature was kept at about 30° throughout the additions. The reaction mixture was stirred at 30° for another 6 hours. The resulting solution was washed with water (3 × 100 mL) and extracted with chloroform (3 × 100 mL). The combined chloroform extracts were concentrated under reduced pressure and the crude residue was purified by distillation to give 5.95 g (56%) of the title product, mp 70–71°; IR (KBr) 1690, 1740 cm^{-1}; ^1H NMR ($CDCl_3$) δ 2.21 (s, 3 H), 5.26 (s, 2 H), 7.3–8.0 (m, 4 H).

2-(Tosyloxy)-3-pentanone [α-Tosyloxylation of a Ketone].[58] To a hot mixture of hydroxy(tosyloxy)iodobenzene (3.92 g, 10 mmol) and 25 mL of acetonitrile was added 10 mL of 3-pentanone. After the mixture was heated at reflux for 10 minutes, the solution was concentrated in vacuo, and the residual material was dissolved in dichloromethane (30 mL). The solution was washed with water (2 × 100 mL), dried, and concentrated on a rotary evaporator. The residual mass was distilled under reduced pressure (at 0.2 mm) to give 2.40 g (94%) of the product as an oil, ^1H NMR ($CDCl_3$) δ 0.98 (t, $J \approx 7$ Hz, 3.0 H), 1.33 (d, $J \approx 6.5$ Hz, 2.9 H), 2.44 and 2.58 (overlapping s and q, $J \approx 7$ Hz for q, 4.7 H), 4.81 (q, $J \approx 7$ Hz, 0.9 H), 7.57 (m AA′BB′, 4-OH).

α-[(10-Camphorsulfonyl)oxy]acetophenone [α-(10-Camphorsulfonyl)-oxylation of a Ketone] (Eq. 52).[62] A mixture of hydroxy[(+)-(10-camphorsulfonyl)oxy]iodobenzene (1.8 g, 4 mmol) and acetophenone (480 mg, 4 mmol) in 20 mL of acetonitrile was heated for 15 minutes at reflux. The resulting clear solution was concentrated in vacuo to give an oil. Column chromatography on silica gel (Et_2O/CH_2Cl_2 9:1) yielded 1.33 g (95%) of pure product, mp 60–61°; IR (neat) 2940, 1735, 1705, 1590, 1450, 1380, 1220, 1170, 1090 cm^{-1}, ^1H NMR ($CDCl_3$) δ 0.9 (s, 3 H), 1.1 (s, 3 H), 1.17–2.67 (m, 7 H). 2.68–3.95 (m, 2 H), 5.58 (s, 2 H), 7.45–8.07 (m, 5 H).

α-Anilinoacetophenone from Acetophenone [α-Amination of a Ketone via α-Tosyloxylation of a Ketone] (Eq. 55c).[64] Hydroxy(tosyloxy)iodobenzene (1.96 g, 5 mmol) was added to a stirred solution of acetophenone (0.6 g; 5 mmol) in acetonitrile (25 mL) and the mixture was refluxed for 2 hours. A solution of aniline (0.93 g, 10 mmol) in 5 mL of acetonitrile was added dropwise to the solution which was then heated under reflux for another 2 hours. After the solution was cooled to room temperature, the reaction mixture was concentrated in vacuo, acidified with 2 M hydrochloric acid, and extracted with ether (3 × 40 mL) to remove iodobenzene and unreacted ketone. The aqueous phase was basified with saturated aqueous sodium bicarbonate solution. The resulting solid was filtered, washed thoroughly with water, and crystallized from ether to give 548 mg (52%) of α-anilinoacetophenone, mp 93–94°; IR (KBr) 3410 (NH), 1685 (C=O) cm^{-1}; ^{1}H NMR ($CDCl_3$, 80 MHz) δ 4.51 (s, 2 H), 4.66 (s, 1 H), 6.57–8.02 (m, 10 H).

2-(2-Benzothiazolyl)-2,2-dimethoxyethanol [α-Hydroxydimethylacetal Formation via α-Tosyloxylation of a Ketone] (Eq. 55).[67]

Step I. 2-(α-Tosyloxyacetyl)benzothiazole. To a stirred solution of 2-acetylbenzothiazole (885 mg, 5 mmol) in dichloromethane (10–15 mL) was added hydroxy(tosyloxy)iodobenzene (1.96 g, 5 mmol) at room temperature in 4–5 portions over a period of 15 minutes. The resulting yellow-orange or sometimes darker mixture was stirred at room temperature for about 1 hour at which point it became homogeneous. The solvent was removed under reduced pressure and an aqueous sodium bicarbonate solution was added slowly to the residue. Extraction with dichloromethane (3 × 20 mL) followed by evaporation of solvent under reduced pressure and crystallization from ethanol gave 954 mg (55%) of the title compound as a colorless crystalline solid, mp 139–140°; IR 1715 (C=O) cm^{-1}, ^{1}H NMR ($CDCl_3$) δ 2.47 (s, 3 H), 5.65 (s, 2 H), 7.26–8.31 (m, 8 H).

Step II. 2-(2-Benzothiazolyl)-2,2-dimethoxyethanol. A suspension of the above α-tosyloxyketone (694 mg, 2 mmol) in methanol was cooled to 0–5° and an ice-cold solution of potassium hydroxide (280 mg, 5 mmol) in methanol (15 mL) was added with stirring. After stirring was continued at 0° for 2 hours, the mixture was allowed to warm to room temperature, diluted with water (25 mL), and extracted with dichloromethane (4 × 10 mL). The organic phase was dried (Na_2SO_4) and concentrated in vacuo to yield 344 mg (72%) of the title compound, mp 80–82°; ^{1}H NMR ($CDCl_3$) δ 3.36 (s, 6 H), 4.10 (s, 2 H), 7.30 8.10 (m, 4 H); MS, m/z 239 (M^+).

α-[Bis(phenoxy)phosphoryloxy]acetophenone [α-Phosphoryloxylation of a Ketone] (Eq. 57).[68] A mixture of hydroxy[bis(phenoxy)phosphoryloxy]-iodobenzene (2.37 g, 5.04 mmol) and acetophenone (1.25 g, 10.4 mmol) in acetonitrile (45 mL) was heated under reflux for 3 hours and 20 minutes. The reaction mixture was concentrated under reduced pressure and the residual oil was dissolved in dichloromethane. This solution was extracted with water and then 5% aqueous sodium bicarbonate solution. The solvent and volatile impurities (iodobenzene, acetophenone) were removed in vacuo, yielding 1.1 g (59%) of the product as an oil, IR 1709, 1292 cm^{-1}; ^{1}H NMR ($CDCl_3$) δ 5.45 (d, J = 10.1 Hz, 1 H); ^{13}C NMR ($CDCl_3$) δ 69.8 (d, J_{CP} = 5.7 Hz, C-α), 191.1 (d, J_{CP} = 5.7 Hz, C=O); ^{31}P NMR ($CDCl_3$) δ -11.6 (t, J_{PH} = 10.2 Hz).

2-(α-Hydroxyacetyl)pyridine [α-Hydroxylation of a Silyl Enol Ether] (Eq. 60).[71,72] Boron trifluoride etherate (2.84 g, 20 mmol) and the silyl enol ether of 2-acetylpyridine (1.93 g, 10 mmol) were added to a stirred and ice-cooled (0–5°) suspension of iodosobenzene (2.42 g, 11 mmol) in water (50 mL). The mixture was stirred for 2 hours, after which the temperature was raised to room temperature. Stirring was then continued for another 2 hours, during which time all of the iodosobenzene went into the solution, indicating completion of the reaction. The solution was neutralized with an excess of aqueous sodium bicarbonate solution and then extracted with dichloromethane (5 × 50 mL). The combined extracts were dried ($MgSO_4$) and concentrated under reduced pressure to yield the crude product. Addition of a mixture of hexane and ether (20 mL each), followed by filtration and cooling of the filtrate at 0° gave 850 mg (62%) of the pure product as a colorless, crystalline solid, mp 70–71°; IR (Nujol) 1720 (C=O), 3510 (OH) cm^{-1}; ^1H NMR ($CDCl_3$) δ 3.30 (br 1 H, exchanged with D_2O), 5.13 (s, 2 H), 7.30–8.72 (m, 4 H); MS (70 eV), m/z 137 (M^+, 40), 107 (88), 106 (35), 79 (95), 78 (100).

2,6-Bis[(methoxymethyl)carbonyl]pyridine (78) [α-Alkoxylation of a Silyl Enol Ether] (Eq. 65).[73] The bis-silyl enol ether of 2,6-diacetylpyridine (6.14 g, 20 mmol) was treated with iodosobenzene (8.80 g, 40 mmol), boron trifluoride etherate (11.36 g, 80 mmol), and 10 mL of methanol in 500 mL of dry dichloromethane at −70° and the temperature was raised to room temperature over a 2-hour period. To the crude mixture (isolated as described in the preceding procedure) was added hexane (50 mL) and the resulting mixture was left undisturbed for a few minutes, then filtered and cooled slowly to about 10°. After 30 minutes, 2.67 g (60%) of colorless crystalline product **78**, mp 100–101°, was collected by filtration and drying. Recrystallization from hexane gave an analytical sample, mp 101–102°. Additional product was isolated from the mother liquor. Combined yield: 3.16 g (71%); IR (Nujol) 1720 cm^{-1}; ^1H NMR ($CDCl_3$) δ 3.55 (s, 6 H), 5.05 (s, 4 H), 8.05–8.40 (m, 3 H); MS, m/z 223 (M^+, 10), 208 (100), 192 (8), 176 (18), 134 (27), 105 (20).

Methyl 2-Phenyl-2-(mesyloxy)acetate [α-Sulfonyloxylation of a Ketene Silyl Acetal] (Eq. 70).[74] Hydroxy(mesyloxy)iodobenzene (3.16 g, 10 mmol) was added to a solution of the methyl trimethylsilyl phenylketene acetal derived from methyl phenylacetate (3.33 g, 15 mmol) in dry dichloromethane (50 mL). The mixture was stirred at room temperature for 2 hours and then washed with aqueous sodium bicarbonate solution (3 × 50 mL). The organic phase was dried ($MgSO_4$) and concentrated in vacuo to yield the crude mesyloxyester which was purified by column chromatography on silica gel (hexane-dichloromethane, 1:1) to give 1.58 g (65%) of the title compound, mp 91–92°; IR (KBr) 1760 cm^{-1} (C=O); ^1H NMR ($CDCl_3$) δ 3.10 (s, 3 H), 3.80 (s, 3 H), 6.00 (s, H), 7.40–7.80 (m, 5 H); ^{13}C NMR ($CDCl_3$) δ 168.2 (s), 132.2 (s), 130.0 (s), 129.0 (s), 127.7 (s), 78.9 (s), 53.0 (s), 39.45 (s); MS, m/z 185 (53), 165 (15), 145 (15), 107 (100), 90 (12), 79 (65), 51 (17).

2-α-(Trifluoromethanesulfonyloxyacetyl)thiophene [An α-Ketotriflate from a Silyl Enol Ether] (Eq. 72).[75] To a cooled (−78°) suspension of iodosobenzene (2.64 g, 12 mmol) in dry dichloromethane (50 mL) was added trimethylsilyl triflate (3.33 g, 15 mmol). After the mixture was stirred for 10–15 minutes under N_2, 2-acetylthiophene trimethylsilyl enol ether (1.98 g, 10 mmol) in dichloromethane (10 mL) was added dropwise and stirring was continued for 1.5 hours at −78°. The mixture was then brought to room temperature, stirred for an additional hour, washed with cold water (2 × 50 mL) and with saturated aqueous sodium bicarbonate solution (25 mL), dried ($MgSO_4$), and evaporated in vacuo. The residue was recrystallized with hexanes/ether to yield 1.89 g (69%) of the title compound, mp 85–86°, ^{1}H NMR ($CDCl_3$) δ 5.50 (s, 2 H), 7.22–7.81 (m, 3 H); MS, m/z 274 (M^+, 2), 110 (100), 83 (12), 69 (12).

2-Trifluoroacetoxycyclohexanone [α-Trifluoroacetoxylation of a Silyl Enol Ether] (Eq. 76).[76] To a stirred solution of bis(trifluoroacetoxy)iodobenzene (4.30 g, 10 mmol) and pyridine (0.84 g, 10 mmol) in choroform (60 mL) was added a solution of cyclohexanone trimethylsilyl enol ether (1.70 g, 10 mmol) in choroform (10 mL). The mixture was stirred at room temperature under argon for 2 hours. The solution was concentrated, and the residue was extracted with pentane. The extract was filtered, the filtrate was evaporated, and the residue was distilled to yield 1.26 g (60%) of α-trifluoroacetoxycyclohexanone, bp 85° (20 mm); mp 35°; IR 1780 ($OCOCF_3$), 1725 (C=O) cm^{-1}; MS, m/z 210(M^+, 76), 166 (66), 96 (20), 84 (16), 69 (10), 68 (48), 55 (100).

Diethyl (2-Oxo-2-phenylethyl)phosphonate [Phosphorylation of a Silyl Enol Ether] (Eq. 77).[77] Boron trifluoride etherate (284 mg, 2 mmol) was added to iodosobenzene (220 mg, 1 mmol) in dry dichloromethane (10 mL) at −40°. The mixture was warmed to 0° until a yellow solution formed, and then was cooled to −40°. To this mixture, acetophenone trimethylsilyl enol ether (192 mg, 1 mmol) and triethyl phosphite (183 mg, 1.1 mmol) were added successively. The resulting solution was stirred at −40° for 1 hour and left to reach room temperature for 1 hour. After stirring was continued for another 30 minutes, the solution was neutralized with aqueous sodium bicarbonate solution and extracted with dichloromethane (2 × 10 mL). The combined organic phases were dried ($MgSO_4$) and concentrated under reduced pressure to give crude product which was purified by flash chromatography on silica gel to yield 192 mg (75%) of the title compound, ^{1}H NMR ($CDCl_3$) δ 1.27 (t, J = 7.0 Hz, 6 H), 3.62 (d, J = 22.6 Hz, 2 H), 4.05–4.20 (m, 4 H), 7.40 (m, 3 H), 7.98–8.03 (m, 2 H); MS, m/z 256 (M^+, 5), 146 (16), 120 (23).

Tri(2-oxo-2-phenylethyl) Phosphate[A Tris-ketol Phosphate from a Silyl Enol Ether] (Eq. 78).[78] To a mixture of crystalline phosphoric acid (100 mg, 1.02 mmol) and *p*-(difluoroiodo)toluene (780 mg, 3.05 mmol) in dry *tert*-butyl alcohol (15 mL) was added under nitrogen the silyl enol ether of acetophenone (1.25 g, 6.5 mmol). The mixture was stirred for 5.75 hours at room temperature and then concentrated in vacuo. The residual semi-solid was crystallized from acetone-hexanes to give 285 mg (63%) of the product, ^{1}H NMR ($CDCl_3$) δ 5.58 (d, J_{HP} = 11.1 Hz, 6 H); ^{13}C NMR ($CDCl_3$) δ 69.6 (d, J_{CP} = 5.6 Hz), 192.3 (d, J_{CP} = 4.7 Hz); ^{31}P NMR ($CDCl_3$) δ 0.0 (septet, J_{PH} = 11.1 Hz).

3,2′:5′,3″-Terthiophene (85) [via Formation of a 1,4-Butanedione from a Silyl Enol Ether] (Eq. 80).[81]

Step I. 1,4-Di(3′-thienyl)-1,4-butanedione. 3-Acetylthiophene silyl enol ether (1.98, 10 mmol) was added to a stirred mixture of iodosobenzene (0.88 g, 4 mmol) and boron trifluoride etherate (2.13 g, 15 mmol) in dry dichoromethane (150 mL) at −50° under nitrogen. The mixture was stirred for 1 hour at −50° and then for an additional hour at room temperature. During this period the color changed from light yellow to dark brown. The solution was washed with water (2 × 25 mL) and aqueous sodium bicarbonate solution, and the combined aqueous washings were extracted with dichloromethane (3 × 40 mL). The organic extracts were combined, dried ($MgSO_4$), concentrated in vacuo, and crystallized from ethanol to yield 600 mg (60%) of the title compound as yellow crystals, mp 129–130°; IR (Nujol) 1660 (C=O) cm^{-1}; ^{1}H NMR ($CDCl_3$) δ 3.35 (s, 4 H), 7.28–8.18 (m, 6 H); MS, m/z 250 (M^+, 10), 139 (7), 111 (100), 83 (16).

Step II. 3,2′:5′,3″-Terthiophene (**85**). To a solution of the above 1,4-diketone (500 mg, 2 mmol) in dichloromethane (10–15 mL) was added with stirring phosphorus pentasulfide (2.20 g, 5 mmol). Solid sodium bicarbonate (840 mg, 10 mmol) was then added in 5–6 portions during 5 minutes. After the reaction mixture was stirred overnight at room temperature, water (50 mL) was added and the aqueous layer was extracted with dichloromethane (2 × 20 mL). The combined organic extracts were washed with water (3 × 25 mL), dried ($MgSO_4$), concentrated in vacuo, and crystallized from dimethoxyethane to yield 370 mg (75%) of pure product **85**, mp 192–193°; MS, m/z 248 (M^+, 100), 203 (9), 171 (6), 140 (6), 127 (16), 121 (6), 83 (2).

2-Phenacylcyclohexanone [Carbon-Carbon Bond Formation] (Eq. 82).[84] To a stirred solution of iodosobenzene (220 mg, 1 mmol) in dichloromethane (5 mL) was added tetrafluoroboric acid-dimethyl ether (0.2 mL) at −50°. The mixture was warmed to 0° until a yellow solution formed, cooled to −78°, and treated with acetophenone silyl enol ether (192 mg, 1 mmol). The color of the reaction mixture changed immediately from light yellow to colorless. The cold solution of α-phenacyl phenyliodonium tetrafluoroborate was added at room temperature to a stirred solution of a 1-(cyclohexenyloxy)trimethylsilane (170 mg, 1 mmol) in dichloromethane (5 mL). The reaction mixture was stirred for 10 minutes, poured into water (50 mL), and extracted with dichloromethane (2 × 10 mL). The organic extract was dried (Na_2SO_4) and concentrated under reduced pressure. Column chromatography on silica gel with ethyl acetate-hexane mixtures as eluant gave 108 mg (50%) of α-phenacylcyclohexanone, ^{1}H NMR

($CDCl_3$) δ 1.4–1.2 (m, 6 H), 2.4 (m, 1 H), 2.6 (m, 1 H), 3.5 (m, 2 H), 7.5–7.9 (m, 5 H); MS, m/z 216 (M^+), 173, 159, 133, 120, 105 (PhCO), 77.

γ,γ,γ-Trifluorobutyrophenone [1*H*, 1*H*-Perfluoroalkylation of a Silyl Enol Ether] (Eq. 83).[85] (1*H*, 1*H*-Perfluoroethyl)phenyliodonium triflate (436 mg, 1 mmol) was added to a mixture of acetophenone silyl enol ether (192 mg, 1 mmol) and spray-dried potassium fluoride (302 mg, 5.2 mmol) in dry dichloromethane (3 mL) under an argon atmosphere. After the mixture was stirred for 1.5 hours at room temperature, filtration through a short column of silica gel, removal of the solvent, and purification by TLC on silica gel gave 176 mg (87%) of γ,γ,γ-trifluorobutyrophenone, mp 59–60°; IR (KBr) 1690 (C=O) cm^{-1}; ^1H NMR ($CDCl_3$) δ 2.20–2.80 (m, 2 H), 3.10–3.35 (m, 2 H), 7.30–7.70 (m, 3 H), 7.83–8.10 (m, 2 H); ^{19}F NMR ($CDCl_3$) δ 65.3 (t, J = 1 Hz, CF_3); MS, m/z 202 (M^+).

2-Phenylcyclohexanone [α-Phenylation of a Silyl Enol Ether] (Eq. 88).[87] A solution of 1-[(trimethylsilyl)oxy]cyclohexene (1.71 g, 10.06 mmol) in tetrahydrofuran (5 mL) was added dropwise during 2.5 minutes under N_2 at −40° to a stirred mixture of diphenyliodonium fluoride (1.51 g, 5.03 mmol) in tetrahydrofuran (15 mL). The reaction mixture was kept below −38° for 3 hours, allowed to warm during 2 hours to 10°, and kept at room temperature for 30 minutes. The resulting solution was treated with water (ca. 3 mL) and concentrated on a rotary evaporator. The residual material was taken up in dichloromethane (100 mL) and the solution was washed with water (2 × 20 mL) and saturated aqueous sodium chloride solution (15 mL), dried ($MgSO_4$), and concentrated to a yellow oil (2.24 g). Flash chromatography on silica gel with hexanes and hexanes/dichloromethane gave 775 mg (88%) of 2-phenylcyclohexanone as a white solid, mp 54–57°; IR (film) 1701 cm^{-1}; ^1H NMR ($CDCl_3$) δ 1.7–2.56 (five closely spaced m, 8 H), 3.59 (dd, J = 11.9, 5.5 Hz, 1 H), 7.10–7.4 (m, 6 H); ^{13}C NMR ($CDCl_3$) δ 24.9, 27.5, 34.8, 41.9, 57.1, 126.9, 128.4, 128.6, 138.9, 210.5.

2-*N*-(*p*-Toluenesulfonyl)aminocyclohexanone [α-Amination of a Silyl Enol Ether] (Eq. 91).[89] $Cu(MeCN)_4{\cdot}ClO_4$ (22 mg, 0.06 mmol) in acetonitrile (5 mL) was added under nitrogen to a suspension of 1-[(trimethylsilyl)oxy]-cyclohexene (0.17 g, 0.20 mL, 1.0 mmol) and *N*-(*p*-toluenesulfonyl)iminophenyliodinane (**97**) (250 mg, 0.67 mmol) in 5 mL of acetonitrile at −20°. The reaction mixture was stirred for 1.5 hours, at which time the mixture became homogeneous. The solution was filtered through a plug of silica gel and eluted with 200 mL of ethyl acetate. The solvent was removed in vacuo to give an oil that was purified by medium performance liquid chromatography using hexane:ethyl acetate (5:1) as eluant to yield 115 mg (64%) of the title compound, mp 133–135°, TLC R_f 0.28 (hexane:ethyl acetate, 2:1); IR ($CHCl_3$), 3350 (NH) 1694 (C=O) cm^{-1}; ^1H NMR ($CDCl_3$, 500 MHz) δ 1.48–1.71 (m, 3 H), 1.86 (m, 1 H), 2.07 (m, 1 H), 2.22 (dt, J = 1.0, 10.6 Hz, 1 H), 2.41 (s, 3 H), 2.42–2.54 (m, 2 H), 3.76 (m, 1 H), 5.79 (bd, J = 4.3 Hz, 1 H), 7.28 (d, J = 8.4 Hz, 2 H), 7.72

(d, J = 8.2 Hz, 2 H); ^{13}C NMR ($CDCl_3$, 126 MHz) δ 21.4, 23.9, 27.4, 36.8, 40.7, 60.6, 126.9, 129.7, 137.0, 143.5, 205.6; HRMS (FAB, MNBA) 268.1003.

3-Azido-2,4-pentanedione [α-Azidonation of a β-Dicarbonyl Compound] (Eq. 93).[90] To a cooled suspension of iodosobenzene (2.20 g, 10 mmol) in dry chloroform (50 mL) under N_2 was added azidotrimethylsilane (2.30 g, 20 mmol). The mixture was stirred for 20 minutes and 2,4-pentanedione (1.0 g, 10 mmol) was added. Stirring was continued at room temperature for 2 hours and then at reflux for 3 hours. The cooled reaction mixture was washed with water (4 × 50 mL), dried ($MgSO_4$), and concentrated under reduced pressure to yield the crude product, which was purified by silica gel column chromatography using hexane-ether (9:1) as eluant to give 1.07 g (76%) of the title product, IR (neat) 2108 (N_3), 1709 (C=O), 1616 (C=C) cm^{-1}; ^{1}H NMR ($CDCl_3$) δ 2.1 (s, 6 H), 5.5 (s, 1 H); MS, m/z 141 (M^+, 4), 113(8), 99(7), 43(100).

2-Mesyloxy-1-phenyl-1,3-butanedione [α-Mesyloxylation of a β-Dicarbonyl Compound] (Eq. 94).[90] To a dry suspension of iodosobenzene (2.20 g, 10 mmol) in dry chloroform (40 mL) was added methanesulfonic acid (0.96 g, 10 mmol). The mixture was stirred for 10 minutes at room temperature followed by addition of 1-phenyl-1,3-butanedione (benzoylacetone, 1.62 g, 10 mmol). The mixture was heated under reflux for 2 hours and then treated with aqueous sodium bicarbonate solution. The organic layer was dried ($MgSO_4$) and concentrated in vacuo to yield crude α-mesyloxy compound, which was purified by column chromatography using hexane-ether as eluant to give 1.87 g (73%) of pure title compound, IR (neat) 1734 (C=O), 1693 (C=O), 1280, 1170 (S=O) cm^{-1}; ^{1}H NMR ($CDCl_3$) δ 2.1 (s, 3 H), 3.1 (s, 3 H), 6.1 (s, 1 H), 7.3–8.2 (m, 5 H); MS, m/z 256 (M^+, 5), 167 (9), 134 (10), 105 (100).

Ethyl 2-Ethoxy-3-oxo-3-phenylpropanoate [α-Ethoxylation of a β-Dicarbonyl Compound] (Eq. 95).[90] To a suspension of iodosobenzene (1.43 g, 6.5 mmol) in dry chloroform (50 mL) was added under nitrogen boron trifluoride etherate (2.34 g, 2.03 mL, 16.5 mmol). The mixture was stirred at room temperature for 10 minutes and ethyl 3-phenyl-3-oxopropanoate (ethyl benzoylacetate, 1.25 g, 6.5 mmol) was added. The reaction mixture was heated under reflux for 3 hours. The resulting solution was basified with a saturated solution of sodium bicarbonate and the aqueous layer was extracted with chloroform (3 × 50 mL). The combined organic extracts were dried ($MgSO_4$) and concentrated under reduced pressure to yield the crude α-ethoxy ester, which was purified by column chromatography to give 0.91 g (59%) of the title product, IR 3480 (OH), 1750 (C=O), 1700 (C=O) cm^{-1}; ^{1}H NMR ($CDCl_3$) δ 1.2 (t, 1 H), 3.5 (q, 2 H), 4.2 (q, 2 H), 5.2 (s, 1 H), 7.3–7.7 & 7.8–8.3 (m, 5 H); MS, m/z 236 (M^+, 2), 163(4), 105(100).

2-Benzoyl-6-methoxybenzofuran-3(2*H*)-one [2-Aroylcoumaran-3-ones from 2-Acetyl-5-aryl Benzoates] (Eq. 107).[98] To a solution of 2-acetyl-5-methoxyphenyl benzoate (1.35 g, 5.0 mmol) in dioxane (25 mL) was added hydroxy(tosyloxy)iodobenzene (1.96 g, 5.0 mmol). After the mixture was refluxed for 2 hours, potassium hydroxide (3.4 g, 60 mmol) was added to the cooled mixture, which was heated under gentle reflux with stirring for another 30 minutes. The resulting mixture was cooled to room temperature, poured into dilute sulfuric acid (40 mL), and extracted with chloroform (4 × 25 mL). The combined organic extracts were dried (Na_2SO_4) and evaporated in vacuo. The residue was crystallized with ethanol to give 0.99 g (74%) of the product, mp 128–130°; IR (Nujol) 1605 cm^{-1}.

5-Benzoyltetrahydrofuran-2-one [Formation of a Five-membered Oxalactone from a 4-Aroylbutyric Acid] (Eq. 109).[100] To a solution of 4-benzoylbutyric acid (1.92 g, 10 mmol) in dry dichloromethane (100 mL) was added hydroxy(tosyloxy)iodobenzene (3.92 g, 10 mmol) with stirring. After the resulting slurry was heated at reflux for 15 hours, the homogeneous solution was cooled to room temperature and washed with cold aqueous sodium bicarbonate solution. The organic phase was dried ($MgSO_4$) and concentrated in vacuo. The residue was crystallized from dichloromethane-hexane to give 1.40 g (74%) of 5-benzoyltetrahydrofuran-2-one, mp 78–79°; IR ($CHCl_3$) 1790 (C=O, lactone), 1700 (C=O, ketone) cm^{-1}; ^{1}H NMR ($CDCl_3$) δ 2.55 (m, 4 H), 5.90 (m, 1 H), 7.35–7.80 (m, 3 H), 7.85–8.10 (m, 2 H).

α-Chloroacetophenone [α-Chlorination of a Ketone] (Eq. 114).[102] Benzyltrimethylammonium dichloroiodate (2.86 g, 8.23 mmol) was added to a solution of acetophenone (0.50 g, 4.16 mmol) in dichloroethane (50 mL) and methanol (20 mL) and the mixture was heated under reflux for 3 hours. The yellow solution gradually changed to brown. The solvent was removed by distillation. To the resulting residue was added 5% aqueous sodium bisulfite solution (20 mL). The mixture was extracted with ether (4 × 40 mL), the ether extracts were dried ($MgSO_4$), filtered, and evaporated at reduced pressure to give 624 mg (97%) of α-chloroacetophenone, mp 53–54°; IR (KBr) 1700 (C=O) cm^{-1}; ^{1}H NMR ($CDCl_3$) δ 4.67 (s, 2 H), 7.27–8.0 (m, 5 H).

2-(α-Saccharinyl)-3-pentanone [α-Imidylation of a Ketone] (Eq. 117).[107] (Disaccharinyliodo)benzene (568 mg, 1 mmol) in 3-pentanone (5 mL) was heated at 65° until the reaction mixture became homogeneous (5 hours). After removal of the pentanone by distillation, the residue was chromatographed on silica gel using light petroleum-chloroform mixtures as eluant. The order of elution was iodobenzene, α-saccharinylpentanone, and saccharin. The title compound (48 mg; 18%) was obtained as a crystalline solid, mp 95°; IR (Nujol) 1730 and 1720 cm^{-1}; ^{1}H NMR ($CDCl_3$) δ 1.2 (t, 3 H), 1.8 (d, 3 H), 2.8 (q, 2 H), 5.8 (q, 1 H), 8.0 (s, 4 H); MS, m/z 267 (M^+, 3), 238 (3), 183 (12).

1-Oxacyclohept-3-en-2-one [Lactones to Higher Homologous α,β-Unsaturated Lactones via Trimethylsilyloxycyclopropanes] (Eq. 118).[108] 1-Trimethylsilyloxy-2-oxabicyclo[4.1.0]heptane (0.94 g, 5.04 mmol) was dissolved in dichloromethane with stirring under a nitrogen atmosphere. The solution was cooled to 0° and iodosobenzene (1.11 g, 5.04 mmol) was added followed by tetra-*n*-butylammonium fluoride in tetrahydrofuran (5.04 mL of a 1.0 M solution, 5.04 mmol). The reaction mixture was warmed to room temperature over 1 hour and the slurry was stirred for an additional 15 hours at room temperature. The resulting homogeneous mixture was washed with water (2 × 25 mL) and dried ($MgSO_4$). Chromatography on silica gel with hexane/dichloromethane (19:1) yielded 425 mg (75%) of 1-oxacyclohept-3-en-2-one as an oil, IR 1720 (C=O), 1626 (C=C) cm^{-1}; ^1H NMR ($CDCl_3$) δ 1.90 (m, 2 H), 2.60 (m, 2 H), 4.33 (t, 2 H), 5.55 (m, 1 H), 6.45 (m, 1 H).

1-[(Triisopropylsilyl)oxy]-3-azidocyclohexene [β-Azidonation of a Triisopropylsilyl Enol Ether] (Eq. 119).[109] Trimethylsilyl azide (3.19 mL, 24 mmol) was added at −15° under argon to a suspension of iodosobenzene (2.64 g, 12 mmol) and 1-[(triisopropylsilyl)]oxy]cyclohexene (2.55 g, 10 mmol) in dichloromethane (100 mL). After the reaction mixture was stirred for one minute, gas evolution (N_2) was observed. The reaction mixture became clear after 10 minutes and was gradually warmed to 25°. The solvent was removed in vacuo, and the resulting yellow oil was kept under high vacuum for 24 hours. Flash chromatography of the oil on silica gel (230–400 mesh, 20 g) with hexane as eluant gave 2.47 g (84%) of the title compound as a colorless oil, IR (film) 2946, 2867, 2093, 1656, 1463 cm^{-1}; ^1H NMR ($CDCl_3$, 300 MHz) δ 1.0–1.25 (m, 21 H), 1.62–1.87 (m, 4 H), 2.05–2.12 (m, 2 H), 3.98–4.04 (m, 1 H), 4.96 (d, J = 4.3 Hz, 1 H).

5-Ethoxy-2-(5H)-furanone [γ-Functionalization of 2-(Trimethylsilyloxy)furan] (Eq. 120).[111] To a slurry of iodosobenzene (2.20 g, 10 mmol) in 20 mL of absolute ethanol was added at 0° 2-(trimethylsilyloxy)furan (1.56 g, 10 mmol) followed by boron trifluoride etherate (1.70 g, 12 mmol). The reaction mixture was stirred at room temperature overnight under nitrogen. The volume of the reaction mixture was reduced to one-third, the resulting solution was treated with a saturated aqueous solution of sodium bicarbonate, and the aqueous layer was extracted with dichloromethane. The combined organic phases were dried ($MgSO_4$) and concentrated under reduced pressure, and the crude product was purified by column chromatography using hexane/dichloromethane as eluant to give 0.70 g (55%) of the title product as an oil; IR (neat) 1796 (C=O) cm^{-1}; ^1H NMR ($CDCl_3$, 200 MHz) δ 1.23 (t, 3 H), 3.8 (m, 2 H), 5.95 (dd, 1 H), 6.2 (dd, 1 H), 7.3 (dd, 1 H).

Methyl *p*-Methoxyphenylacetate [Oxidative Rearrangement of an Aryl Methyl Ketone to a Methyl Arylacetate] (Eq. 121).[115] To a stirred solution of *p*-methoxyacetophenone (1.50 g, 10 mmol) in methanol was added solid hydroxy-(tosyloxy)iodobenzene (4.31 g, 11 mmol) and the reaction mixture was left at room temperature for 30 hours. Most of the methanol was removed in vacuo, water (25 mL) was added, the resulting mixture was basified with saturated aqueous sodium bicarbonate solution and extracted with dichloromethane (4 × 25 mL). The combined organic extracts were dried ($MgSO_4$) and evaporated in vacuo to yield the crude ester which was purified by column chromatography to give 1.22 g (68%) of pure title product as an oil, bp 120–122° (5 mm); IR (Nujol) 1740 cm^{-1}; ^1H NMR ($CDCl_3$) δ 3.48 (s, 2 H), 3.59 (s, 3 H), 3.67 (s, 3 H), 6.6–7.2 (AA′BB′, 4 H). Vacuum distillation of the crude mixture also gave the pure product in almost identical yield.

Methyl 2-Phenylpropanoate [Oxidative Rearrangement of an Aryl Ethyl Ketone to a Methyl 2-Arylpropanoate] (Eq. 125).[113] To a stirred solution of iodobenzene diacetate (386 mg, 1.2 mmol) and propiophenone (161 mg, 1.2 mmol) in trimethyl orthoformate (3 mL) was added sulfuric acid (0.1 ml, 2 mmol) dropwise at room temperature. The reaction mixture was stirred for 10 minutes at 60°, quenched with water (10 mL), and extracted with ether (2 × 10 mL). The organic extracts were washed with water (20 mL) and dried ($MgSO_4$). The solvent was evaporated under reduced pressure and the residue was purified by column chromatography on silica gel to give 160 mg (81%) of pure methyl 2-phenylpropanoate, bp 104–105°(18 mm); IR ($CHCl_3$) 1730 cm^{-1}; ^1H NMR ($CDCl_3$) δ 1.48 (d, J = 7 Hz, 3 H), 3.62 (s, 3 H), 3.70 (q, J = 7 Hz, 1 H), 7.24 (s, 5 H).

Isoflavone [Oxidative Rearrangement of a Flavanone] (Eq. 133).[122] To a solution of flavanone (1.12 g, 5 mmol) in acetonitrile or propionitrile (15–20 mL) was added hydroxy(tosyloxy)iodobenzene (1.96 g, 5 mmol). After the mixture was heated under reflux for about 12 hours, the solvent was removed in vacuo and water (50–70 mL) was added. The resulting mixture was extracted with dichloromethane (3 × 50 mL). The combined organic extracts were washed with water and dried ($MgSO_4$). Removal of the solvent and crystallization from either petroleum ether or ethanol gave 721 mg (65%) of isoflavone, mp 131–132°; IR (KBr) 1640 cm^{-1}; ^1H NMR ($CDCl_3$) δ 7.32–7.78 (m, 8 H), 8.01 (s, 1 H), 8.31 (dd, J = 7.5 and 11.8 Hz).

1,2,3,4-Tetrahydroxanthone [Oxidative Rearrangement of a 2-Spirochromanone] (Eq. 135).[125] *Method A (Thermal).* To a stirred solution of 2-spiro-(cyclopentane)chromanone (505 mg, 2.5 mmol) in acetonitrile (25 mL) containing a crystal of *p*-toluenesulfonic acid was added hydroxy(tosyloxy)iodobenzene (1.18 g, 3.0 mmol). The reaction mixture was heated under reflux for 10 hours. The solvent was removed under reduced pressure, water (50 mL) was added, and the mixture was extracted with dichloromethane (3 × 50 mL). The combined organic extracts were washed with aqueous sodium bicarbonate solution and water, and dried (Na_2SO_4). The solvent was evaporated under reduced pressure and the residue was purified by column chromatography on silica gel using benzene as eluant to give 400 mg (80%) of the title compound, mp 88–89°; ^{13}C

NMR ($CDCl_3$) δ 21.0, 21.6, 21.9, 28.1, 117.4, 118.2, 124.1, 125.0, 125.5, 132.7, 155.7, 163.5, 177.3. *Method B (Ultrasound).* To a solution of 2-spiro(cyclopentane)chromanone (505 mg, 2.5 mmol) in acetonitrile (10 mL) containing a crystal of *p*-toluenesulfonic acid was added hydroxy(tosyloxy)iodobenzene (1.18 g, 3.0 mmol). The mixture kept in an ultrasonic bath at 45° and the progress of the reaction was monitored by TLC. After completion of the reaction (5 minutes), the mixture was worked up as described above to afford 450 mg (90%) of the product, mp 88–89°.

TABULAR SURVEY

The tables are arranged in parallel with the text and in the order of increasing complexity of the substrates. Open-chain compounds are cited before cyclic molecules, where applicable, in the following order: aliphatic, aromatic, and heterocyclic. For each category, the order of citation of the various carbonyl compounds, where applicable, is: aldehydes, ketones, esters, lactones, β-diketones, diesters, and mixed β-dicarbonyl compounds. Within each class of compounds, attempts have been made to arrange various substituents in order of increasing carbon chain, increasing atomic number of the element, or further complexity of a group.

Numbers in parentheses are yields of isolated pure products, whereas a dash indicates that no yield is reported. Where isolated yields and yields by GLC or NMR are reported, we give only the former. Where isolated yields are not reported and yields based on NMR or GLC are available, we give the latter along with a footnote. Numbers without parentheses are ratios of products.

The following abbreviations are used in the tables:

Ac	acetyl
BTMG	*N-tert*-butyl-*N′,N″*-tetramethylguanidine
Bn	benzyl
HCIB	hydroxy[(+)-(10-camphorsulfonyl)oxy]iodobenzene
HMIB	hydroxy(mesyloxy)iodobenzene
HTIB	hydroxy(tosyloxy)iodobenzene
IBD	iodobenzene diacetate
IBTA	iodobenzene bis(trifluoroacetate)
Mes	mesityl
MOM	methoxymethyl
Ms	mesyloxy
Py	pyridine
R_f	perfluoroalkyl
TBDMS	*tert*-butyldimethylsilyl
TEBA	benzyltrimethylammonium chloride
Tf	trifluoromethanesulfonyl

TFA	trifluoroacetic acid
THF	tetrahydrofuran
TIPS	triisopropylsilyl
TMOF	trimethyl orthoformate
TMS	trimethylsilyl
Ts	*p*-toluenesulfonyl
((((	ultrasound

TABLE IA. α-HYDROXYLATION OF ACYCLIC KETONES VIA α-HYDROXYDIMETHYLACETAL FORMATION

Substrate	Reagent-Conditions	Product(s) and Yield(s) (%)	Refs.
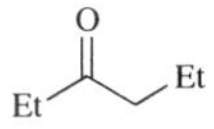	$PhI(OAc)_2$, KOH/MeOH	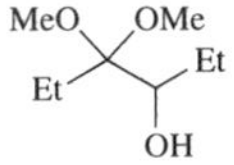 (65)	27
	o-$OIC_6H_4CO_2H$, KOH/MeOH	" (62)	27
	1. $PhI(OAc)_2$ or $(PhIO)_n$ or o-$OIC_6H_4CO_2H$, KOH/MeOH 2. HCl, H_2O	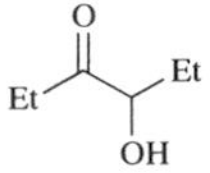 [a] (50)	27
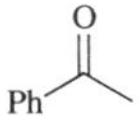	$PhI(OAc)_2$, KOH/MeOH	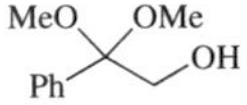 (81)	27
	o-$OIC_6H_4CO_2H$, KOH/MeOH	" (65)	193
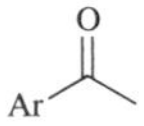	1. $PhI(OAc)_2$ or $(PhIO)_n$ or o-$OIC_6H_4CO_2H$, KOH/MeOH 2. HCl, H_2O	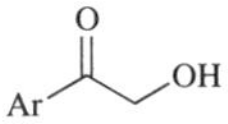 Ar — % Ph (60) p-FC_6H_4 (70) p-ClC_6H_4 (63) p-BrC_6H_4 (70) p-IC_6H_4 (71) p-$O_2NC_6H_4$ (48) p-MeC_6H_4 (45) p-$MeOC_6H_4$ (50) 3,4-$(MeO)_2C_6H_4$ (40)	23
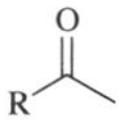	$PhI(OAc)_2$, KOH/MeOH	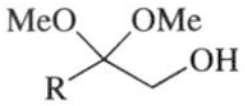 R — % 2-pyridyl (61) 3-pyridyl (40-45) 4-pyridyl (58)	30

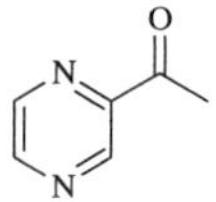	$PhI(OAc)_2$, KOH/MeOH	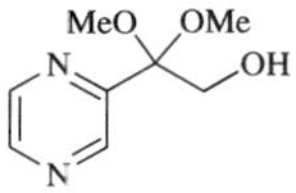	(62)	30
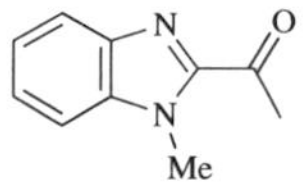	$PhI(OAc)_2$, KOH/MeOH	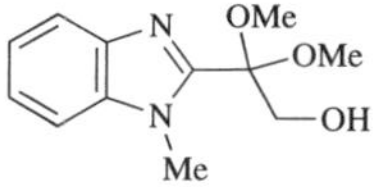	(65)	30
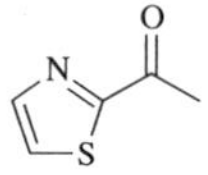	$PhI(OAc)_2$, KOH/MeOH	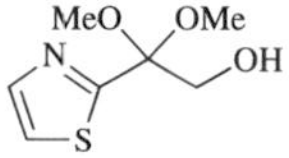	(65)	31a
	1. $PhI(OAc)_2$ or $(PhIO)_n$ or *o*-$OIC_6H_4CO_2H$, KOH/MeOH 2. HCl, H_2O	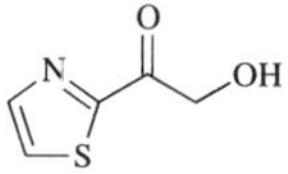	(57)	31a
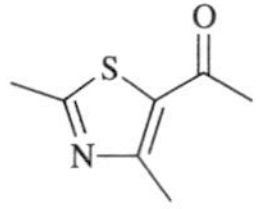	$PhI(OAc)_2$, KOH/MeOH	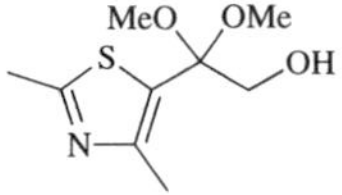	(69)	31a
	1. $PhI(OAc)_2$ or $(PhIO)_n$ or *o*-$OIC_6H_4CO_2H$, KOH/MeOH 2. HCl, H_2O	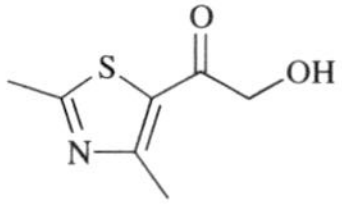	(66)	31a
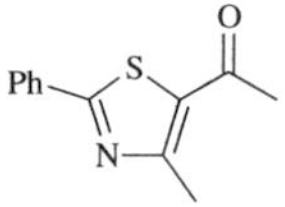	$PhI(OAc)_2$, KOH/MeOH	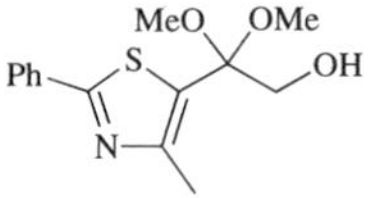	(59)	31a
	1. $PhI(OAc)_2$ or $(PhIO)_n$ or *o*-$OIC_6H_4CO_2H$, KOH/MeOH 2. HCl, H_2O	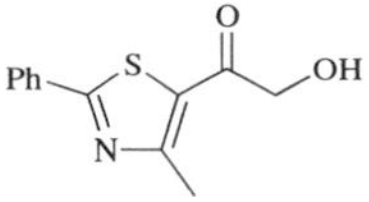	(46)	31a

TABLE IA. α-HYDROXYLATION OF ACYCLIC KETONES VIA α-HYDROXYDIMETHYLACETAL FORMATION (*Continued*)

Substrate	Reagent-Conditions	Product(s) and Yield(s) (%)	Refs.
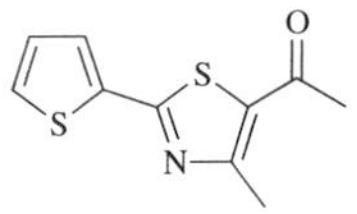	$PhI(OAc)_2$, KOH/MeOH	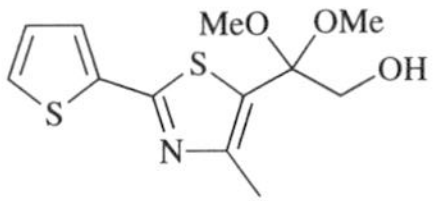 (58)	31a
	1. $PhI(OAc)_2$ or $(PhIO)_n$ or *o*-$OIC_6H_4CO_2H$, KOH/MeOH 2. HCl, H_2O	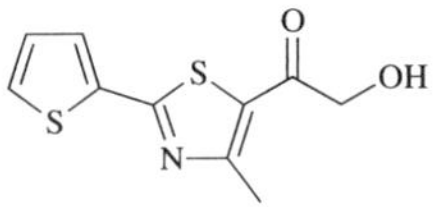 (45)	31a
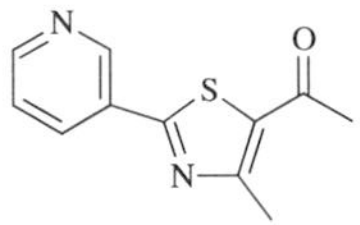	$PhI(OAc)_2$, KOH/MeOH	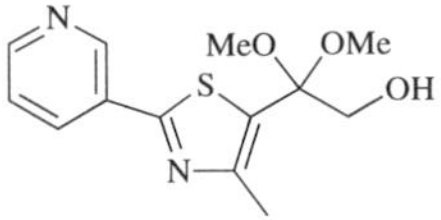 (56)	31a
	1. $PhI(OAc)_2$ or $(PhIO)_n$ or *o*-$OIC_6H_4CO_2H$, KOH/MeOH 2. HCl, H_2O	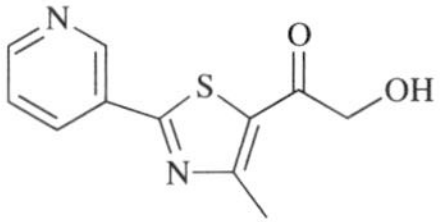 (44)	31a
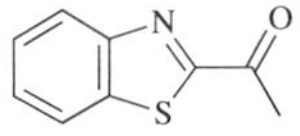	$PhI(OAc)_2$, KOH/MeOH	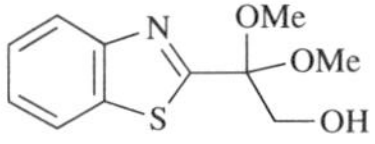 (50)	31a
	1. $PhI(OAc)_2$ or $(PhIO)_n$ or *o*-$OIC_6H_4CO_2H$, KOH/MeOH 2. HCl, H_2O	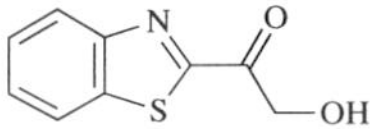 (45)	31a
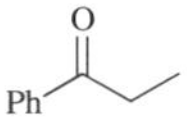	$PhI(OAc)_2$, KOH/MeOH	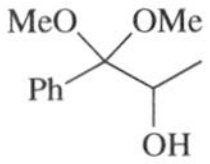 (71)	27
	o-$OIC_6H_4CO_2H$, KOH/MeOH	" (70)	27

	1. $PhI(OAc)_2$ or $(PhIO)_n$ or o-$OIC_6H_4CO_2H$, KOH/MeOH 2. HCl, H_2O	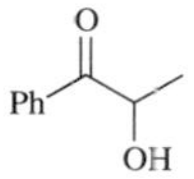	(57)	27
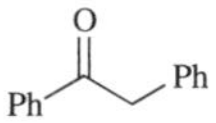	$(PhIO)_n$, KOH/MeOH	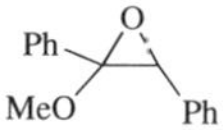	(53)	23
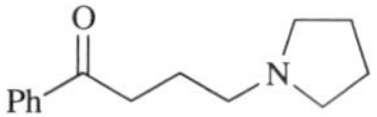	$PhI(OAc)_2$, KOH/MeOH	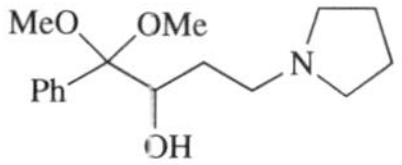	(25)	29, 30
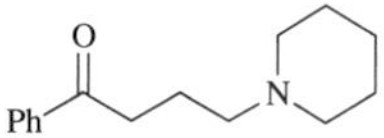	$PhI(OAc)_2$, KOH/MeOH	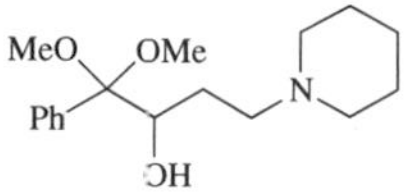	(50)	29, 30
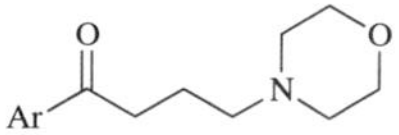	$PhI(OAc)_2$, KOH/MeOH	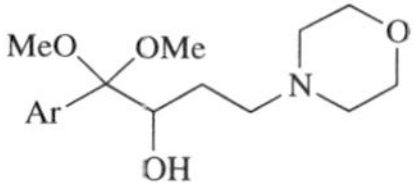	Ar / % Ph (60) p-$MeOC_6H_4$ (44)	29, 30
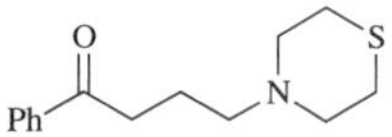	$PhI(OAc)_2$, KOH/MeOH	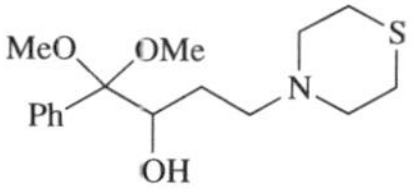	(65)	29, 30
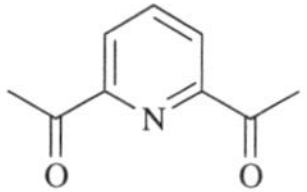	$PhI(OAc)_2$, KOH/MeOH; or $(PhIO)_n$, KOH/MeOH	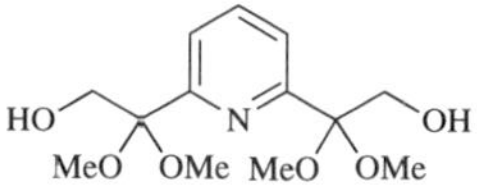	(72)	195
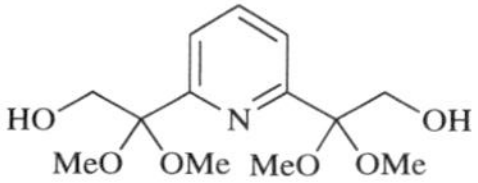	p-TsOH/H_2O, acetone, 4 d	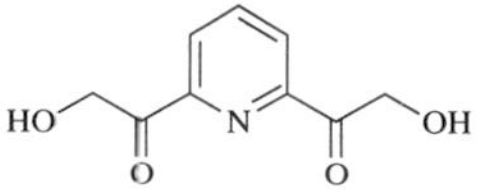	(90)	23, 195

[a] A mixture of this product and its isomer, 3-hydroxy-2-pentanone, was isolated in the ratio 5.8:1.

TABLE IB. α-HYDROXYLATION OF CYCLIC KETONES VIA α-HYDROXYDIMETHYLACETAL FORMATION

Substrate	Reagent-Conditions	Product(s) and Yield(s) (%)	Refs.
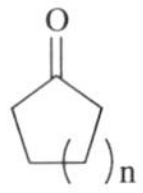	Reagent (see table), KOH/MeOH	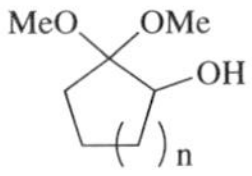 n / Reagent / %: 1 IBD (78); 1 o-$OIC_6H_4CO_2H$ (83); 2 IBD (75); 2 o-$OIC_6H_4CO_2H$ (74); 3 IBD (68); 3 o-$OIC_6H_4CO_2H$ (71); 6 IBD (56); 6 o-$OIC_6H_4CO_2H$ (61)	27
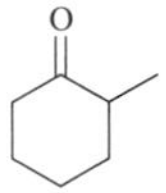	IBD, KOH/MeOH	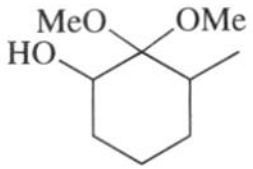 + 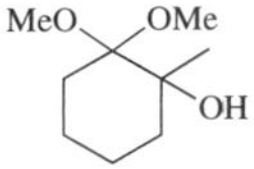 6:1 (62)	27
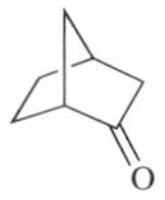	o-$OIC_6H_4CO_2H$, KOH/MeOH	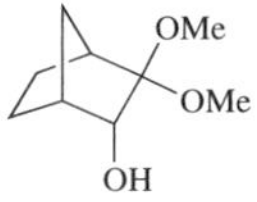 (67)	27
	IBD, KOH/MeOH	" (47)	27
	1. IBD, KOH/MeOH 2. 3N HCl	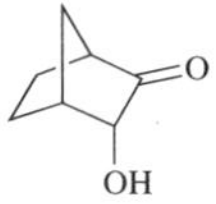 (33)	27
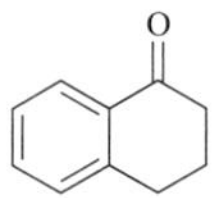	1. IBD, KOH/MeOH, 0-5°, 1 h 2. 23-25°, 20 h	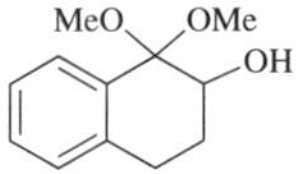 (70)	36
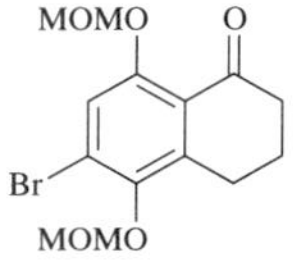	IBD, KOH/MeOH	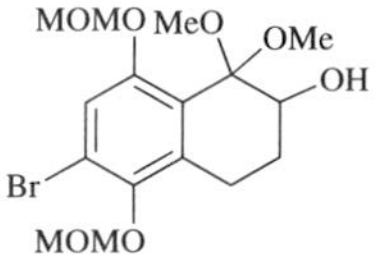 (—)	196

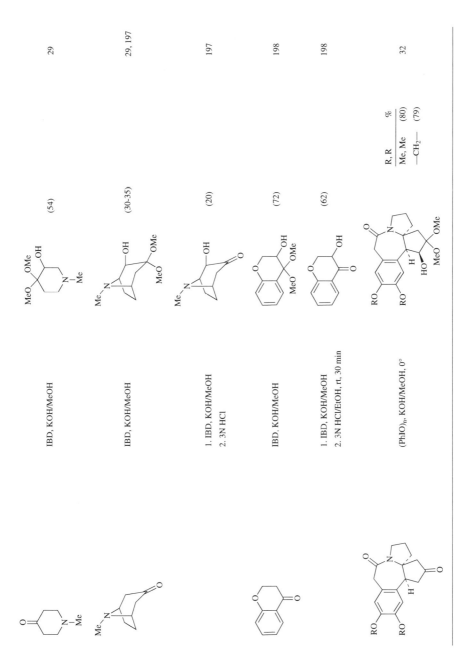

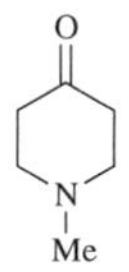	IBD, KOH/MeOH	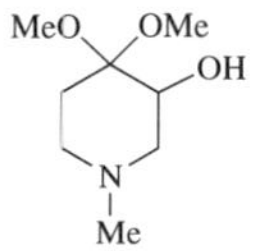	(54)	29
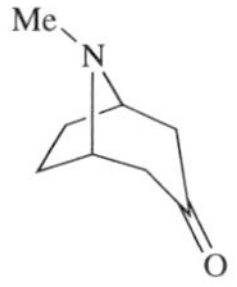	IBD, KOH/MeOH	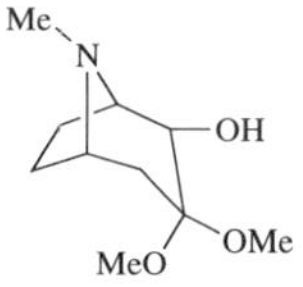	(30-35)	29, 197
	1. IBD, KOH/MeOH 2. 3N HCl	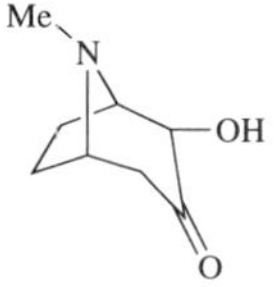	(20)	197
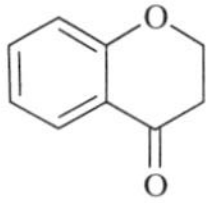	IBD, KOH/MeOH	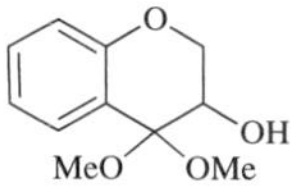	(72)	198
	1. IBD, KOH/MeOH 2. 3N HCl/EtOH, rt, 30 min	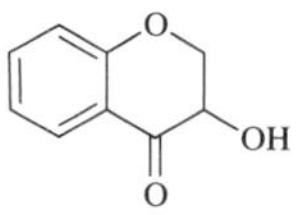	(62)	198
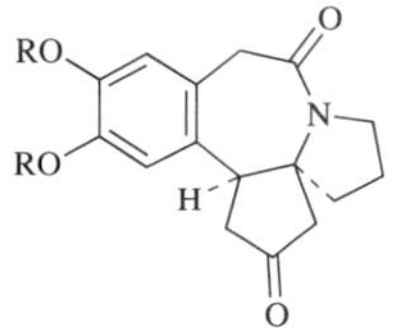	$(PhIO)_n$, KOH/MeOH, 0°	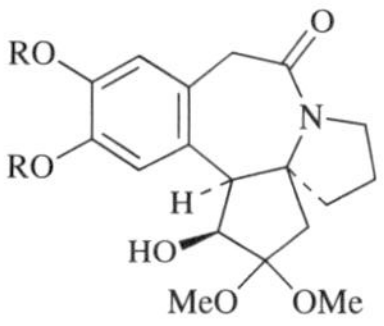	R, R — % Me, Me — (80) —CH_2— — (79)	32

TABLE IC. α-HYDROXYLATION OF STEROIDAL KETONES VIA α-HYDROXYDIMETHYLACETAL FORMATION

Substrate	Reagent-Conditions	Product(s) and Yield(s) (%)	Refs.
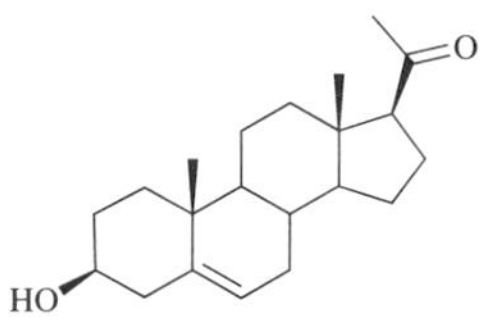	$(PhIO)_n$, KOH/MeOH, 0°-rt	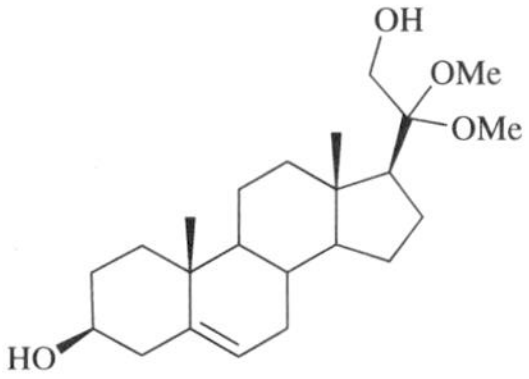 (67)	33
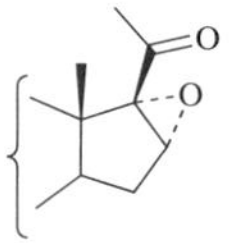	IBD, NaOH/MeOH, 20°, 3 h	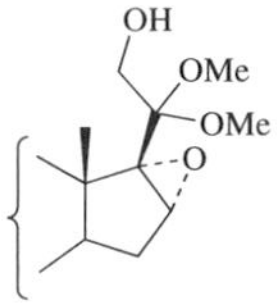 (70)	199
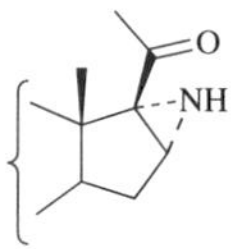	IBD, NaOH/MeOH, 20°, 2 h	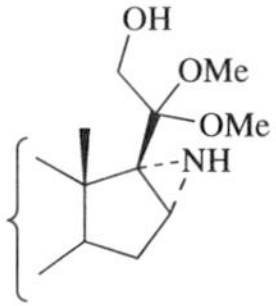 (63)	199
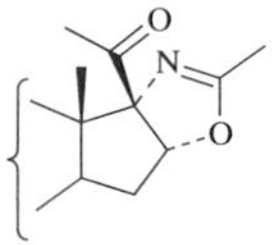	IBD, NaOH/MeOH, 20°, 3.5 h	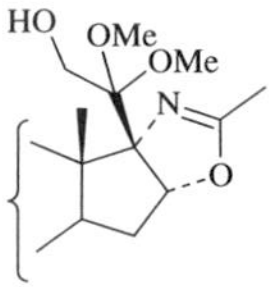 (60)	199
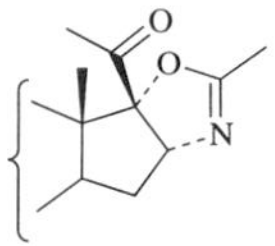	IBD, NaOH/MeOH, 20°, 120 h	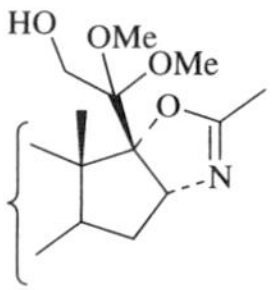 (82)	199

	IBD, NaOH/MeOH, 20°, 7 h	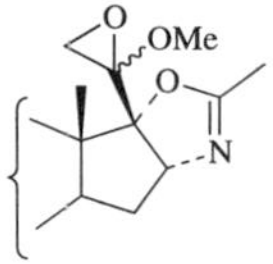	(80)	199
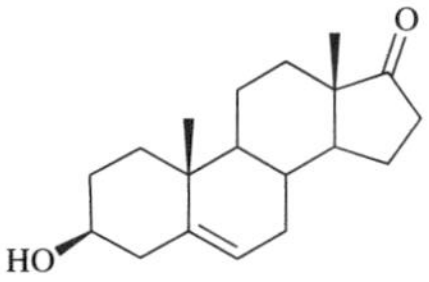	o-OIC$_6$H$_4$CO$_2$H (2.4 eq), KOH (6.6 eq)/MeOH , 70°, 8 h or rt, 15 d	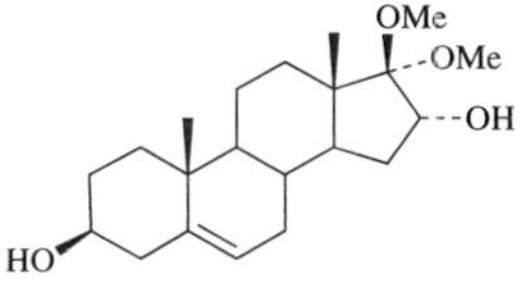	(19)	42
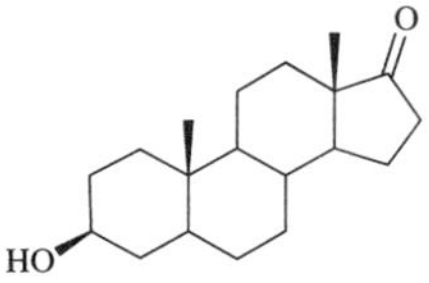	"	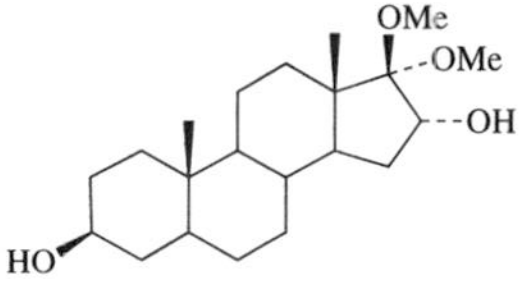	(18)	42
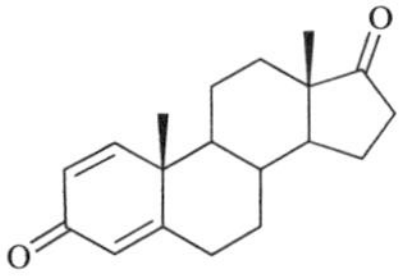	"	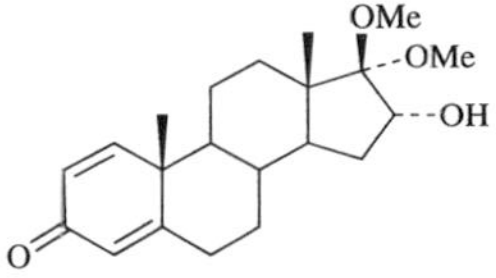	(26)	42
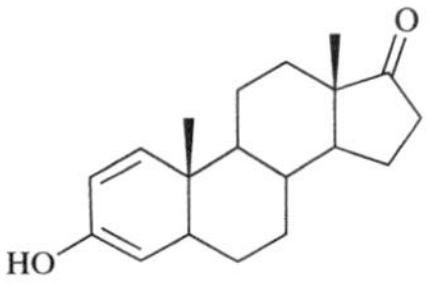	"	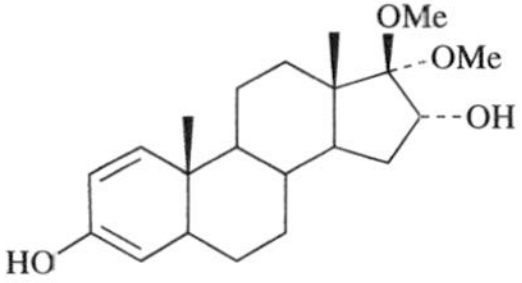	(22)	42

TABLE ID. α-HYDROXYLATION OF METAL-COMPLEXED KETONES VIA DIMETHYLACETAL FORMATION

Substrate	Reagent-Conditions	Product(s) and Yield(s) (%)	Refs.
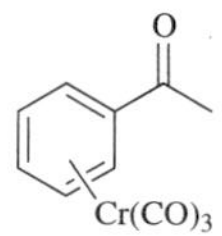	IBD, KOH/MeOH, 0°-rt	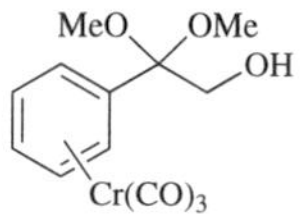 (60)	35, 36
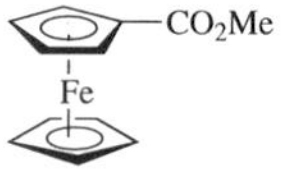	1. IBD, KOH/MeOH 2. HCl, H_2O	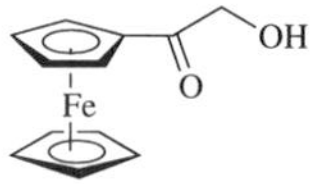 (60)	23
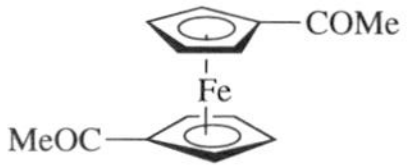	1. IBD, KOH/MeOH 2. HCl, H_2O	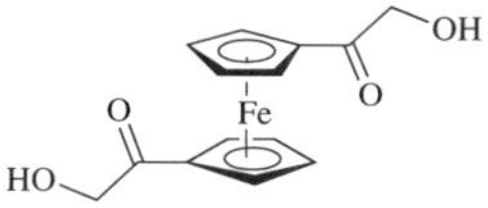 (34)	23
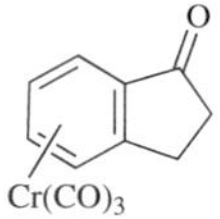	IBD, KOH/MeOH, 0°-rt	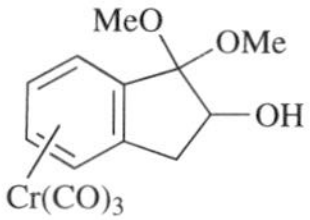 (60)	35, 36
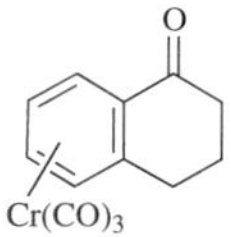	IBD, KOH/MeOH, 0°-rt	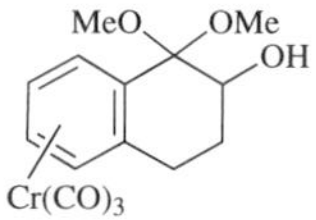 (80)	35, 36
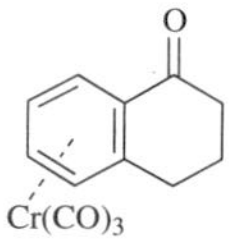 (+)	IBD, KOH/MeOH, 0°-rt	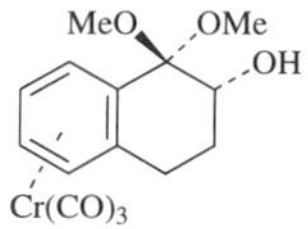 (2*R*)-(—) (80)	35, 36

TABLE II. FUNCTIONALIZATION OF α,β-UNSATURATED CARBONYL COMPOUNDS

Substrate	Reagent-Conditions	Product(s) and Yield(s) (%)	Refs.
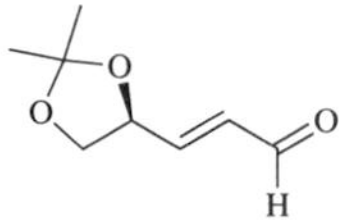O O O H	IBD (2 eq), KOH/MeOH, rt, 40 min	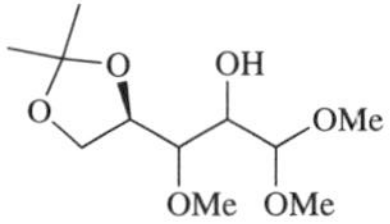O O OH OMe OMe OMe (67) 15:5:5:3 diastereomeric mixture	40
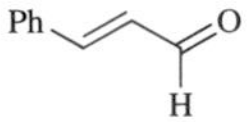Ph O H	IBD (2 eq), KOH/MeOH, rt, 30 min	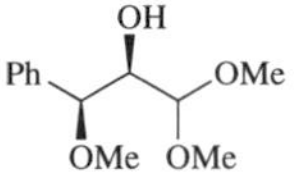OH Ph OMe OMe OMe + 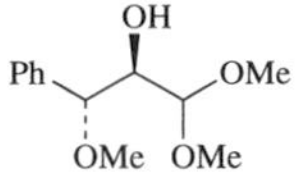OH Ph OMe OMe OMe (62) 53:47 diastereomeric mixture	40
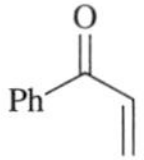O Ph	IBD (1.1 eq), KOH/MeOH, rt, 20 min	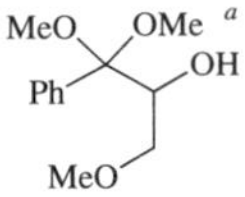MeO OMe OH Ph MeO [a] (76) + 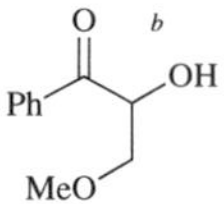O Ph OH MeO [b] (69)	40
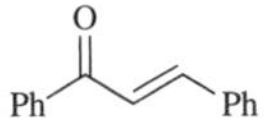O Ph Ph	IBD (1.1 eq), KOH/MeOH, 0°-rt overnight	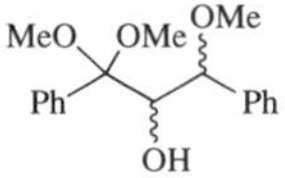MeO OMe OMe Ph Ph OH (60)	37
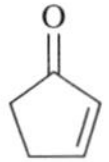O	IBD (1.1 eq), KOH/MeOH, 0°, 40 min	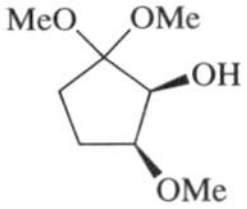MeO OMe OH OMe (55)	40
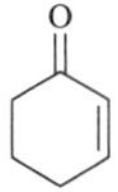O	IBD (1.1 eq), KOH/MeOH, 0°, 1 h	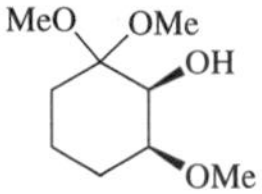MeO OMe OH OMe (52)	40
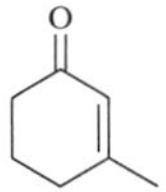O	IBD (3 eq), KOH/MeOH, rt, 1 h	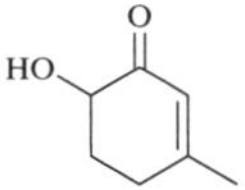HO O (32) + 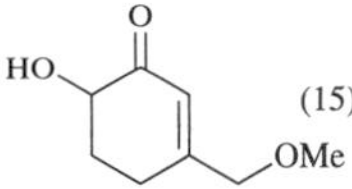HO O OMe (15)	40

TABLE II. FUNCTIONALIZATION OF α,β-UNSATURATED CARBONYL COMPOUNDS (*Continued*)

Substrate	Reagent-Conditions	Product(s) and Yield(s) (%)	Refs.
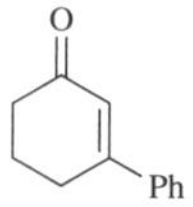	IBD (2 eq), KOH/MeOH, rt, 30 min	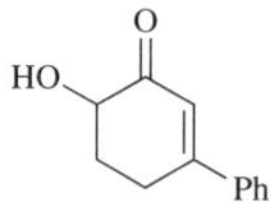 (58) + 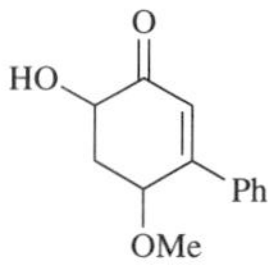 (15)	40
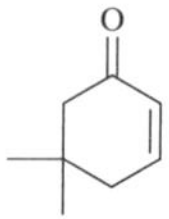	IBD (1.1 eq), KOH/MeOH, rt, 30 min	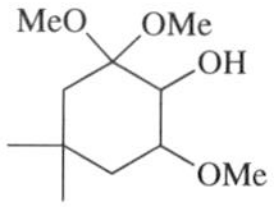 (23)	40
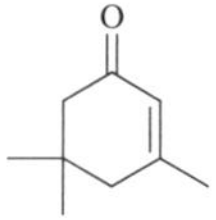	IBD (3 eq), KOH/MeOH, rt, 30 min	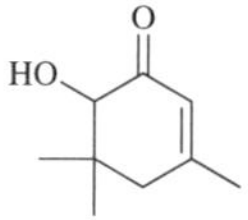 (15) + 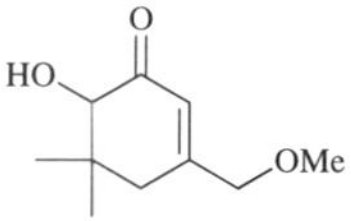 (8)	40
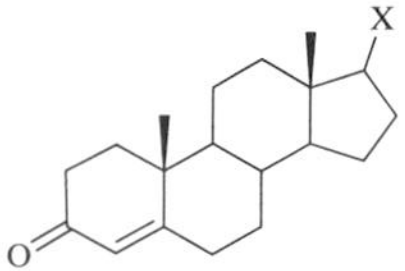	*o*-$OIC_6H_4CO_2H$, KOH/MeOH, 70°, 1 h	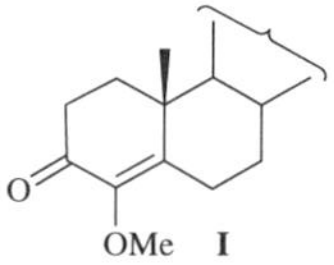 **I** + 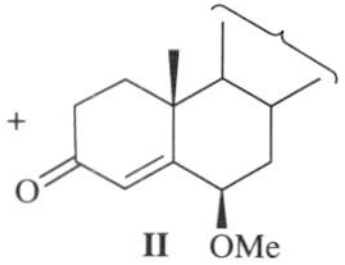 **II** + 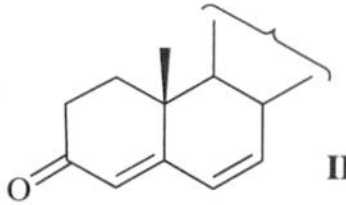 **III**	41, 42

X	% **I**	% **II**	% **III**
=O	(25)	(33)	(20)
OH	(23)	(47)	(22)

[a] The general workup mixture contained acetal and ketone in a ratio of 1:2.

[b] The reaction mixture was concentrated and subjected directly to column chromatography without workup with saturated NaCl. The mixture contained acetal and ketone in a ratio of 2:1.

TABLE III. C(3)-HYDROXYLATION OF CHROMONE AND FLAVONES

Substrate	Reagent-Conditions	Product(s) and Yield(s) (%)	Refs.
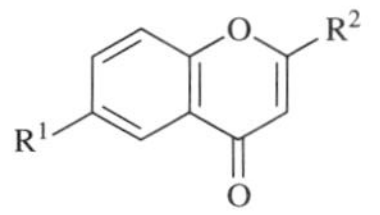	IBD, KOH/MeOH, 5°-rt	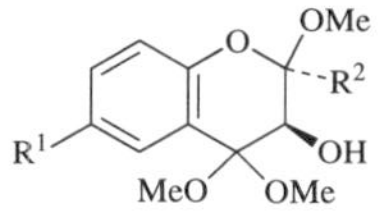 R^1 R^2 %	
		H H (53)	37, 38
		H Ph (67)	37, 38
	IBD, KOH/MeOH-THF, rt	" R^1 R^2 %	
		H p-ClC_6H_4 (69)	39
		H p-MeC_6H_4 (60)	39
		H p-$MeOC_6H_4$ (67)	39
		Cl Ph (71)	39
		Cl p-ClC_6H_4 (68)	39
		Cl p-$MeOC_6H_4$ (70)	39
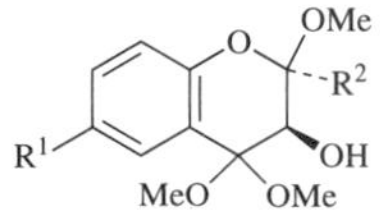	Conc HCl, acetone	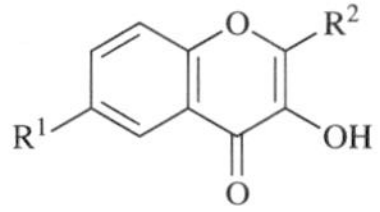 R^1 R^2 %	
		H H (90)	38
		H Ph (85)	38
		H p-ClC_6H_4 (85)	39
		H p-MeC_6H_4 (82)	39
		H p-$MeOC_6H_4$ (84)	39
		Cl Ph (87)	39
		Cl p-ClC_6H_4 (89)	39
		Cl p-$MeOC_6H_4$ (90)	39
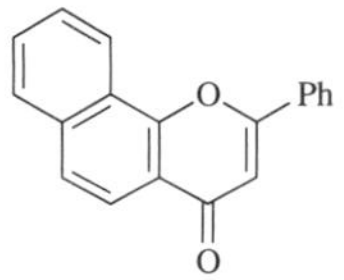	IBD, KOH/MeOH, rt, overnight or Conc HCl/acetone	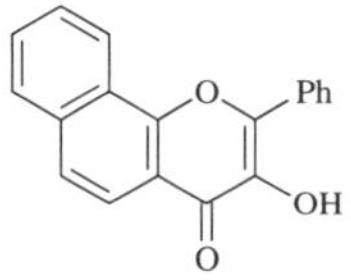 (40)	38

TABLE IV. SYNTHESIS OF *CIS*-3-HYDROXYFLAVANONES AND RELATED COMPOUNDS

Substrate	Reagent-Conditions	Product(s) and Yield(s) (%)				Refs.
			R^1	R^2	%	
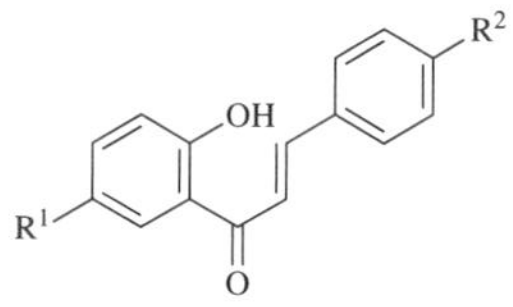	IBD, KOH/MeOH, 5-10°	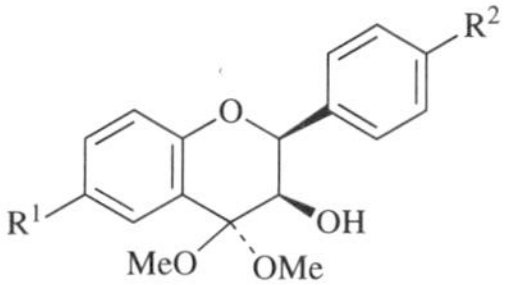	H	H	(65)	37, 43
			H	Cl	(70)	44
			H	Me	(71)	44
			H	OMe	(65)	44
			Cl	H	(69)	44
			Cl	OMe	(68)	44
			Cl	Cl	(71)	44
			Me	H	(68)	44
			Me	Cl	(70)	44
			R^1	R^2	%[a]	
	1. IBD, KOH/MeOH, 5-10° 2. 50% aq AcOH	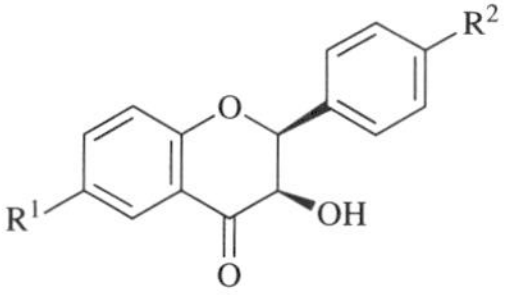	H	H	(71)	43
			H	Cl	(75)	44
			H	Me	(74)	44
			H	OMe	(76)	44
			Cl	H	(76)	44
			Cl	OMe	(76)	44
			Cl	Cl	(78)	44
			Me	H	(75)	44
			Me	Cl	(77)	44
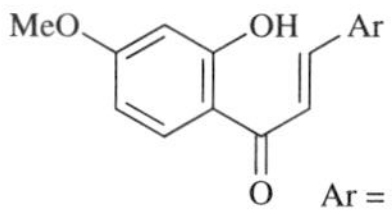 Ar = 3,4-$MeOC_6H_3$	IBD, KOH/MeOH, 10-15°, overnight	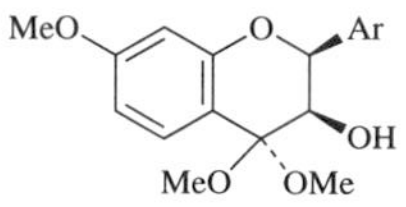 (25) + 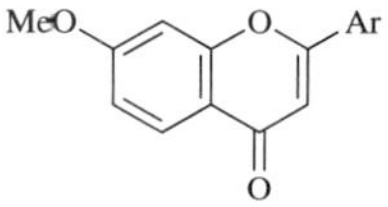 (50)				44
	1. IBD, KOH/MeOH, 5-10° 2. 50% aq AcOH	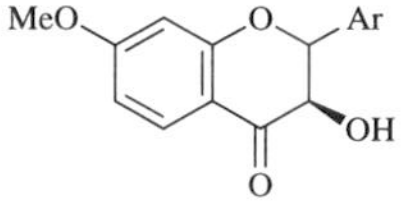 (72)				44

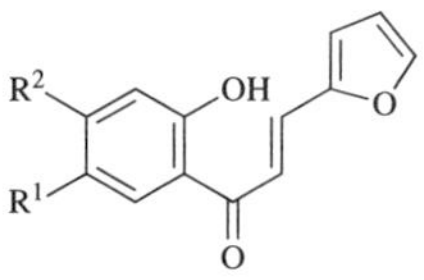

1. IBD, KOH/MeOH, CH_2Cl_2, 0-5°
2. See table

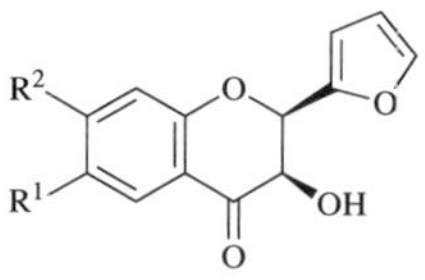

45

R^1	R^2	Cond	%
H	H	A	(41)
Cl	H	B	(51)
Cl	Me	B	(45)

A = Conc HCl/EtOH (1:99), pet ether
B = 50% aq AcOH, 3 h, rt

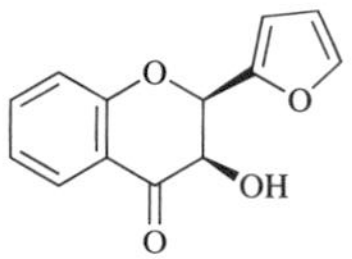

5% NaOH, rt, 30 min

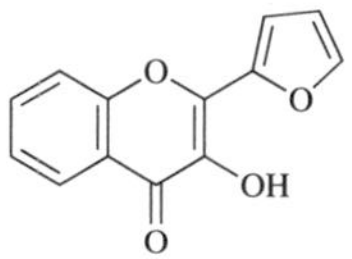

(90)

45

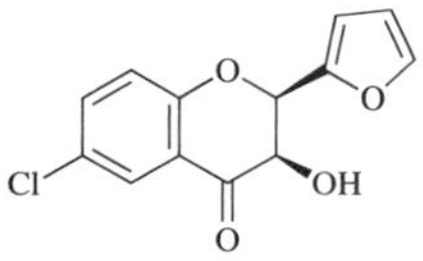

5% NaOH, rt, 30 min

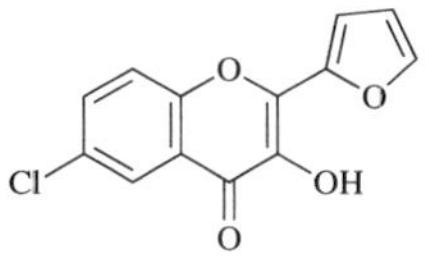

(94)

45

[a] The yields of this product are reported with respect to the product (starting material for step 2 in conditions) in the previous entry.

TABLE V. ARYLATIONS OF CARBONYL COMPOUNDS

Substrate	Reagent-Conditions	Product(s) and Yield(s) (%)	Refs.
	Ph_2IBr, K/liq NH_3	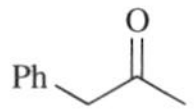 (71)	200
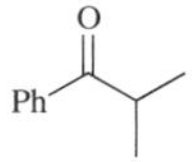	Ph_2ICl, Na/*t*-BuOH, 26°	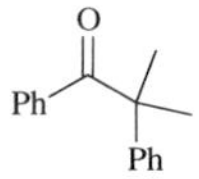 (57)	46
	Ph_2ICl, Na/EtC(Me)$_2$OH, 0-5°	" (81)	46
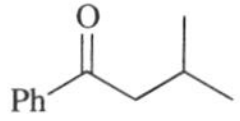	Ph_2ICl, Na/EtC(Me)$_2$OH, 0-5°	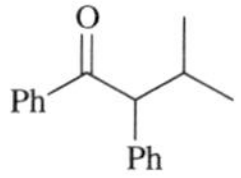 (23)	46
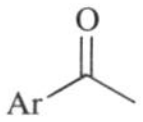 Ar = mesityl	Ph_2ICl, Na/*t*-BuOH, reflux, 6 h	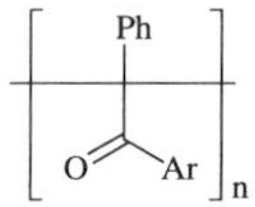 (—)	46
	Ph_2ICl, $NaNH_2$/liq NH_3	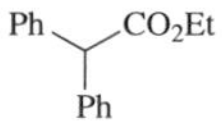 (57) + 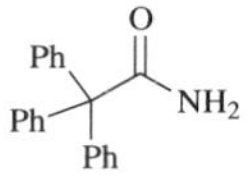 (31)	47
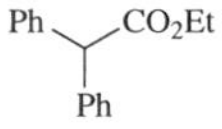	Ph_2ICl, K/*t*-BuOH	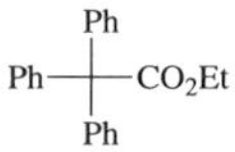 (28)	47
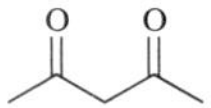	Ph_2ICl, Na/liq NH_3 (2 eq)	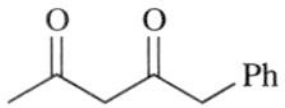 (92)	52
	(p-ClC_6H_4)$_2$ICl, Na/liq NH_3	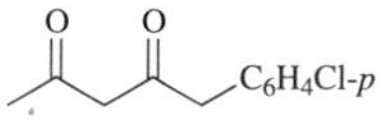 (44)	52

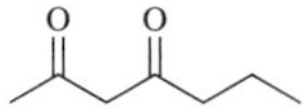	Ph_2ICl, Na/liq NH_3 (2 eq)	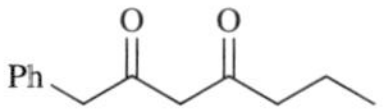 (98)	52
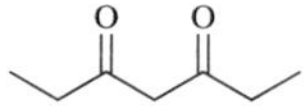	Ph_2ICl, Na/liq NH_3 (2 eq)	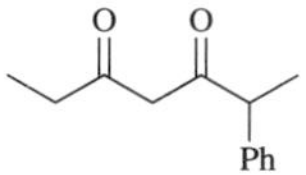 (50)	52
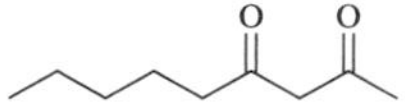	Ph_2ICl. Na/liq NH_3 (2 eq)	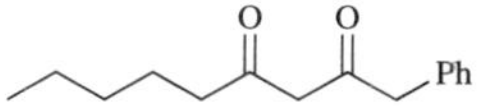 (78)	52
	(p-$MeC_6H_4)_2ICl$, Na/liq NH_3 (2 eq)	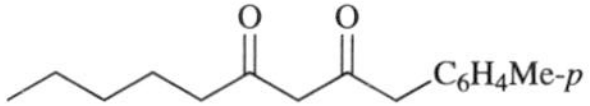 (21)	52
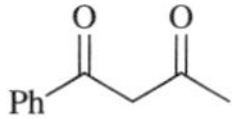	Ph_2ICl, Na/liq NH_3 (2 eq)	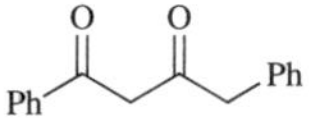 (61)	52
	(p-$MeC_6H_4)_2ICl$, Na/liq NH_3 (2 eq)	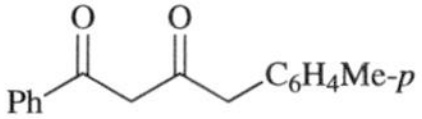 (44)	52
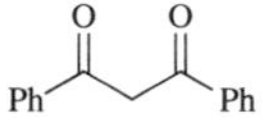	Ph_2ICl, K/t-BuOH	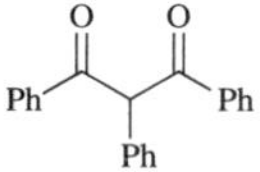 (31) + 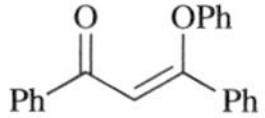 (9)	48
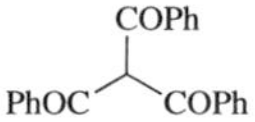	Ph_2ICl, K/t-BuOH	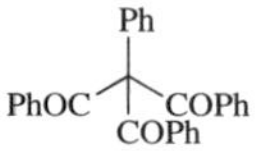 (10) + 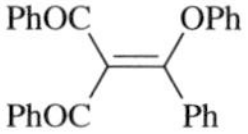 (68)	48
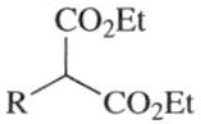	Ph_2ICl, K/t-BuOH	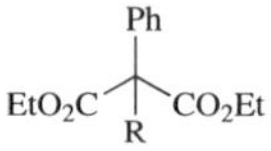 R: H (—); Et (55); Ph (37); NHCOMe (34)	47

R	%
H	(—)
Et	(55)
Ph	(37)
NHCOMe	(34)

TABLE V. ARYLATIONS OF CARBONYL COMPOUNDS (*Continued*)

Substrate	Reagent-Conditions	Product(s) and Yield(s) (%)	Refs.
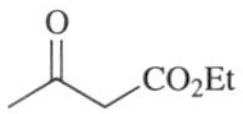	Ph_2IBF_4, K/*t*-BuOH	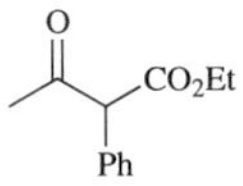 (—)	47
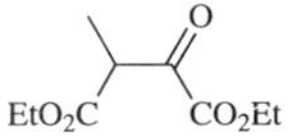	Ph_2ICl, K/*t*-BuOH	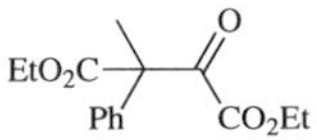 (70)	47
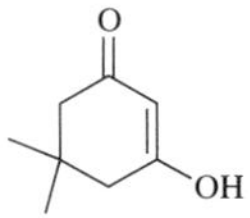	Ph_2IX, *t*-BuONa	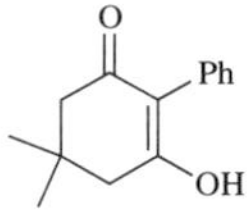 **I** + 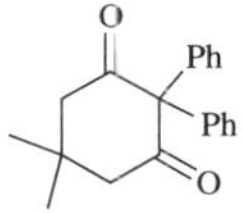 **II**	

X	Solvent	Temp	Time	**I**	**II**	Refs.
Br	dioxane:water (1:1)	100°	2 h	(8.5)	(4)	201
Cl	*t*-BuOH	83°	4 h	(22)	(23)	48
OTs	*t*-BuOH	83°	4 h	(18)	(19)	48
Cl	H_2O	100°	4 h	(—)	(14)	48

Substrate	Reagent-Conditions	Product(s) and Yield(s) (%)	Refs.
	$(o\text{-}O_2NC_6H_4)_2IBr$, *t*-BuONa/*t*-BuOH 60°, 4 h	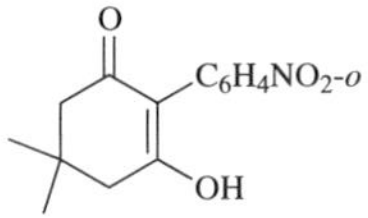 (29)	48
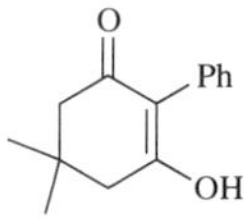	Ph_2ICl, *t*-BuONa/*t*-BuOH, reflux	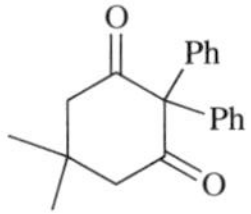 (71)	48
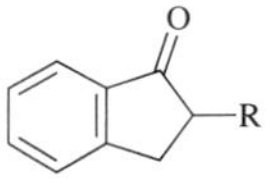	Ph_2ICl, Na/*t*-BuOH, reflux, 12 h	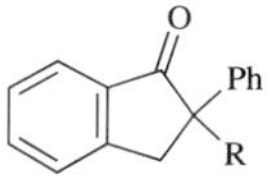 R: Me (68); Ph (74)	46

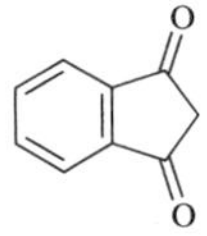	Ph_2ICl. Na/*t*-BuOH	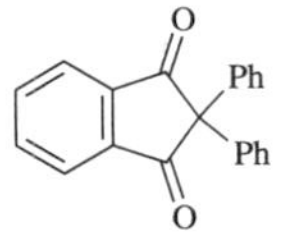	(28)	51
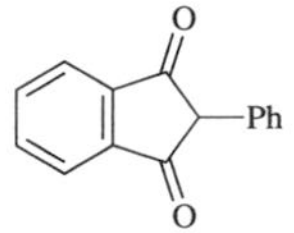	Ph_2ICl, Na/*t*-BuOH	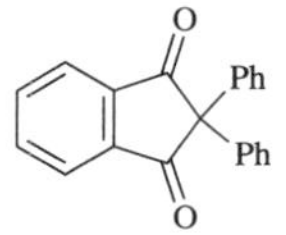	(52)	51
	$(Mes)_2ICl$, Na/*t*-BuOH	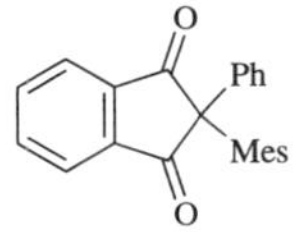	(23)	51
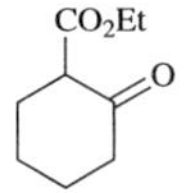	Ph_2ICl, K/*t*-BuOH	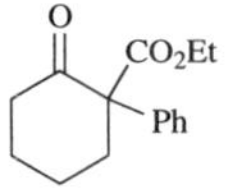	(60)	47
	Ph_2IOAc, BTMG, *t*-BuOH, reflux, 2 h	"	(55)	50
	Ph_2IOAc, BTMG, 1,1-diphenylethylene *t*-BuOH	"	(80)	50
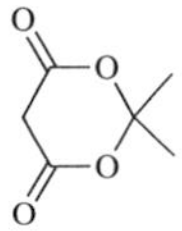	Ar_2IX, *t*-BuOK/*t*-BuOH, 70°	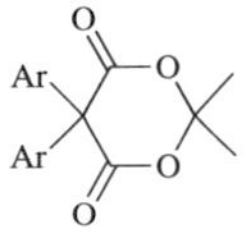		49

Ar	X	Time	%
Ph	Cl	16 h	(73)
p-ClC_6H_4	Cl	72 h	(72)
m-$O_2NC_6H_4$	Cl	16 h	(65)
p-MeC_6H_4	Br	72 h	(95)
p-$MeOC_6H_4$	I	72 h	(59)

TABLE V. ARYLATIONS OF CARBONYL COMPOUNDS (*Continued*)

Substrate	Reagent-Conditions	Product(s) and Yield(s) (%)	R	Ar	X	Time	%	Refs.
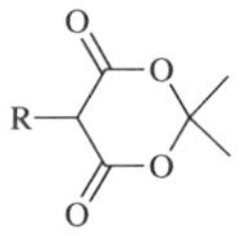	Ar_2IX, *t*-BuOK/*t*-BuOH, 70°	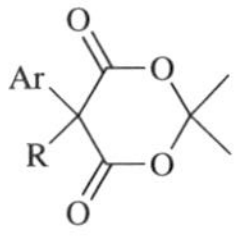	*i*-Pr	Ph	Cl	26 h	(56)	49
			i-Pr	*m*-$O_2NC_6H_4$	Cl	20 h	(55)	
			Bn	Ph	Cl	18 h	(88)	
			Bn	*p*-ClC_6H_4	Cl	72 h	(82)	
			Bn	*m*-$O_2NC_6H_4$	Cl	8 h	(79)	
			Bn	*p*-MeC_6H_4	Br	72 h	(84)	
			Bn	*p*-$MeOC_6H_4$	I	72 h	(60)	
			PhCH=$CHCH_2$	Ph	Cl	12 h	(95)	
			PhCH=$CHCH_2$	*p*-ClC_6H_4	Cl	72 h	(87)	
			PhCH=$CHCH_2$	*m*-$O_2NC_6H_4$	Cl	15 h	(64)	
			PhCH=$CHCH_2$	*p*-MeC_6H_4	Br	72 h	(81)	
			PhCH=$CHCH_2$	*p*-$MeOC_6H_4$	I	72 h	(59)	
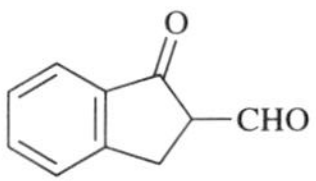	Ph_2ICl (2 eq)	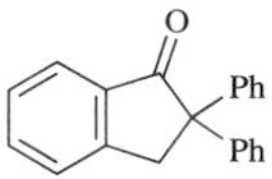 (54)						46
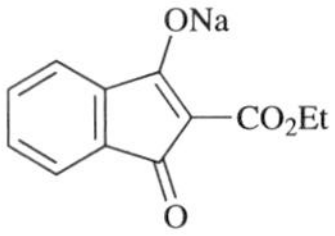	Ph_2ICl, *t*-BuOH, reflux, 3 d	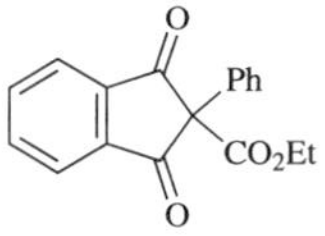 (—) + several products						200
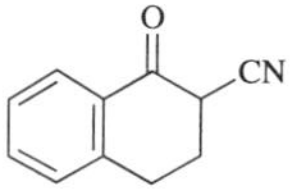	Ph_2ICl, *t*-BuOK/*t*-BuOH, reflux overnight	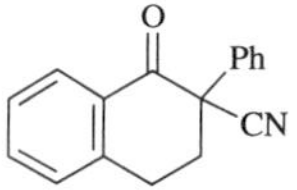 (50)						46

TABLE VI. α-FUNCTIONALIZATION OF ESTERS UNDER BASIC CONDITIONS

Substrate	Reagent-Conditions	Product(s) and Yield(s) (%)	Refs.
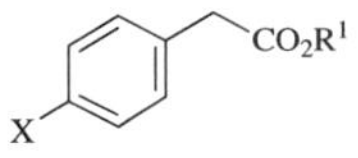	IBD, see table	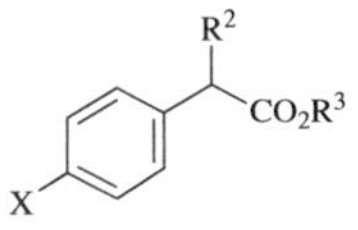 (see table below)	53
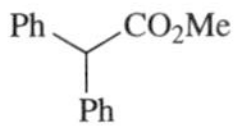	IBD, NaOMe/MeOH	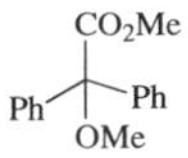 (65)	53
	IBD, NaOEt/EtOH	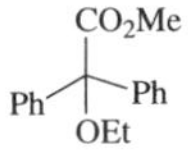 (65)	53
	IBD, $KOH/H_2O-C_6H_6$	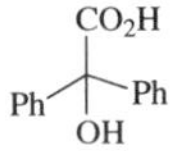 (75)	53
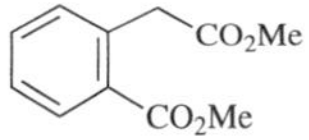	IBD, NaOMe/MeOH	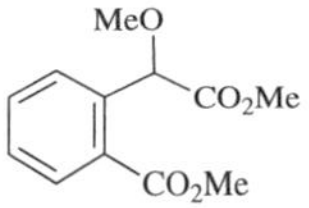 (50)	54

X	R^1	R^2	R^3	Base	Temp	Time	%
H	Me	OMe	Me	NaOMe/MeOH	rt	3 h	(70)
H	Et	OMe	Me	NaOMe/MeOH	rt	3 h	
H	Me	OEt	Et	NaOEt/EtOH	rt	3 h	(45)
H	Et	OEt	Et	NaOEt/EtOH	—	—	(50)
H	Me	OH	H	$KOH/H_2O-C_6H_6$	—	—	(50)
H	Et	OH	H	$KOH/H_2O-C_6H_6$	—	—	(75)
OMe	Me	OMe	Me	NaOMe/MeOH	—	3 h	(80)
OMe	Me	OEt	Et	NaOEt/EtOH	—	—	(50)
OMe	Me	OH	H	$KOH/H_2O-C_6H_6$	—	—	(66)
Cl	Et	OMe	Me	NaOMe/MeOH	—	—	(80)
Cl	Et	OEt	Et	NaOEt/EtOH	—	—	(70)
Cl	Et	OH	H	$KOH/H_2O-C_6H_6$	—	—	(60)
Br	Et	OMe	Me	NaOMe/MeOH	—	—	(75)
Br	Et	OEt	Et	NaOEt/EtOH	—	—	(60)
Br	Et	OH	H	$KOH/H_2O-C_6H_6$	—	—	(60)

TABLE VII. α-ACETOXYLATION OF CARBONYL COMPOUNDS

Substrate	Reagent-Conditions	Product(s) and Yield(s) (%)				Refs.
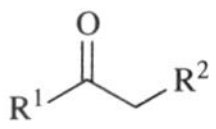	IBD, H_2SO_4/AcOH, Ac_2O, 30°	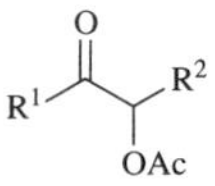	R^1	R^2	%	55a
			Ph	H	(25)	
			p-ClC_6H_4	H	(56)	
			p-$O_2NC_6H_4$	H	(50)	
			p-MeC_6H_4	H	(22)	
			Ph	$COCF_3$	(63)	

TABLE VIII. α-HYDROXYLATION OF KETONES UNDER ACIDIC CONDITIONS

Substrate	Reagent-Conditions	Product(s) and Yield(s) (%)					Refs.
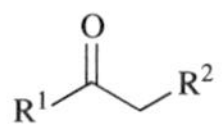	IBTA-TFA, MeCN, H_2O, reflux	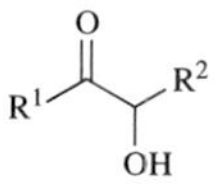	R^1	R^2	Time	%	57
			cyclopropyl	H	3 h	(74)	
			t-Bu	H	3 h	(41)	
			1-adamantyl	H	3 h	(70)	
			Ph	H	3 h	(69)	
			4-FC_6H_4	H	3 h	(67)	
			4-$O_2NC_6H_4$	H	3 h	(29)[a]	
			2-MeC_6H_4	H	3 h	(70)	
			4-MeC_6H_4	H	3 h	(72)	
			4-$MeOC_6H_4$	H	3 h	(58)	
			2-furyl	H	4 h	(69)	
			2-thienyl	H	3 h	(73)	
			2,5-Me_2-3-thienyl	H	3 h	(84)	
			Ph	Me	3 h	(36)	
			4-BrC_6H_4	Me	4 h	(21)	
			—$(CH_2)_4$—		2 h	(47)[b]	
			—$(CH_2)_5$—		3 h	(94)	

[a] *p*-Nitrobenzoic acid was also obtained in 29% yield.
[b] The product was isolated as a dimer.

TABLE IX. α-TOSYLOXYLATION OF CARBONYL COMPOUNDS UNDER NEUTRAL CONDITIONS

Substrate	Reagent-Conditions	Product(s) and Yield(s) (%)	R^1	R^2	Solvent	Temp	Time	%	Refs.
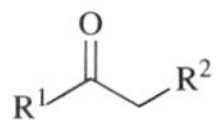	HTIB	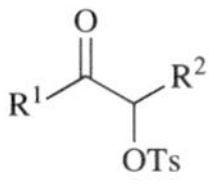	Me	H	MeCN	reflux	20 min	(71)	58
			Me	H	MeCN	55°,))))	15 min	(74)	202
			Me	Me	MeCN	reflux	—	(—)[b]	58
			Me	Me	MeCN	))))	10 min	(91)[c]	202
			Et	Me	MeCN	reflux	10 min	(94)	58
			Et	Me	MeCN	))))	10 min	(92)	202
			c-C_3H_5	H	MeCN	reflux	20 min	(80)[a]	58
			c-C_3H_5	H	MeCN	))))	15 min	(86)	202
			Ph	H	MeCN	reflux	45 min	(73)	58
			Ph	H	MeCN	))))	30 min	(55)	202
			4-ClC_6H_4	H	MeCN	reflux	1-2 h	(68)	64
			4-BrC_6H_4	H	MeCN	reflux	1-2 h	(80)	64
			4-$O_2NC_6H_4$	H	MeCN	reflux	1-2 h	(—)	203
			4-$MeOC_6H_4$	H	MeCN	reflux	1-2 h	(75)	64
			2-$PhCO_2C_6H_4$	H	MeCN	reflux	2 h	(85)	98
			Ph	Ph	CH_2Cl_2	reflux	1 d	(52)[a]	58
			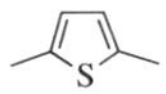	H	CH_2Cl_2	rt	3-4 d	(80)	58
			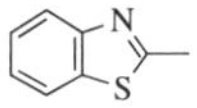	H	CH_2Cl_2	rt	—	(55)	67
			Me	COMe	MeCN	75°	10 min	(73)[a]	58
			Me	COMe	MeCN	55°,))))	10 min	(74)[a]	202
			Ph	COPh	MeCN	reflux	20 min	(~100)[a]	58
			Ph	CO_2Et	MeCN	75°	10 min	(75)	58
			—$(CH_2)_3$—		MeCN	))))	10 min	(42)	202
			—$(CH_2)_4$—		CH_2Cl_2	rt	3 h	(40)	58

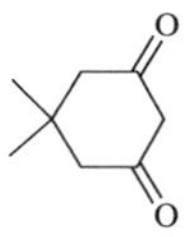	HTIB, MeCN reflux, 20 min	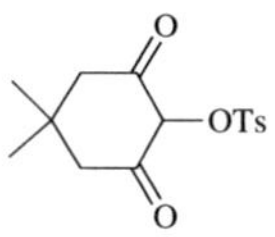 (86)	58

[a] The product is a mixture of $TsOCH_2COEt$ and $CH_3COCH(OTs)CH_3$ in the ratio 1:1.6 as determined by ^1H NMR data.

[b] The product is a mixture of $TsOCH_2COEt$ and $CH_3COCH(OTs)CH_3$; the ratio is unknown.

[c] The yield given is of crude product.

TABLE X. α-MESYLOXYLATION OF CARBONYL COMPOUNDS UNDER NEUTRAL CONDITIONS

Substrate	Reagent-Conditions	Product(s) and Yield(s) (%)						Refs.
			R^1	R^2	Temp	Time	%	
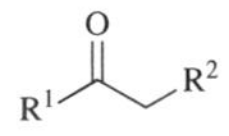	HMIB, MeCN	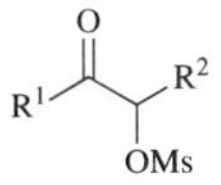	Me	H	20°	18 h	(75)	59
			Me	H	reflux	12 min	(76)	60
			Et	Me	reflux	12 min	(87)	60
			c-C_3H_5	Me	reflux	35 min	(91)	60
			Ph	H	reflux	18.5 h	(62.5)	60
			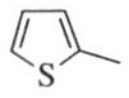	H	rt	18 d	(72)	60
			Me	COMe	rt	15 min	(60)	60
			Ph	COPh	reflux	15 min	(96)	60
			OEt	CO_2Et	reflux	142 min	(64.5)	60
			Me	CO_2Et	reflux	8 min	(81)	60
			Ph	CO_2Et	reflux	15 min	(98.5)	60
			—$(CH_2)_4$—		rt	70 min	(49)	60
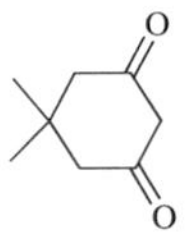	HMIB, MeCN reflux, 20 min	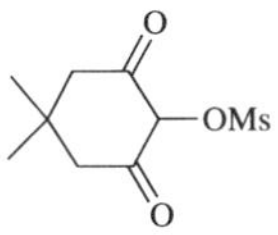 (81)						60

TABLE XI. α-(10-CAMPHORSULFONYL)OXYLATION OF CARBONYL COMPOUNDS

For this table, R = 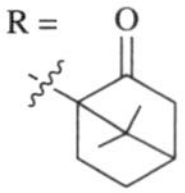

Substrate	Reagent-Conditions	Product(s) and Yield(s) (%)	Refs.
	HCIB, MeCN, 80°, 0.75 h	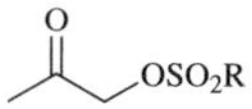 (87)	62
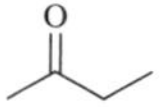	HCIB, MeCN, 80°, 0.75 h	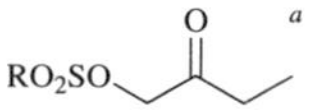 [a] (67) + 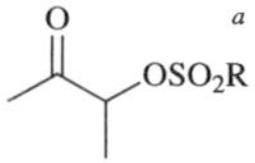 [a] (28)	62
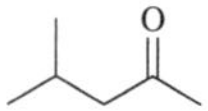	HCIB, MeCN. 80°, 0.25 h	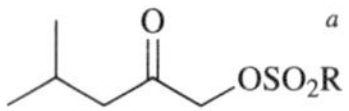 [a] (60) + $RO_2SO-OSO_2R$ (19)	62
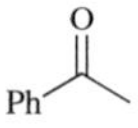	HCIB, MeCN, 80°, 0.25 h	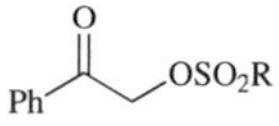 (95)	62
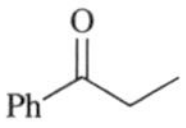	HCIB, MeCN, 80°, 0.75 h	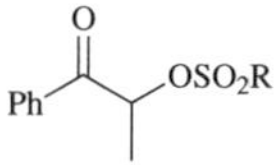 (90)	62
	HCIB, MeCN, 80°, 0.25 h	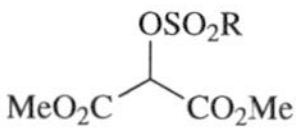 (70)	62
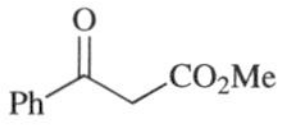	HCIB, MeCN, 20°, 24 h	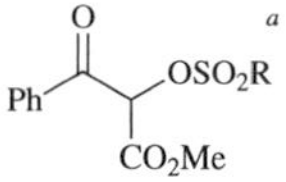 [a] (95)	62
	HCIB, MeCN, 80°, 3 h	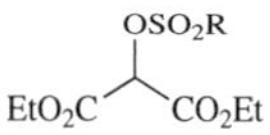 (40)	62

TABLE XI. α-(10-CAMPHORSULFONYL)OXYLATION OF CARBONYL COMPOUNDS (*Continued*)

Substrate	Reagent-Conditions	Product(s) and Yield(s) (%)	Refs.
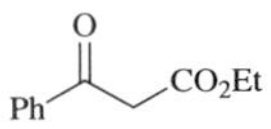	HCIB, CH_2Cl_2, 20°, 8 h	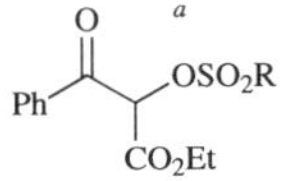 [a] (95)	62
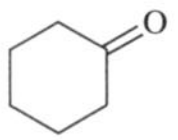	HCIB, CH_2Cl_2, 20°, 3 h	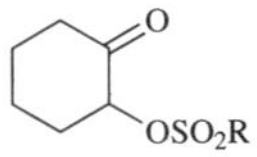 (38)	62
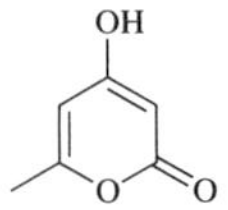	HCIB, CH_2Cl_2, 20°, 72 h	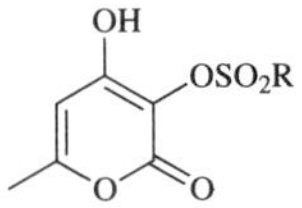 (42)	62
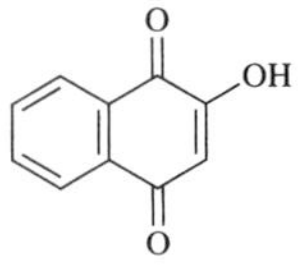	HCIB, CH_2Cl_2, 20°, 24 h	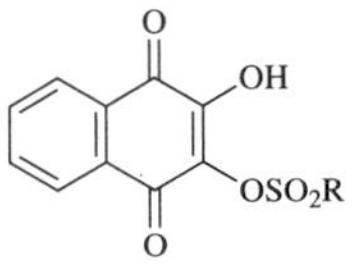 (65)	62

[a] The product yields are based on ^1H NMR data.

TABLE XII. SYNTHESIS OF α-FUNCTIONALIZED CARBONYL COMPOUNDS VIA α-TOSYLOXYLATION

Substrate	Reagent-Conditions	Product(s) and Yield(s) (%)	Refs.
α-Hydroxylation			
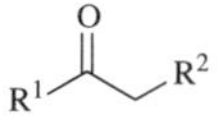	1. HTIB, MeCN, reflux, 1 h 2. $NaNO_2/H_2O$, reflux, 8-9 h	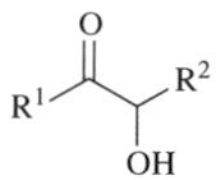 R^1 / R^2 / % Et / Me / (86) 1-adamantyl / H / (72) Ph / H / (80) 4-$MeOC_6H_4$ / H / (78)	66
	1. HTIB, MeCN, reflux, 1 h 2. KOH/MeOH, 0-5°, 2 h	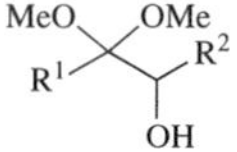 R^1 / R^2 / % Ph / H / (55) 4-ClC_6H_4 / H / (74) 4-BrC_6H_4 / H / (65) 4-$O_2NC_6H_4$ / H / (73) Ph / Ph / (83) —$(CH_2)_4$— / (83)	67
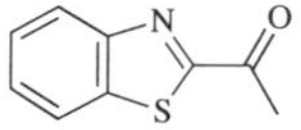	1. HTIB, CH_2Cl_2, rt 2. KOH/MeOH, 0-5°, 2 h	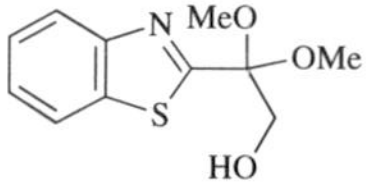 (72)	67
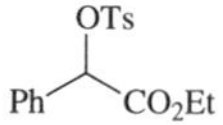	$NaNO_2$, H_2O, reflux, 8 h	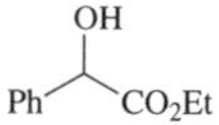 (60)	66
α-Aryloxylation			
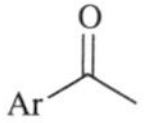	1. HTIB, MeCN, reflux, 2 h 2. Ar'OH, K_2CO_3, dry acetone	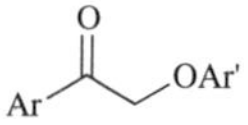 Ar / Ar / % Ph / Ph / (17) Ph / 4-$O_2NC_6H_4$ / (31) 4-BrC_6H_4 / 4-$O_2NC_6H_4$ / (30) 4-BrC_6H_4 / 4-MeC_6H_4 / (47) 4-MeC_6H_4 / Ph / (35) 4-MeO_6H_4 / 4-MeC_6H_4 / (36)	61

TABLE XII. SYNTHESIS OF α-FUNCTIONALIZED CARBONYL COMPOUNDS VIA α-TOSYLOXYLATION (*Continued*)

Substrate	Reagent-Conditions	Product(s) and Yield(s) (%)	Refs.
α-Aryloxylation			
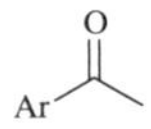	1. HTIB, MeCN, reflux, 2 h 2. Ar'CO_2H, Et_3N, reflux, 1-3 h	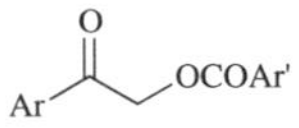 Ar / Ar / % Ph / Ph / (67) 4-ClC_6H_4 / Ph / (67) 4-$O_2NC_6H_4$ / Ph / (56) 4-MeO_6H / Ph / (60) Ph / 4-$O_2NC_6H_4$ / (62)	61
α-Amination			
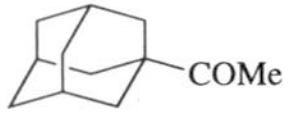	1. HTIB, MeCN, reflux, 1-2 h 2. Piperidine, reflux, 1 h	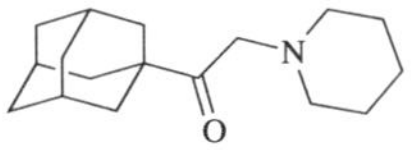 (77)	66
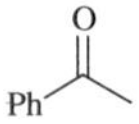	1. HTIB, MeCN, reflux, 1-2 h 2. Et_2NH, MeCN, reflux, 1 h	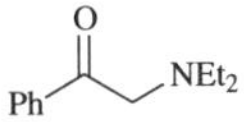 (65)	63, 66
	1. HTIB, MeCN, reflux, 1-2 h 2. Piperidine, MeCN, reflux, 1 h	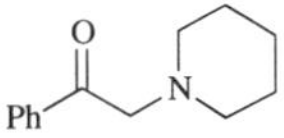 (79)	63, 66
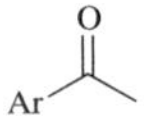	1. HTIB, MeCN, reflux, 1-2 h 2. Ar'NH_2 (2 eq), reflux, 2 h	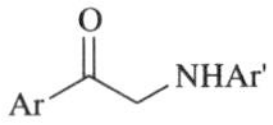 (52-70) Ar / Ar' Ph / Ph 4-ClC_6H_4 / Ph 4-BrC_6H_4 / Ph 4-$MeOC_6H_4$ / Ph Ph / 4-ClC_6H_4 Ph / 4-$MeOC_6H_4$ 4-BrC_6H_4 / 4-$MeOC_6H_4$ 4-$MeOC_6H_4$ / 4-$MeOC_6H_4$	64

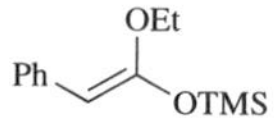	1. HTIB, CH_2Cl_2, rt, 1 h 2. Piperidine, reflux, 1 h	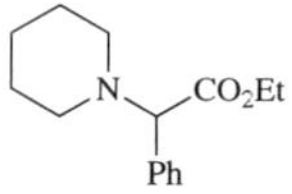	(61)	66
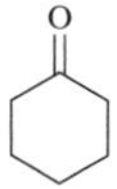	1. HTIB, MeCN, rt, 1 h 2. Piperidine, MeCN, reflux, 2 h	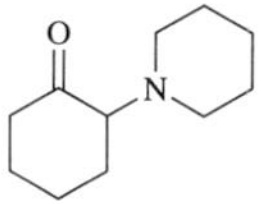	(50)	66
Miscellaneous				
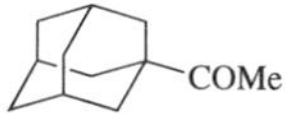	1. HTIB, MeCN, reflux, 1 h 2. NaCN, H_2O, reflux, 1 h	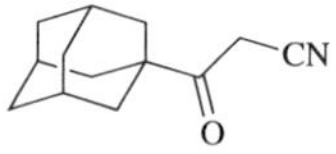	(64)	66
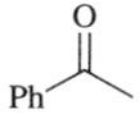	1. HTIB, MeCN, reflux, 1 h 2. NaCN, H_2O, reflux, 1 h	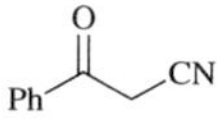	(72)	63, 66
	1. HTIB, MeCN, reflux, 1 h 2. NaN_3, H_2O, reflux, 1 h	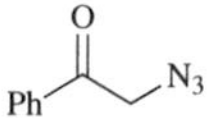	(80)	63, 66
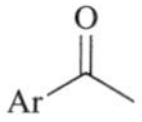	1. HTIB, MeCN, reflux, 2 h 2. KSCN, MeCN, reflux, 5-10 min	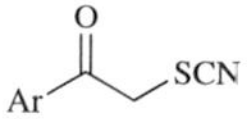	Ar / % Ph (71) 4-ClC_6H_4 (75) 4-BrC_6H_4 (78) 4-MeC_6H_4 (52) 4-$MeOC_6H_4$ (68)	65
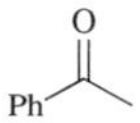	1. HTIB, MeCN, reflux, 2 h 2. 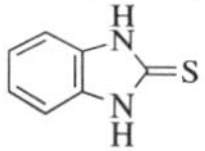 , reflux, 6-7 h	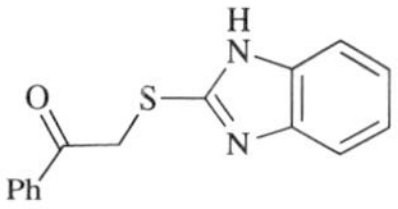	(—)	203

TABLE XII. SYNTHESIS OF α-FUCNTIONALIZED CARBONYL COMPOUNDS VIA α-TOSYLOXYLATION (*Continued*)

Substrate	Reagent-Conditions	Product(s) and Yield(s) (%)	Refs.
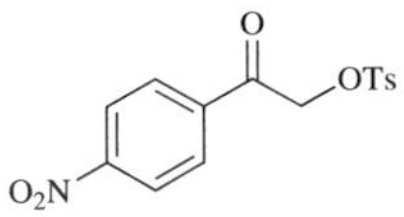	 , acetone, reflux, 1 h	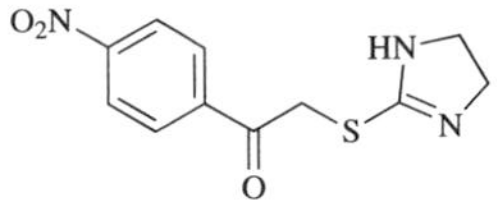 (67)	203
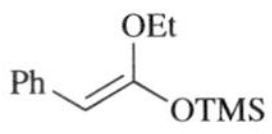	1. HTIB, MeCN, rt, 2 h 2. NaN_3, H_2O, reflux, 1 h	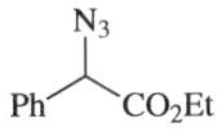 (40)	66

TABLE XIII. α-PHOSPHORYLOXYLATION AND α-PHOSPHONYLOXYLATION OF CARBONYL COMPOUNDS

Substrate	Reagent-Conditions	Product(s) and Yield(s) (%)	Refs.
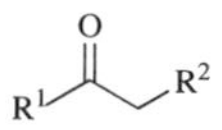	PhI(OH)OPO(OPh)$_2$	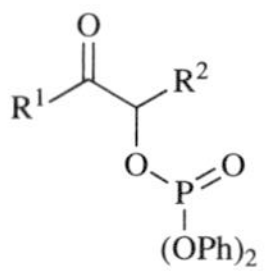 R[1] | R[2] | Solv | Temp | Time | % Me | H | MeCN | reflux | 30 min | (81) c-C_3H_5 | H | MeCN | reflux | 3 h | (59) Ph | H | MeCN | reflux | 3.3 h | (59) Ph | COPh | CH_2Cl_2 | rt | 15 min | (90) —$(CH_2)_4$— | | CH_2Cl_2 | rt | 7.5 h | (62)	68
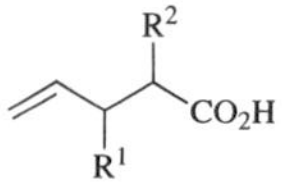	PhI(OH)OPO(OPh)$_2$, CH_2Cl_2, rt	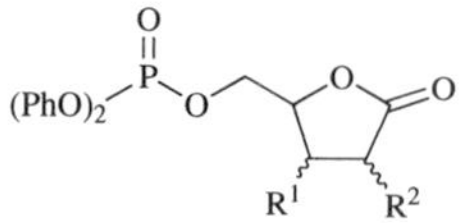 R[1] | R[2] | Time | % H | H | 2 h, 40 min | (55) H | Me | 8 h, 43 min | (64) OH | H | 24 h | (12.5)	68
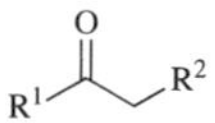	PhI(OH)OPO(Me)OPh	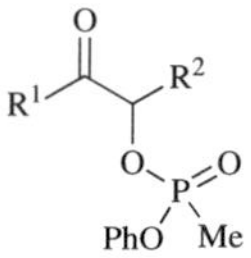 R[1] | R[2] | Solv | Temp | Time | % Me | H | acetone | reflux | 3 h | (75) Ph | H | MeCN | reflux | 1 h | (48) Ph | COPh | CH_2Cl_2 | rt | 20 min | (68) —$(CH_2)_4$— | | MeCN | reflux | 3 h | (45)	69
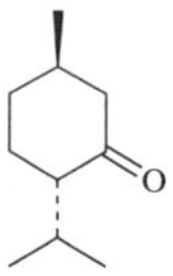	PhI(OH)OPO(Me)OPh, MeCN, reflux	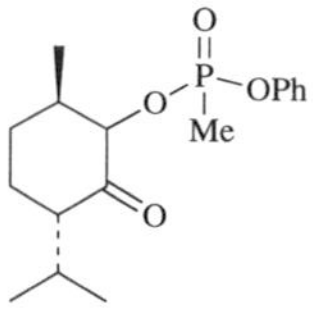 (30)	69

TABLE XIV. α-HYDROXYLATION OF SILYL ENOL ETHERS

Substrate	Reagent-Conditions	Product(s) and Yield(s) (%)			Refs.
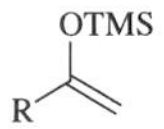	1. $(PhIO)_n$, $BF_3•OEt_2$, H_2O, 0-5°, 2 h 2. rt, 2 h	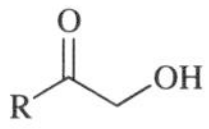	R	%	71
			t-Bu	(83)	
			Ph	(65)	
			4-ClC_6H_4	(68)	
			4-$O_2NC_6H_4$	(70)	
			4-$MeOC_6H_4$	(72)	
			2-furyl	(78)	
			2-thienyl	(50)	
			2-pyridyl	(62)	
			3-pyridyl	(54)	
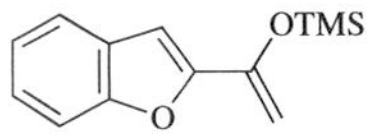	1. $(PhIO)_n$, $BF_3•OEt_2$, H_2O, 0-5°, 2 h 2. rt, 2 h	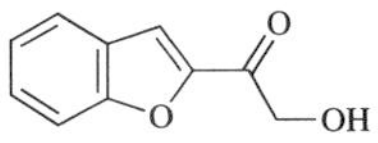 (59)			71
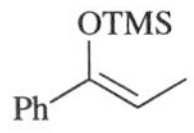	1. $(PhIO)_n$, $BF_3•OEt_2$, H_2O, 0-5°, 2 h 2. rt, 2 h	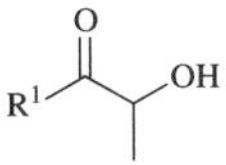 (74)			71
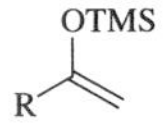	1. $(PhIO)_n$, $BF_3•OEt_2$, CH_2Cl_2, –40°, 1 h 2. rt, 0.5 h	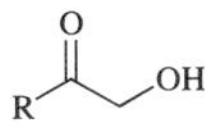	R	%	70
			Ph	(57)	
			4-ClC_6H_4	(63)	
			4-$MeOC_6H_4$	(65)	
			2-pyridyl	(45)	
			3-pyridyl	(38)	
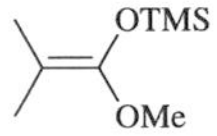	$(PhIO)_n$, H_2O, rt 16 h	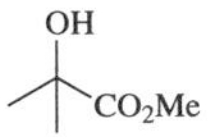 (71)			72
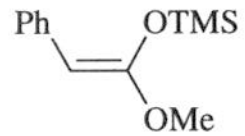	1. $(PhIO)_n$, $BF_3•OEt_2$, CH_2Cl_2, 0°, 1 h 2. rt, 3 h	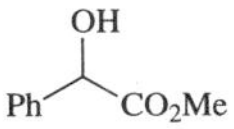 (64)			72
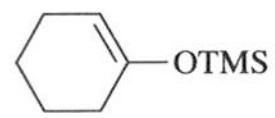	1. $(PhIO)_n$, $BF_3•OEt_2$, H_2O, 0-5°, 2 h 2. rt, 2 h	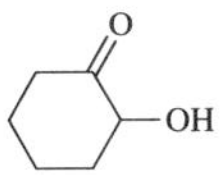 (80)			71

TABLE XV. α-ALKOXYLATION OF SILYL ENOL ETHERS

Substrate	Reagent-Conditions	Product(s) and Yield(s) (%)	R^1	R^2	R^3	%	Refs.
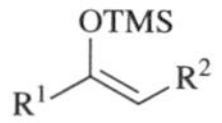	1. $(PhIO)_n$, BF_3•OEt_2 (2 eq), R^3OH, CH_2Cl_2, –70°, 1 h 2. rt, 2 h	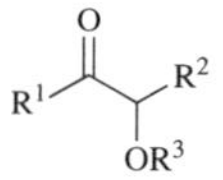	*t*-Bu	H	Me	(85)	73
			Ph	H	Me	(78)	
			4-ClC_6H_4	H	Me	(76)	
			4-$O_2NC_6H_4$	H	Me	(68)	
			4-$MeOC_6H_4$	H	Me	(71)	
			2-furyl	H	Me	(60)	
			2-thienyl	H	Me	(54)	
			2-pyridyl [a]	H	Me	(70)	
			Ph	H	Et	(80)	
			Ph	H	*i*-Pr	(45)	
			Ph	Me	Me	(75)	
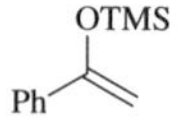	$(PhIO)_n$, MeOH	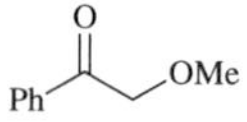 (78)					72
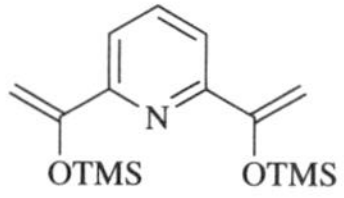	1. $(PhIO)_n$ (2.2 eq), BF_3•OEt_2 (6 eq), MeOH, CH_2Cl_2, –70°, 1 h 2. rt, 0.5 h	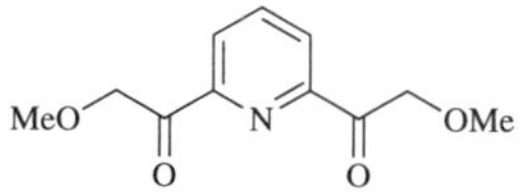 (71)					73
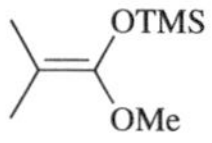	$(PhIO)_n$, MeOH, reflux, 3 d	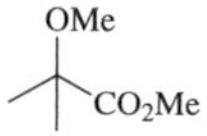 (64)					72
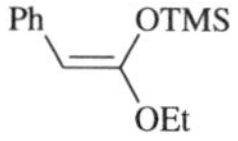	$(PhIO)_n$, MeOH, rt, 3 d	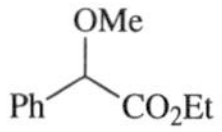 (51)					72
	$(PhIO)_n$, EtOH, rt, 3 d	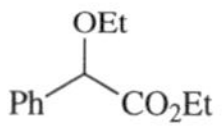 (62)					72

TABLE XV. α-ALKOXYLATION OF SILYL ENOL ETHERS (*Continued*)

Substrate	Reagent-Conditions	Product(s) and Yield(s) (%)	Refs.
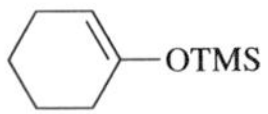 OTMS	1. $(PhIO)_n$, BF_3•OEt_2 (2 eq), R^3OH, CH_2Cl_2, –70°, 1 h 2. rt, 2 h; or $(PhIO)_n$, MeOH	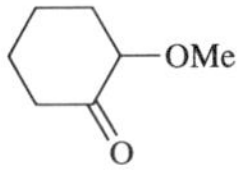 OMe O (78)	72, 73
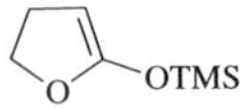 O OTMS	$(PhIO)_n$, MeOH	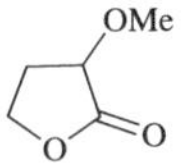 OMe O O (56)	72
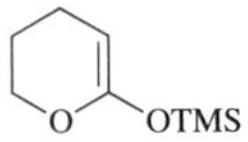 O OTMS	$(PhIO)_n$, MeOH	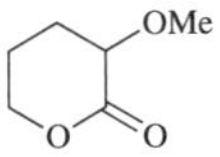 OMe O O (63)	72

[a] For this reaction, 3 eq of reagent were used, and the reaction remained at rt for only one hour.

TABLE XVI. α-SULFONYLOXYLATION OF SILYL ENOL ETHERS

Substrate	Reagent-Conditions	Product(s) and Yield(s) (%)		Refs.
Ketones				
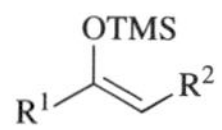	HTIB, CH_2Cl_2, rt, 2 h	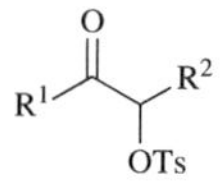	R^1 / R^2 / % Ph / H / (92) 2-furyl / H / (88) 2-thienyl / H / (90) 2-pyridyl / H / (78) —$(CH_2)_4$— / (85)	74
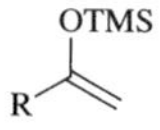	HMIB, CH_2Cl_2, rt, 2 h	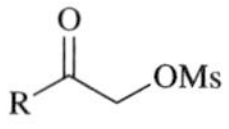	R / % Ph / (89) 2-furyl / (90)	74
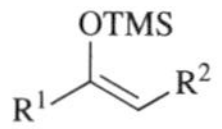	$(PhIO)_n$, TMSOTf, CH_2Cl_2, 2 h	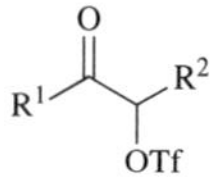	R^1 / R^2 / % Ph / H / (70) 4-ClC_6H_4 / H / (53) 2-furyl / H / (70) 2-thienyl / H / (69) Ph / Me / (77) —$(CH_2)_4$— / (64) —$(CH_2)_5$— / (74)	75
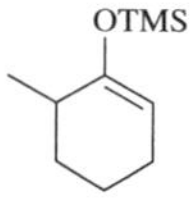	HTIB, CH_2Cl_2, rt, 2 h	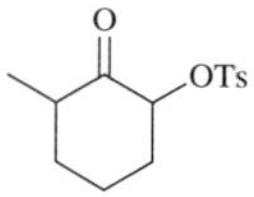	(80)	74

TABLE XVI. α-SULFONYLOXYLATION OF SILYL ENOL ETHERS (*Continued*)

Substrate	Reagent-Conditions	Product(s) and Yield(s) (%)	Refs.
Esters			
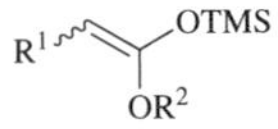	HTIB, CH_2Cl_2, rt, 2 h	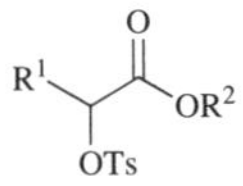 R^1 / R^2 / %: Et / Me / (65); Ph / Me / (81); Ph / Et / (60)	74
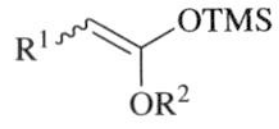	HMIB, CH_2Cl_2, rt, 2 h	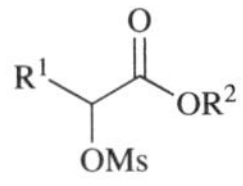 R^1 / R^2 / %: Et / Me / (65); Ph / Me / (65); Ph / Et / (85)	74
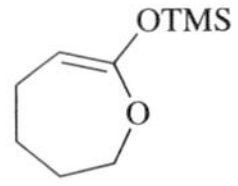	HTIB, CH_2Cl_2, rt, 2 h	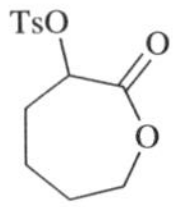 (69)	74

TABLE XVII. α-ACETOXYLATION AND α-TRIFLUOROACETOXYLATION OF SILYL ENOL ETHERS

Substrate	Reagent-Conditions	Product(s) and Yield(s) (%)	Refs.
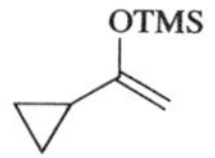	1. IBD, CH_2Cl_2, rt 2. $BF_3 \bullet OEt_2$	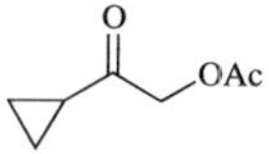 (67)	76
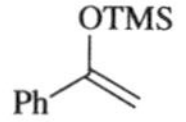	1. IBD, CH_2Cl_2, rt 2. $BF_3 \bullet OEt_2$	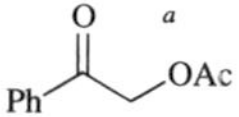 [a] (90)	76
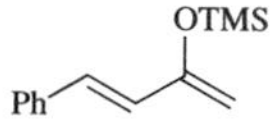	1. IBD, CH_2Cl_2, rt 2. $BF_3 \bullet OEt_2$	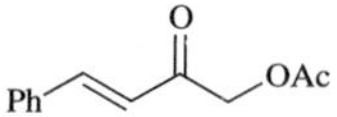 (78)	76
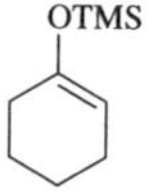	1. IBD, CH_2Cl_2, rt 2. $BF_3 \bullet OEt_2$	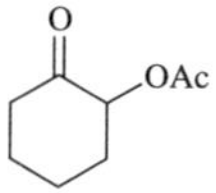 (81)	76
	IBTA, Py, $CHCl_3$, 2 h	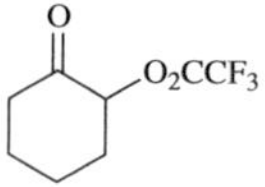 (60)	76

[a] α-Acetoxyacetophenone is produced in 87% yield when enol acetate of acetophenone is treated with IBD in acetic acid.

TABLE XVIII. α-PHOSPHORYLATION OF SILYL ENOL ETHERS OF KETONES

Substrate	Reagent-Conditions	Product(s) and Yield(s) (%)	Refs.
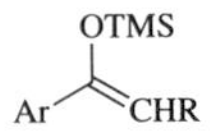	1. $(PhIO)_n$, BF_3•OEt_2, CH_2Cl_2, –40° 2. $P(OEt)_3$, –40° to rt	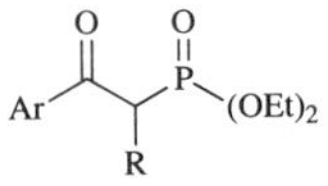 Ar / R / % Ph / H / (76) *m*-BrC_6H_4 / H / (62) *p*-MeC_6H_4 / H / (81) *p*-$MeOC_6H_4$ / H / (83) Ph / Me / (75)	77

TABLE XIX. α-PHOSPHORYLOXYLATION OF SILYL ENOL ETHERS OF KETONES

Substrate	Reagent-Conditions	Product(s) and Yield(s) (%)			Refs.
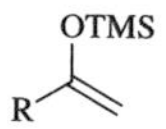	p-$MeC_6H_4IF_2$, H_3PO_4, t-BuOH	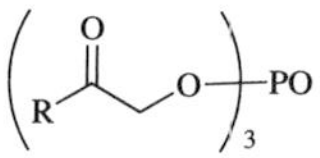	R	%	78
			Me	(64)	
			t-Bu	(73)	
			Ph	(63)	
			2-furyl	(60)	
			2-pyridyl	(60)	

TABLE XX. CARBON-CARBON BOND FORMATION VIA SILYL ENOL ETHERS

Substrate	Reagent-Conditions	Product(s) and Yield(s) (%)	R	%	Refs.
Self Coupling					
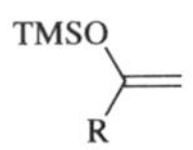	1. $(PhIO)_n$, BF_3•OEt_2, CH_2Cl_2, –40°, 1 h 2. rt, 1 h	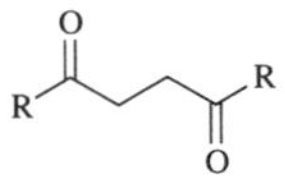	*t*-Bu	(55)	80
			Ph	(48)	79, 80
			o-HOC_6H_4	(43)	79, 80
			p-FC_6H_4	(60)	80
			p-ClC_6H_4	(62)	79, 80
			p-$O_2NC_6H_4$	(50)	80
			p-MeC_6H_4	(57)	80
			p-$MeOC_6H_4$	(58)	79, 80
			2-furyl	(50)	80
			2-thienyl	(56)	80
			4-Cl-2-thienyl	(54)	80
			4-Me-2-thienyl	(59)	80, 81
			2-benzofuryl	(63)	80
	$(PhI)_2O(BF_4)_2$, CH_2Cl_2, rt or $(PhI)_2O(PF_6)_2$, CH_2Cl_2, rt	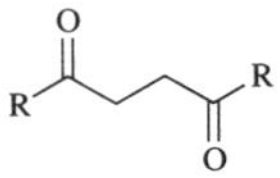	Ph	(—)	204
			p-ClC_6H_4	(—)	
			p-MeC_6H_4	(—)	
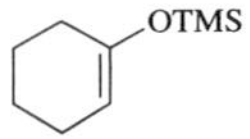	$PhIF_2$, BF_3•OEt_2, –78° to –10°	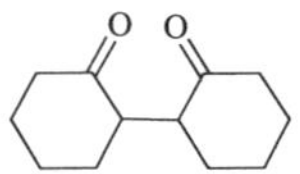 (82)			82

Cross Coupling

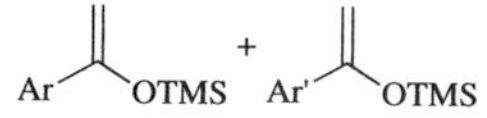

Ar
OTMS
+
Ar'
OTMS

1. $(PhIO)_n$, $HBF_4 \cdot OMe_2$, CH_2Cl_2, −50° to 0°
2. −78°

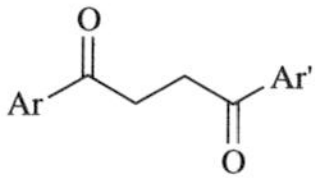

O
Ar
Ar'
O

Ar	Ar'	%	Refs.
Ph	p-ClC_6H_4	(42)	84
Ph	p-$O_2NC_6H_4$	(27)	84
Ph	p-MeC_6H_4	(40)	84
Ph	p-$MeOC_6H_4$	(34)	84
p-ClC_6H_4	Ph	(46)	84
p-ClC_6H_4	p-MeC_6H_4	(43)	84
p-ClC_6H_4	p-$MeOC_6H_4$	(51)	84
p-ClC_6H_4	p-ClC_6H_4	(38)	84
p-$O_2NC_6H_4$	p-$MeOC_6H_4$	(61)	84
p-$O_2NC_6H_4$	p-$MeOC_6H_4$	(27)[a]	80
p-MeC_6H_4	Ph	(31)	84

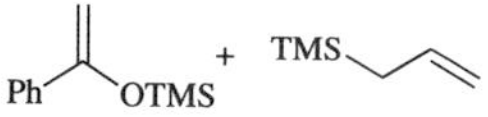

Ph
OTMS
+
TMS

"

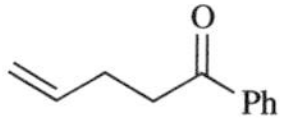

O
Ph

(63) 84

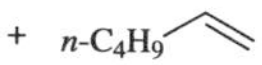

+
n-C4H9

"

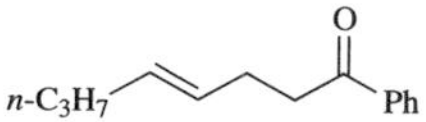

O
n-C3H7
Ph

(20) 83, 84

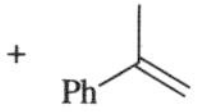

+
Ph

"

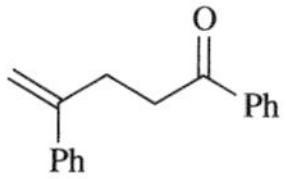

O
Ph
Ph

(59) 84

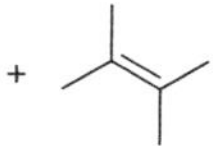

+

"

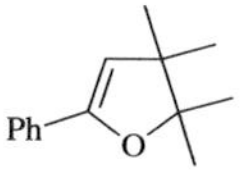

Ph
O

(80) 84

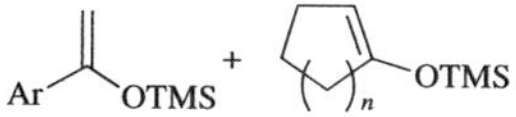

Ar
OTMS
+
OTMS
n

"

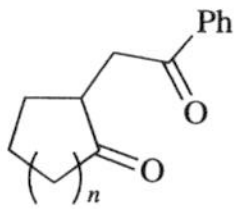

Ph
O
O
n

Ar	n	%	Refs.
Ph	1	(40)	84
Ph	2	(50)	83,84
p-ClC_6H_4	1	(44)	84
p-$O_2NC_6H_4$	1	(75)	84
p-MeC_6H_4	1	(55)	84
p-MeC_6H_4	2	(50)	84
p-$MeOC_6H_4$	1	(30)	84

[a] The conditions of this reaction were the same as that of the first entry of this table, Table XX.

TABLE XXI. α-PERFLUOROALKYLATION OF SILYL ENOL ETHERS

Substrate	Reagent-Conditions	Product(s) and Yield(s) (%)	Refs.
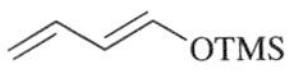	n-$C_8F_{17}I(Ph)OSO_3H$, CH_2Cl_2, rt, 4 h	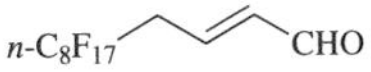 (54)	85
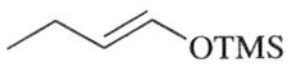	n-$C_8F_{17}I(Ph)OSO_2CF_3$, CH_3CN, rt, 1 h	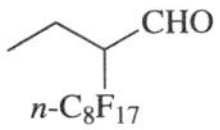 (65)	85
	$R_fI(Ph)OSO_2CF_3$, Py, rt	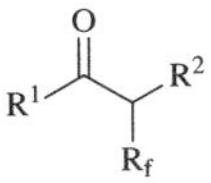	85

R^1	R^2	R_f	Solv	Time	%
Me	H	n-C_8F_{17}	CH_2Cl_2	1 h	(88)
n-C_6H_{13}	H	C_2F_5	MeCN	12 h	(77)
t-Bu	H	n-C_8F_{17}	MeCN	0.5 h	(82)
Ph	H	i-C_3F_7	CH_2Cl_2	2 h	(79)
—$(CH_2)_4$—		n-C_6F_{13}	MeCN	0.7 h	(71)

Substrate	Reagent-Conditions	Product(s) and Yield(s) (%)	Refs.
	$R_fI(Ph)OSO_3H$, Py	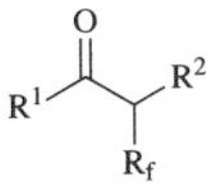	85

R^1	R^2	R_f	Solv	Temp	Time	%
Me	Ph	C_2F_5	CH_2Cl_2	reflux	1 h	(59)
Ph	H	C_2F_5	CH_2Cl_2	reflux	0.7 h	(85)
—$(CH_2)_3$—		n-C_8F_{17}	MeCN	45°	0.5 h	(76)
—$(CH_2)_3$—		n-$C_{10}F_{21}$	MeCN	40°	1 h	(80)
—$(CH_2)_5$—		n-C_6F_{13}	MeCN	45°	0.5 h	(83)

Substrate	Reagent-Conditions	Product(s) and Yield(s) (%)	Refs.
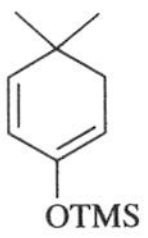	$C_2F_5I(Ph)OSO_2CF_3$, Py, CH_2Cl_2	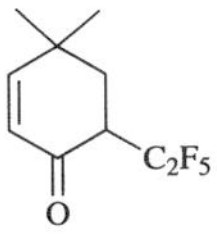 (72)	205
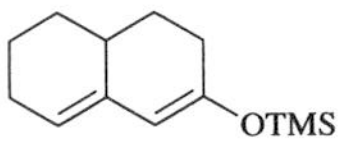	n-$C_8F_{17}I(Ph)OSO_2CF_3$, CH_2Cl_2, rt, 2 h	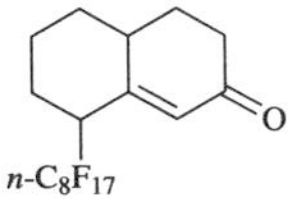 (85)	85

TABLE XXII. 1H,1H–PERFLUOROALKYLATION OF SILYL ENOL ETHERS

Substrate	Reagent-Conditions	Product(s) and Yield(s) (%)	R^1	R^2	R_f	Solv	Temp	Time	%	Refs.
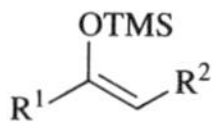	$R_fCH_2I(Ph)OSO_2CF_3$, KF (0-5 eq)	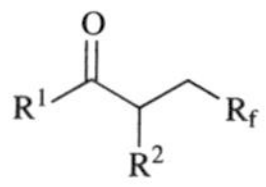	n-C_6H_{13}	H	CF_3	CH_2Cl_2	rt	1.5 h	(80)	86
			n-C_6H_{13}	H	C_2F_5	CH_2Cl_2	rt	4 h	(63)	
			n-C_6H_{13}	H	n-C_7F_{15}	CH_2Cl_2	rt	8 h	(45)	
			Ph	H	CF_3	CH_2Cl_2	rt	1.5 h	(87)	
			Ph	H	CF_3	CH_2Cl_2	rt	3 h	(40)	
			Ph	H	CF_3	$CHCl_3$	rt	4 h	(45)	
			Ph	H	CF_3	THF	rt	5 h	(38)	
			Ph	H	CF_3	Et_2O	rt	4 h	(39)	
			Ph	H	CF_3	CH_2Cl_2	rt	5 h	(16)	
			Ph	H	C_2F_5	CH_2Cl_2	rt	4 h	(67)	
			Ph	H	C_2F_5	CH_2Cl_2	reflux	3 h	(87)	
			Ph	H	i-C_3F_7	CH_2Cl_2	rt	8 h	(18)	
			Ph	H	i-C_3F_7	CH_2Cl_2	reflux	7 h	(32)	
			Ph	H	n-C_7F_{15}	CH_2Cl_2	rt	12 h	(50)	
			Ph	H	n-C_7F_{15}	CH_2Cl_2	reflux	10 h	(54)	
			PhCH=CH	H	CF_3	CH_2Cl_2	rt	1.5 h	$(75)^a$	
			Ph	Me	CF_3	CH_2Cl_2	rt	1.5 h	(55)	
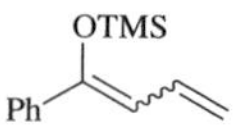	$CF_3CH_2I(Ph)OTf$-KF (5.2 eq) CH_2Cl_2, rt, 1.5 h	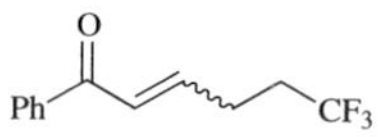 (42)								86
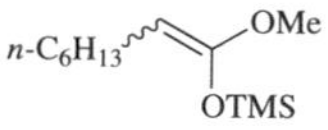	"	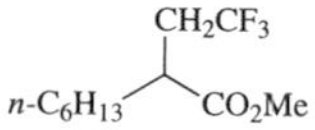 (76)								86
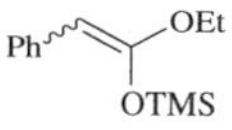	"	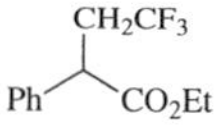 (92)								86

TABLE XXII. 1*H*,1*H*–PERFLUOROALKYLATION OF SILYL ENOL ETHERS (*Continued*)

Substrate	Reagent-Conditions	Product(s) and Yield(s) (%)	Refs.
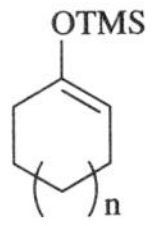	$R_fCH_2I(Ph)OTf$-KF (5.2 eq) CH_2Cl_2, rt, 1.5h	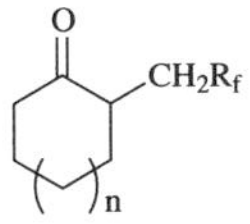 n / R_f / % 1 / n-C_7H_{15} / (21) 2 / CF_3 / (49) 3 / CF_3 / (71)	86
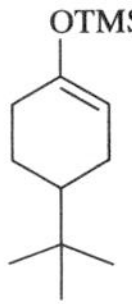	$CF_3CH_2I(Ph)OTf$-KF (5.2 eq) CH_2Cl_2, rt, 1.5h	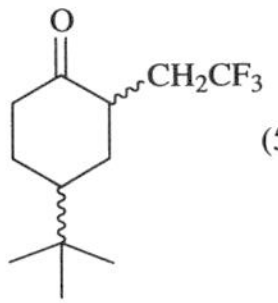 (53)	86

[a] The product is the *trans* isomer

TABLE XXIII. α-PHENYLATION OF SILYL ENOL ETHERS OF KETONES

Substrate	Reagent-Conditions	Product(s) and Yield(s) (%)	Refs.
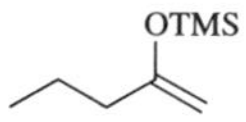	1. Ph_2IF, THF, –40°, 2 h 2. 10°, 2 h	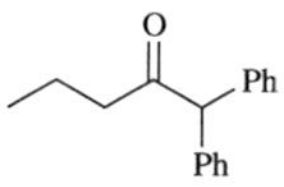 (24)	87
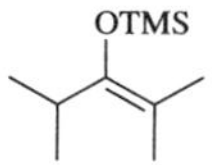	"	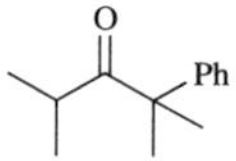 (20)	87
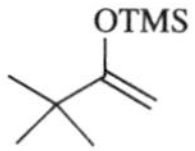	"	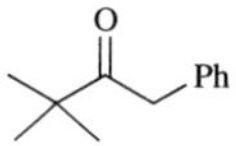 (37)	87
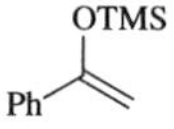	"	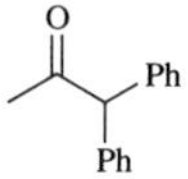 (47)	87
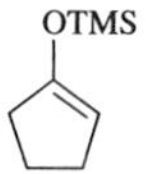	"	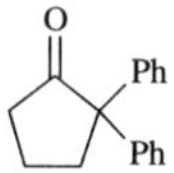 (51)	87
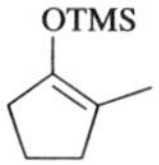	"	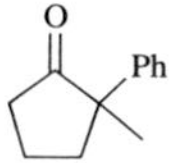 (45)	87
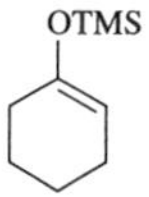	"	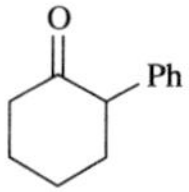 (88)	87
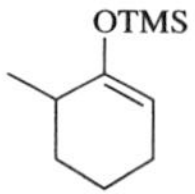	"	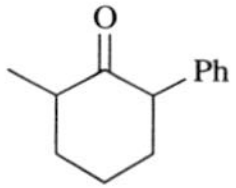 (78)	87

TABLE XXIII. α-PHENYLATION OF SILYL ENOL ETHERS OF KETONES (*Continued*)

Substrate	Reagent-Conditions	Product(s) and Yield(s) (%)	Refs.
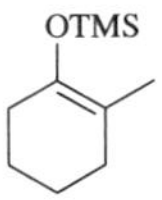	1. Ph_2IF, THF, –40°, 2 h 2. 10°, 2 h	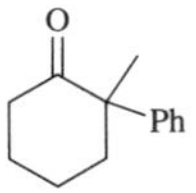 (77)	87
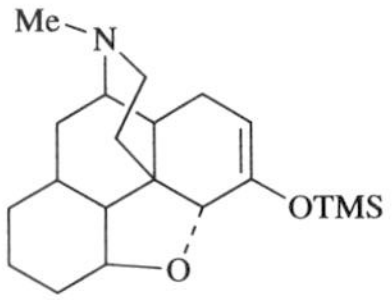	Ph_2IF, THF, CH_2Cl_2	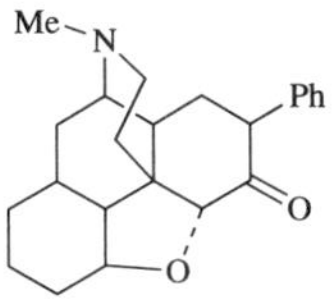 (11)	88
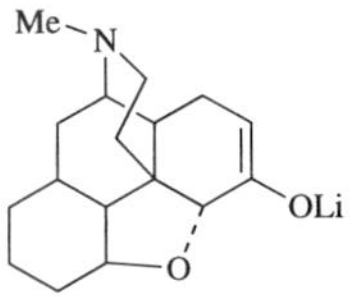	1. $Ph_2I^+I^-$, THF, DMF, –45°, 2 h 2. rt, overnight	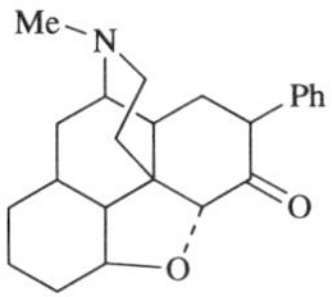 (71)	88

TABLE XXIV. α-AMINATION OF SILYL ENOL ETHERS

Substrate	Reagent-Conditions	Product(s) and Yield(s) (%)	Refs.
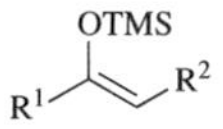	PhI=NTs, MeCN $CuClO_4$ (5-10 mol %), –20°	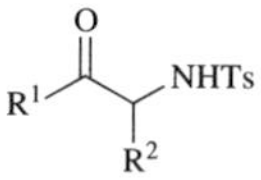 R^1 / R^2 / % n-Bu / H / (70) Ph / H / (75) Ph / Me / (58)	89
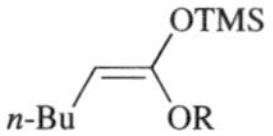	PhI=NTs, MeCN, $CuClO_4$ (5-10 mol %)	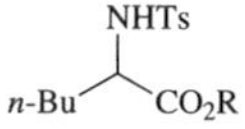 R / Temp / % Me / 25° / (27) Me / –20° / (10) Ph / 25° / (43) Ph / –20° / (50)	89
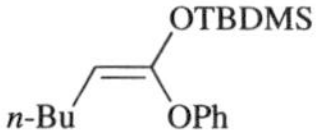	PhI=NTs, MeCN $CuClO_4$ (5-10 mol %), 25°	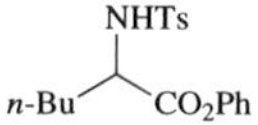 (45)	89
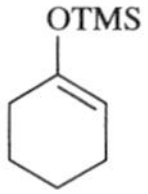	PhI=NTs, MeCN $CuClO_4$ (5-10 mol %), –20°	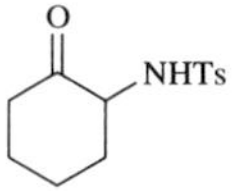 (64)	89
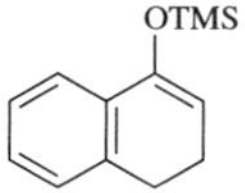	PhI=NTs, MeCN $CuClO_4$ (5-10 mol %), –20°	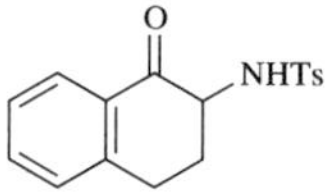 (53)	89

TABLE XXV. α-FUNCTIONALIZATION OF β–DICARBONYL COMPOUNDS WITH IODOSOBENZENE

Substrate	Reagent-Conditions	Product(s) and Yield(s) (%)	Refs.
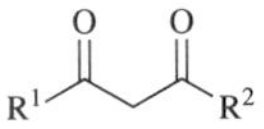	1. $(PhIO)_n$, $CHCl_3$, $TMSN_3$ (2 eq), rt, 2 h 2. Reflux, 3 h	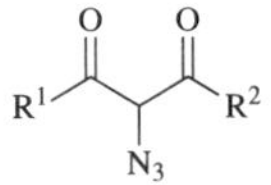 R^1 R^2 % Me Me (76) Ph Me (70) Me OMe (52) Ph OEt (70)	90
	$(PhIO)_n$, $CHCl_3$, $MeSO_3H$, reflux, 2 h	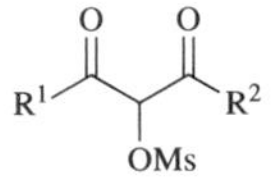 R^1 R^2 % Me Me (83) Ph Me (73) Me OMe (76) Ph OEt (76)	90
	$(PhIO)_n$, BF_3•OEt_2, MeOH, rt, 5 h	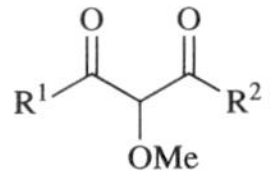 R^1 R^2 % Me Me (67) Me OMe (63)	90
	$(PhIO)_n$, BF_3•OEt_2 (0.65 eq), $CHCl_3$, reflux, 3 h	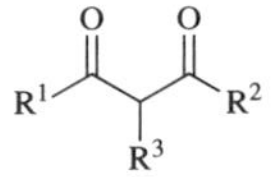 R^1 R^2 R^3 % Me Me $CH(COMe)_2$ (74) Ph Me CH(COPh)COMe (76) Me OMe $CH(COMe)CO_2Me$ (46) Ph OEt $CH(COPh)CO_2Et$ (68)	90
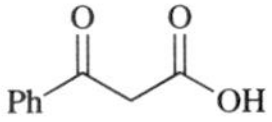	$(PhIO)_n$, BF_3•OEt_2, EtOH, rt	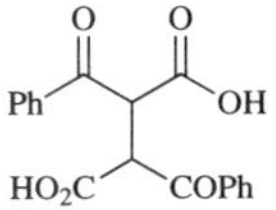 (46)	91
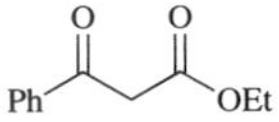	$(PhIO)_n$, BF_3•OEt_2, EtOH	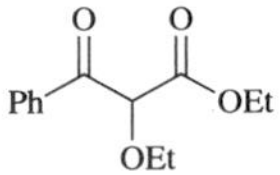 (59)	90
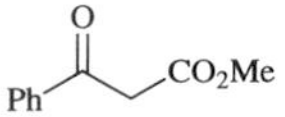	$(PhIO)_n$, BF_3•OEt_2, MeOH, rt, 8 h	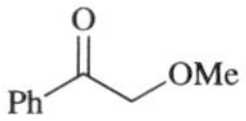 (63) + 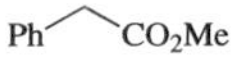 (24)	90

TABLE XXVI. α-FUNCTIONALIZATION OF β–DICARBONYL COMPOUNDS WITH IODOBENZENE DIACETATE

Substrate	Reagent-Conditions	Product(s) and Yield(s) (%)	Refs.
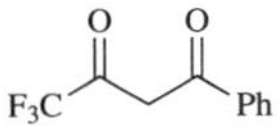	IBD (3 eq), AcOH, H_2O	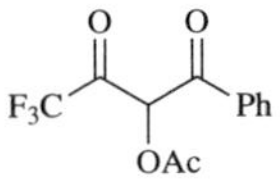 (22)	55
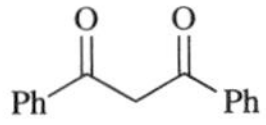	IBD (3 eq), AcOH, H_2O	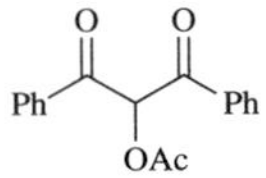 (24)	55
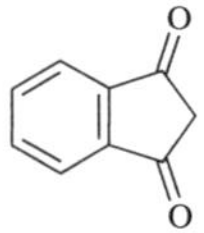	IBD (2 eq), MeOH, H_2SO_4, rt, 2 h	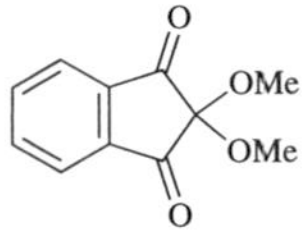 (44)	92
	IBD (2 eq), EtOH, H_2SO_4, 40°, 2 h	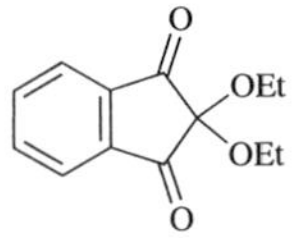 (49)	92
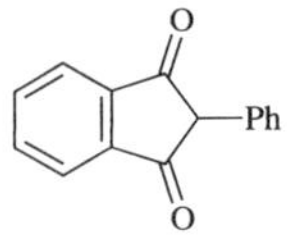	IBD (3 eq), AcOH, H_2O	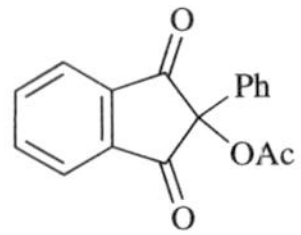 (35)	56
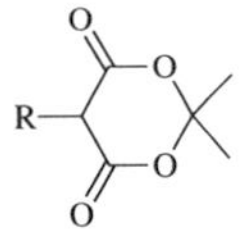	IBD, K_2CO_3, TEBA, $CHCl_2$	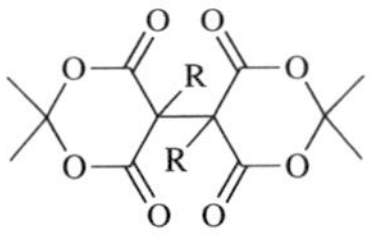	93

R	Time	%
Me	7 h	(45)
$(CH_2)_2CN$	11 h	(39)
Bn	10 h	(47)
p-$ClC_6H_4CH_2$	8 h	(52)
p-$MeC_6H_4CH_2$	11 h	(42)
p-$MeOC_6H_4CH_2$	10 h	(48)

TABLE XXVII. PERFLUOROALKYLATION OF ENOLATE ANIONS OF β–DICARBONYL COMPOUNDS WITH PERFLUOROALKYLIODONIUM TRIFLATES

Substrate	Reagent-Conditions	Product(s) and Yield(s) (%)	Refs.
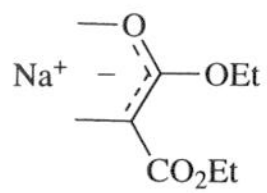	$n\text{-}C_3F_7I(Ph)OTf$, DMF, rt, 1 h	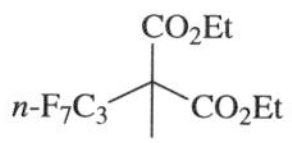 $n\text{-}F_7C_3$ — CO_2Et / CO_2Et (20)	94
	$i\text{-}C_3F_7I(Ph)OTf$, DMF, rt, 1 h	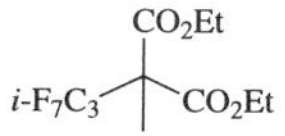 $i\text{-}F_7C_3$ — CO_2Et / CO_2Et (3)	94
	$n\text{-}C_8F_{17}I(Ph)OTf$, DMF, rt, 1 h	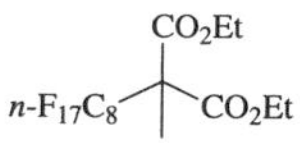 $n\text{-}F_{17}C_8$ — CO_2Et / CO_2Et (24)	94
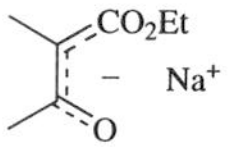	$n\text{-}C_8F_{17}I(Ph)OTf$, DMF, 0°, 3 h	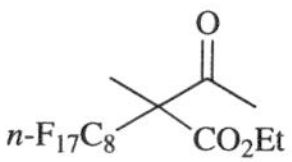 $n\text{-}F_{17}C_8$ — CO_2Et (28) + EtO_2C ... $OC_8F_{17}\text{-}n$ 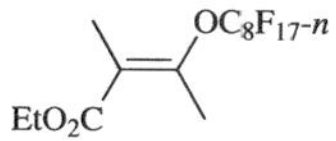(8)	94
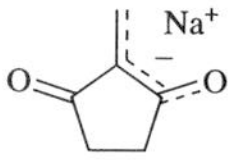	$n\text{-}C_8F_{17}I(Ph)OTf$, DMF, 0°, 3 h	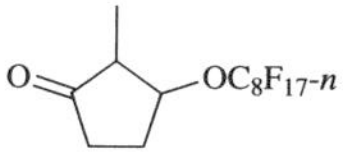 $OC_8F_{17}\text{-}n$ (26) + $C_8F_{17}\text{-}n$ 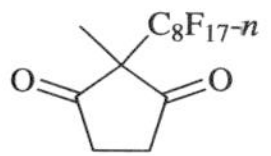(7)	94
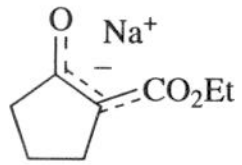	$n\text{-}C_8F_{17}I(Ph)OTf$, DMF, 0°, 3 h	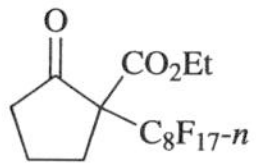 CO_2Et / $C_8F_{17}\text{-}n$ (20) + $OC_8F_{17}\text{-}n$ / CO_2Et 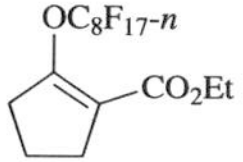(8)	94

TABLE XXVIII. SYNTHESIS OF SPIROOXETAN-3-ONES VIA INTRAMOLECULAR PARTICIPATION OF HYDROXY GROUP

Substrate	Reagent-Conditions	Product(s) and Yield(s) (%)	Refs.
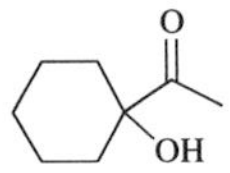	IBD, KOH/MeOH, 20°	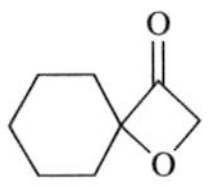 (~10)	95
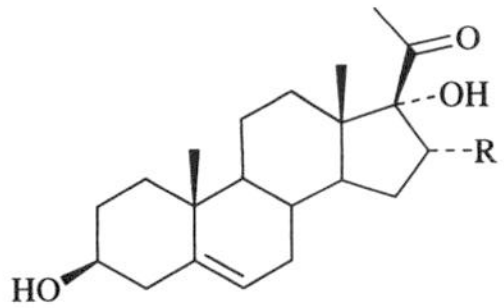	IBD, KOH/MeOH, 20°	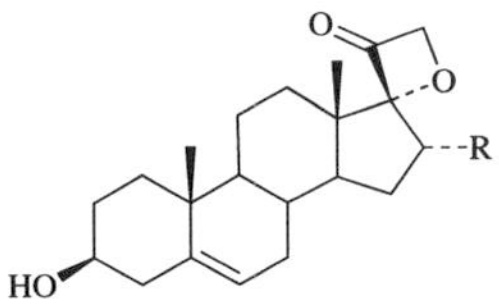 R / Time / % H / 5 h / (70) Me / 4.5 h / (75) NHAc / 1 h / (70)	95

TABLE XXIX. SYNTHESIS OF COUMARAN-3-ONES VIA INTRAMOLECULAR PARTICIPATION OF PHENOLIC GROUP

Substrate	Reagent-Conditions	Product(s) and Yield(s) (%)	Refs.
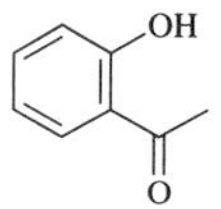	1. IBD, KOH/MeOH, 0-5° 2. rt, 2 h	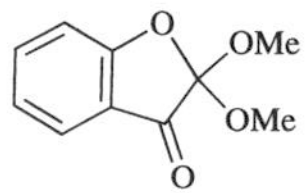 (20)	26
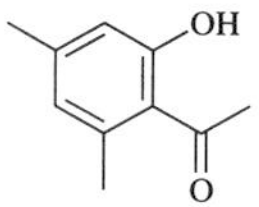	"	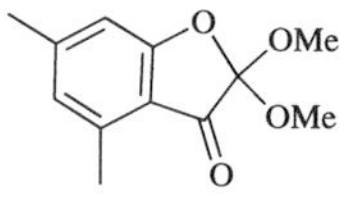 (35)	26
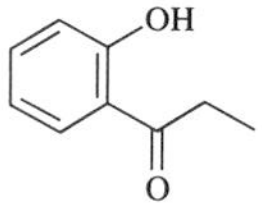	"	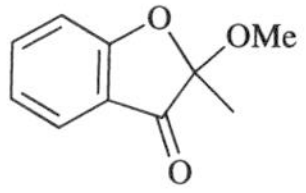 (21)	26
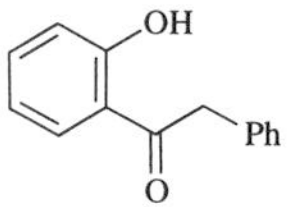	"	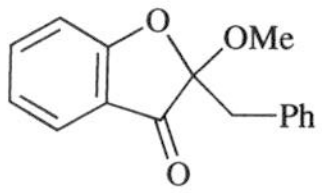 (40)	26
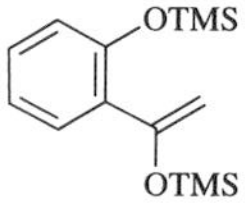	1. $(PHIO)_n$, BF_3•OEt_2, ether, H_2O, –40°, 1 h 2. –40° to rt, 1 h 3. rt, 0.5 h	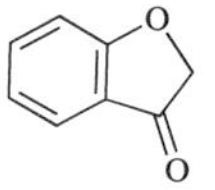 (31)[a]	99

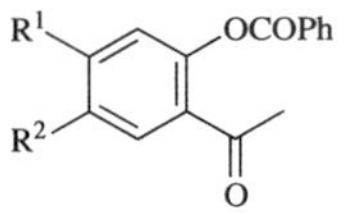

1. HTIB, dioxane, reflux, 2 h
2. KOH, rt to reflux, 0.5 h

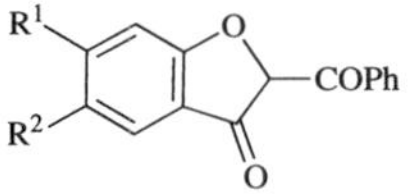

R^1	R^2	%	Refs.
H	H	(80)	98
H	Cl	(82)	
Me	Cl	(75)	
OMe	H	(74)	

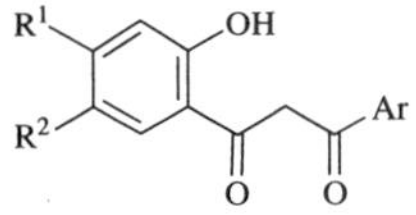

1. IBD, KOH/MeOH, 0-5°, 1 h
2. rt, 2 h
3. 6N HCl

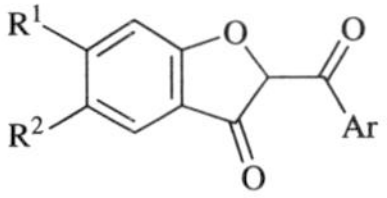

R^1	R^2	Ar	%	Refs.
H	H	Ph	(75)	97
H	Cl	Ph	(—)	97
Me	Cl	Ph	(—)	97
MeO	H	Ph	(—)	97
H	COMe	Ph	(82)	206
H	COMe	p-MeC_6H_4	(80)	206
H	COMe	p-$MeOC_6H_4$	(86)	206
H	COEt	p-MeC_6H_4	(79)	206
H	COEt	p-$MeOC_6H_4$	(82)	206

[a] In addition to coumaran-3-one, α-hydroxy-*o*-hydroxyacetophenone and 1,4-di(*o*-hydroxyphenyl)butane-1,4-dione (self-coupling) products were also obtained in 25 and 25% yields, respectively.

TABLE XXX. SYNTHESIS OF LACTONES VIA INTRAMOLECULAR PARTICIPATION OF CARBOXY GROUP

Substrate	Reagent-Conditions	Product(s) and Yield(s) (%)	Refs.
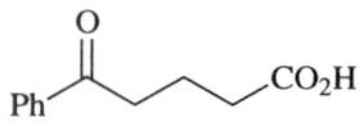	HTIB, CH_2Cl_2, reflux, 15 h	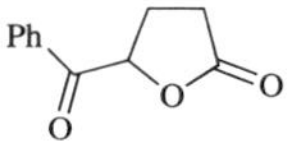 (74)	100
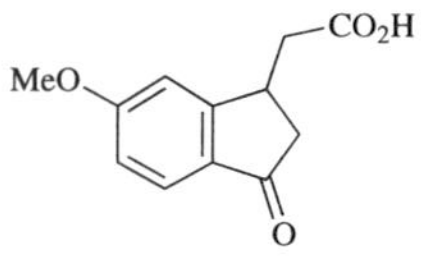	HTIB, CH_2Cl_2, reflux	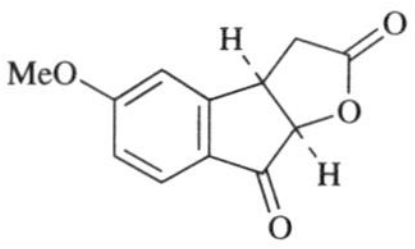 (78)	100
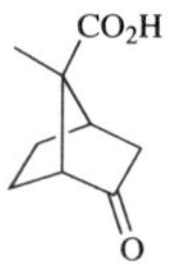	"	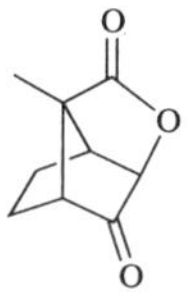 (76)	100
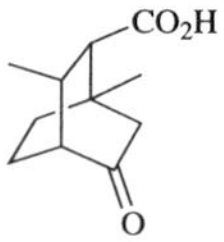	"	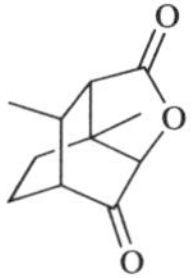 (81)	100

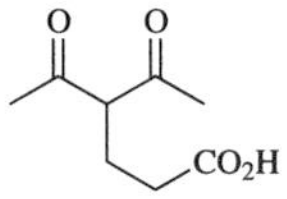	"	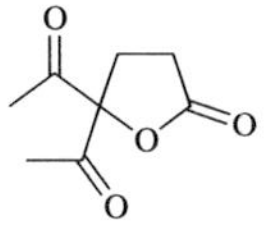	(48)	100
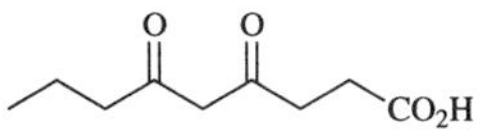	"	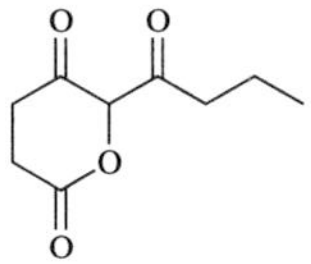	(68)	100
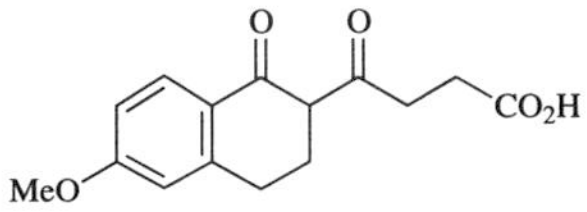	"	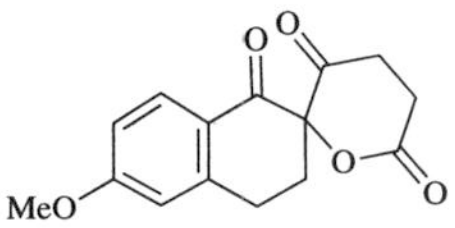	(36)	100
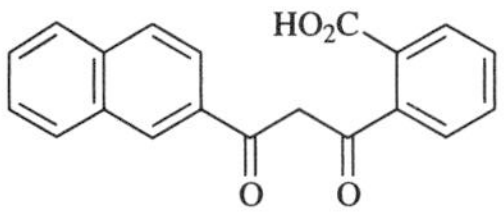	"	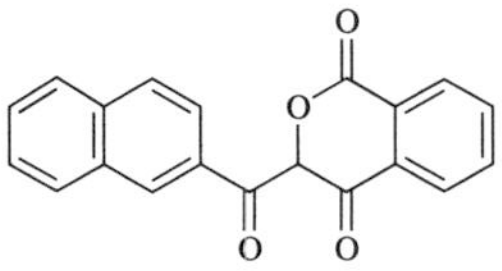	(79)	100

TABLE XXXI. CHLORINATION OF KETONES AND β-DIKETONES

Substrate	Reagent-Conditions	Product(s) and Yield(s) (%)				Refs.
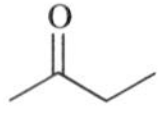	$PhICl_2$, AcOH		80:20 (—)			101
	$PhICl_2$, $h\nu$, C_6H_6		98:2 (—)			101
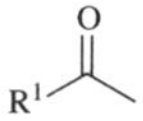	$BnMe_3N^+Cl_2I^-$ (2 eq) $Cl(CH_2)_2Cl$, MeOH	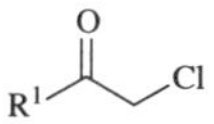	R^1	Time	%	
			Ph	3 h	(97)	102
			4-ClC_6H_4	6 h	(97)	102
			3-BrC_6H_4	5 h	(66)	102
			4-BrC_6H_4	5 h	(98)	101
			2-HOC_6H_4	10 h	(73)	102
			4-HOC_6H_4	10 h	(95)	102
			4-MeC_6H_4	10 h	(99)	102
			3-$MeOC_6$H$	3 h	(95)	102
			4-$MeOC_6H_4$	3 h	(97)	102
			4-EtC_6H_4	5 h	(98)	102
			2,5-$(MeO)_2C_6H_3$	4 h	(95)	102
			2-naphthyl	5 h	(99)	102
			2-thienyl	3 h	(95)	102
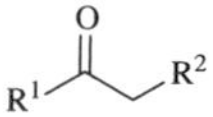	$PhICl_2$, AcOH	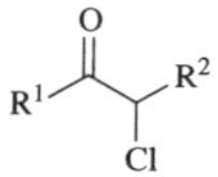	R^1	R^2	%	
			4-BrC_6H_4	H	(70)	102
			Ph	Ph	(30)	101
			4-MeC_6H_4	Ph	(35)	101
			1,3,5-$Me_3C_6H_2$	Ph	(35)	101
			Ph	COPh	(30)	101

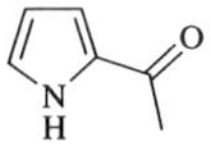	$BnMe_3N^+Cl_2I^-$ (2 eq), THF, rt, 12-16 h	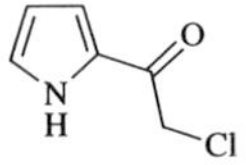	(85)	103
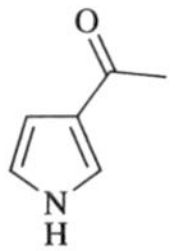	"	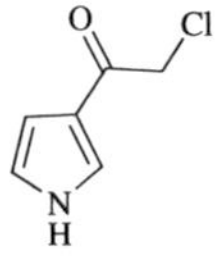	(62)	103
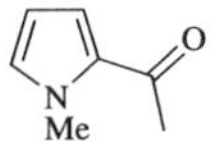	"	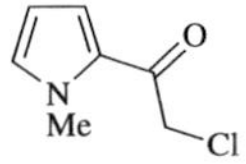	(95)	103
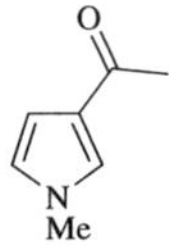	"	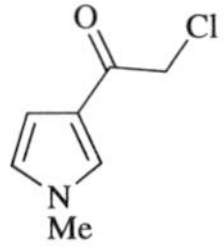	(75)	103
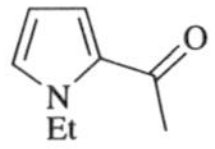	"	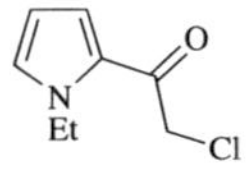	(85)	103
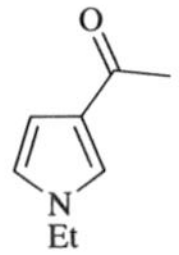	"	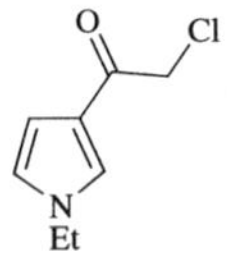	(81)	103
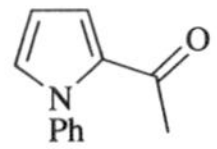	"	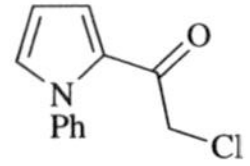	(78)	103
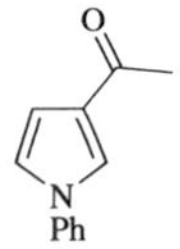	"	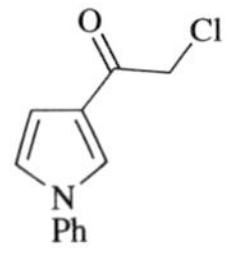	(93)	103

TABLE XXXI. CHLORINATION OF KETONES AND β-DIKETONES (*Continued*)

Substrate	Reagent-Conditions	Product(s) and Yield(s) (%)	Refs.
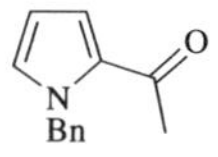	"	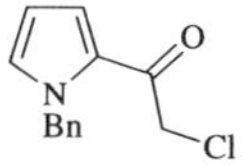 (82)	103
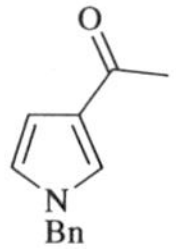	"	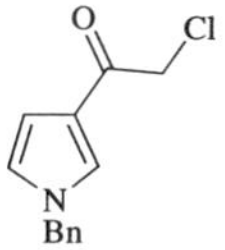 (76)	103
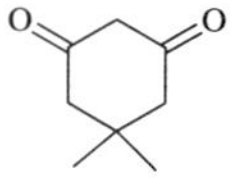	$PhICl_2$, AcOH	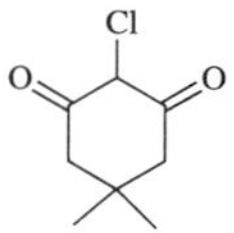 (—)	101

TABLE XXXII. OXIDATION OF STEROIDAL SILYL ENOL ETHERS BY *p*-(DIFLUOROIODO)TOLUENE

Substrate	Reagent-Conditions	Product(s) and Yield(s) (%)	Refs.
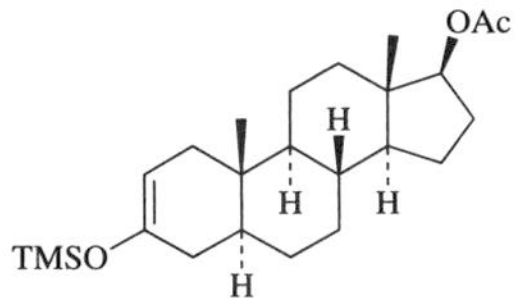	*p*-$MeC_6H_4IF_2$ $CF_2ClCFCl_2$, MeCN (1:1)	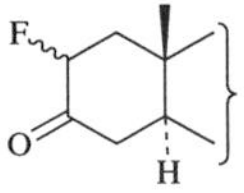 (37.5) + 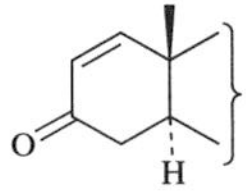 (17) 4:6, 2α + 2β	104
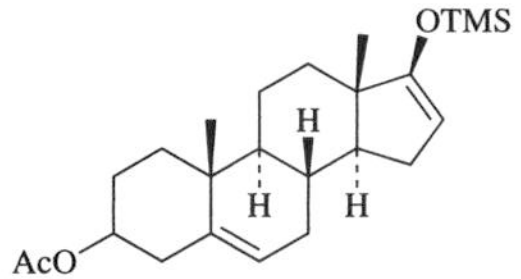	"	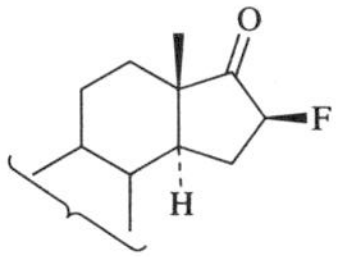 (22.5) 16β (no 16α)	104
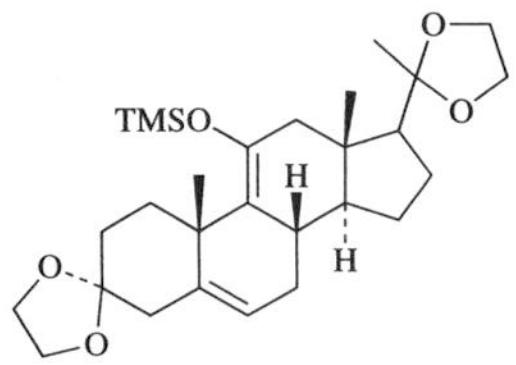	"	No reaction	104
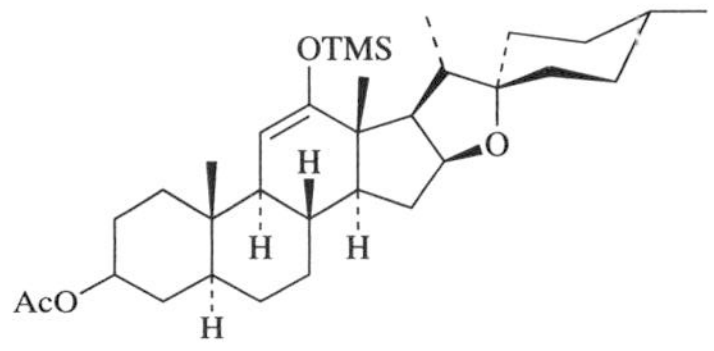	"	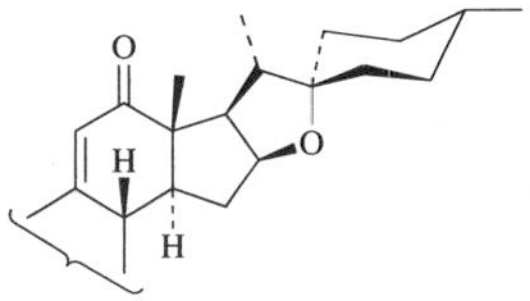 (67.5)	104

TABLE XXXIII. α-IMIDYLATION OF KETONES

Substrate	Reagent-Conditions	Product(s) and Yield(s) (%)	Refs.
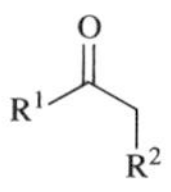	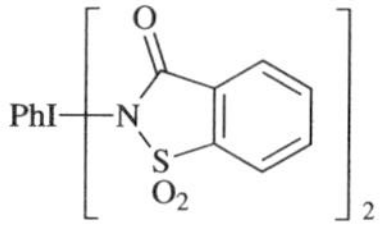	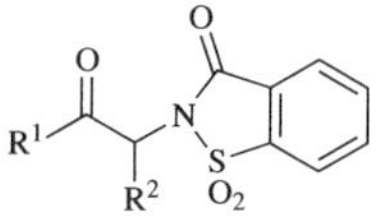	107
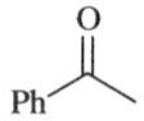	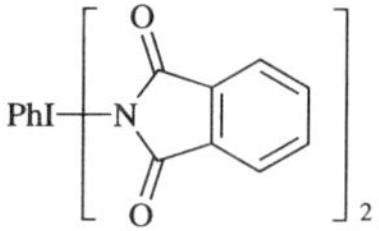, 70°, 30 min	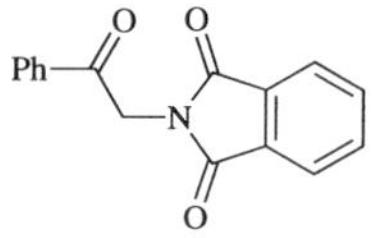 (20)	108

R^1	R^2	Solv	Temp	Time	%
H	Me	acetone	reflux	2 h	(57)
H	Et	butanone	reflux	6 h	(18)
Me	Et	—	65°	5 h	(18)
H	Ph	—	70°	5 h	(70)
H	p-MeC_6H_4	—	70°	5 h	(75)
—$(CH_2)_3$—		—	70°	4 h	(45)
—$(CH_2)_4$—		—	55°	4 h	(51)

TABLE XXXIV. LACTONES TO HIGHER HOMOLOGOUS α,β-UNSATURATED LACTONES VIA TRIMETHYLSILYLOXYCYCLOPROPANOLS

Substrate	Reagent-Conditions	Product(s) and Yield(s) (%)	Refs.
	$(PhIO)_n$, $(n\text{-}Bu)_4NF$, THF, CH_2Cl_2	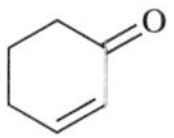 (72)	108
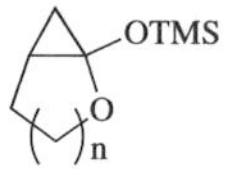	$(PhIO)_n$, $(n\text{-}Bu)_4NF$, THF, CH_2Cl_2	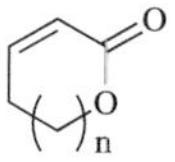 n / %: 1 (72); 2 (75); 3 (62); 9 (78)	108
	$(PhIO)_n$, $(n\text{-}Bu)_4NF$, THF, CH_2Cl_2	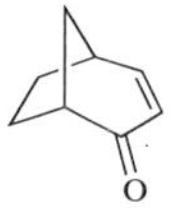 (73)	108

TABLE XXXV. β-FUNCTIONALIZATION OF TRIISOPROPYLSILYL ENOL ETHERS

Substrate	Reagent-Conditions	Product(s) and Yield(s) (%)	Refs.
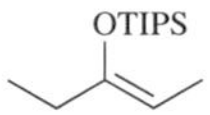	$(PhIO)_n$, $TMSN_3$, CH_2Cl_2	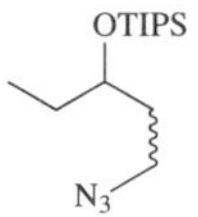 %: Z (42), E (32) + 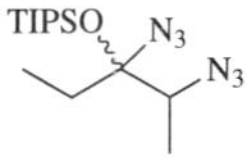 (28)	109
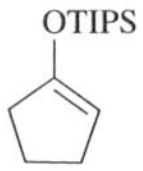	$(PhIO)_n$, $TMSN_3$, CH_2Cl_2	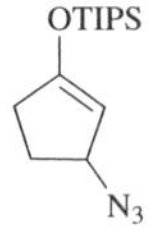 (73)	109
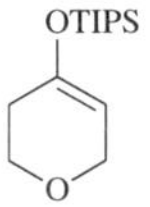	1. $(PhIO)_n$, $TMSN_3$, CH_2Cl_2 2. CH_2=$CHCH_2SnBu_3$, Me_2AlCl	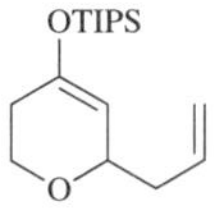 (62)	109
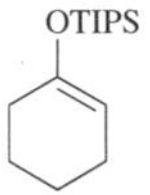	$(PhIO)_n$, $TMSN_3$, CH_2Cl_2	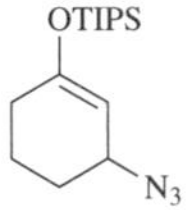 (83)	109
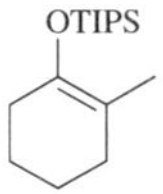	$(PhIO)_n$, $TMSN_3$, CH_2Cl_2	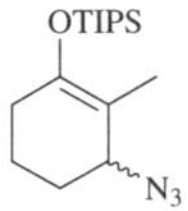 (95)[a]	109
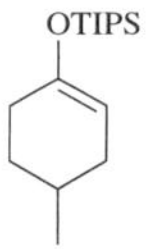	$(PhIO)_n$, $TMSN_3$, CH_2Cl_2	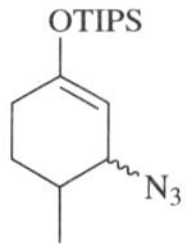 (95)[a]	109

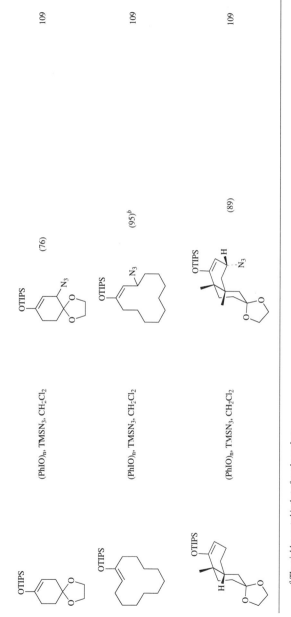

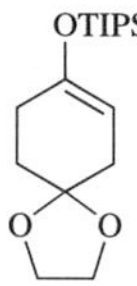	$(PhIO)_n$, $TMSN_3$, CH_2Cl_2	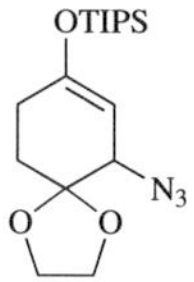	(76)	109
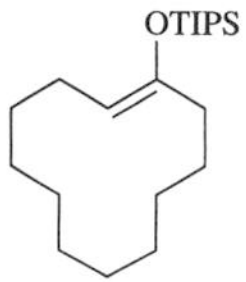	$(PhIO)_n$, $TMSN_3$, CH_2Cl_2	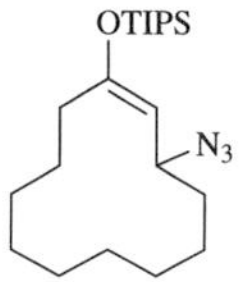	(95)[b]	109
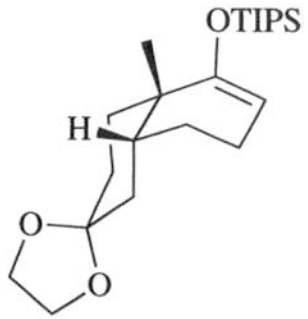	$(PhIO)_n$, $TMSN_3$, CH_2Cl_2	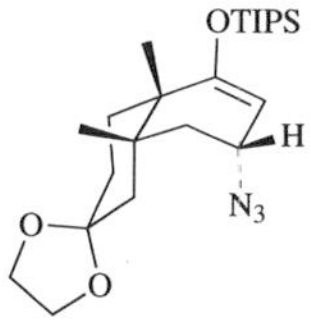	(89)	109

[a] The yield reported is that of crude product.

[b] The *E*/*Z* ratio = 4:1.

TABLE XXXVI. 5-SUBSTITUTED-2(5H)-FURANONES VIA γ-FUNCTIONALIZATION OF SILYL ENOL ETHERS

Substrate	Reagent-Conditions	Product(s) and Yield(s) (%)				Refs.
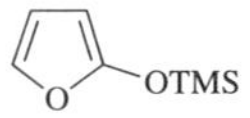	$(PhIO)_n$, ROH, BF_3•OEt_2, solvent	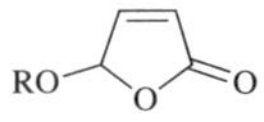	R	Sclv	%	
			Et	—	(55)	111
			Ac	CH_2Cl_2	(72)	
			Ms	CH_2Cl_2	(67)	
			p-Ts	CH_2Cl_2	(66)	
	$(PhIO)_n$, $TMSN_3$, BF_3•OEt_2, CH_2Cl_2	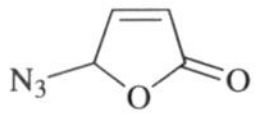	(51)			111
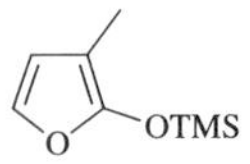	$(PhIO)_n$, MsOH, BF_3•OEt_2, CH_2Cl_2	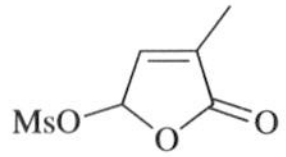	(74)			111
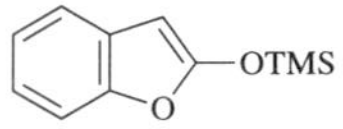	$(PhIO)_n$, AcOH, BF_3•OEt_2, CH_2Cl_2	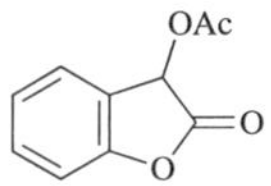	(65)			111
	$(PhIO)_n$, *p*-TsOH, BF_3•OEt_2, CH_2Cl_2	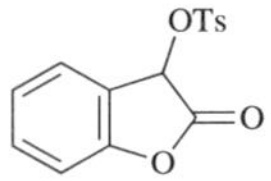	(62)			111

TABLE XXXVII. 2-ARYL (AND HETEROARYL)ALKANOATES BY 1,2-ARYL MIGRATION OF ALKYL ARYL KETONES

Substrate	Reagent-Conditions	Product(s) and Yield(s) (%)	Refs.
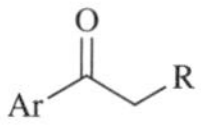	Reagent, acid, solvent (See table)	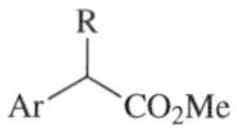	

Ar	R	Reagent	Acid	Solv	Temp	Time	%	Refs.
Ph	H	$(PhIO)_n$	H_2SO_4	MeOH	rt	2-3 h	(61)[a]	115
Ph	H	$(PhIO)_n$	FSO_3H	MeOH	rt	12 h	(61)[a]	114
Ph	H	IBD	H_2SO_4	MeOH	60°	1 h	(59)[b]	113
Ph	H	HTIB	—	MeOH	rt	36 h	(57)	114
p-FC_6H_4	H	HTIB	—	MeOH	rt	45 h	(51)	115
p-ClC_6H_4	H	$(PhIO)_n$	H_2SO_4	MeOH	rt	2-3 h	(53)[a]	115
p-ClC_6H_4	H	$(PhIO)_n$	FSO_3H	MeOH	rt	12 h	(50)[a]	114
p-ClC_6H_4	H	HTIB	—	MeOH	rt	40 h	(55)	115
p-BrC_6H_4	H	IBD	H_2SO_4	MeOH	60°	5 h	(44)[b]	113
p-MeC_6H_4	H	$(PhIO)_n$	H_2SO_4	MeOH	rt	2-3 h	(63)[a]	115
p-MeC_6H_4	H	IBD	H_2SO_4	MeOH	60°	50 min	(68)[b]	113
p-MeC_6H_4	H	HTIB	—	MeOH	rt	30 h	(62)	115
o-MeC_6H_4	H	IBD	H_2SO_4	MeOH	60°	1 h	(62)[b]	113
p-$MeOC_6H_4$	H	$(PhIO)_n$	FSO_3H	MeOH	rt	12 h	(71)[a]	114
p-$MeOC_6H_4$	H	$(PhIO)_n$	H_2SO_4	MeOH	rt	2-3 h	(70)[a]	115
Ph	Me	$(PhIO)_n$	H_2SO_4	TMOF	rt	10 h	(78)	115
Ph	Me	HTIB	—	TMOF	rt	20 h	(75)	115
Ph	Me	IBD	H_2SO_4	TMOF	60°	10 min	(81)	113
p-FC_6H_4	Me	IBD	H_2SO_4	TMOF	60°	5 min	(82)	112, 113
p-MeC_6H_4	Me	IBD	H_2SO_4	TMOF	60°	5 min	(85)	113
p-$MeOC_6H_4$	Me	$(PhIO)_n$	H_2SO_4	TMOF	rt	3 h	(81)	115
p-$MeOC_6H_4$	Me	HTIB	—	TMOF	rt	15 h	(77)	115
p-$MeOC_6H_4$	Me	IBD	H_2SO_4	TMOF	rt	1 h	(88)	113

TABLE XXXVII. 2-ARYL (AND HETEROARYL)ALKANOATES BY 1,2-ARYL MIGRATION OF ALKYL ARYL KETONES (*Continued*)

Substrate	Reagent-Conditions	Product(s) and Yield(s) (%)	Refs.
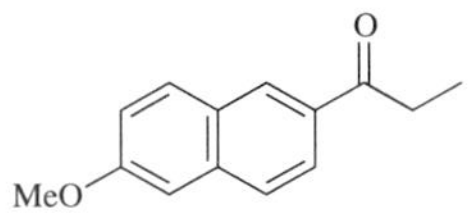	$(PhIO)_n$, H_2SO_4, TMOF, rt, 3 h	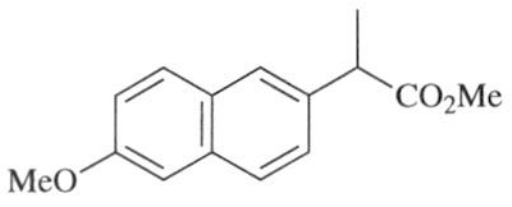 (80)	115
	HTIB, TMOF, rt, 12 h	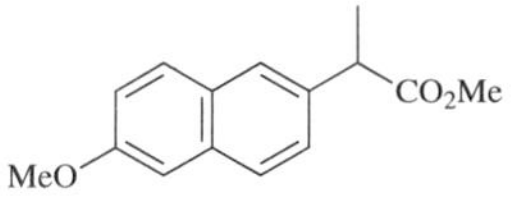 (73)	115
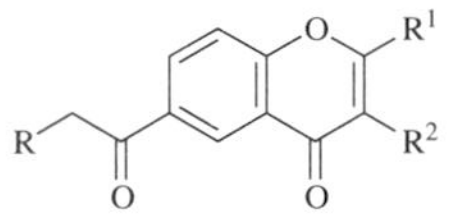	Reagent, acid, solvent (See table)	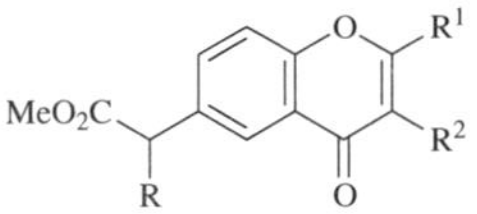	116

R	R[1]	R[2]	Reagent	Acid	Solv	Temp	%	Refs.
H	Me	Me	$(PhIO)_n$	BF_3•Et_2O	MeOH	rt	(70)	116
H	Ph	Me	$(PhIO)_n$	BF_3•Et_2O	MeOH	rt	(70)	116
Me	Me	Me	IBD	H_2SO_4	TMOF	50-60°	(95)	116
Me	Ph	Me	IBD	H_2SO_4	TMOF	rt	(95)	116
Me	Ph	OMe	IBD	H_2SO_4	TMOF		(94)	207
Me	Ph	OBn	IBD	H_2SO_4	TMOF	rt	(95)	207
Me	*p*-ClC_6H_4	OMe	IBD	H_2SO_4	TMOF		(95)	207
Me	*p*-ClC_6H_4	OBn	IBD	H_2SO_4	TMOF	rt	(96)	207
Me	*p*-MeC_6H_4	OMe	IBD	H_2SO_4	TMOF	rt	(96)	207
Me	*p*-MeC_6H_4	OBn	IBD	H_2SO_4	TMOF	rt	(94)	207
Me	*p*-$MeOC_6H_4$	OBn	IBD	H_2SO_4	TMOF	rt	(92)	207
Et	Me	Me	IBD	H_2SO_4	TMOF		(90)	207
Et	Ph	OBn	IBD	H_2SO_4	TMOF	rt	(90)	208
Et	*p*-ClC_6H_4	OMe	IBD	H_2SO_4	TMOF	rt	(90)	208
Et	*p*-ClC_6H_4	OBn	IBD	H_2SO_4	TMOF	rt	(80)	208

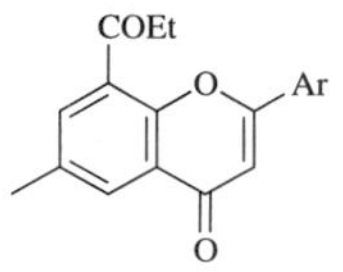

IBD, (2 eq), H_2SO_4, TMOF

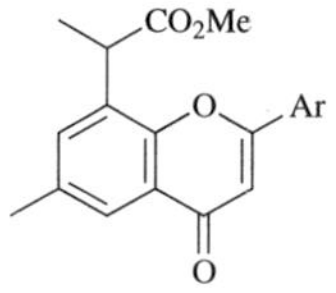

Ar	%
Ph	(90-95)
p-ClC_6H_4	(90-95)
p-MeC_6H_4	(90-95)
p-$MeOC_6H_4$	(90-95)

209

[a] The yield is based on ^{1}H NMR of crude ester.

[b] The corresponding α-methoxyacetophenones were isolated in 13-18% yields.

[c] The reaction time was 5 h.

TABLE XXXVIII. OXIDATIVE REARRANGEMENT OF 3-AROYLPROPIONIC ACIDS

Substrate	Reagent-Conditions	Product(s) and Yield(s) (%)	Refs.
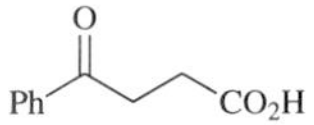	IBD, H_2SO_4, TMOF, 60°, 2 h	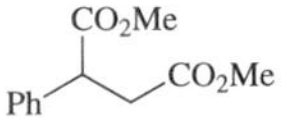 (87)	113
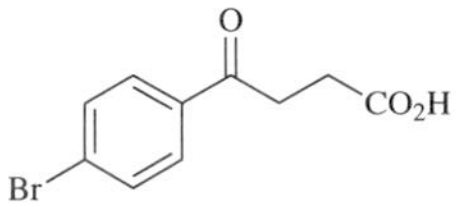	IBD, H_2SO_4, TMOF, 60°, 2 h	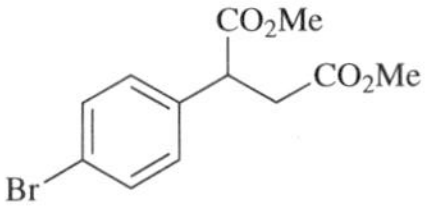 (76)	113
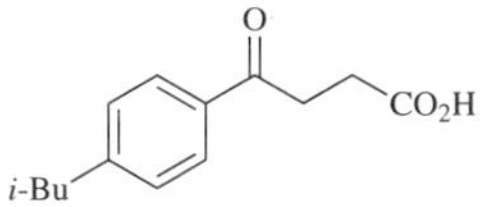	IBD, H_2SO_4, TMOF, 60°, 1 h	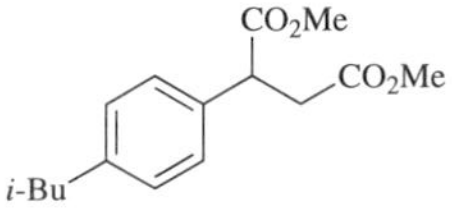 (81)	113
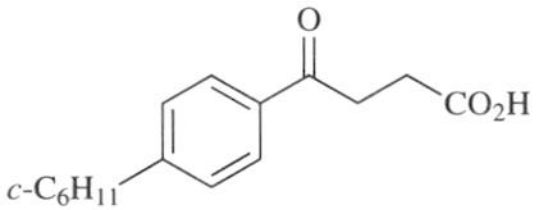	IBD, H_2SO_4, TMOF, 60°, 2 h	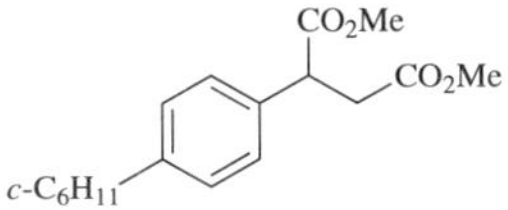 (82)	113
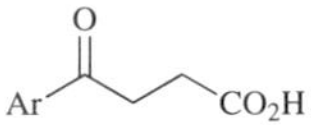 Ar = 2-naphthyl	IBD, H_2SO_4, TMOF, 60°, 1 h	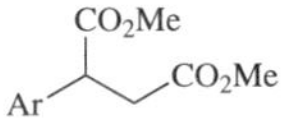 (74)	113
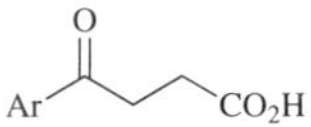 Ar = 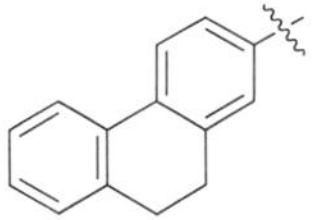	IBD, H_2SO_4, TMOF, 60°, 5 h	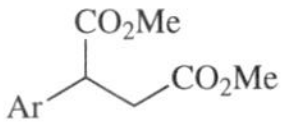 (75)	113

TABLE XXXIX. METHYL α-SUBSTITUTED ARYLACETATES FROM THE OXIDATIVE REARRANGEMENT OF ARYL METHYL KETONES

Substrate	Reagent-Conditions	Product(s) and Yield(s) (%)	Refs.
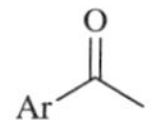	IBD (2.2 eq), H_2SO_4, TMOF, overnight	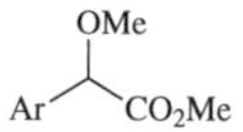 Ar / %: Ph (82); *p*-ClC_6H_4 (80); *p*-BrC_6H_4 (84); *p*-MeC_6H_4 (84); *p*-$MeOC_6H_4$ (86)	118
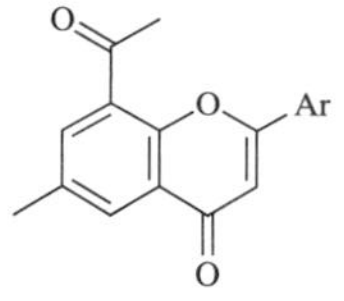	IBD (2.2 eq), H_2SO_4, TMOF	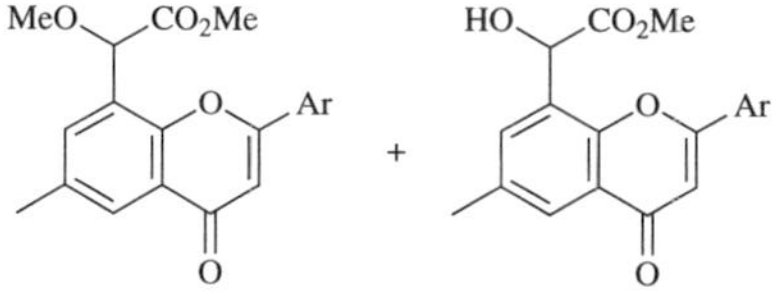 Ar / %: Ph (90); *p*-ClC_6H_4 (85); *p*-MeC_6H_4 (83); *p*-$MeOC_6H_4$ (87)	209

TABLE XL. 1,2-ARYL MIGRATION OF CHALCONES

Substrate	Reagent-Conditions	Product(s) and Yield(s) (%)	Refs.
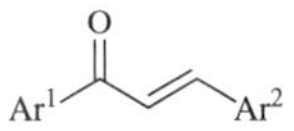	Reagent, acid, MeOH or TMOF	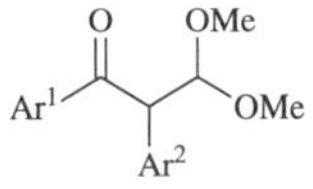	114

Ar^1	Ar^2	Reagent	Acid	Time	%
Ph	Ph	$(PhIO)_n$	FSO_3H	6 h	(48)
Ph	Ph	$(PhIO)_n$	CF_3SO_3H	12 h	(20)
Ph	Ph	$(PhIO)_n$	$BF_3{\bullet}OEt_2$	12 h	(25)
Ph	Ph	HTIB	—	40 h	(58)
Ph	p-ClC_6H_4	$(PhIO)_n$	FSO_3H	12 h	(64)
Ph	p-ClC_6H_4	HTIB	MeOH	40 h	(60)
Ph	p-$MeOC_6H_4$	$(PhIO)_n$	FSO_3H	12 h	(18)
Ph	p-$MeOC_6H_4$	$(PhIO)_n$	CF_3SO_3H	12 h	(18)
Ph	p-$MeOC_6H_4$	$(PhIO)_n$	$BF_3{\bullet}OEt_2$	2 h	(49)
Ph	p-$MeOC_6H_4$	HTIB	—	48 h	(55)
p-$MeOC_6H_4$	Ph	$(PhIO)_n$	FSO_3H	1 h	(43)
p-$MeOC_6H_4$	Ph	$(PhIO)_n$	CF_3SO_3H	12 h	(59)
p-$MeOC_6H_4$	Ph	$(PhIO)_n$	$BF_3{\bullet}OEt_2$	12 h	(67)
p-$MeOC_6H_4$	Ph	HTIB	—	36 h	(65)
p-$MeOC_6H_4$	p-$MeOC_6H_4$	$(PhIO)_n$	FSO_3H	12 h	(42)
p-$MeOC_6H_4$	p-$MeOC_6H_4$	$(PhIO)_n$	CF_3SO_3H	12 h	(42)
p-$MeOC_6H_4$	p-$MeOC_6H_4$	HTIB	—	55 h	(46)

Substrate	Reagent-Conditions	Product(s) and Yield(s) (%)	Refs.
	IBD, H_2SO_4, TMOF, rt	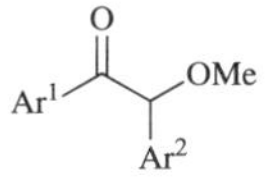	119

Ar^1	Ar^2	Time	%
Ph	Ph	24 h	(80)
Ph	p-ClC_6H_4	20 h	(92)
p-MeC_6H_4	Ph	18 h	(90)
p-MeC_6H_4	p-ClC_6H_4	18 h	(90)
p-$MeOC_6H_4$	Ph	18 h	(88)
p-$MeOC_6H_4$	p-ClC_6H_4	18 h	(94)

TABLE XLI. OXIDATIVE REARRANGEMENT OF FLAVANONES

Substrate	Reagent-Conditions	Product(s) and Yield(s) (%)	Refs.
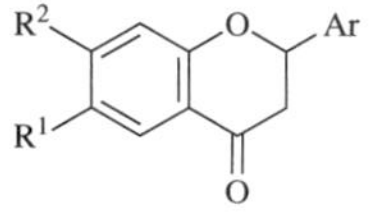	HTIB, CH_3CN, reflux	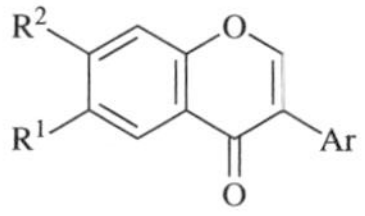	122
		R¹ / R² / Ar / %: H / H / Ph / (75); H / H / p-ClC_6H_4 / (72); H / H / p-$MeOC_6H_4$ / (76); H / OMe / p-$MeOC_6H_4$ / (74); Cl / H / Ph / (75); Cl / H / p-$MeOC_6H_4$ / (80); Me / H / p-ClC_6H_4 / (78)	
	$(PhIO)_n$, p-MsOH, CH_2Cl_2	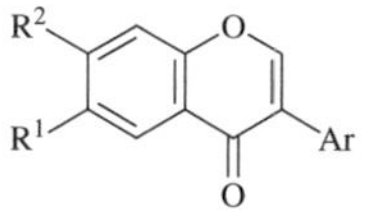	123
		R¹ / R² / Ar / %: H / H / Ph / (80); H / H / p-ClC_6H_4 / (84); H / H / p-$MeOC_6H_4$ / (75); Cl / H / Ph / (70); Cl / H / p-ClC_6H_4 / (70); Cl / H / p-$MeOC_6H_4$ / (80); Me / H / p-ClC_6H_4 / (75)	
	IBD, H_2SO_4, CH_3CN, reflux	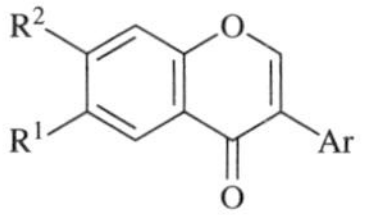	123
		R¹ / R² / Ar / %: H / H / Ph / (38); Cl / H / p-ClC_6H_4 / (66)	
	IBD, H_2SO_4, AcOH	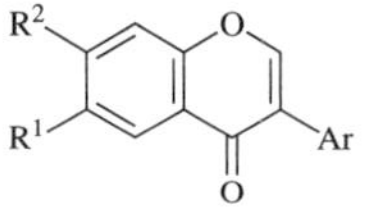	123
		R¹ / R² / Ar / %: H / H / p-ClC_6H_4 / (80); H / H / p-$MeOC_6H_4$ / (75); Cl / H / p-ClC_6H_4 / (66); Cl / H / p-$MeOC_6H_4$ / (70); Me / H / p-ClC_6H_4 / (60)	

TABLE XLI. OXIDATIVE REARRANGEMENT OF FLAVANONES (*Continued*)

Substrate	Reagent-Conditions	Product(s) and Yield(s) (%)	Refs.
	IBD, H_2SO_4, TMOF	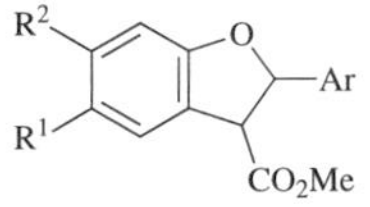	124

R^1	R^2	Ar	%
H	H	Ph	(40)
H	H	p-ClC_6H_4	(60)
Cl	H	Ph	(43)
Cl	H	p-ClC_6H_4	(35)
Cl	Me	Ph	(47)
Me	H	Ph	(75)
Me	H	p-ClC_6H_4	(75)

TABLE XLII. OXIDATIVE REARRANGEMENT OF 2,2-DIALKYLCHROMANONES TO CHROMONES AND TETRAHYDROXANTHONES

Substrate	Reagent-Conditions	Product(s) and Yield(s) (%)	R^1	R^2	R^3	Time	%	Refs.
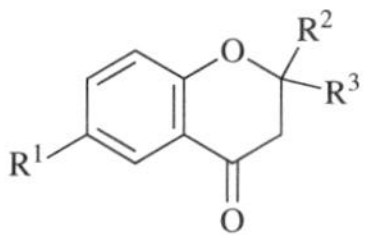	PhI(OH)OTs, MeCN, reflux	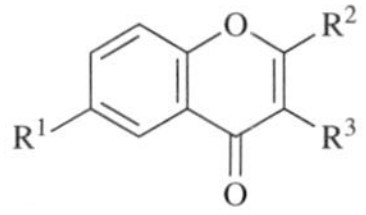	H	H	H	20 h	(60)	125
			H	H	H	25 h	(58)	
			H	Me	H	18 h	(65)	
			H	Me	Me	16 h	(70)	
			H	Me	Et	16 h	(45)	
			Cl	Me	Et	16 h	(48)	
			Cl	Me	*i*-Pr	18 h	(61)	
			H	Me	*i*-Pr	18 h	(56)	
			Cl	—$(CH_2)_4$—		10 h	(80)	
			H	—$(CH_2)_4$—		10 h	(85)	
			Cl	—$(CH_2)_5$—		10 h	(78)	
				—$(CH_2)_5$—		10 h	(82)	
	PhI(OH)OTs, MeCN, ((((, 45°		R^1	R^2	R^3	Time	%	125
			H	H	H	10 h	(84)	
			H	H	H	35 h	(60)	
			H	Me	H	10 h	(90)	
			H	Me	Me	10 h	(92)	
			H	Me	Et	10 h	(60)	
			Cl	Me	Et	10 h	(62)	
			H	Me	*i*-Pr	10 h	(75)	
			Cl	Me	*i*-Pr	10 h	(72)	
			H	—$(CH_2)_4$—		5 h	(90)	
			Cl	—$(CH_2)_4$—		5 h	(92)	
			H	—$(CH_2)_5$—		5 h	(90)	
			Cl	—$(CH_2)_5$—		5 h	(94)	

TABLE XLIII. DEHYDROGENATION OF FLAVANONES AND 2-ARYL-1,2,3,4-TETRAHYDROQUINOLONES

Substrate	Reagent-Conditions	Product(s) and Yield(s) (%)				Refs.
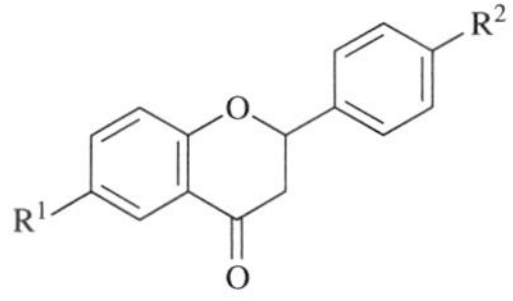	HTIB, MeOH, rt	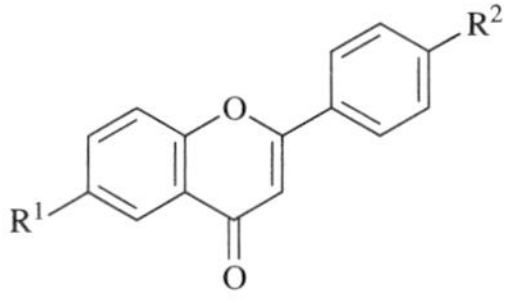	R^1	R^2	%	128
			H	H	(68)	
			H	Cl	(65)	
			Cl	H	(70)	
			Cl	Cl	(75)	
			Cl	OMe	(71)	
			OMe	OMe	(65)	
	IBD, MeOH, reflux	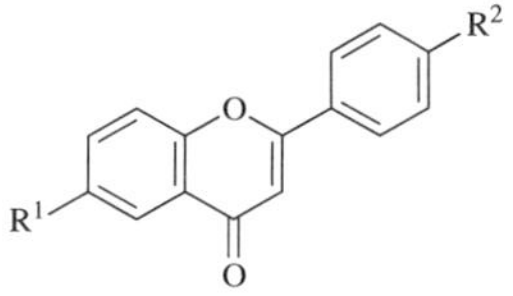	R^1	R^2	%	123
			H	H	(70)	
			H	Cl	(72)	
			H	OMe	(68)	
			Cl	H	(72)	
			Cl	Cl	(80)	
			Cl	OMe	(72)	
			Me	Cl	(75)	
	IBD, AcOH	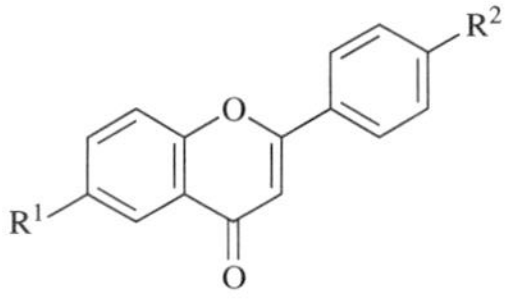	R^1	R^2	%	123
			H	H	(70)	
			H	Cl	(70)	
			Cl	H	(75)	
			Cl	Cl	(80)	
			Cl	OMe	(60)	

IBD, MeCN, reflux

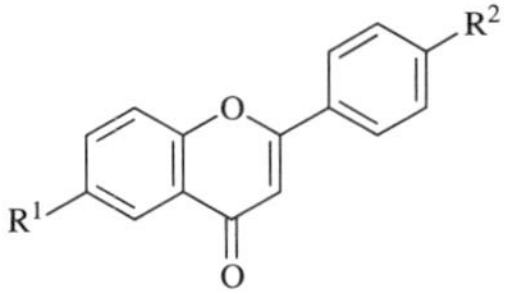

R^1	R^2	%
H	Cl	(60)
H	OMe	(75)
Cl	H	(60)
Cl	Cl	(85)
Cl	OMe	(75)

123

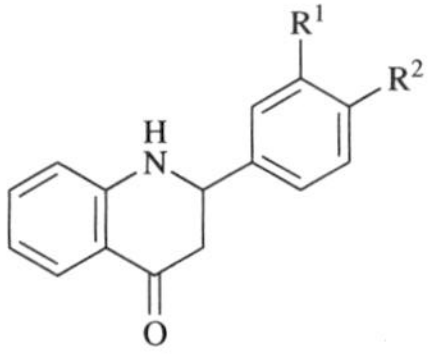

IBD, KOH/MeOH, 60°, 12-18 h

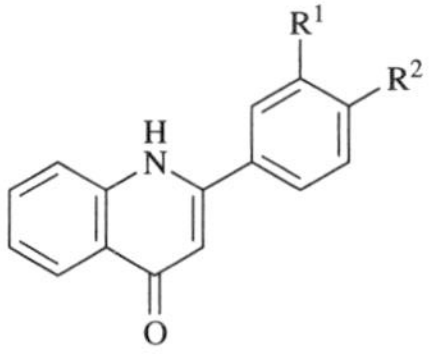

R^1	R^2	%
H	H	(85)
Cl	H	(89)
Me	H	(80)
OMe	H	(75)

129

REFERENCES

[1] Varvoglis, A. *Chem. Soc. Rev.* **1981**, *10*, 377.
[2] Umemoto, T. *Yuki, Gosei Kagaku Kyokai Shi.* **1983**, *41*, 251 [*Chem. Abstr.* **1983**, *98*, 214835y].
[3] Koser, G. F. In *The Chemistry of Functional Groups,* Suppl. *D*; Patai, S.; Rappoport, Z., Eds., Wiley: New York, 1983, p. 721, p. 1265.
[4] Nguyen, T. T.; Martin, J. C. In *Comprehensive Heterocyclic Chemistry*; Katritzky, A. R.; Rees, C. W., Eds.; Pergamon: Oxford, Vol. 1, 1984, p. 563.
[5] Varvoglis, A. *Synthesis* **1984**, 709.
[6] Ochiai, M.; Nagao, Y. *Yuki Gosei Kagaku Kyokai Shi.* **1986**, *44*, 660 [*Chem. Abstr.* **1987**, *106*, 84682s].
[7] Moriarty, R. M.; Prakash, O. *Acc. Chem. Res.* **1986**, *19*, 244.
[8] Merkushev, E. B. *Russ. Chem. Rev. (Engl. Transl.)* **1987**, *56*, 826.
[9] Moriarty, R. M.; Vaid, R. K.; Koser, G. F. *Synlett* **1990**, 365.
[10] Moriarty, R. M.; Vaid, R. K. *Synthesis* **1990**, 431.
[11] Varvoglis, A. *The Organic Chemistry of Polycoordinated Iodine,* VCH: New York, 1992.
[12] Stang, P. J. *Angew. Chem., Int. Ed. Engl.* **1992**, *31*, 274.
[13] Moriarty, R. M.; Prakash, O. In *Advances in Heterocyclic Chemistry*; Katritzky, A. R., Ed.; Academic: New York, Vol. 69, 1998, p. 1.
[14] Prakash, O.; Saini, N.; Sharma, P. K. *Heterocycles* **1994**, *38*, 409.
[15] Prakash, O.; Singh, S. P. *Aldrichimica Acta* **1994**, *27*, 15.
[16] Stang, P. J. In *The Chemistry of Triple-Bonded Functional Groups, Supplement C2, Vol. 2*; Patai, S., Ed.; Wiley: New York, 1994, p. 1164.
[17] Stang, P. J. In *Modern Acetylene Chemistry*; Stang, P. J.; Diederich, F., Eds.; VCH: Weinheim, 1995, p. 67.
[18a] Koser, G. F. In *The Chemistry of Halides, Pseudo-Halides and Azides, Suppl. D*; Patai, S.; Rappoport, Z., Eds.; Wiley: New York, 1995, p. 1173.
[18b] Ochial, M.; Kitagawa, Y.; Yamamoto, Y. *J. Am. Chem. Soc.* **1997**, *119*, 11598.
[19] Prakash, O.; Saini, N.; Tanwar, M. P.; Moriarty, R. M. *Cont. Comp. Org. Synth.* **1995**, *2*, 121.
[20] Prakash, O. *Aldrichimica Acta* **1995**, *28*, 63.
[21] Stang, P. J.; Zhdankin, V. V. *Chem. Rev.* **1996**, *96*, 1123.
[22] Jones, A. B. In *Comprehensive Organic Synthesis*; Trost, B. M.; Fleming, I., Eds.; Pergamon: Oxford, Vol. 7, 1991, p. 151.
[23] Moriarty, R. M.; Hu, H.; Gupta, S. C. *Tetrahedron Lett.* **1981**, *22*, 1283.
[24] Schardt, B. C.; Hill, C. L. *Inorg. Chem.* **1983**, *22*, 1563.
[25] Moriarty, R. M.; Prakash, I.; Musallam, H. A. *Tetrahedron Lett.* **1984**, *25*, 5867.
[26] Moriarty, R. M.; Prakash, O.; Prakash, I.; Musallam, H. A. *J. Chem. Soc., Chem. Commun.* **1984**, 1342.
[27] Moriarty, R. M.; Hou, K. C. *Tetrahedron Lett.* **1984**, *25*, 691.
[28] Koser, G. F.; Wettach, R. H.; Troup, J. M.; Frenz, B. A. *J. Org. Chem.* **1976**, *41*, 3609.
[29] Moriarty, R. M.; Prakash, O.; Karalis, P.; Prakash, I. *Tetrahedron Lett.* **1984**, *25*, 4745.
[30] Moriarty, R. M.; Prakash, O.; Thachet, C. T.; Musallam, H. A. *Heterocycles* **1985**, *23*, 633.
[31a] Prakash, O.; Goyal, S.; Sehgal, S.; Singh, S. P.; Moriarty, R. M. *Indian J. Chem.* **1988**, *27B*, 929.
[31b] Ray, S.; Pal, S. K.; Saha, C. K. *Indian J. Chem.* **1995**, *34B*, 112.
[32] Yasuda, S.; Yamada, T.; Hanaoka, M. *Tetrahedron Lett.* **1986**, *27*, 2023.
[33] Moriarty, R. M.; John, L. S.; Du, P. C. *J. Chem. Soc., Chem. Commun.* **1981**, 641.
[34] Daum, S. J. *Tetrahedron Lett.* **1984**, *25*, 4725.
[35] Moriarty, R. M.; Engerer, S. C.; Prakash, O.; Prakash, I.; Gill, U. S.; Freeman, W. A. *J. Chem. Soc., Chem. Commun.* **1985**, 1715.
[36] Moriarty, R. M.; Engerer, S. C.; Prakash, O.; Prakash, I.; Gill,, U. S.; Freeman, W. A. *J. Org. Chem.* **1987**, *52*, 153.
[37] Moriarty, R. M.; Prakash O.; Freeman, W. A. *J. Chem. Soc., Chem. Commun.* **1984**, 927.
[38] Moriarty, R. M.; Prakash, O.; Musallam, H. A. *J. Heterocycl. Chem.* **1985**, *22*, 583.
[39] Prakash, O.; Pahuja, S.; Tanwar, M. P. *Indian J. Chem.* **1994**, *33B*, 272.
[40] Tamura, Y.; Yakura, T.; Terashi, H.; Haruta, J.; Kita, Y. *Chem. Pharm. Bull.* **1987**, *35*, 570.

[41] Numazawa, M.; Ogata, M. *J. Chem. Soc., Chem. Commun.* **1986**, 1092.
[42] Numazawa, M.; Mutsumi, A.; Ogata, M. *Chem. Pharm. Bull.* **1988**, *36*, 3381.
[43] Moriarty, R. M.; Prakash, O. *J. Org. Chem.* **1985**, *50*, 151.
[44] Prakash, O.; Pahuja, S.; Sawhney, S. N. *Indian J. Chem.* **1991**, *30B*, 1023.
[45] Prakash, O.; Mendiratta, S. *Synth. Commun.* **1992**, *22*, 327.
[46] Beringer, F. M.; Daniel, W. J.; Galton, S. A.; Rubin, G. *J. Org. Chem.* **1966**, *31*, 4315.
[47] Beringer, F. M.; Forgione, P. S. *J. Org. Chem.* **1963**, *28*, 714.
[48] Beringer, F. M.; Forgione, P. S.; Yudis, M. D. *Tetrahedron* **1960**, *8*, 49.
[49] Chen, Z. C.; Jin, Y. Y.; Stang, P. J. *J. Org. Chem.* **1987**, *52*, 4115.
[50] Barton, D. H. R.; Finet, J. P.; Gianotti, C.; Halley, F. *J. Chem. Soc., Perkin Trans. 1* **1987**, 241.
[51] Beringer, F. M.; Galton, S. A. *J. Org. Chem.* **1963**, *28*, 3417.
[52] Hampton, K. G.; Harris, T. M.; Hauser, C. R. *J. Org. Chem.* **1964**, *29*, 3511.
[53] Moriarty, R. M.; Hu, H. *Tetrahedron Lett.* **1981**, *22*, 2747.
[54] Tamura, Y.; Sasho, M.; Akai, S.; Kishimoto, H.; Sekihachi, J.; Kita, Y. *Chem. Pharm. Bull.* **1987**, *35*, 1405.
[55a] Mizukami, F.; Ando, M.; Tanaka, T.; Imamura, J. *Bull. Chem. Soc. Jpn.* **1978**, *51*, 335.
[55b] Andrews, I-P.; Lewis, N. J.; McKillop, A.; Wells, A. S. *Heterocycles* **1994**, *38*, 713.
[56] Podolesov, B. *J. Org. Chem.* **1984**, *49*, 2644.
[57] Moriarty, R. M.; Berglund, B. A.; Penmasta, R. *Tetrahedron Lett.* **1992**, *33*, 6065.
[58] Koser, G. F.; Relenyi, A. G.; Kalos, A. N.; Rebrovic, L.; Wettach, R. H. *J. Org. Chem.* **1982**, *47*, 2487.
[59] Zefirov, N. S.; Zhdankin, V. V.; Dan'kov, Y. V.; Koz'min, A. S.; Chizhov, O. S. *J. Org. Chem. USSR (Engl. Transl.)* **1985**, *21*, 2252.
[60] Lodaya, J. S.; Koser, G. F. *J. Org. Chem.* **1988**, *53*, 210.
[61] Prakash, O.; Saini, N.; Sharma, P. K. *J. Indian Chem. Soc.* **1995**, *72*, 129.
[62] Hatzigrigoriou, E.; Varvoglis, A.; Christianopoulou, M. B. *J. Org. Chem.* **1990**, *55*, 315.
[63] Moriarty, R. M.; Awasthi, A. K.; Penmasta, R. *194th ACS National Meeting, New Orleans*, ORGN, 1987, 58.
[64] Prakash, O.; Rani, N.; Goyal, S. *Indian J. Chem.* **1992**, *31B*, 349.
[65] Prakash, O.; Rani, N. *Synth. Commun.* **1993**, *23*, 1455.
[66] Penmasta, R., *The Applications of Hypervalent Iodine Oxidation to Problems in Organic Synthesis*, Ph.D. Thesis, University of Illinois at Chicago, 1988.
[67] Prakash, O.; Saini, N.; Sharma, P. K. *J. Chem. Res. (S)* **1993**, 430.
[68] Koser, G. F.; Lodaya, J. S.; Ray, III, D. G.; Kokil, P. B. *J. Am. Chem. Soc.* **1988**, *110*, 2987.
[69] Moriarty, R. M.; Condeiu, C.; Prakash, O. *20th IUPAC Meeting on Natural Products, Chicago*, 1996, Sy 39.
[70] Moriarty, R. M.; Prakash, O.; Duncan, M. P. *Synthesis* **1985**, 943.
[71] Moriarty, R. M.; Duncan, M. P.; Prakash, O. *J. Chem. Soc., Perkin Trans. 1* **1987**, 1781.
[72] Duncan, M. P., *New Bond Forming Reactions Using Hypervalent Organoiodine*, Ph.D. Thesis, University of Illinois at Chicago, 1987.
[73] Moriarty, R. M.; Prakash, O.; Duncan, M. P.; Vaid, R. K.; Musallam, H. A. *J. Org. Chem.* **1987**, *52*, 150.
[74] Moriarty, R. M.; Penmasta, R.; Awasthi, A. K.; Epa, R. W.; Prakash, I. *J. Org. Chem.* **1989**, *54*, 1101.
[75] Moriarty, R. M.; Epa, R. W.; Penmasta, R.; Awasthi, A. K. *Tetrahedron Lett.* **1989**, *30*, 667.
[76] Brunovlenskaya, I. I.; Kusainova, K. M.; Kashin, A. K. *J. Org. Chem. USSR (Engl. Transl.)* **1988**, *24*, 316.
[77] Kim, D. Y.; Mang, J. Y.; Oh, D. Y. *Synth. Commun.* **1994**, *24*, 629.
[78] Koser, G. F.; Chen, K.; Huang, Y.; Summers, C. A. *J. Chem. Soc., Perkin Trans. 1* **1994**, 1375.
[79] Moriarty, R. M.; Prakash, O.; Duncan, M. P. *J. Chem. Soc., Chem. Commun.* **1985**, 420.
[80] Moriarty, R. M.; Prakash, O.; Duncan, M. P. *J. Chem. Soc., Perkin Trans. 1* **1987**, 559.
[81] Moriarty, R. M.; Prakash, O.; Duncan, M. P. *Synth. Commun.* **1985**, *15*, 789.
[82] Zefirov, N. S.; Samoniya, N. S.; Kutateladze, T. G.; Zhdankin, V. V. *J. Org. Chem. USSR (Engl. Transl.)* **1991**, *27*, 220.

[83] Zhdankin, V. V.; Tykwinski, R.; Caple, R.; Berglund, B.; Koz'min, A. S.; Zefirov, N. S. *Tetrahedron Lett.* **1988**, *29*, 3703.
[84] Zhdankin, V. V.; Mullikin, M.; Tykwinski, R.; Berglund, B.; Caple, R.; Zefirov, N. S.; Koz'min, A. S. *J. Org. Chem.* **1989**, *54*, 2605.
[85] Umemoto, T.; Kuriu, Y.; Nakayama, S.; Miyano, O. *Tetrahedron Lett.* **1982**, *23*, 1471.
[86] Umemoto, T.; Gotoh, Y. *Bull. Chem. Soc. Jpn.* **1987**, *60*, 3823.
[87] Chen, K.; Koser, G. F. *J. Org. Chem.* **1991**, *56*, 5764.
[88] Gao, P.; Portoghese, P. S. *J. Org. Chem.* **1995**, *60*, 2276.
[89] Evans, D. A.; Faul, M. F.; Bilodeau, M. T. *J. Am. Chem. Soc.* **1994**, *116*, 2742.
[90] Moriarty, R. M.; Vaid, R. K.; Ravikumar, V. T.; Vaid, B. K.; Hopkins, T. E. *Tetrahedron* **1988**, *44*, 1603.
[91] Bregant, N.; Matijevic, J.; Sirola, I.; Balenovic, K. *Bull. Sci. Acad. Sci. Arts, RSF Youg., Sect. A.* **1972**, *17*, 148 [*Chem. Abstr.* **1973**, *78*, 4047d].
[92] Prakash, O.; Sharma, P. K.; Saini, N. *Indian J. Chem.* **1995**, *34B*, 632.
[93] Yan, J.; Zhong, L. R.; Chen, Z. C. *J. Org. Chem.* **1991**, *56*, 459.
[94] Umemoto, T.; Gotoh, Y. *Bull. Chem. Soc. Jpn.* **1986**, *59*, 439.
[95] Taruta, A. M.; Kamernitzky, A. V.; Fadeeva, T. M.; Zhulin, A. V. *Synthesis* **1985**, 1129.
[96] Prakash, O.; Tanwar, M. P.; Goyal, S.; Pahuja, S. *Tetrahedron Lett.* **1992**, *33*, 6519.
[97] Prakash, O.; Goyal, S.; Pahuja, S.; Singh, S. P. *Synth. Commun.* **1990**, *20*, 1409.
[98] Prakash, O.; Goyal, S. *Synthesis* **1992**, 629.
[99] Moriarty, R. M.; Prakash, O.; Duncan, M. P. *Synth. Commun.* **1986**, *16*, 1239.
[100] Moriarty, R. M.; Vaid, R. K.; Hopkins, T. E.; Vaid, B. K.; Prakash, O. *Tetrahedron Lett.* **1990**, *31*, 201.
[101] Dneprovskii, A. S.; Krainyuchenko, I. V.; Temnikova, T. I. *J. Org. Chem. USSR (Engl. Transl.)* **1978**, *14*, 1414.
[102] Kajigaeshi, S.; Kakinami, T.; Moriwaki, M.; Fujisaki, S.; Maeno, K.; Okamoto, T. *Synthesis* **1988**, 545.
[103] Croce, P. D.; Ferraccioli, R.; Ritieni, A. *Synthesis* **1990**, 212.
[104] Tsushima, T.; Kawada, K.; Tsuji, T. *Tetrahedron Lett.* **1982**, *23*, 1165.
[105] Papadopoulou, M.; Varvoglis, A. *J. Chem. Res. (S)* **1983**, 66.
[106] Hadjiarapoglou, L.; Spyroudis, S.; Varvoglis, A. *Synthesis* **1983**, 207.
[107] Papadopoulou, M.; Varvoglis, A. *J. Chem. Res. (S)* **1984**, 166.
[108] Moriarty, R. M.; Vaid, R. K.; Hopkins, T. E.; Vaid, B. K.; Prakash, O. *Tetrahedron Lett.* **1990**, *31*, 197.
[109] Magnus, P.; Lacour, J. *J. Am. Chem. Soc.* **1992**, *114*, 767 and 3993.
[110a] Magnus, P.; Roe, M. B.; Hulme, C. *J. Chem. Soc., Chem. Commun.* **1995**, 263.
[110b] Magnus, P.; Lacour, J.; Weber, W. *J. Am. Chem. Soc.* **1993**, *115*, 9347.
[111] Moriarty, R. M.; Vaid, R. K.; Hopkins, T. E.; Vaid, B. K.; Tuncay, A. *Tetrahedron Lett.* **1989**, *30*, 3019.
[112] Tamura, Y.; Shirouchi, Y.; Haruta, J. *Synthesis* **1984**, 231.
[113] Tamura, Y.; Yakura, T.; Shirouchi, Y.; Haruta, J. *Chem. Pharm. Bull.* **1985**, *33*, 1097.
[114] Moriarty, R. M.; Khosrowshahi, J. S.; Prakash, O. *Tetrahedron Lett.* **1985**, *26*, 2961.
[115] Prakash, O.; Goyal, S.; Moriarty, R. M.; Khosrowshahi, J. S. *Indian J. Chem.* **1990**, *29B*, 304.
[116] Singh, O. V.; Prakash, O.; Garg, C. P.; Kapoor, R. P. *Indian J. Chem.* **1989**, *28B*, 814.
[117] Prakash, O.; Kumar, J.; Sadana, A.; Saini, N. Kurukshetra University, Kurukshetra, Haryana, India, unpublished results.
[118] Singh, O. V. *Tetrahedron Lett.* **1990**, *31*, 3055.
[119] Singh, O. V.; Garg, C. P.; Kapoor, R. P. *Synthesis* **1990**, 1025.
[120] Rebrovic, L.; Koser, G. F. *J. Org. Chem.* **1984**, *49*, 2462.
[121] Prakash, O.; Kumar, D.; Saini, R. K.; Singh, S. P. *Tetrahedron Lett.* **1994**, *35*, 4211.
[122] Prakash, O.; Pahuja, S.; Goyal, S.; Sawhney, S. N.; Moriarty, R. M. *Synlett* **1990**, 337.
[123] Prakash, O.; Tanwar, M. P. *J. Chem. Res. (S)* **1995**, 1429, *(M)* 213.
[124] Prakash, O.; Tanwar, M. P. *Bull. Chem. Soc. Jpn.* **1995**, *68*, 1168.
[125] Kumar, D.; Singh, O. V.; Singh, S. P.; Prakash, O. *Synth. Commun.* **1994**, *24*, 2637.

[126] Moriarty, R. M.; Enache, L. A.; Zhao, L.; Gilardi, R.; Mattson, M. V.; Prakash, O. *J. Med. Chem.* **1998**, *41*, 468.
[127] Grunewald, G. L.; Ye, Q. *J. Org. Chem.* **1988**, *53*, 4021.
[128] Prakash, O.; Pahuja, S.; Moriarty, R. M. *Synth. Commun.* **1990**, *20*, 1417.
[129] Prakash, O.; Kumar, D.; Saini, R. K.; Singh, S. P. *Synth. Commun.* **1994**, *24*, 2167.
[130] Vedejs, E. *J. Am. Chem. Soc.* **1974**, *96*, 5944.
[131] Vedejs, E.; Engler, D. A.; Telschow, J. E. *J. Org. Chem.* **1978**, *43*, 188.
[132] Vedejs, E.; Larsen, S. *Org. Synth.* **1985**, *64*, 127.
[133] Davis, F. A.; Sheppard, A. C. *J. Org. Chem.* **1987**, *52*, 954.
[134] Davis, F. A.; Chen, B. C. *Chem. Rev.* **1992**, *92*, 919.
[135] Gardner, J. N.; Carlon, F. E.; Gnoj, O. *J. Org. Chem.* **1968**, *33*, 3294.
[136] Masui, M.; Ando, A.; Shioiri, T. *Tetrahedron Lett.* **1988**, *29*, 2835.
[137] Bailey, E. J.; Barton, D. H. R.; Elks, J.; Templeton, J. F. *J. Chem. Soc.* **1962**, 1578.
[138] Jefford, C. W.; Rimbault, C. G. *Tetrahedron Lett.* **1977**, 2375.
[139] Rubottom, G. M.; Vasquez, M. A.; Pelegrina, D. R. *Tetrahedron Lett.* **1974**, 4319.
[140] Brook, A. G.; Macrae, D. M. *J. Organomet. Chem.* **1974**, *77*, C19.
[141] Hassner, A.; Reuss, R. H.; Pinnick, H. W. *J. Org. Chem.* **1975**, *40*, 3427.
[142] Rubottom, G. M.; Marrcro, R. *J. Org. Chem.* **1975**, *40*, 3783.
[143] Rubottom, G. M.; Gruber, J. M. *Tetrahedron Lett.* **1978**, 4603.
[144] Rubottom, G. M.; Gruber, J. M. *J. Org. Chem.* **1978**, *43*, 1599.
[145] Horiguchi, Y.; Nakamura, E.; Kuwajima, I. *Tetrahedron Lett.* **1989**, *30*, 3323.
[146] McCormick, J. P.; Tomasik, W.; Johnson, M. W. *Tetrahedron Lett.* **1981**, *22*, 607.
[147] Lee, T. V.; Toczek, J. *Tetrahedron Lett.* **1982**, *23*, 2917.
[148] Rubottom, G. M.; Gruber, J. M.; Kincaid, K. *Synth. Commun.* **1976**, *6*, 59.
[149] Davis, F. A.; Weismiller, M. C. *J. Org. Chem.* **1990**, *55*, 3715.
[150] Yamakawa, K.; Satoh, T.; Ohba, N.; Sakaguchi, R. *Chem. Lett.* **1979**, 763.
[151] Adam, W.; Prechtl, F. *Chem. Ber.* **1991**, *124*, 2369.
[152] Guertin, K. R.; Chan, T. *Tetrahedron Lett.* **1991**, *32*, 715.
[153] McKillop, A.; Hunt, J. D.; Taylor, E. C. J. *Org. Chem.* **1972**, *27*, 3381.
[154] Willgerodt, O. *Chem. Ber.* **1887**, *20*, 2467.
[155] Kindler, K. *Justus Liebigs Ann. Chem.* **1923**, *431*, 193.
[156] Cavalieri, L.; Pattison, D. B.; Carmack, M. *J. Am. Chem.Soc.* **1945**, *67*, 1783.
[157] Carmack, M.; DeTar, D. F. *J. Am. Chem. Soc.* **1946**, *68*, 2025.
[158] Maeyer, R.; Wehl, J. *Angew. Chem.* **1964**, *76*, 861.
[159] Wolff, E.; Folkers, K. *Org. React.* **1951**, *6*, 439.
[160] Hundt, R. H. *Chem. Rev.* **1961**, *61*, 52.
[161] Brown, E. V. *Synthesis* **1975**, 358.
[162] Giordano, C.; Castaldi, G.; Uggeri, F. *Angew. Chem., Int. Ed. Engl.* **1984**, *23*, 413.
[163] McKillop, A.; Swann, B. P.; Taylor, E. C. *J. Am. Chem. Soc.* **1971**, *93*, 4919.
[164] McKillop, A.; Swann, B. P.; Taylor, E. C. *J. Am. Chem. Soc.* **1973**, *95*, 3340.
[165] Taylor, E. C.; Chiang, C. S.; McKillop, A.; White, J. F. *J. Am. Chem. Soc.* **1976**, *98*, 6750.
[166] Walker, J. A.; Pillai, M. D. *Tetrahedron Lett.* **1977**, 3707.
[167] Higgins, S. D.; Thomas, C. B. *J. Chem. Soc., Perkin Trans. 1* **1982**, 235.
[168] McKillop, A.; Taylor, E. C. In *Comprehensive Organometallic Chemistry*; Wilkinson, G.; Stone, F. G. A.; Abel, E. W., Eds.; Pergamon: Oxford, Vol. 7, 1983, p. 465.
[169] Wang, J.; Zhang, G. *Hecheng Huaxue* **1993**, *18*, 1 [*Chem. Abstr.* **1994**, *120*, 322289e].
[170] Myrboh, B.; Ila, H.; Junjappa, H. *Synthesis* **1981**, 126.
[171] Fujii, K.; Nakao, K.; Yamauchi, T. *Synthesis* **1982**, 456.
[172] Giordano, C.; Castaldi, G.; Casagrande, R.; Abis, L. *Tetrahedron Lett.* **1982**, *23*, 1385.
[173] Giordano, C.; Castaldi, G.; Casagrande, F.; Belli, A. *J. Chem. Soc., Perkin Trans. 1* **1982**, 2575.
[174] Verhe, R.; Dekimple, N. In *The Chemistry of Functional Groups, Suppl. D*; Patai, S.; Rappoport, Z., Eds.; Wiley: London, 1983, p. 813.
[175] Sharefkin, J. G.; Saltzman, H. *Org. Synth.* **1963**, *43*, 62; **1973**, *Coll. Vol. V*, 660.
[176] Pausaker, K. H. *J. Chem. Soc.* **1953**, 107.

[177] Lucas, H. J.; Kennedy, E. R. *Org. Synth.* **1955**, *Coll. Vol. III*, 482.
[178] McKillop, A.; Kemp, D. *Tetrahedron* **1989**, *45*, 3299.
[179] Spyroudis, S.; Varvoglis, A. *Synthesis* **1975**, 445.
[180] Neiland, O.; Karele, B. *J. Org. Chem. USSR (Engl. Transl.)* **1970**, *6*, 889.
[181] Boeseken, J; Wicherling, E. *Rec. Trav. Chim. Pay-Bas* **1936**, *55*, 936.
[182] Askenasy, P.; Meyer, V. *Chem. Ber.* **1893**, *26*, 1354.
[183] Beringer, F. M.; Geering, E. J.; Kuntz, I.; Mausner, M. *J. Phys. Chem.* **1956**, *60*, 141.
[184] Zupan, M.; Pollak, A. *J. Fluorine Chem.* **1976**, *7*, 445.
[185] Carpenter, W. *J. Org. Chem.* **1966**, *31*, 2688.
[186] Bockemuller, W. *Chem. Ber.* **1931**, *64*, 522.
[187] Garvey, B. S.; Halley, L. F.; Allen, C. F. H. *J. Am. Chem. Soc.* **1937**, *59*, 1827.
[188] Umemoto, T.; Gotoh, Y. *J. Fluorine Chem.* **1985**, *28*, 235.
[189] Umemoto, T.; Gotoh, Y. *Bull. Chem. Soc. Jpn.* **1987**, *60*, 3307.
[190] Umemoto, T.; Kuriu, Y.; Shuyanra, H.; Miyano, O.; Nakayama, S-L. *J. Fluorine Chem.* **1982**, *20*, 695.
[191] Emeleus, H. J.; Heal, H. G. *J. Chem. Soc.* **1946**, 1126.
[192] Kajigaeshi, S.; Kakinami, T.; Yamasaki, H.; Fujisaki, S.; Kondo, M.; Okamoto, T. *Chem. Lett.* **1987**, 2109.
[193] Moriarty, R. M.; Hou, K. C.; Prakash, I.; Arora, S. K. *Org. Synth.* **1986**, *64*, 138.
[194] Beringer, F. M.; Galton, S. A.; Huang, S. J. *J. Am. Chem. Soc.* **1962**, *84*, 2819.
[195] Hu, H. *Hypervalent Iodine(III) in Organic Synthesis*, Ph.D. Thesis, University of Illinois at Chicago, 1982.
[196] Tamura, Y., Annoura, H.; Yamamoto, H.; Kondo, H.; Kita, Y.; Fujioka, H. *Tetrahedron Lett.* **1987**, *28*, 5709.
[197] Moriarty, R. M.; Prakash, O.; Vavilidolanu, P. R.; Vaid, R. K.; Freeman, W. A. *J. Org. Chem.* **1989**, *54*, 4008.
[198] Moriarty, R. M.; Prakash, O.; Thachet, C. T. *Synth. Commun.* **1984**, *14*, 1373.
[199] Kamernitzky, A. V.; Taruta, A. M.; Fadeeva, T. M.; Istomina, Z. I. *Synthesis* **1985**, 326.
[200] Rossi, R. A.; Bunnett, J. F. *J. Am. Chem. Soc.* **1974**, *96*, 112.
[201] Neilands, O.; Vavags, G.; Gurdriniece, E. *J. Gen. Chem. USSR* **1958**, *28*, 1201.
[202] Tuncay, A.; Dustman, J. A.; Fisher, G.; Tuncay, C. I.; Suslick, K. S. *Tetrahedron Lett.* **1992**, *33*, 7647.
[203] Prakash, O.; Rani, N.; Goyal, S. *J. Chem. Soc., Perkin Trans. 1* **1992**, 707.
[204] Zhdankin, V.V.; Tykwinski, R.; Berglund, B.; Mullikin, M.; Caple, R. *J. Org. Chem.* **1989**, *54*, 2609.
[205] Cruciani, G.; Semisch, C.; Margaretha, P. *J. Photochem. Photobiol.* **1988**, *44A*, 219.
[206] Khanna, M. S.; Sangeeta; Garg, C. P.; Kapoor, R. P. *Synth. Commun.* **1992**, *22*, 2555.
[207] Singh, O. V.; Garg, C. P.; Kapoor, R. P.; Kapil, A.; Moza, N. *Indian J. Chem.* **1992**, *31B*, 248.
[208] Bhatti, S. P.; Garg, C. P.; Kapoor, R. P.; Sharma, S.; Kapil, A. *Indian J. Chem.* **1995**, *34B*, 879.
[209] Kapoor, R. P.; Garg, C. P.; Bhatti, S. P. Kurukshetra University, Kurukshetra, Haryana, India, unpublished results.

CUMULATIVE CHAPTER TITLES BY VOLUME

Volume 1 (1942)

1. **The Reformatsky Reaction**: Ralph L. Shriner
2. **The Arndt-Eistert Reaction**: W. E. Bachmann and W. S. Struve
3. **Chloromethylation of Aromatic Compounds**: Reynold C. Fuson and C. H. McKeever
4. **The Amination of Heterocyclic Bases by Alkali Amides**: Marlin T. Leffler
5. **The Bucherer Reaction**: Nathan L. Drake
6. **The Elbs Reaction**: Louis F. Fieser
7. **The Clemmensen Reduction**: Elmore L. Martin
8. **The Perkin Reaction and Related Reactions**: John R. Johnson
9. **The Acetoacetic Ester Condensation and Certain Related Reactions**: Charles R. Hauser and Boyd E. Hudson, Jr.
10. **The Mannich Reaction**: F. F. Blicke
11. **The Fries Reaction**: A. H. Blatt
12. **The Jacobson Reaction**: Lee Irvin Smith

Volume 2 (1944)

1. **The Claisen Rearrangement**: D. Stanley Tarbell
2. **The Preparation of Aliphatic Fluorine Compounds**: Albert L. Henne
3. **The Cannizzaro Reaction**: T. A. Geissman
4. **The Formation of Cyclic Ketones by Intramolecular Acylation**: William S. Johnson
5. **Reduction with Aluminum Alkoxides (The Meerwein-Ponndorf-Verley Reduction)**: A. L. Wilds

Volume 7 (1953)

1. **The Pechmann Reaction**: Suresh Sethna and Ragini Phadke
2. **The Skraup Synthesis of Quinolines**: R. H. F. Manske and Marshall Kulka
3. **Carbon-Carbon Alkylations with Amines and Ammonium Salts**: James H. Brewster and Ernest L. Eliel
4. **The von Braun Cyanogen Bromide Reaction**: Howard A. Hageman
5. **Hydrogenolysis of Benzyl Groups Attached to Oxygen, Nitrogen, or Sulfur**: Walter H. Hartung and Robert Simonoff
6. **The Nitrosation of Aliphatic Carbon Atoms**: Oscar Touster
7. **Epoxidation and Hydroxylation of Ethylenic Compounds with Organic Peracids**: Daniel Swern

Volume 8 (1954)

1. **Catalytic Hydrogenation of Esters to Alcohols**: Homer Adkins
2. **The Synthesis of Ketones from Acid Halides and Organometallic Compounds of Magnesium, Zinc, and Cadmium**: David A. Shirley
3. **The Acylation of Ketones to Form β-Diketones or β-Keto Aldehydes**: Charles R. Hauser, Frederic W. Swamer, and Joe T. Adams
4. **The Sommelet Reaction**: S. J. Angyal
5. **The Synthesis of Aldehydes from Carboxylic Acids**: Erich Mosettig
6. **The Metalation Reaction with Organolithium Compounds**: Henry Gilman and John W. Morton, Jr.
7. **β-Lactones**: Harold E. Zaugg
8. **The Reaction of Diazomethane and Its Derivatives with Aldehydes and Ketones**: C. David Gutsche

Volume 9 (1957)

1. **The Cleavage of Non-enolizable Ketones with Sodium Amide**: K. E. Hamlin and Arthur W. Weston
2. **The Gattermann Synthesis of Aldehydes**: William E. Truce
3. **The Baeyer-Villiger Oxidation of Aldehydes and Ketones**: C. H. Hassall
4. **The Alkylation of Esters and Nitriles**: Arthur C. Cope, H. L. Holmes, and Herbert O. House

Volume 19 (1972)

1. **Conjugate Addition Reactions of Organocopper Reagents**: Gary H. Posner

2. **Formation of Carbon-Carbon Bonds via π-Allylnickel Compounds**: Martin F. Semmelhack

3. **The Thiele-Winter Acetoxylation of Quinones**: J. F. W. McOmie and J. M. Blatchly

4. **Oxidative Decarboxylation of Acids by Lead Tetraacetate**: Roger A. Sheldon and Jay K. Kochi

Volume 20 (1973)

1. **Cyclopropanes from Unsaturated Compounds, Methylene Iodide, and Zinc-Copper Couple**: H. E. Simmons, T. L. Cairns, Susan A. Vladuchick, and Connie M. Hoiness

2. **Sensitized Photooxygenation of Olefins**: R. W. Denny and A. Nickon

3. **The Synthesis of 5-Hydroxyindoles by the Nenitzescu Reaction**: George R. Allen, Jr.

4. **The Zinin Reaction of Nitroarenes**: H. K. Porter

Volume 21 (1974)

1. **Fluorination with Sulfur Tetrafluoride**: G. A. Boswell, Jr., W. C. Ripka, R. M. Scribner, and C. W. Tullock

2. **Modern Methods to Prepare Monofluoroaliphatic Compounds**: Clay M. Sharts and William A. Sheppard

Volume 22 (1975)

1. **The Claisen and Cope Rearrangements**: Sara Jane Rhoads and N. Rebecca Raulins

2. **Substitution Reactions Using Organocopper Reagents**: Gary H. Posner

3. **Clemmensen Reduction of Ketones in Anhydrous Organic Solvents**: E. Vedejs

4. **The Reformatsky Reaction**: Michael W. Rathke

Volume 23 (1976)

1. **Reduction and Related Reactions of α,β-Unsaturated Compounds with Metals in Liquid Ammonia**: Drury Caine

2. **The Acyloin Condensation**: Jordan J. Bloomfield, Dennis C. Owsley, and Janice M. Nelke

3. **Alkenes from Tosylhydrazones**: Robert H. Shapiro

Volume 24 (1976)

1. **Homogeneous Hydrogenation Catalysts in Organic Solvents**: Arthur J. Birch and David H. Williamson

2. **Ester Cleavages via S_N2-Type Dealkylation**: John E. McMurry

3. **Arylation of Unsaturated Compounds by Diazonium Salts (The Meerwein Arylation Reaction)**: Christian S. Rondestvedt, Jr.

4. **Selenium Dioxide Oxidation**: Norman Rabjohn

Volume 25 (1977)

1. **The Ramberg-Bäcklund Rearrangement**: Leo A. Paquette

2. **Synthetic Applications of Phosphoryl-Stabilized Anions**: William S. Wadsworth, Jr.

3. **Hydrocyanation of Conjugated Carbonyl Compounds**: Wataru Nagata and Mitsuru Yoshioka

Volume 26 (1979)

1. **Heteroatom-Facilitated Lithiations**: Heinz W. Gschwend and Herman R. Rodriguez

2. **Intramolecular Reactions of Diazocarbonyl Compounds**: Steven D. Burke and Paul A. Grieco

Volume 27 (1982)

1. **Allylic and Benzylic Carbanions Substituted by Heteroatoms**: Jean-François Biellmann and Jean-Bernard Ducep

2. **Palladium-Catalyzed Vinylation of Organic Halides**: Richard F. Heck

Volume 28 (1982)

1. **The Reimer-Tiemann Reaction**: Hans Wynberg and Egbert W. Meijer

2. **The Friedländer Synthesis of Quinolines**: Chia-Chung Cheng and Shou-Jen Yan

3. **The Directed Aldol Reaction**: Teruaki Mukaiyama

Volume 29 (1983)

1. **Replacement of Alcoholic Hydroxy Groups by Halogens and Other Nucleophiles via Oxyphosphonium Intermediates**: Bertrand R. Castro

Volume 36 (1988)

1. **The [3 + 2] Nitrone-Olefin Cycloaddition Reaction**: Pat N. Confalone and Edward M. Huie

2. **Phosphorus Addition at *sp*^{2} Carbon**: Robert Engel

3. **Reduction by Metal Alkoxyaluminum Hydrides. Part II. Carboxylic Acids and Derivatives, Nitrogen Compounds, and Sulfur Compounds**: Jaroslav Málek

Volume 37 (1989)

1. **Chiral Synthons by Ester Hydrolysis Catalyzed by Pig Liver Esterase**: Masaji Ohno and Masami Otsuka

2. **The Electrophilic Substitution of Allylsilanes and Vinylsilanes**: Ian Fleming, Jacques Dunoguès, and Roger Smithers

Volume 38 (1990)

1. **The Peterson Olefination Reaction**: David J. Ager

2. **Tandem Vicinal Difunctionalization: β-Addition to α,β-Unsaturated Carbonyl Substrates Followed by α-Functionalization:** Marc J. Chapdelaine and Martin Hulce

3. **The Nef Reaction**: Harold W. Pinnick

Volume 39 (1990)

1. **Lithioalkenes from Arenesulfonylhydrazones**: A. Richard Chamberlin and Steven H. Bloom

2. **The Polonovski Reaction**: David Grierson

3. **Oxidation of Alcohols to Carbonyl Compounds via Alkoxysulfonium Ylides: The Moffatt, Swern, and Related Oxidations**: Thomas T. Tidwell

Volume 40 (1991)

1. **The Pauson-Khand Cycloaddition Reaction for Synthesis of Cyclopentenones**: Neil E. Schore

2. **Reduction with Diimide**: Daniel J. Pasto and Richard T. Taylor

3. **The Pummerer Reaction of Sulfinyl Compounds**: Ottorino DeLucchi, Umberto Miotti, and Giorgio Modena

4. **The Catalyzed Nucleophilic Addition of Aldehydes to Electrophilic Double Bonds**: Hermann Stetter and Heinrich Kuhlmann

Volume 48 (1996)

1. **Asymmetric Epoxidation of Allylic Alcohols: The Katsuki–Sharpless Epoxidation Reaction**: Tsutomu Katsuki and Victor S. Martin

2. **Radical Cyclization Reactions**: B. Giese, B. Kopping, T. Göbel, J. Dickhaut, G. Thoma, K. J. Kulicke, and F. Trach

Volume 49 (1997)

1. **The Vilsmeier Reaction of Fully Conjugated Carbocycles and Heterocycles**: Gurnos Jones and Stephen P. Stanforth

2. **[6 + 4] Cycloaddition Reactions**: James H. Rigby

3. **Carbon–Carbon Bond-Forming Reactions Promoted by Trivalent Manganese**: Gagik G. Melikyan

Volume 50 (1997)

1. **The Stille Reaction**: Vittorio Farina, Venkat Krishnamurthy, and William J. Scott

Volume 51 (1997)

1. **Asymmetric Aldol Reactions Using Boron Enolates**: Cameron J. Cowden and Ian Paterson

2. **The Catalyzed α-Hydroxylation and α-Aminoalkylation of Activated Olefins (The Morita–Baylis–Hillman Reaction)**: Engelbert Ciganek

3. **[4 + 3] Cycloaddition Reactions**: James H. Rigby and F. Christopher Pigge

Volume 52 (1998)

1. **The Retro–Diels–Alder Reaction. Part I. C—C Dienophiles**: Bruce Rickborn

2. **Enantioselective Reduction of Ketones**: Shinichi Itsuno

Volume 53 (1998)

1. **The Oxidation of Alcohols by Modified Oxochromium(VI)-Amine Reagents**: Frederick A. Luzzio

2. **The Retro–Diels–Alder Reaction. Part II. Dienophiles with One or More Heteroatoms**: Bruce Rickborn

AUTHOR INDEX, VOLUMES 1–54

Volume number only is designated in this index.

CHAPTER AND TOPIC INDEX, VOLUMES 1–54

Many chapters contain brief discussions of reactions and comparisons of alternative synthetic methods related to the reaction that is the subject of the chapter. These related reactions and alternative methods are not usually listed in this index. In this index, the volume number is in **boldface**, the chapter number is in ordinary type.